The First
International Scientific Consensus
Workshop Proceedings



# DEGRADABLE MATERIALS

## *Perspectives, Issues, and Opportunities*

Editors

**Sumner A. Barenberg**
Baxter Healthcare Corporation
Round Lake, IL

**John L. Brash**
McMaster University
Chemical Engineering
Hamilton, Ontario

**Ramani Narayan**
Michigan Biotech Institute
Michigan State University
East Lansing, MI

**Anthony E. Redpath**
EcoPlastics
Willowdale, Ontario

**CRC Press**
**Boca Raton Ann Arbor Boston**

Acquiring Editor: Russ Hall
Production Director: Sandy Pearlman
Coordinating Editor: Kristen Peterson
Cover Design: Chris Pearl
Cover Photo: Frans Lanping, Minden Pictures
Composition: CRC Press, Inc.

**Library of Congress Cataloging-in-Publication Data**

Degradable materials : perspectives, issues and opportunities : the first international scientific consensus workshop proceedings / editors, S.A. Barenberg . . . [et al.].
p. cm.
"The First International Scientific Consensus Workshop on Degradable Materials was convened on November 2, 1989 in Toronto Canada"--Introd.
Includes bibliographical references.
ISBN 0-8493-4274-0
1. Polymers--Biodegradation--Congresses. I. Barenberg, S. A. (Sumner A.) II. International Scientific Consensus Workshop on Degradable Materials (1st : 1989 : Toronto, Ont.)
QP801.P64D44 1990
628.4'45--dc20

90-1562
CIP

This book represents information obtained from authentic and highly regarded sources. Reprinted material is quoted with permission, and sources are indicated. A wide variety of references are listed. Every reasonable effort has been made to give reliable data and information, but the authors and the publisher cannot assume responsibility for the validity of all materials or for the consequences of their use.



Direct all inquiries to CRC Press, Inc., 2000 Corporate Blvd., N.W., Boca Raton, Florida 33431.



International Standard Book Number 0-8493-4274-0

Library of Congress Card Number 90-1562
Printed in the United States

## INTRODUCTION

The First International Scientific Consensus Workshop on Degradable Materials was convened on November 2, 1989 in Toronto Canada, with 165 participants representing over 130 companies, research institutes and/or governments from 14 different countries. The Workshop addressed the global use of degradable materials ranging from "HI TECH" applications such as biomedical implants to commodity applications such as plastic packaging and in agriculture farming materials. To date no effort has been made on a global basis to standardize terminology and characterization techniques, understand the various mode of degradation as a function of application, define secondary effects of degradable materials, develop standard reference materials,or define the issues and needs associated with degradable materials. To this end an international group of concerned scientists on an ad hoc basis, decided to put in place a consensus workshop to develop a preliminary set of international scientific guidelines.

The objectives of the Workshop were to:

- Define State of the Art
- Propose/Agree to Standard Terminology
- Define Characterization Techniques & Protocols
- Define Mechanisms of Degradation
- Propose Standard Reference Materials
- Identify Issues and Needs

These Proceedings document the findings of the Workshop, chapters 1 & 26, and the scientific papers presented, chapters 2-25. The question and answer, and discussion sessions were documented by court stenographers and appear at the end of each of the scientific papers.

The results of the Workshop indicate that the field of degradable materials is emerging as an interdisciplinary technology with a defined market niche that complements recycling and solid land fill. It was also recognized that the science and technology of degradable materials is embryonic on a global scale given the state of the art and the interdisciplinary nature of the technology,as is discussed in chapters 1 and 26.

We would like to acknowledge the financial and in kind support which we received that made this Workshop and Proceedings possible:

- Canadian Natural Sciences and Engineering Research Council
- United States Food and Drug Administration
- Ontario Centre for Materials Research

- Abbott Laboratories
- Amoco Chemical Co
- Baxter Healthcare
- Becton Dickinson
- B.F. Goodrich
- Cambridge Consulting LTD
- Ciba Geigy
- Council for Solid Waste Solutions
- E.I. DuPont
- Himont
- International Paper
- Johnson & Johnson
- 3M
- Sam Yang Company
- Society Plastics Industry
- Union Carbide

The Second International Scientific Consensus Workshop on Degradable Materials has been scheduled to be held in France in the spring of 1991 with Michele Vert, Ann Christine Albertsson, Gerald Scott, and David Williams serving as the Workshop organizing committee.

FIRST INTERNATIONAL SCIENTIFIC CONSENSUS WORKSHOP ON DEGRADABLE MATERIALS

HILTON INTERNATIONAL HOTEL

Toronto, Ontario, Canada
2-4 November, 1989

PLENARY SESSION

Thursday, 2 November, 1989

| | | |
|---|---|---|
| Welcome | John Brash | McMaster University |
| Introduction | Ramani Narayan | Michigan Biotechnology |
| Biodegradation | Michele Vert | University Rouen |
| Environmental | Roger Lloyd | ICI |
| Environmental | Mark Matlock | ADM |
| Photochemcial | Jim Guillet | University Toronto |

MECHANISM & CHARACTERIZATION: CASE STUDIES

| | | |
|---|---|---|
| Chairperson | Jim Anderson | Case Western Reserve |
| Alpha-Hydroxy Acid Based Biodegradable Polymers Poly (L-Lactide) and Polydepsipeptides | Jan Feijen | University Twente |
| Photo-Biodegradation of Plastics. A Systems Approach to Plastic Waste & Litter | Gerald Scott | University Aston |
| Photo-Degradable Plastics in Agriculture | Dan Gilead | Plastophil |
| Photodegradative Additives | George S. Upton | Ampacet |
| Degradation Studies of Polyolefins | Richard G. Austin | Exxon Chemical Co. |
| Modified Starch Based Environmentally Degradable Plastics | Peter D. Campbell | St. Lawrence Starch |
| Biodegradation & Test Methods for Environmental & Biomedical Applications of Polymers | A. C. Albertsson | Royal Institute |

## MECHANISMS & CHARACTERIZATION: CASE STUDIES

(Continued)

Thursday, 2 November, 1989

| | | |
|---|---|---|
| Long Term In Vitro Enzymatic Biodegradation of Pellethane 2363-80A | James M. Anderson | Case Western Reserve |

Friday, 3 November, 1989 - AM Session

| | | |
|---|---|---|
| Chairperson | Gary Loomis | DuPont |
| The Role of Active Species Within Tissue in Degradation Processes | David F. Williams | University Liverpool |
| Evaluating Biodegradable Plastics with In-Vitro Enzyme Assays | Paul Allenza | Allied-Signal |
| Plastic Degradation & Recycling Through Selective Solubility | Melissa Bouzianis | Belland Plastics |
| Biodegradation of a Poly (Glycolic Acid) Epineural Tubulization Device | Don F. Gibbons | 3M |
| Studies on the Environmental Degradation of Starch-Based Plastics | Gene Iannotti | University Missouri |
| Preliminary Results of Screening Studies on Effectiveness of Cerium Salts in Accelerating Photodegradability of Thermoplastics | Lucinda Ballinger | Rhone Poulenc |

## MECHANISMS & CHARACTERIZATION: CASE STUDIES

(Continued)

### Friday, 3 November 1989 - PM Session

| | | |
|---|---|---|
| Chairperson | Bill Bailey | University Maryland |
| Biodegradation of Poly (Beta-Hydroxylakanoates) | David Gilmore | U of Massachusetts |
| Degradation Mechanisms in Polyethylene-Starch Blends | Dick Wool | University Illinois |
| Environmental | Tony Andrady | Research Triangle |
| Design of Polyester-Based Biodegradable Materials | Yutaka Tokiwa | MITI |
| Effects of Photodegradants on the Environmental Fate of Linear Low Density Polyethylene | Fred C. Schwab | Mobil Chemical |
| The Evaluation of Degradation Rates in Photodegradable Ecolyte Polystyrene | Tony Redpath | Ecoplastics |

## CONSENSUS BREAKOUT SESSIONS

| | | |
|---|---|---|
| Biodegradation | Shalaby Shalaby<br>Michele Vert | Johnson & Johnson<br>University of Rouen |
| Environmental | Ramani Narayan<br>Richard Wool | Michigan Biotechnology<br>University Illinois |
| Photochemical | Tony Redpath<br>Sam Mitra | Ecoplastics Limited<br>3M |

CONSENSUS PRESENTATIONS

| | | |
|---|---|---|
| Chairperson | Jan Feijen<br>Dave Williams | University Twente<br>University Liverpool |
| Biodegradation | Michele Vert<br>Shalaby Shalaby | University Rouen<br>Johnson & Johnson |
| Environmental | Ramani Narayan<br>Richard Wool | Michigan Biotechnology<br>University Illinois |
| Photochemical | Tony Redpath<br>Sam Mitra | Ecoplastics Limited<br>3M |
| Issues & Summary | Ramani Narayan | Michigan State |

IN MEMORY

BILL BAILEY, Ph.D.

DEVOTED SCIENTIST AND PROFESSOR

## TABLE OF CONTENTS

# INTRODUCTION

RAMANI NARAYAN
Michigan Biotechnology Institute
&
Michigan State University
Lansing, Michigan 48909

Good morning. Welcome to the First International Scientific Consensus Workshop on Degradable Materials.

I would like to start off by emphasizing that this is a workshop and not a conference where you just listen to the speakers and go home. We (the Organizing Committee) expect everybody to participate in the various sessions by way of questions, comments, critiques, etc. The success of this workshop will be dictated by how many people participate and the number and quality of questions, answers and issues.

With that in mind, a set of goals that this workshop would address has been put together. The technical sessions are structured to give you the current status of the experimental work in the area and where the technology stands. This workshop puts together a unique combination of researchers in differing diciplines and, for the first time, it brings together the well-established and advanced biomaterials research community with the burgeoning environmentally degradable polymers group. Figure 1 illustrates this classification. As shown in the figure, the environmentally degradable materials are further classified into biodegradable, photodegradable and chemically degradable.

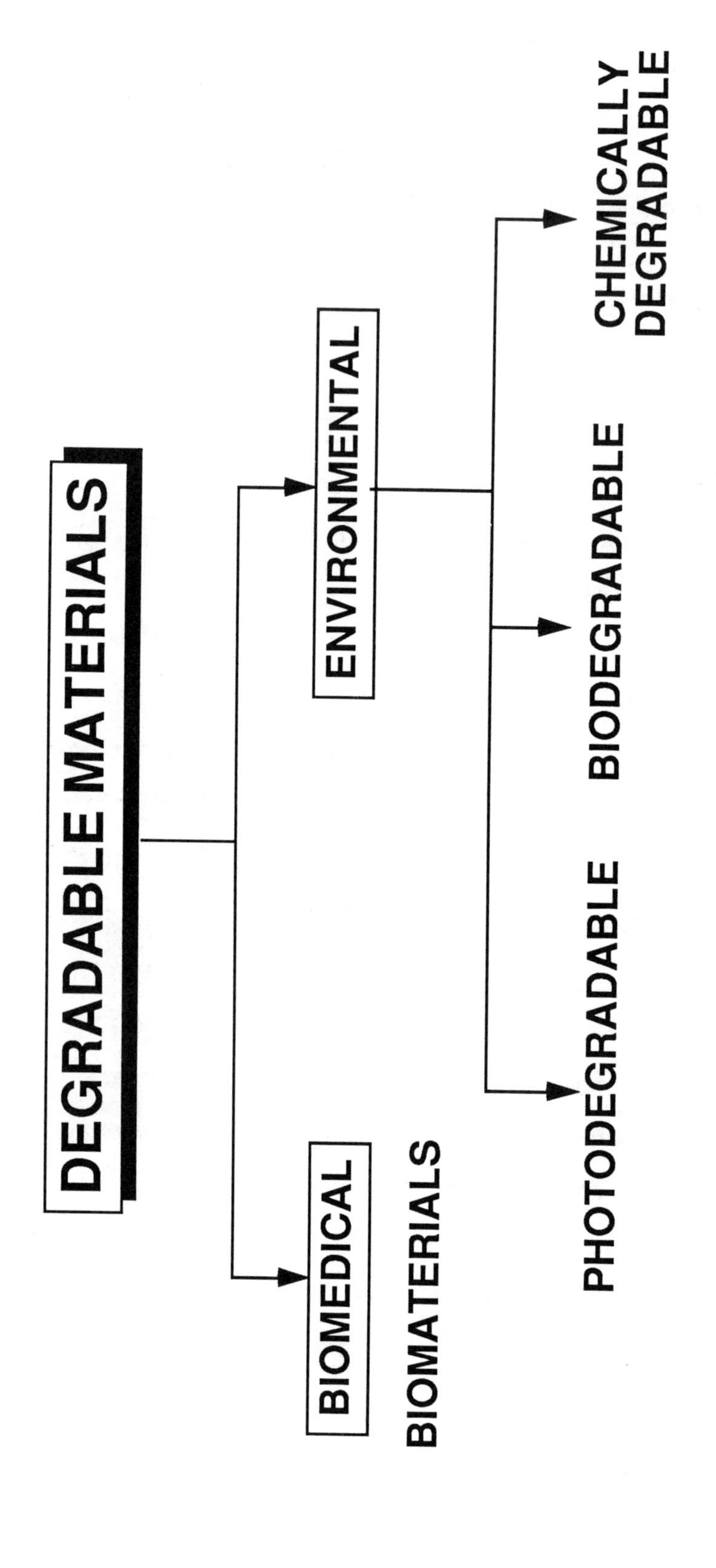


Figure 1. Classification of Degradable Materials

While both applications (biomedical and environmental) invoke the concept of degradability, test procedures and protocols, design strategies, manufacturing approaches and processing parameters will be different. Moreover, an understanding of the chemical and physical properties of a polymer which are responsible for the degradation in one medium will be relevant in large measure to the second. It must be recognized that materials need not be exclusively bio or photo or chemically degradable, but involve a combination of degradation mechanisms.

## WORKSHOP GOALS

The first goal is "Issues-the Research Agenda". We need to identify research needs and unresolved issues and put them in order of priority. There are a number of issues, and some of them will be addressed by speakers at this workshop, for example:

- What do we know about the biochemical mechanisms of breakdown of hydrocarbon polymer chains?
- Can microbial metabolism of "plastics" (induced by a photo, bio or chemical trigger) results in toxic products? - Aren't fatty acids the major (if not only) product.
- Degradation in anaerobic environment (landfills) - What's the state of knowledge - rates of degradation?

In short, with the hindsight of the presentations made by the various speakers we need to identify research needs and have all the unresolved issues tabled.

More specifically relating to mechanisms of degradation of environmentally degradable materials, we do not know what is the primary or predominant mechanism. Is it biological, chemical, or photochemical? Unfortunately nobody likes to use the word "chemical" because of the stigma associated with it as preceived by the public, but in scientific context if that is the mechamism, then we would so identify it.

Do all of the above mechanisms occur synergistically, concertedly, or consecutively?

The second set of goals are:

- Definitions
- Test procedures and protocols
- Reference standards.

We need to agree or disagree on a set of definitions for degradable, bio, and photodegradable materials. These is an urgent need to identify standard test procedure and protocols for measuring degradability and comparing different degradable materials. Reference standards need to be identified and developed.

Again, as in the case of Goal 1, we need to come up with research needs and unresolved issues in our discussions at the breakout sessions. We would then have to list them in order of priority.

With reference to Goal #2, ASTM (American Society for Testing and Materials) has taken a lead in trying to come up with the definitions, test procedures, and protocols, and many of the people involved in the ASTM activity are participants at this workshop. I would like to give you an overview of thic activity.

ASTM committee D-20, the Committee on Plastics, has formed a subcommittee "Environmentally Degradable Plastics", D20.96. The scope of this subcommittee is the promotion of knowledge and the development of standards (classification, guide, practice, specification, terminology, test method) for plastics which are intended to environmentally degrade. I chair the subcommittee and the subcommittee is divided into five sections:

- Biodegradable (D-20.96.01)
- Photodegradable (D-20.96.02)
- Chemically degradable (D-20.96.03)
- Environmental Fate (D-20-96.04)
- Terminology (D-20-96.05).

The terminology section after much deliberations came up with two sets of definitions-one general and broad in scope and the other more specific. Both definitions were balloted at the subcommittee level. The suggested definitions were:

General Definition -

Degradable plastics - Plastic materials that disintegrate under environmental conditions in a reasonable and demonstrable period of time.

Biodegradable plastics - Plastic materials that disintegrate under environmental conditions in a reasonable and demonstrable period of time, where the primary mechanism is through the action of microorganisms such as bacteria, yeast fungi and algae.

Photodegradable plastics - Plastic materials that disintegrate under environmental conditions in a reasonable and demonstrable period of time, where the primary mechanism is through the action of sunlight.

Specific Definition

Degradable plastics - Plastic materials that undergo bond scission in the backbone of a polymer through chemical, biological and/or physical forces in the environment at a rate which is reasonably accelerated, as compared to a control, and which leads to fragmentation or disintegration of the plastics.

Biodegradable plastics - Those degradable plastics where the primary mechanism of degradation is through the action of microorganisms such as bacteria, fungi, algae, yeasts.

Photodegradable plastics - Those degradable plastics where the primary mechanism of degradation is through the action of sunlight.

This may end up being modified at the next ASTM meeting and in all probability will. However, they are good starting points for hammering out a consensus definition at the breakout sessions.

The phot and bio sections have identified a number of standard test environments for evaluating degradable materials in those environments. In terms of test methods, mechanical property loss is being considered as a simple test method to evaluate degradation of plastics, particularly for photodegradable plastics. A biometer assay involving $CO_2/CH_4$ evolution (aerobic/anaerobic environments) is being worked on as a test method for evaluation of biodegradation. Molecular weight analysis is important because reduction in molecular weight represents breakdown of the polymer. The Environmental Fate Section has been charged with designing test methods to establish the toxicity of the degraded products.

All of these documents are being prepared by the various sections of ASTM D-20.96. These will have to go through the ASTM approval process, which I will not describe here, but those interest can obtain it from ASTM. Figure 3 gives an overview of these activities.

In summary, the technical presentations to follow present the state of the art in the area of biomaterials, Environmental-photodegradable and Environmental-biodegradable materials. The consensus breakout sessions are divided into the same three areas and the goals for the breakout sessions are to come up with:

- Definitions/Terminology
- Testing of materials - specifically test methods, test environments, protocols and other issues.
- Research needs and Issues - degradation mechanisms, reference standards, environmental fate of degradation products and other issues and needs that the sessions bring out.

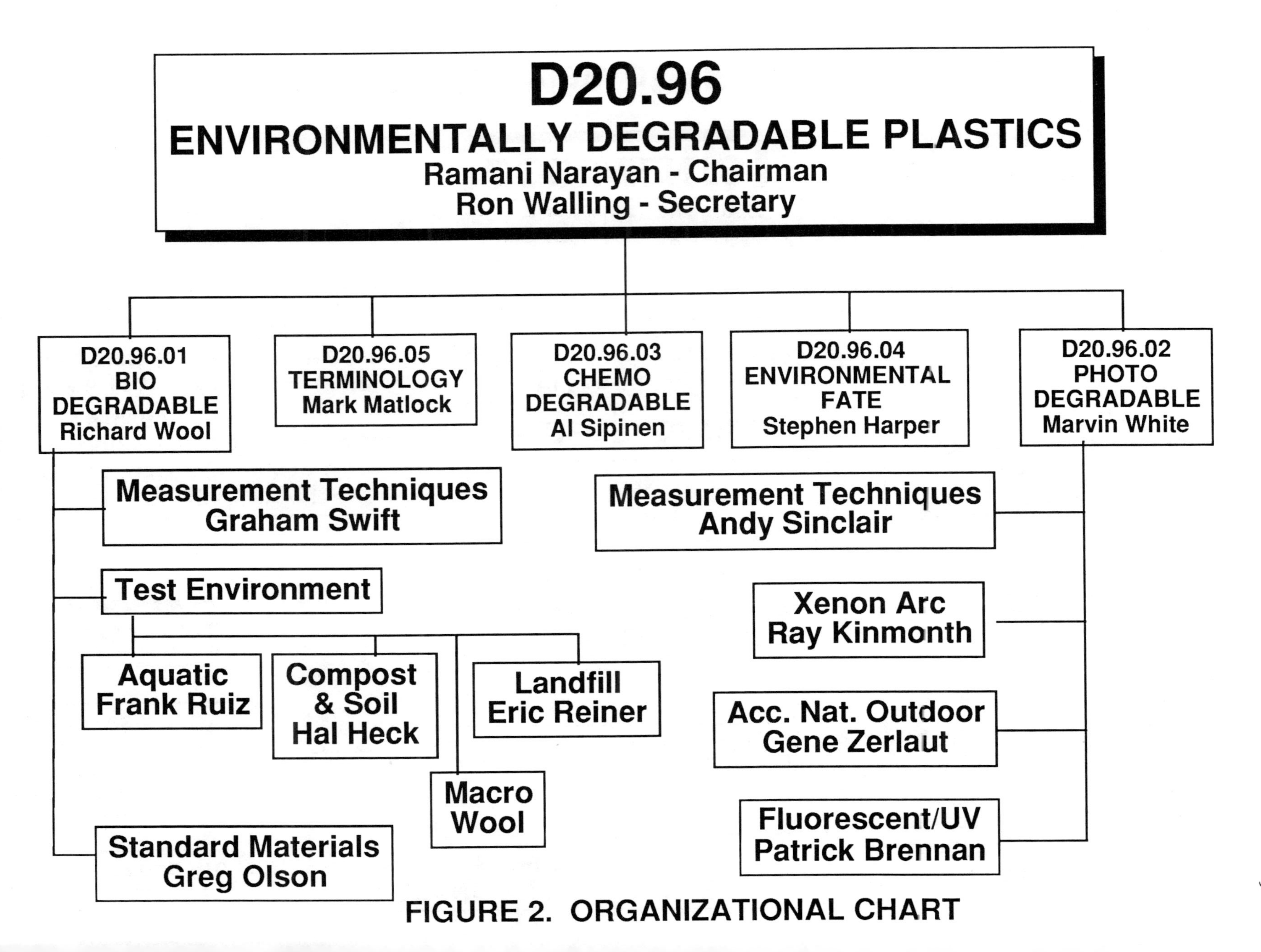


FIGURE 2. ORGANIZATIONAL CHART

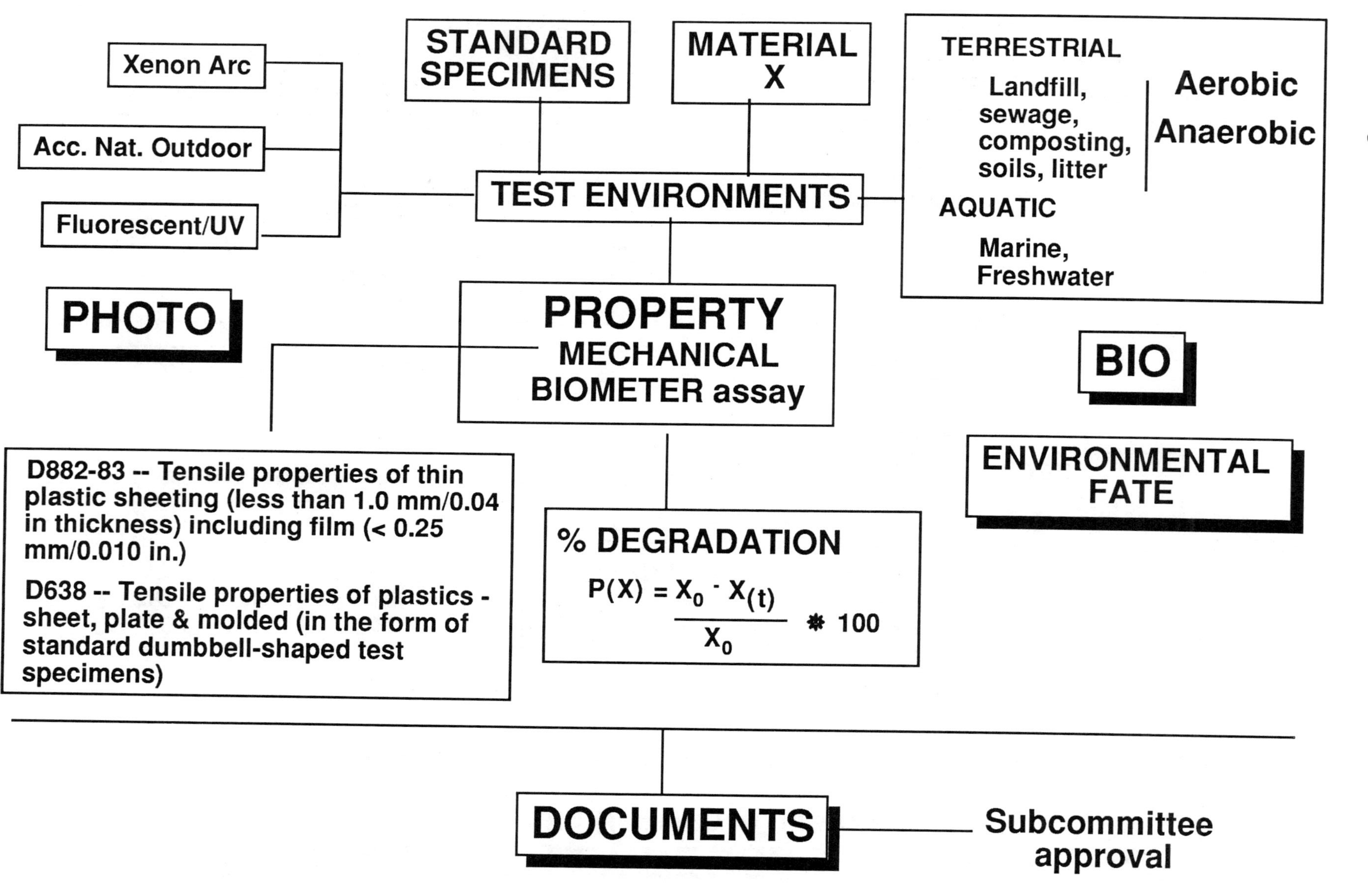


Figure 3. Overview of ASTM Test Activity

# DEGRADATION OF POLYMERIC BIOMATERIALS WITH RESPECT TO TEMPORARY THERAPEUTIC APPLICATIONS: TRICKS AND TREATS

Michele Vert, URA CNRS D500 - University of Rouen,
Laboratory of Macromolecular Substances, INSA Rouen, BP 08
76131 Mont-Saint-Aignan Cedex, France

## INTRODUCTION

Degradable materials are compounds of increasing interest and are the source of new scientific and technological problems in fields as different as computer hardware, environment, medicine, pharmacology, etc.... . The connection between degradable materials and life is becoming more and more important as humanity is taking increasing care of the preservation of both its environment and its health. Although the term "Biomaterials" has been selected as reflecting non-viable materials used in a medical device, intended to interact with biological systems (1), the term "Biodegradable" has been used for both environmental and human lives, so far. The lack of accepted nomenclature to differentiate environmental and mammalian biological systems is partly responsible for the confusions. Another reason is that differences between environmental degradation by microorganisms like bacteria or fungi and degradation under physiological conditions imposed by mammalian bodies can hardly be perceived in their totality by one discipline-contributors forced to deal with complex problems relevant of interdisciplinary research and efforts.

Instead of solely reviewing the current literature with its

confusing remarks and conclusions, I would like to take the chance of this introductory contribution to focus the attention of the participants on several aspects of degradable biomaterials which seem to be of great importance regardless of the goals: basic knowledge or applications. Indeed, in the particular field of biomaterials where a degradable material has to age in a living animal organism, all the research efforts must take into account, from the very beginning, the largest number of prerequisites and of specifications which are typical of each application. In other words, it seems to me useless to spend too much time on one particular problem such as _in vitro_ investigations of degradation rates in aqueous media taken as model of _in vivo_ conditions if one has no preliminary information on the _in vivo_ behavior of the candidate degradable material to anchor the simulation data on the scale of life effects and if aggressive conditions like sterilization are not taken into account soon. Therefore, this talk has been built up so as to draw your attention on several main problems in the field (the tricks) even if there are many reasons to be optimistic for future applications (the treats).

First of all, let us point out that no usual material is really stable in an animal body, regardless of its inorganic or organic nature. There is an increasing number of therapaeutic problems which find solutions based on the use of biomaterials. It is only in the last three decades that attention has been paved to the fact that some of these problems concern organs or body parts which have to be replaced for the rest of the patient's life whereas others request the aid of a prosthetic device for a limited period of time, namely the healing period. Table 1 gives a list of such problems which are relevant of time limited prosthetic therapy.

All these time-limited disorders are related to the ability of living systems to self-repair. In most cases, the present solutions are based on the use of biomaterials which have not been designed purposely but result from adaptations of domestic or industrial materials. Such materials often request removal after healing as it is the case for metallic osteosynthetic parts, for example. However, in time-limited therapy, materials which could provide safe healing conditions and then degrade _in situ_ are more and more regarded as of great interest. Of course, each of the therapeutic problems listed in Table 1 raises its own list of specifications to be fulfilled and requests degradable materials with adapted life-time. Therefore, we are forced to conclude that the concept of degradable biomaterial is directly connected to therapeutic applications. In a mammalian body, all the common materials are more or less attacked by living cells or biological chemicals. However, only those which can exhibit suitable life-times should be referred to as biodegradable. All the others should be referred to as unstable. Accordingly, degradation whose effects can be considered as shortcomings is obviously excluded from this contribution.

TABLE 1: TIME LIMITED BIOLOGICAL DISORDERS AND RELATED TEMPORARY AIDS

| | |
|---|---|
| Soft tissue injuries | Suture, staples, surgical adhesions, wound dressings, skin substitutes, vascular prostheses |
| Bone fractures | Bone plates, screws, nails, pins |
| Perturbed biological processes | Drug delivery systems, polymeric drugs, macromolecular prodrugs |
| Altered body fluids | Artificial plasma, synthetic blood, artificial vitreous |

Let us consider the materials which are degradable for the benefits of therapy, including diagnosis aspects. In literature, one can find degradable materials in the field of inorganic materials (porous calcium phosphate (2-3), corals (4),) and in that of organic compounds either from natural origin (proteins (5-6), polysaccharides (7)) or totally man-made (synthetic polymers). As it has been the case for industrial or domestic applications of synthetic polymers, it is because of their outstanding properties and especially because of the possibility of covering very large ranges of properties and characteristics that synthetic polymers are regarded as compounds of choice in the field of therapy. Inorganic materials like degradable ceramics or corals can provide suitable solutions to a limited number of temporary therapeutic problems. The limitation is obviously due to their intrinsic properties which make these compounds useable for hard tissue repair only. In contrast, synthetic polymers which can degrade advantageously in the body offer potential solutions to most of the recognized problems. Synthetic polymers which have been shown or only proposed as degradable when in contact with living systems are presented in Table 2.

TABLE 2: POLYMERS REFERRED TO AS DEGRADABLE POLYMERS IN LITERATURE

| | |
|---|---|
| Polyanhydrides (8) | Polyorthoesters (9) |
| Polycyanoacrylates (10-11) | Polyaminotriazoles (12) |
| Polydihydropyranes (13) | Polyurethanes (14-15) |
| Polypeptides (16) | Polydepsipeptides (17) |
| Polyphosphazenes (18) | Aliphatic polyesters (19) |

Actually, in most cases, no clear distinction was made between degradation due to microorganisms living in the environment and degradation in living bodies. Furthermore, when degradation was clearly referred to mammalian living bodies, the use of different terms such as "bioabsorbable", "biodegradable", "bioerodible" and "bioresorbable" in the absence of specific definitions maintained some confusion. Whereas one should reasonably take the chance of using these terms to defined scientific phenomena in the open literature, their involvement in the definition of commerical devices makes the move more difficult. Several years ago, we raised the problem of distinguishing degradable compounds which disappear in the body and are eliminated through natural pathways from those which also disappear from the implantation site but remain stored somewhere either because they cannot be filtrated nor metabolized or because molecules interact strongly with living elements (molecules, cells, tissues,). Therefore, we proposed to take advantages of the existence of different but closely related terms to express some nuances based on observed phenomena and personal experiences. In particular, we decided to use "biodegradable" to qualify those polymers or devices which break down at the contact of living systems so that the integrity of the biomaterial, and in some cases but not necessarily of the macromolecules themselves, is affected and lead to the loss of the desired properties. Whether molecules are cleaved is not the problem here. We are considering only the fact that fragments can move away from their site of action but not necessarily from the body. We reserve the term "bioresorbable" to those biomaterials which can degrade in such a way that low molecular weight compounds which are formed during the degradation, can be eliminated from the body through natural pathways, namely filtration or metabolization. The use of the term "bioabsorbable" is limited to those compounds which disappear from their initial implantation location with no macromolecule degradation, a definition which agrees well with the acceptance of the term "absorption" in physics. This definition fits well with a slow dissolution of water-soluble implants in body fluids for example. Accordingly, a bioabsorbable biomaterial is also a degrdable biomaterial but it can be a bioresorbable biomaterial only if the dispersed macromolecules are excreted or metabolized. Bioerodible fits well with a degradation occurring at the surface only.

Accordingly, a bioresorbable material is the top-brand of degradable biomaterial to be achieved, in particular if the final goal is therapy.

Anyhow, the concept of bioresorption applies to any no-longer desired material present in the body regardless of its natural or artificial origin (blood clot or therapeutic device) and, in the field of biomaterials, bioresorbability is just one more prerequisite to be fulfilled with respect to those properties required of regular biomaterials, namely, <u>biocompatibility</u>, which includes non-toxicity, non-carcinogenicity, non-immunogenicity and non-thrombogenicity, <u>functionality</u>, <u>sterilizability</u>, <u>adjustability</u> which is often related to the achievement of functionality, <u>manipulability</u> for practical reasons and <u>storability</u> which is necessary for economical reasons. Biodegradable and thus bioresorbable biomaterials raise specific problems. For instance, <u>stability of macromolecules</u> is required for processing, sterilization and storage stages whereas <u>unstability</u> is required to fulfill degradation requirements. Fulfilling such opposite properties can sometimes be problematic and compromise is not always feasible. This point is one of the major problems for biodegradable polymeric materials. On the other hand, biocompatibility must be extended to degradation by-products (20). This is the reason why we have decided to focus our efforts on polymers derived from metabolites from the very beginning of our research on bioresorbable polymers. The key point in the achievement of a degradable polymer is the introduction of cleavable bonds in the main chain. Many chemical groups are known as cleavable either by simple hydrolysis or by enzymatic reactions (esters, amides, ...). The amount of cleavable bonds present in the polymer main chair will determine the size of the fragments and thus, in some extents, the fulfillment of bioresorption requirements. For the last fifteen years, we have been working with polymers derived from glycolic acid, L- and D-lactic acids (21), L- and D-malic acids (22) and citric acid (23) with the goal of covering needs ranging from solid state to hydrogels and water-soluble bioresorbable polymers.

Among all the polymers which have been recognized as degradable in animal bodies, the family of aliphatic polyesters seem to be the richer and the most promising, in particular because these aliphatic polyesters break down to metabolites. Although there is always some arguments about the possibility for enzymatic hydrolysis (19), a majority of specialists are now agreeing on simple hydrolytic degradation as far as rigid polyesters derived from lactis and glycolic acids are concerned (21, 25-26). As early as 1981, we have pointed out that discrepancies, which are often found from one author to the other in the series of lactic/glycolic homo and copolymers, were mostly due to the fact that usually, factors which can affect the intrinsic properties of degradable are considered separately (20). In particular, wrong statements can be made because of a lack of careful characterization and definition of the compounds which are studied. This statement applies to all the polymeric biomaterials recognized as biodegradable so far. The situation is improving because research teams specialize on specific families. Everybody can easily associate research groups and polymers (Allcock and polyphosphazenes, Langer and polyanhydrides, Heller and polyorthoesters, Schindler and polycaprolactones, etc...). In contrast, aliphatic polyesters have been studied by enough different groups to prevent such association except, maybe, for the pioneers of the so-called bioabsorbable sutures (27).

According to our experience in the field of aliphatic polyesters, one can say that the degradation of biodegradable biomaterials depends on a great number of factors and phenomena. A tentative list is given in in Table 3.

THE TRICKS

When aging, biodegradable materials have all their characteristics which change during degradation, and this is especially true for polymers. We just saw that a great number of factors can affect the degradation processes and thus can modify

TABLE 3: FACTORS WHICH CAN AFFECT THE IN VIVO BEHAVIOR OF DEGRADABLE POLYMERS

Chemical structure, gross composition, distribution of repeat units in multimers, presence of ionic groups, presence of unexpected units or chain defects,

Configurational structure

Molecular weight

Molecular weight distribution (polydispersity)

Presence of low molecular weight compounds (monomer, oligomers solvents, initiators, drugs, etc...)

Processing conditions

Shape

Sterilizing process

Morphology (amorphous vs. semi-crystalline, presence of microstructures, presence of residual stresses)

Annealing

Storage history

Site of implantation

Adsorbed and absorbed compounds (water, lipids, ions etc...)

Physico-chemical factors (ion exchange, ionic strength, pH,)

Physical factors (shape and size changes, variations of diffusion coefficients, of HHB, mechanical stresses, stress and solvent-induced crackings, etc...)

Mechanism of hydrolysis (enzymes vs. water)

the fate of a degradable material. One must keep in mind that the properties of a polymer can be appreciated through samples only, and that polymers are not well defined compounds in contrast to low molecular weight organic molecules which have defined melting points, boiling points, solubility, etc... . If for any reason, polymeric samples bearing the same name have different characteristics, and it is often the case, one will obviously find different degradation behaviors which will be regarded as discrepancies or misinterpreted if their origins are ignored. This is a first trick.

A second trick, which is worse than the first one, is related to the fact that factors identified in Table 3 are not independent. For example, we have now evidences which show that ionic strength alters very much the uptake of water in the case of poly (a-hydroxy acid)-type aliphatic polyesters whereas pH is a secondary factor provided degradation is carried out in different isoosmolar buffer solutions (28). The effect of ionic strength is probably due to exchanges of chemical species via osmotic phenomena. Figure 1 shows the uptakes of water and the losses of mechanical properties found for compression molded specimens of PLA 37.5 GA 25 amorphous copolymer in pure water and in initially isoosmolar pH=7.4 phosphate buffer. This finding is issued from a four year investigation of the in vitro behavior of poly(a-hydroxy acids) in aqueous media which is being published. Accordingly, isoosmolarity appears to be an important factor for in vitro investigations and characterizations. People ought to take care of it when selecting a buffering system. Neglecting the control of isoosmolarity can be the source of misinterpretations when comparison of _in vivo_ and _in vitro_ characteristics are used to show contribution specific of living media. Indeed, _in vivo_ isoosmolarity is physiologically fixed, whereas it is sometimes difficult to avoid changes of osmolarity _in vitro_ when identification and quantification of compounds released from a degradable material is desired.

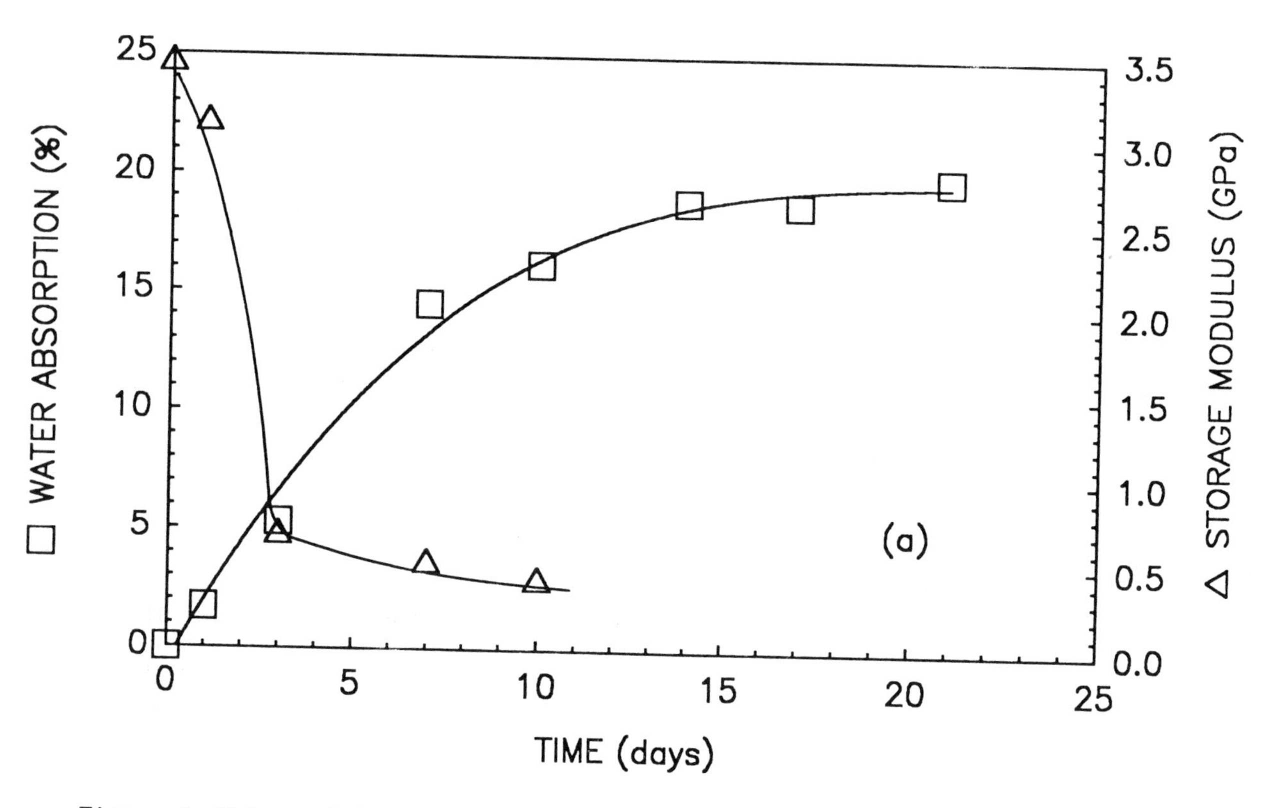


Figure 1: Water-uptake and loss of active modulus E' as a function of time for PLA 37.5 GA 25 copolymer specimens in water (a) and in isoosmolar pH=7.4 phosphate buffer (b)

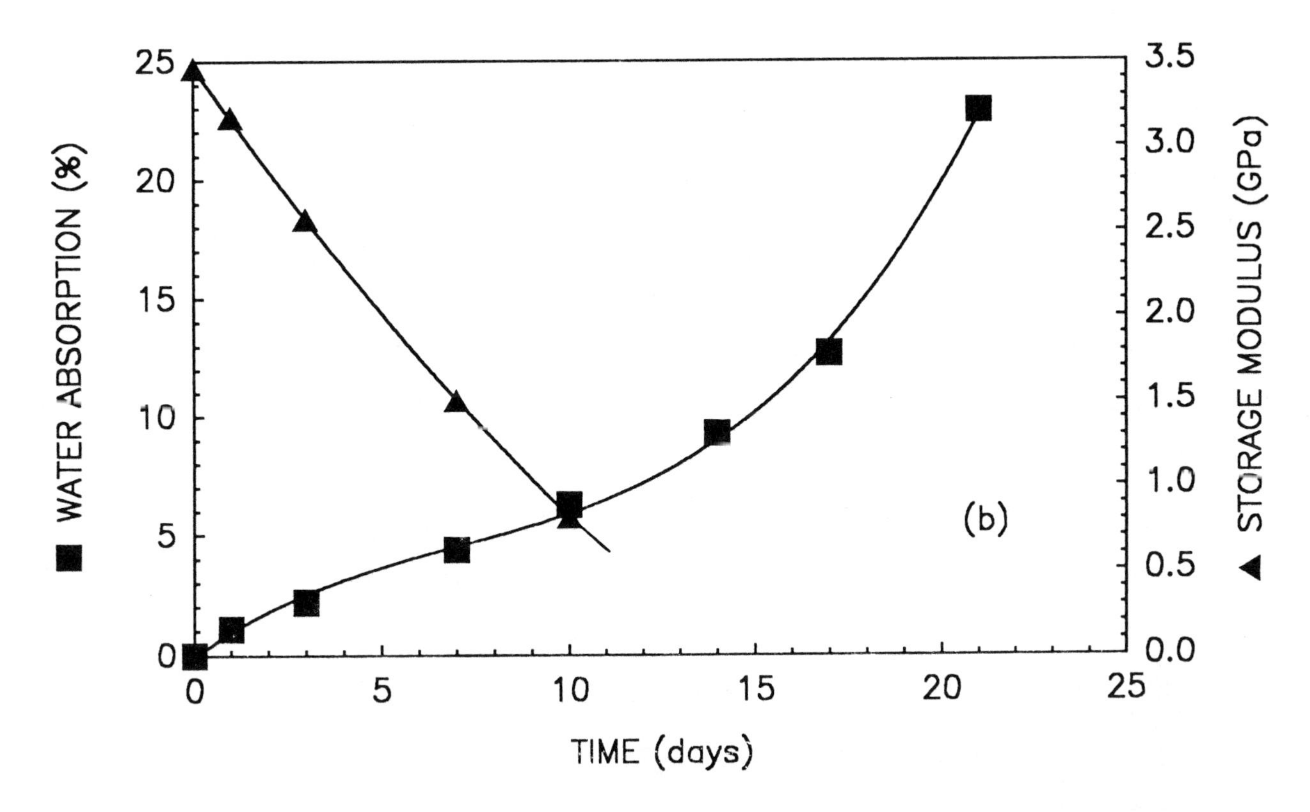


Figure 1: Water-uptake and loss of active modulus E' as a function of time for PLA 37.5 GA 25 copolymer specimens in water (a) and in isoosmolar pH=7.4 phosphate buffer (b)

In an attempt to define a method suitable for the monitoring of the fate of poly(a-hydroxy acids) in vitro, and for the same of avoiding the discrepancies due to uncontrolled sampling or experimentation, we have retained the principle of using specimens fabricated and processed similarly and allowed to age simultaneously at 37°C in aqueous media taken as model of biological fluids. Distilled water, isoosmolar saline and isoosmolar pH=7.4 phosphate buffer were selected as suitable media to detect the effects of pH and ionic strength by comparison. The specimens were given a parallelepipedic shape with sharp edges as no biocompatibility evaluation was planned. The dimensions (2 x 10 x 15 mm) were selected to allow the monitoring of aging consequences by various techniques including weighing (water absorption and weight loss), SEC (molecular weight and polydispersity), DSC and X-ray scattering (crystallinity), potentiometry and cryometry and enzymatic assay (pH changes and L-lactic acid release), and viscoelasticimetry (mechanical properties). Data collection was organized to be achieved from identified specimens allowed to age at 37°C in the aqueous media for periods as long as 2 years in order to include the case of long-lasting PLA 100, the most stable polymer of the LA/GA series (21). The principle of data averaging was also selected when necessary to data consistency. When it was acceptable, the same specimen was split in parts to perform measurements with different techniques in order to limit discrepancies related to synthesis and processing histories. Altogether, we had to use up to 30 specimens for each investigated polymer. We decided to let the specimens age without stirring since body fluids move slowly in smooth or hard tissues. Accordingly, these specimens were placed in vessels containing three floor glass holders immersed in the aging medium, the floors being made of pieces of a polyethylene mesh. Only one vessel was used to accommodate all the specimens of a given polymer. All the vessels corresponding to the various studied polymers were accommodated in the same water bath thermostated at 37°C. To fulfill this requirement, we had to use a one meter long water bath where the several sets of specimens corresponding do different polymers were allowed to age simultaneously (29).

Besides the confirmation that degradation of poly(a-hydroxy acids) depends on acid-catalyzed hydrolytic cleavage of main chain ester bonds as already shown (30), one of the main results of this study was the demonstration that degradation does not proceed homogeneously but is faster in the center of the specimens than at the surface. Figure 2a shows the aspect of PLA 50 specimens allowed to age for 5 weeks in an initially isoosmolar pH=7.4 phosphate buffer (29). SEC measurements of molecular weights in the center and at the surface revealed that molecular weights became rapidly smaller, c.a. 2-3,000, than at the surface c.a. 20,000, thus explaining that at 12 weeks only a crust was remaining (29). Figure 2b shows the aspect of similar specimens implanted subcutaneously on the back of rabbits for 2 weeks. Here again, degradation occurred faster in the center than at the surface, blistering being due to the effects of stresses on the plasticized, partially degraded material (31).

The similarity between pictures in figures 2a and 2b has

validated our in vitro approach and the method has been used for many members of the poly(a-hydroxy acid) family and for molded devices (28). The presence of both low molecular weight compounds in the central part and high molecular weights ones at the surface well accounted for the bimodal platterns observed in many instances for partially degraded, intrinsically amorphous poly(a-hydroxy acids) which could not be related to the differences in degradation rates between amorphous and crystalline microdomains as it is the case for the semicrystalline members of the family. Figure 3 shows the situation found for a biodegradable screw made from a PLA 96 semicrystalline polymer after 40 weeks in initially isoosmolar pH=7.4 phosphate buffer (28).

a

b

Figure 2: a) Cross-sections of PLA 50 specimens allowed to age for 5 weeks in isoosmolar pH=7.4 phosphate buffer.(upper)

b) Cross-sections of PLA 50 specimens allowed to age for 2 weeks after subcutaneous implantation in rabbits (deformation is due to the effects of in vivo stresses on the weakened specimen

Figure 3: Visual aspect of a PLA 96 degradable screw after 6 months in isoosmolar pH=7.4 phosphate buffer

In this case, both the center and the surface zones were solid because of the crystallinity. Degradation was faster in amorphous microdomains than in crystalline ones as it is now well known but it was also faster in the center than at the surface.

Before to end up this talk, I would like to draw your attention towards another trick which was found after sterilization of LA/GA polymers by $\gamma$-rays (32). Molecular weights of a $\gamma$-ray sterilized LA-based polymer, namely PLA 50, was measured and compared with the results obtained for various LA/GA copolymers. All these polymers were drastically degraded by the radiations as shown in figure 4.

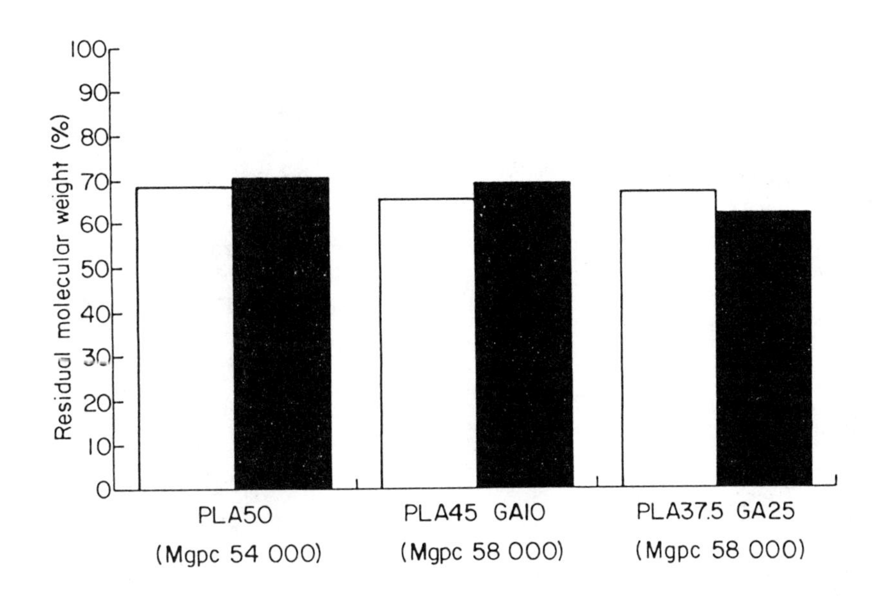


Figure 4: Histograms of the variations of molecular weights due to $\gamma$-ray sterilization for various LA/GA polymers (open: 3.77 MRads, full: .84 MRads)

However, we found that GA-containing copolymers continued to degrade on storage whereas the GA-free samples did not (figure 5). As degradation is very much dependent on molecular weights, one can imagine easily the possible consequences of such behaviors. Degradation of radiation sensitive degradable materials can be the source of many other tricks and I hope that these examples will be helpful to those who will join the field in the future.

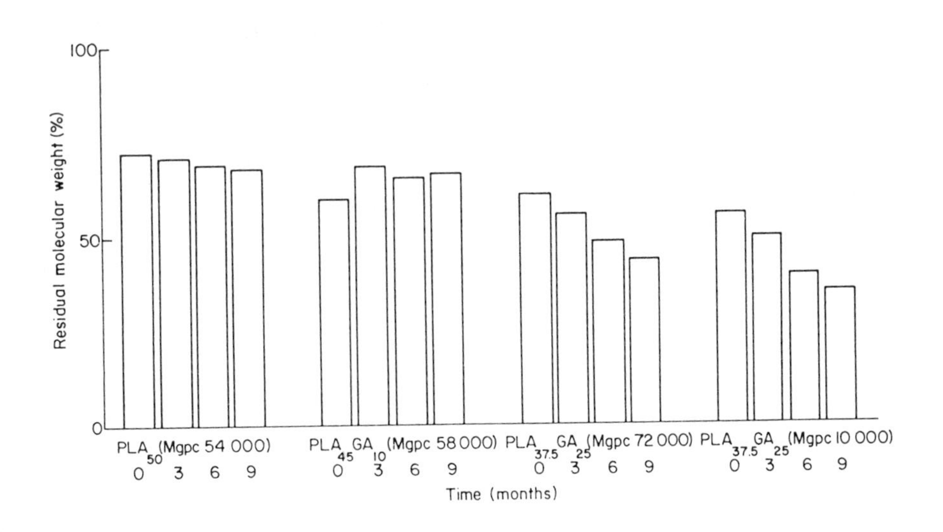


Figure 5: Histograms of the variations of molecular weights of various LA/GA-containing polymers during storage after $\gamma$-ray sterilization (3.77 MRads).

THE TREATS

In spite of the tricks mentioned before, degradable materials are now receiving great attention with respect to many therapeutic applications. Several groups in the world are now investigating applications for which degradable corals, ceramics and polymers are processed to different form and size.

TABLE 5 : biodegradable devices and some related applications

| | |
|---|---|
| LARGE SIZE IMPLANTS | sutures,<br>osteosynthesis parts<br>plates, screws, nails, cords, pins<br>bone reconstruction and augmentation<br>vascular grafts |
| MICROPARTICLES | sustain release, transport,<br>microembolization |
| NANOPARTICLES | sustain release, transport, |
| MACROMOLECULES | polymeric microemulsions<br>macromolecular prodrugs<br>polymeric drugs |

Table 5 presents some general devices which are currently mentioned in literature dealing with polymers. Even if only a limited number of compounds or systems have reached the stage of commercialization, the future seems to be widely opened to new biodegradable and bioresorbable materials (33). Polymers like polyanhydrides, polyphosphazenes, polydepsipeptides, polyorthoesters, etc... are attractive candidates. However, the effects of factors listed in Table 3 are still far from being known for most of them.

ACKNOWLEDGEMENTS

The illustration of this lecture was made possible by essential contributions of many of my coworkers and partners whose names appear in the reference list. I am very much indebted to them.

REFERENCES

1) "Definitions in Biomaterials", D.F. Williams Ed., Elsevier, Amsterdam, 1987
2) M.P. Levin, L. Getter, D.E. Cutright and N.S. Bashar, Oral Sur., 38, 344, 1974
3) A. Al-Arabi and P.K. BAJPAI, Clin. Res., 30, 294A, 1982
4) G. Guillemin, J. Fournié, J.L. Patat and M. Chétail, C.R. Acad. Sci., 293, 371, 1981
5) H.A. Tucker, in "Absorbable Gelatin (Gelfoam) Sponges, C.C. Thomas Ed., Springfield, 1965, 111
6) M. Gerondas, US Pat. 3,523,807, 19707) R.L. Kronenthal, in "Polymers in Medicine and Surgery", Polym. Sci. Tech., 8, 119, 1975
8) K.W. Leong, B.C. Brott and R. Langer, Polymer Prep., 25, 201, 1984
9) N.S. Choi and J. Heller, US Pat. Appl. 544,808, 1975
10) F. Leonard, R.K. Kulkarni, G. Brandes, J. Nelson and J.J. Cameron, J. Appl. Polym. Sci., 10, 259, 1966
11) C.W.R. Wade and F. Leonard, J. Biomed. Mat. Res., 6, 215, 1972
12) E.E. Schmitt and R.A. Polystina, US pat. 3,809,683, 1974
13) N.N. Graham, Br. Polym. J., 10, 260, 1978
14) T.E. Lipatova, T.T. Alekseeva, L.A. Bakalo and T.L. Tereschenko, Polim. Med., 10, 19, 1980
15) V. Markovitch, A. Tetsuko and K.J. Kolff, Trans. Am. Soc. Art. Intern. Organs, 8, 79, 1972
16) W.G. Miller and J. Monroe, Biochemistry, 7, 253, 1968
17) M. Goodman and K.S. Kirshenbaum, US Pat. 3,773,737, 1973
18) H.R. Allcock and T.J. Fuller, J. Am. Chem. Soc., 103, 223, 1981
19) S.J. Holland, B.J. Tighe, and P.L. Gould, J. Control. Rel., 4, 155, 1986
20) M. Vert, F. Chabot, J. Leray and P. Christel, Makromol. Chem., Suppl. 5, 30, 1981
21) M. Vert, P. Christel, F. Chabot and J. Leray, in "Macromolecular Materials", G.W. Hastings and P. Ducheyne Eds., CRC Press, Boca Raton, 1984, 119
22) R.W. Lenz and M. Vert, US Pat. 4,265,247, 1981 and US Pat.4,320,753, 1982
23) M. Boustta, J. Huguet and M. Vert, Polymer Prepr., 30, 366, 1989
24) T.N. Salthouse and B.F. Matlaga, Surg. Gynecol. Obst., 142, 544, 1976
25) J.M. Brady, D.E. Cutright, R.A. Miller and G.C. Battistone, J. Biomed. Mat. Res., 7, 155, 1973
26) T. Nakamura, S. Hitomi, S. Watanabe, Y. Shimizu, K. Jamshidi, S.-H. Hyon and Y. Ikada, J. Biomed. Mat. Res., 23, 1115, 1989
27) E.J. Frazza and E.E. Schmitt, Biomed. Mat. Res. Symp., 1, 43, 1971
28) Li Su Ming, Thesis of the Universty of Rouen, France, june 1989
29) Li Su Ming, H. Garreau and M. Vert, J. Biomed. Mat. Res., submitted
30) C.G. Pitt and A. Schindler, in "Progress in Contraceptive Delivery Systems, E.S.E. Hafez Ed., M.T.P. Publishers, Lancaster, 1980, 17
31) M. Thérin, P. Christel, Li Su Ming and M. Vert, to be published
32) G. Spenlehauer, M. Vert, J.P. Benoit and A. Boddaert, Biomaterials, 10, 57, 1989
33) M. Vert, Angew. Makromol. Chem., 166/167, 155, 1989

QUESTIONS AND ANSWERS

DR. NARAYAN: The floor is open for questions.

Before you ask a question, please identify yourself so that the court stenographers can actually get your name and affiliation.

DR. SHALABY: I am Shalaby Shalaby with Johnson & Johnson. Just a comment.

Since we are developing a consensus, it appears that from two speakers it looks like it may be unwise to combine nomenclature between the biodegradable and the environment and the bioabsorbable or biodegradable in mammalians, so it is something to think about. We should think from now on to separate mammalian, polymer and mammalians, and polymers that are not intended for any particular use outside.

The second comment I would like to make, when you talk about post-irradiation, degradation that Professor Vert alluded to, we have to be very careful to define under what condition we have treated the system with gamma and under what systems we are storing these polymers post-irradiation. That can make a tremendous difference in the profile of the material.

DR. LOOMIS: Gary Loomis, DuPont.

Michele, did you say that when your amorphous polymer that was DL and 25 percent GA, or whatever -- statistically that should not be possible to have runs of 10 lactic acid units of the same configuration, yet our work has shown that you need 10 lactic acids units of the same configuration to get crystallinity due to the 3(1) helix.

So I wonder if from your x-ray work if you are sure that the x-ray defraction pattern you are seeing is due to polymer crystallinity or could it just be due to lactide from a backbiting, or whatever?

DR. VERT: No. No, No. No, because it occurs in many intrinsically amorphous polymers, or copolymers depending on the composition.

No, but look at the 75 percent L-lactic 25 percent GA, okay? That's an amorphous polymer, initially. But degradation occurs preferentially in the glycolide units level so that slowly you rise the content in lactic acid, and after a couple of weeks you end up with 90 percent of lactic acid.

DR. LOOMIS: I understand that, but in that polymer, the 75/25, it is statistically possible to have runs of 10-L. What I'm saying is in the DL+ glycolic, in that polymer, it's not statistically possible to have runs of 10 lactic acid units in the same.

DR. VERT: No, no. For sure it's not a random copolymer, but the problem is that initially when you have 25 GA, the short sequences of LA cannot crystallize. Even if you need 10 LA you can find them.

DR. LOOMIS: I guess what I'm questioning is: Do you really have 10 LA units in that one?

DR. VERT: I think so as we did it with other copolymers, and the more you have LA, the higher crystallinity.

And it's also true in the case of the annealed polylactic acid which crystallizes easily.

It's connected with some kind of plasticizing effect, and as you increase the amount of LA, the molecular weight is decreasing, of course, so the crystallization is more and more easy, I would say.

But the crystallization occurs on low molecular weight compounds. So it's what you are saying. You need just short sequences. You know, oligomers with less than 10 LA can crystallize if you make -- no, no, if you make condensation -- well, normal esterifications of lactic acid. You end up with a crystalline material, and the number of units is very small. 10 LA means 1,000 molecular weight or so.

DR. D. WILLIAMS: I am David Williams from Liverpool in England.

Could I say I don't agree at all with the suggestion that we separate the medical from the other environmental?

I think that is closing a door which we really don't want to close at the moment because we don't understand exactly what is going on in some of these situations. We don't know whether they are the same mechanisms but just mediated by different phenomenon or they are entirely different mechanisms, and I think it would be very, very wrong.

I think that we have already seen a couple of cases where there is some confusion this morning in terms of nomenclature which would suggest that we should keep these separate.

Chairman, you said at the beginning, in one of your slides, you separated I think the biomedical from the environmental and referred to one as being in vivo and other as being in vitro. We have to be exactly clear, very clear what we mean by "in vitro" because that is not the same as environmental.

In vitro specifically means carrying out studies where there are cells present, cells in culture, and that is not environmental. We have got a really big problem if we do that.

I also am very, very concerned about the definition of "biodegradation" that you had as far as ASTM is concerned. I myself don't think I am a microorganism; I think I am a little bit bigger than that, and I am very worried about a definition which says that biodegradation is needed only by microorganisms.

We have got huge problems with that sort of definition and I think we have got to keep these two very close together.

DR. VERT: Is that to me? Or was that a comment to the Chairman?

DR. D. WILLIAMS: That was a comment for the Chairman.

DR. VERT: Oh. Thank you. I was worrying.

DR. HENDERSON: Alex Henderson, Ethicon Canada.

For biomaterials it also is dependent on the environment where the implant is. That's going to have to be considered too, or defined. You get different rates of degradation depending on blood supply and other factors such as that, so if we are going to define "biomaterial degradation" you are going to have to pick maybe standard implant sites or things like that. It's dependent on the area of implantation.

DR. VERT: Do you have a suggestion?

DR. HENDERSON: No, not right now.

DR. FEIJEN: My name is Jan Feijen.

I would like to ask a question: How to classify in your terminology polymers like copolymers or vinyl ethers and, for instance, unseparated anhydrides?

You are changing the structure after exposure to the surroundings. At least you hydrolyzed the anhydrides, but you are not changing the molecular rate more or less, not the main chain. Do you then propose to make a combination of biodegradable/ bioabsorbable, or bioresorbable/biodegradable, or whatever?

DR. VERT: That's a good question.

DR. FEIJEN: Or how do we solve this problem?

DR. VERT: That's a good question. That's a good question. I guess the key of the answer will be let's say the molecular weight, for example, because for sure this system, they are water soluble.

DR. FEIJEN: Yes, they turn into water soluble compounds, but they will not apparently further degrade.

DR. VERT: Yes, yes.

DR. FEIJEN: So they dissolve?

DR. VERT: So the terminology, as far as I am concerned, would be for low molecular weight compounds which can be excreted by filtration at the level of the keto. Let's assume this. That would be a biodegradable system.

But for high molecular weight compounds - and this was well-known with polyvinyl a long time ago - if the molecular weight is very high you don't have any filtration so you have a storage in the body. In that case the degradation occurs, but not the degradation of micromolecules. This would be bioabsorbable.

DR. FEIJEN: So we further have to specify there is degradation in the molecule? There is a change --

DR. VERT: But you have to accept --

DR. FEIJEN: So we have to come to some conclusion for that?

DR. VERT: You have to accept some precision, I would say.

DR. NARAYAN: Thank you.

# PHBV$^{TM}$ Biodegradable Polyester

T.J. Galvin

ICI Americas Inc.

Wilmington, DE 19897

## INTRODUCTION

Naturally occurring, biodegradable polymers have been known to be present in the environment of many, many years. The polymer family ICI has concentrated on, however, were first identified by the Pasteur Institute, Paris in 1925. They worked with a homopolymer of hydroxy butyric acid (PHB) and generally classified its physical properties and characteristics. This work was expanded on by U.S. researchers in the early 50's. The homopolymer was seen to have some serious deficiencies in thermal stability, brittleness and, most importantly, in ease of manufacture.

ICI's work in this area began in the mid 70's, about the time of the first oil shock. The objective of the work was to produce a polymer or polymers from naturally occurring raw materials, not to develop a bio-degradable polymer. Like the prior workers, ICI found the same serious deficiencies in PHB which eventually led to a variety of studies investigating both the type of bacteria employed and the feedstocks fed to the bacteria. Eventually they zeroed in on the bacteria Alcaligenes Eutrophus and found by carefully controlling a mixed feed stock of organic acid and sugar, they could recover a copolymer of hydroxy butyric and hydroxy valeric acid-polyhydroxybutyrate-valerate of PHBV.

PHBV's process of manufacture, the general properties of the copolymers, biodegradation in various media and possible application areas will be discussed further in this paper.

## I. Manufacturing process

The general process schematic is as follows:

Feedstocks

Bacterial Fermentation

Polymer Extraction

Polymer Drying/Pelletizing

Glucose is currently used as the principal feedstock but virtually any fermentable sugar can be used; ethyl alcohol can also be used. The percent valerate can be controlled simply by adjusting the amount of aliphatic organic utilized.

Unlike prior efforts to obtain the polymer, the ICI process does not use a solvent extraction but a simple aqueous extraction. This decreases the amount of cell debris left in the polymer although some molecular weight is sacrificed. For example, if the copolymer is carefully removed from the cell by solvent extraction, molecular weights of 1 million are obtained; the aqueous method yields polymer of molecular weights in the 600,000 range.

Accordingly, the key elements of the ICI process are:

- control of the valerate content,
- control of the molecular weight, and
- the aqueous extraction procedure.

All these contribute to the production of a consistent, clean product while the last item, in isolation, is key to an economical, hazardous free, environmentally sound process.

GENERAL PROPERTIES

As noted previously, the homopolymer of hydroxybutyric acid (PHB), is a rather brittle, thermally unstable material. The brittleness is due to its very high crystallinity - 80%. The thermal instability comes about because the PHB polymer begins to decompose roughly ten degrees above its 350°F. melting point.

Adding valerate to the polymer leads to a number of improvements including:

- reduction of the crystallinity
- drop in melting point, and
- increased flexibility

Current thinking is that the introduction of valerate leads to the introduction of amorphous regions which in turn leads to a reduced melting point and an increase in flexibility.

PHBV is currently produced with valerate contents ranging from 5 to 20%, either in powder or pellet form. The copolymers can be plasticized, filled, nucleated, etc. much as conventional polymers. They also can be recycled much like PET either into re-usable second generation products or thru use of scrap into first generation items.

PHBV polymers will blend very well with polymers containing some type of polar group such as PVC, chlorinated polyethyline, polycarbonate and low melt nylons and alloys of these resins. The copolymers are not compatible, however, with nonpolar systems as represented by the polyolefins and polystyrene.

PHBV copolymers' general property profile most closely resembles that of polypropylene except in two places where it resembles that of PET. The copolymers are all heavier than water with densities of approximately 1.25 $gm./cm^3$ and all have gas barrier properties comparable to PET.

PHBV copolymers have an excellent toxicological profile. For example the LD50 is greater than 5000 mg. kg. and there is no sign of skin or eye irritation. PHBV copolymers are also biocompatible; various studies have demonstrated that the copolymers are well tolerated by the study and lose strength and weight over time. The rate of absorption by the body is dependant on surface area, molecular weight and valerate content.

## RESIN PROCESSING

PHBV copolymers, either plasticized or as virgin resin, have been processed on conventional plastic equipment to produce films, bottles, and injection molded articles.

With injection molding processes, standard machines with polyethylene type screws (LD = 20/1) are preferred. Barrel temperatures should be about 150-160°C. The PHBV range of copolymers have a very low melt viscosity, so low that conventional melt index measurements do not apply.

Processing of PHBV films has yet to be finalized. The low melt viscosity coupled with a slow recrystalization rate relative to other polymers have proven difficult hurdles to overcome. Progress has been made, however, principally thru careful control of temperatures and thru the use of selected nucleating agents. Work is ongoing in this area.

PHBV copolymers have been successfully spun into fibers thru a dry spinning process utilizing chlorinated solvents. A non-woven floss with specific fiber denier has been made which can be further processed into non-woven articles that range from light tissues to fairly tough fabrics.

The low melting point of plasticized PHBV formulations lends itself readily to conventional paper coating operations. Film thickness is dependant upon both machine speed and the character of the substrate. Solvent solutions of PHBV have also been used to cast film on to a variety of cellulosic substates.

Biodegradation

PHBV is seen as an energy source by the natural microorganisms present in the soil. These microorganisms be they fungi, bacteria or whatever, colonize on the surface of the PHBV article and gradually digest the polymer. The products of this digestion are carbon dioxide and water.

Biodegradation, of course, is a complex process with the rate dependant upon a number of factors including:

- microorganism intensity
- temperature
- moisture
- surface area

Accordingly, similar articles made from PHBV would be expected to degrade in the medias noted as follows:

| | | |
|---|---|---|
| Aneorabic Sewage | - | Fastest |
| Estuarine Sediment | - | |
| Aerobic Sewage | - | |
| Soil | - | |
| Seawater | - | Slowest |

PHBV is indefinitely stable in air. It must be placed in an environment where microbes exist before the degradation occurs; the higher the microbial activity, the faster the degradation.

POSSIBLE APPLICATION AREAS

The key properties of PHBV is its biodegradability, biocompatability and naturalness; its general physical properties are about equivalent to other conventional polymers. Accordingly, it is expected that the copolymers will find niche applications that must have requirements for the key properties.

Some of these include:

- Ortheopedic devices: the biocompatability of PHBV and its ability to gradually degrade in the body over a long period of time makes the copolymer an ideal candidate for orthopedic pins, screws, bone augmentation, etc. PHBV's breakdown products; ie, hydroxy butyric and/or valeric acids, are normally present in the body and excreted via body fluids.

- Personal hygiene: PHBV in either film or non-woven fiber form would seem to be appropriate for this area which is as broad as wound care to disposable diapers.

- Specialty packaging: PHBV bottles, films, paper coatings, etc. are all candidates for this general area. PHBV's excellent grease resistance, in particular, is noteworthy for fast food wrap application.

- Controlled release: PHBV's ability to both breakdown in the body or in the soil makes it a candidate for both pharmaceutical and/or agriculture applications. Its relatively slow breakdown in the body may allow dosage forms of pharmaceuticals not heretofore possible. The same could be said for a variety of herbicides and or insecticides.

- Polmer blends: PHBV's compatability with polar type polymers offers a degradation route for these polymers without the major sacrifice in properties experienced with other solid additive approaches.

## CURRENT STATUS / FUTURE POSSIBILITIES

PHBV is currently produced in Billingham, England at a rate equivalent to 50 tonnes per year. All products are sold thru a wholly owned ICI affiliate company, Marlborough Biopolymers Ltd.. The small pilot plant has been ideal in developing a robust production process. Product from this unit, which is undergoing expansion to 500 tons, is $15 per pound; when the expansion is completed the price will drop to about $7 per pound.

The development prices noted above are far removed from the target price of product from a full scale unit - $2 per pound. Granted, this target price is still more expensive than bulk commodity resins but it does represent a very significant improvement over current pricing. Nevertheless, at the $2 level, PHBV will probably first be employed in the niche markets that demand either biodegradability or biocompatability.

It is difficult to see past this stage at present but as demand grows, as petrochemical raw materials fluctuate versus commodity, agriculture types, as the demand for biodegradable packaging increases, PHBV's price may not seem excessive in the 90's.

ICI, as the inventor of polyethylene in the late 30's, has had a grand stand seat in watching the growth of this polymer over the past fifty years. PHBV is still on its growth curve, but has come a long way in the past 10 years toward meeting the demands of a truly biodegradable thermosplastic in terms of performance, cost and safety.

QUESTIONS AND ANSWERS

DR. FEIJEN: I would like to ask you something in particular about this bacterial degradation. I am more from the biomaterials area so I don't know too much about this bacterial degradation process.

In particular your kind of polymers, how does this process go? We have different bacteria, we have, what, the aerobic/anaerobic type of material. Can you tell us anything about the mechanism or part of the mechanism, how bacteria attack your biodegradable polymers?

DR. GALVIN: How long is the string? That's a tough question to answer because of the reasons that I have put up there before, but it's probably a combination of esterase enzymes and hydrolysis occurrring at the same time.

DR. FEIJEN: but do we know how the bacteria, are they excreting the enzymes or anre certain enzymes at the surface? How is this process going? Is anybody in the audience going to tell us later about this in detail? Or are there any comments?

DR. GALVIN: I see a hand in the corner. I hope they can tell you because I don't believe I can.

DR. GILMORE: David Gilmore, University of Massachusetts. I will be speaking a little bit on this topic tomorrow, at least as far as what I know about the process and what we're doing to study it, and can enlighten you more at that time.

DR. GALVIN: Thank you, sir.

DR. ANDERSON: Jim anderson from Cleveland.

I have two short questions. One refers to the data base that you put up that you talked about or at least you showed the in vivo tissue response.

My question is: What assay system permits you to say that your polymer is much, much less in terms of its response to nylon or polypropylene?

DR. GALVIN: You are asking details I don't have, quite frankly. I am giving this paper from Dr. Lloyd who might have been able to answer you.

DR. ANDERSON: The second question may refer to Dr. Feijen's question, and you showed I believe at three days the surface pitting effect of fungi on the material? What's the dimension of those pits? I wasn't able to pick it up. And is there a direct relationship between the fungal adherence of the organism and the pits in the early stages of the surface degradation phenomenon?

DR. GALVIN: I will have to get back to you.

DR. ANDERSON: Thank you.

DR. D. WILLIAMS: I can follow up that question, if I may.

Could you tell us what material variables control the degradability in vivo? I ask that question because you are probably aware that this matter of some controversy in the literature, that there are several people, myself included, who have implanted PHB and its copolymers into experimental animals and seen no degradation whatsoever in the time period you are saying responds to its half life. There are other people who have shown degradation but in materials of a totally different form.

DR. GALVIN: I think you are probably aware, that has been done on these polymers has not been done in vivo.

DR. D. WILLIAMS: We have material from ICC supplied by ICI for us. We have published that in vivo, published in the literature.

DR. GALVIN: I'm sure you know more about it than I do.

DR. D. WILLIAMS: I should also add that you mentioned esterase enzymes produced by bacteria. We have studied the effects of esterase, specifically nonspecific esterase, in vitro on PHB and its copolymers and saw no degradation whatsoever by esterase unless, unless we pre-degraded the PHB by gamma radiation to such a point the molecular weight made it a rather unviable material anyway.

DR. VERT: You know, nothing is stable in the body, so degradation is a general process, I would say. There is no polymer absolutely stable. The half lifetime of PET is something like 30 years or so. But anyway, it does degrade.

So I guess that biodegradation must be regard as a positive factor and it cannot be a negative factor. If there is a negative factor, it is the unstability with respect to the application.

This means that the biodegradation is directly related, directly related to the applications, okay? And I guess here we are facing an example where we would need two different expressions to reflect the biodegradation because for sure PHB does biodegrade in the environment, but as far as I understand, and I guess David has just pointed it out, it doesn't degrade very much, let's say, and in such a way that we can take advantage of it in vivo, unless you have some other results in this field.

DR. GALVIN: Well, there are a couple things that I don't know that have been investigated thoroughly. You touched on some of the things in polymer work that you are doing. One is the molecular weight. We are talking about molecular weights up around 600- to 7000,000.

Too, there is the effect of the valerate content and how much valerate you can put in, and one thing that I didn't mention previously, which is the polymers can be plasticized with things like glycerol triacetate which should go out faster and increase surface area and so forth.

DR. LOOMIS: I think there was an implication you mode there when you showed some what you called biodegradation from -- I think you went from anaerobic innoculant to soil to aerobic innoculant to sea water, and you showed that the weight loss changes.

But in going from -- if you look down that list I think all you are doing is dropping the pH. You are going from a lower pH to a higher pH.

You can make the same argument that pH has nothing to do with bacteria whatsoever.

One other thing on the cost, you are aware of course there is a German company now that has expressed the gene in ecoli, and it is now producing PHB extra-cellular, which means they don't have to get rid of cell debris and so forth, and they claim that they are going to be selling it for I think 75 cents a pound. Do you have any comment?

DR. GALVIN: No comment.

DR. LOOMIS: No comment.

DR. SHALABY: Actually, I have a few comments and a question.

First, when you talk about nine to twelve months' half life, we should define exactly how do we measure to form and the medium, and I agree with Dr. Williams that the form can make a great deal of a difference.

Secondly, when we talk degradation we should define that degradation could be molecular, volume. It could be change in mechanical property. It could also be change in mass. So this is one of the issues we have to address in our consensus: What do we mean by "degradation"?

My question is: PHB and copolymer is indeed sensitive or degradable in the presence of microflora. For ICI, what would be the most impressive issue that you have to address to make it most useful?

DR. GALVIN: Could you say that again, please?

DR. SHALABY: What would be the most important issue that ICI needs to address to make PHB and its copolymer a useful material and a good substitute for traditional plastics?

DR. SHALABY: What would be the most important issue that ICI needs to address to make PHB and its copolymer a useful material and a good substitute for traditional plastics?

DR. GALVIN: I think the gentleman on the righthand side said what we had to do. What we have got to do is get the volume up and get the price down as far as packaging applications are concerned.

DR. LOOMIS: Technically, what we have to do is to develop the package of additives and what I will call mouse milk that you have to sprinkle into a resin to get a formulation that is neither a film-forming resin or formulation, which has all the nucleating agents and the thermal anti-oxidants and so forth as required. We also have to do more work on formulating products for injection molding to get the impact up or to get the anti-stats or to get the, you know, the formulation that the customer can use.

We have spent, up until this time, most of our effort in trying to get the process right because when we were developing the process before we went to our 50 ton plant, our molecular weights were going anywhere from 200- to 800,000, and we were at the same time trying to develop the applications. It was like shooting at a moving target from a moving platform.

We have now have the process, 95 per cent right, so we can start developing the formulations that will enable the converters to convert it into useful articles. But there still is that question of price.

DR. NARAYAN: Thank you, Tom.

# PHOTODEGRADABLE PLASTICS

James E. Guillet
Department of Chemistry, University of Toronto
Toronto, Canada M5S 1A1

## INTRODUCTION

It has become popular in certain circles to blame many of our contemporary problems on advances in technology. In keeping with this trend, we often blame the packaging industry for increasing amounts of garbage and litter, and the disposal problems associated with it. The fact that the development of the modern packaging industry has had a major effect on reducing the cost of consumer goods both in manufacture and distribution, and the critical role it plays in protecting our health is often ignored. However, it must in fairness be admitted that until very recently the packaging industry has been so concerned with the problem of providing adequate and attractive packaging that it neglected to concern itself about the disposal of these packages once they have served their initial functions. Ecologists have made us fully aware that no human system can be considered in isolation. Each activity must be contemplated in relation to our total environment. We should therefore extend our consideration of packaging practice to the point of ultimate disposal or recycling of the material from which the package is made.

At this point it is helpful to distinguish between two kinds of disposal problems relevant to our modern society. The first of these is garbage disposal. Garbage can be defined as the discarded solid waste products of household or industry which are collected and disposed of in some central facility such as a dump, landfill, or incinerator. Litter, on the other hand, may be defined as a man-made object in a place where it should not be. For example, a fallen tree in the forest is not litter, but a discarded wooden box made from the same material in the same place is. Paper in a garbage can is not litter. The same piece of paper blowing along the side of a road definitely is. Surveys of litter show that by far the greatest proportion is containers or packages of various kinds used for food, beverages or tobacco.

Plastics have one major advantage over glass and metal in packaging applications in that they are inherently organic materials, just like banana skins and coconut shells, and it is therefore possible in principle to make them degrade by natural mechanisms once they have performed their primary function as a temporary container. This paper describes ways in which this can be accomplished.

In consideration of these principles, it is possible to draw up a list of the desirable characteristics for a packaging material.

1. It must be resistant to the material which it is to contain and not contribute to the taste, odor, or toxicity, particularly if it is a food product.

2. It must be light in weight and easily formable into an attractive package.

3. It must be cheap and represent a minimal expenditure of natural resources in its manufacture.

4. It must be resistant to microorganisms which might otherwise attack the materials which it contains.

5. It must be stable and maintain its desirable physical properties for at least the lifetime of the product which it contains.

6. It must be disposable or recyclable by conventional garbage disposal technology.

7. It should degrade by some natural mechanism if it becomes litter.

Plastics, as currently manufactured, fulfil all of these characteristics except the last, and until recently the latter requirement has been considered to be inconsistent with the fourth since it was felt that if the plastic were biologically degradable, it would no longer afford adequate protection against the attack of microorganisms on the product which the plastic is intended to contain. Now, however, it is clear that these two requirements need not be mutually exclusive. It has been found that the resistance of conventional plastics to microorganisms is primarily due to two factors: (1) the low surface area and relative impermeability of plastic films and molded objects and (2) the very high molecular weight of the plastic material. Microorganisms tend to attack the ends of large carbon-chain molecules and the number of ends is inversely proportional to the molecular weight. In order to make plastics degradable, it is necessary first to break them down into very small particles with large surface area, and secondly to reduce their molecular weight.
This can be done utilizing the energy of natural sunlight, and several

commercial processes have been developed for the manufacture of photodegradable plastics.
This can be done utilizing the energy of natural sunlight, and several commercial processes have been developed for the manufacture of photodegradable plastics.

## TECHNOLOGY OF DEGRADABLE PLASTICS

Recently a number of processes have been disclosed using additives to make "unstable" plastics. But packaging materials require not only a polymer that will degrade, but one that will degrade at a controlled and predictable rate. One approach to this is to make a change in the plastic molecule itself. Small quantities of a sensitizing group is exposed to natural sunlight, the sensitizing group absorbs radiation which causes the chain to break at that point and thus form smaller segments. Since the physical properties of a plastic depend on the length of the chain, if the chain is broken in enough places, it becomes biologically degradable.

### The Ecolyte Process

The Ecolyte process developed at the University of Toronto involves inclusion in the backbone of the chain of a polymer, a group of the general structure.

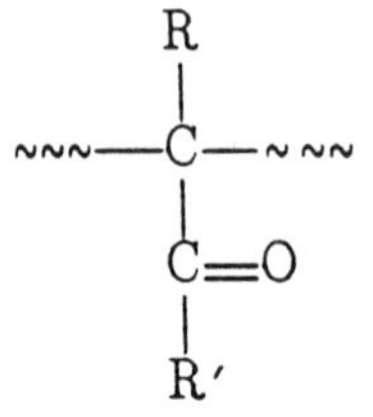

$$
\begin{array}{c}
R \\
| \\
\sim\sim\sim - C - \sim\sim\sim \\
| \\
C{=}O \\
| \\
R'
\end{array}
$$

where R and R′ are various alkyl and aryl substituents. By changing the nature of these two groups, one can control the rate of the degradation process. When the carbonyl group absorbs a quantum of ultraviolet light, the classical photochemical reactions which occur are (a) the Norrish type I reaction, a free-radical split occurring at the carbonyl group to give two free radicals, and (b) the type II process, an intermolecular roarrangement resulting in a scission of the main chain to give a methyl ketone and a terminal double bond.

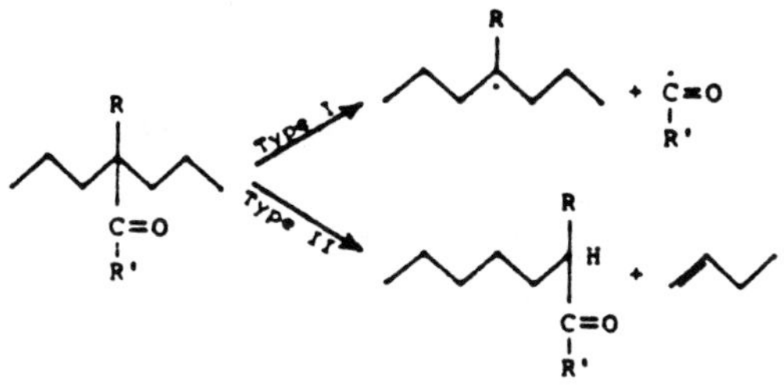

The type I process gives two free radicals, one polymeric and one small acyl radical. The polymeric radical can undergo a rearrangement known as ß-scission which results in a break of C-C bond in the backbone of the polymer and a consequent reduction in molecular weight. The type II reaction, however, is the major photodecomposition process that causes chains to break. In the presence of oxygen, both radical sites can induce photooxidation processes which cause chain degradation over a longer time scale.

Alkyl ketones have absorbances with a maximum of around 280-290 nm and they cut off rather sharply at about 300 nm. Figure 1 shows the solar spectrum along with that of other light sources. the emission spectrum of the sun has an approximate Boltzmann distribution of radiation which cuts off rather sharply at 300 nm. As a result, ketones are stable compounds photochemically as far as visible light is concerned and undergo photochemical reactions only if they are irradiated with light of wavelength less than 330 nm, such as occurs in natural sunlight. This gives the possibility of producing a packaging material which is stable in visible light, but which degrades when thrown away in a outdoor environment. The critical portion of the sun's spectrum, between 290 and 330 nm is called the "erythmal region" and is the radiation responsible for tanning and sunburn of the human skin.

Table I shows the relative intensity of sunlight in various regions of the spectrum. For the erythmal radiation, noon sunlight in Arizona is about 300 or 400 times the intensity of an ordinary fluorescent light. This means that you would have to sit a long time under a fluorescent lamp in order to get much of a tan. In the near UV range there is more radiation in artificial sources although still rather small compared to solar radiation. It is quite obvious that there will be a much larger effect in sunlight than there is under any of these normal lighting conditions.

TABLE I. Output of Artificial Lighting Compared to Solar Radiation ($\mu$watt/cm$^2$)

| Type of lamp | Erythmal $\lambda$=280–320 nm | Near UV $\lambda$=320–400 nm | Visible and IR $\lambda$>400 nm |
|---|---|---|---|
| Incandescent | | | |
| 40 watt | 0 | 0.21 | 21 |
| 100 watt | 0 | 0.89 | 71 |
| 500 watt | 0 | 6.55 | 409 |
| Fluorescent | | | |
| 40 watt | 0.8 | | |
| Noon sunlight, Arizona | 259 | 4640 | 88,000 |

The other factor which is important is that ordinary window glass filters out the erythmal radiation of the sun.

It is not necessary for the plastic to be in direct sunlight in order for the degradation process to occur. Over 50% of the total amount of ultraviolet radiation comes from the sky rather than from the sun itself. Consequently, even if the plastic is in the shade it will still be receiving skylight and hence will degrade. In fact, as long as the plastic can be seen outdoors, it will be undergoing degradation. The rate at which the chains will be broken depends only upon the intensity of the UV light absorbed by the sample. In northern latitudes such as in Canada the intensity of this light will vary with the time of year. This means that in the winter the rate of degradation will be rather slow, while during the summer, the rate will be considerably more rapid. In equatorial regions the intensity of UV radiation does not vary appreciably throughout the year.

Surprisingly, the total amount of UV radiation does not vary much over the surface of the globe. During the arctic summer, for example, the amount of UV radiation is comparable to that of more temperate regions simply because of the longer daylight hours. This then provides a mechanism for degradation in very cold arctic regions where biological processes are either very slow or non-existent.

Although the chain-breaking process begins as soon as the plastic is exposed to solar radiation, there is a certain period of time necessary before an appreciable change in the physical properties occurs. The reason for this is that above a certain molecular weight, which is sometimes called the critical molecular weight, there is only a small change in the physical properties of the polymer as the molecular weight changes. Once the critical molecular weight is reached however, any subsequent decrease in molecular weight will cause a drastic change in the properties of the material. This means that even after exposure to solar radiation, the plastic material will still retain its useful properties for a certain period of time, and this time can be controlled at will in the manufacturing process.

It is found that in order to provide an acceptable rate of degradation for polystyrene, for example, it is necessary to include less than 1% of these carbonyl groups in the polystyrene molecule.

Ecolyte plastics are made by copolymerizing ketone-containing comonomers in small amounts with ethylene, styrene, or other monomers used in the manufacture of commerical plastics. The process is covered by a number of issued US and foreign patents.[1] Condensation polymers such as Nylon and polyesters can also be made photodegradable by this method.[2] Even polymers such as poly(vinylchloride)[3] and poly(acrylonitrile)[4] which normally do not degrade by chain scission are made photodegradable by this technique.

In most commerical applications, where rapid degradation is not required, it is convenient to prepare a concentrate containing 2 to 5% of the ketone monomer in polyethylene or polystyrene. This concentrate, which may also be colored with selected pigments, can then be blended with natural resin in ratios of 1:9 to 1:20 and extruded or molded to provide products with the desired rate of degradation. Control of the rate is provided by changes in concentrate and/or pigment concentration.

In some cases, the ketone groups can be introduced by a chemical post-treatment. Because of the minor amount of modification required, the physical properties of the photosensitive resin are almost identical with those of the untreated plastic.

Since these photodegradable plastics will be used for food packaging, the question of food approval for packages made from them is important. Essentially all of the ketone has been introduced into the polymer chain and is chemically attached to it. For this reason, the ketone groups cannot be extracted from the polymer film or package and hence can have no effect on the toxicity or the taste of the packaged product. This represents a particular advantage of the Ecolyte system in that other processes invariably make use of additives which are merely dissolved and are not chemically bonded into the plastic and there fore will migrate from the plastic into any foodstuff packaged in it.

An alternative method of introducing ketone groups into polymers, useful so-far only in low-density polyethylene made by the high pressure process, is copolymerization with carbon monoxide. Although such copolymers have been known for many years, their UV photochemistry (and hence usefulness in this application) was first reported by Hartley and Guillet[5] in 1968. These copolymers are manufactured on a large scale in the U.S. by several major plastics producers and are particularly useful in applications such as the plastic rings used for six-pack beverage containers.

The photochemistry E-CO resins is similar to that of the Ecolyte vinyl ketone copolymers except that the ketone carbonyl group is in the backbone of the polymer chain. A a result, the yield of free

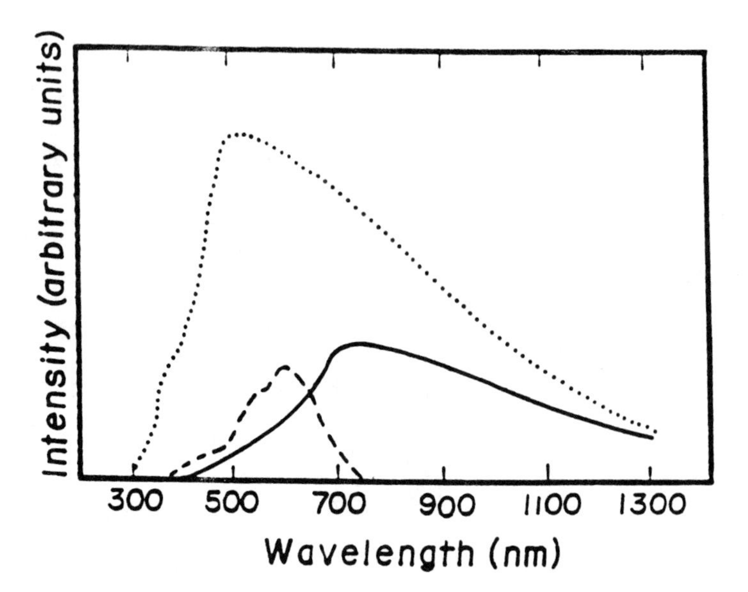


Figure 1. Wavelength distribution of light sources: ...the sun, ---- incandescent lamp, --- fluorescent lamp.

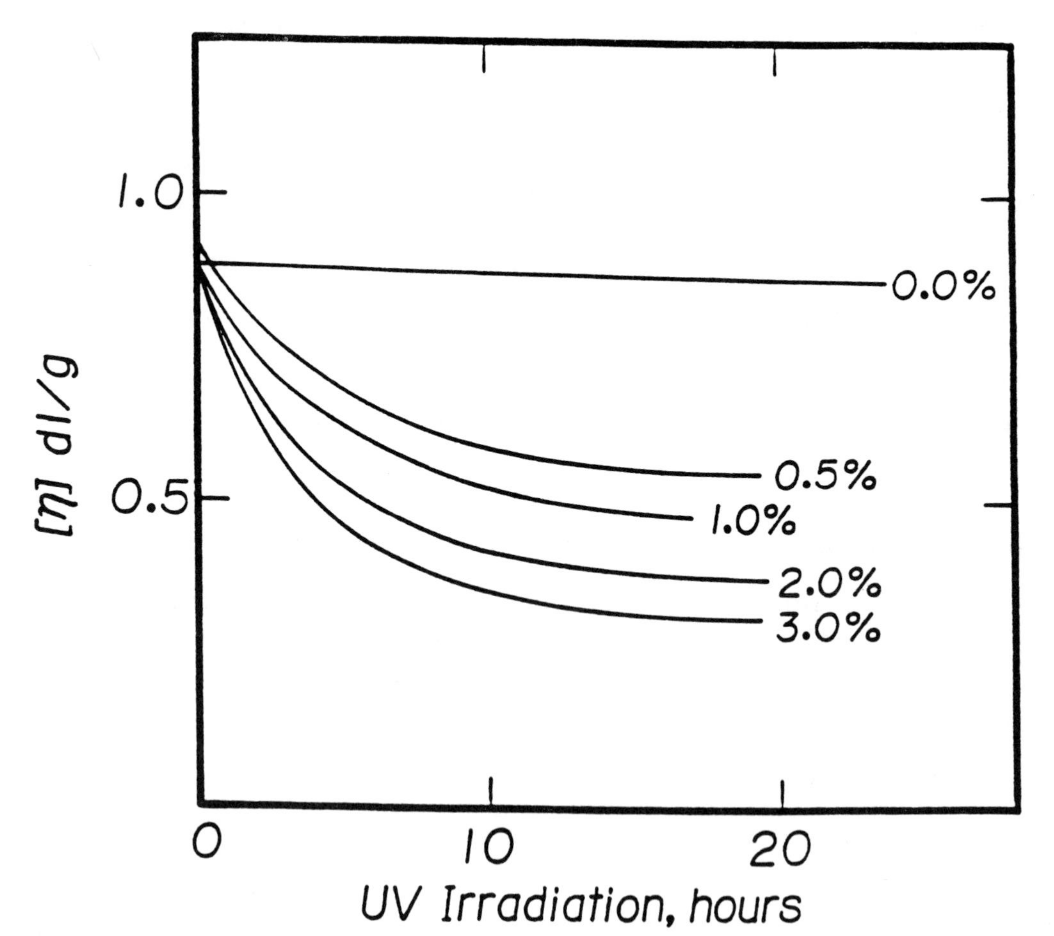


Figure 2. Effect of benzophenone concentration on degradation of polystyrene (from ref. 6).

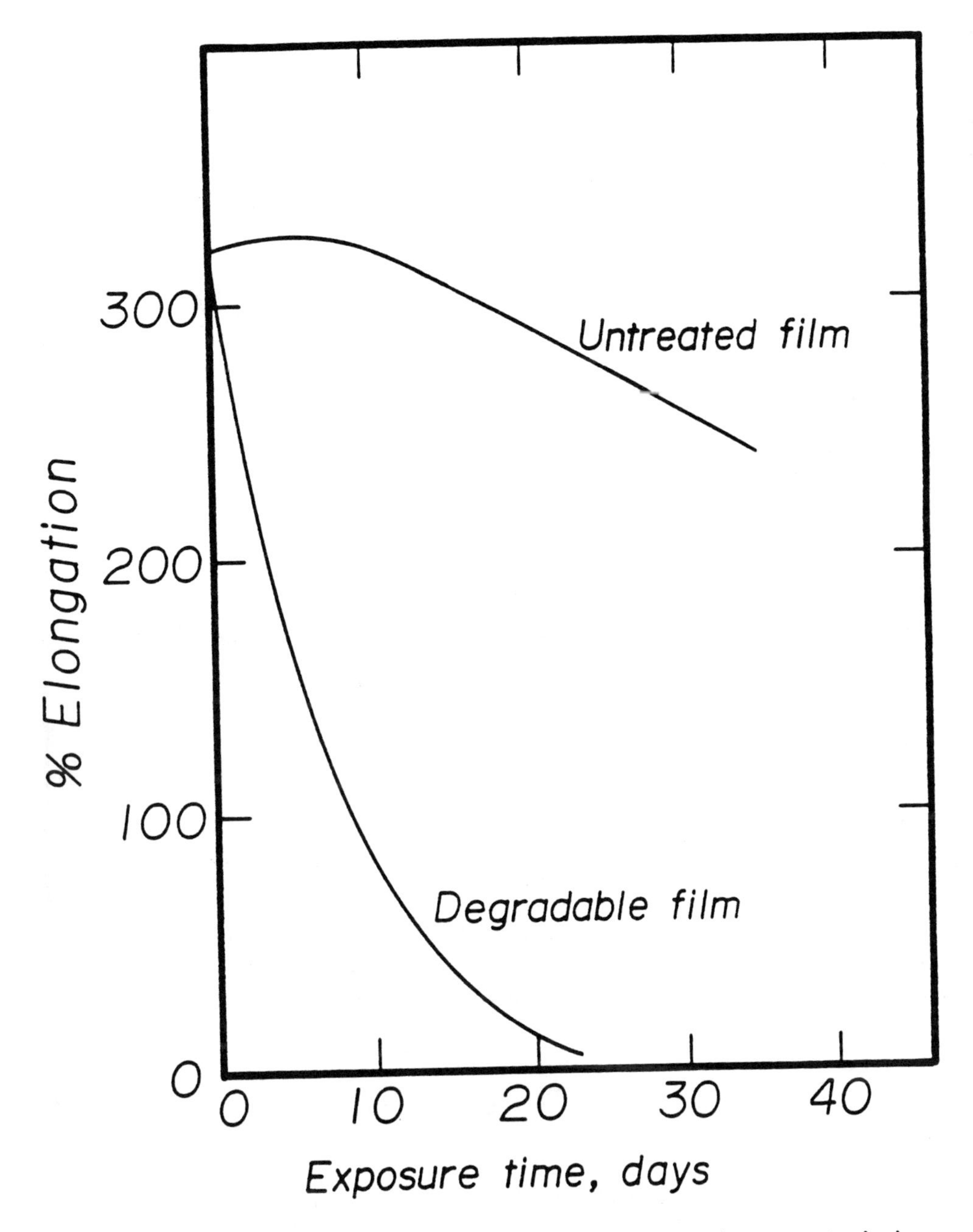


Figure 3. Loss of elongation of polyethylene film containing Akerlund and Rausing additive: outdoor exposure at Jonkoping, Sweden (from ref. 7).

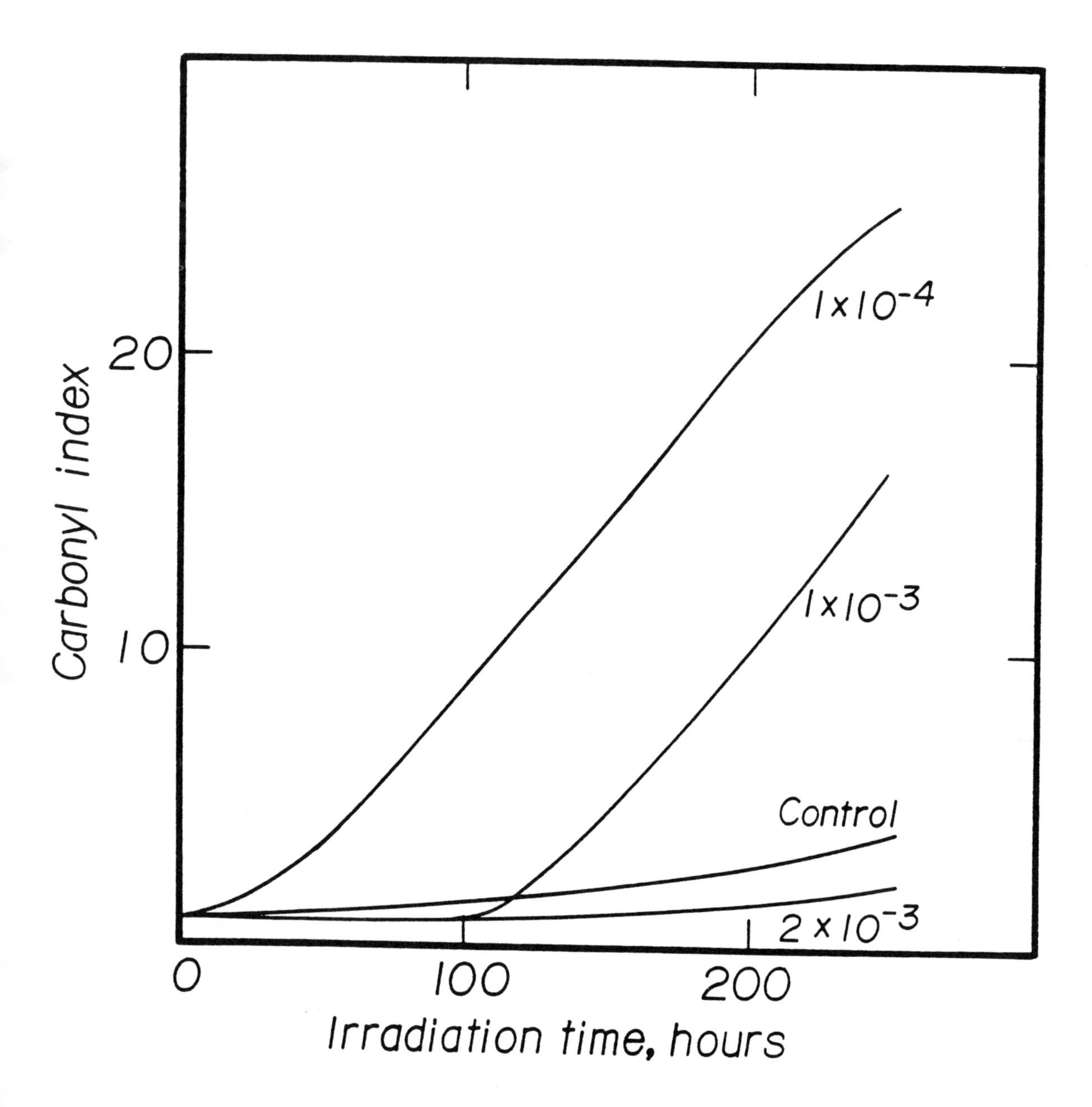


Figure 4. Effect of artificial UV irradiation on polyethylene containing ferric dinonyldithiocarbamate (from ref. 8).

radicals by the type I process is very low, hence the polymer does not photooxidize very rapidly and is less useful in a "masterbatch" application.

Photodegradable Additives

In contrast to the copolymer approach described above, photodegradable plastics can be made by the addition of additives, prior to molding or extrusion, usually in the form of a masterbatch. Additives which have been used in commercial practice include benzophenone, metal salts such as iron stearate, halogenated compounds and ferric dinonyl dithiocarbamate. All of these compounds sensitize the photooxidation of hydrocarbon polymers by classical free radical mechanisms. They are particularly effective in polyethylene and polypropylene, but because of the possibility of migration from the package into food, they are not yet in general use where food contact is indicated.

Typical examples of the performance of additive systems are shown in Figures 2,[6] 3,[7] and 4.[8]

Biodegradation

In general, the photodegradation process is considered to be successful when the litter has disappeared as a visible man-made object. This means that at this point a package has broken up into small particles comparable to sand and is dispersed in the surrounding soil. The rate of this process can be accurately controlled using the principles discussed previously. However, the biological degradation which now ensues may take considerably more time.

The term "biodegradable" has been open to a variety of interpretations. We believe that the proper definition in relation to packaging applications should be, "capable of being chemically transformed by biological agents, including enzymes and microorganisms".

Early studies of the biodegradability of photodegraded Ecolyte polystyrene and polyethylene used classical oxygen uptake procedures using high activity media such as sewage sludge and also a variety of natural soil samples.[9,10] Typical results for Ecolyte polyethylene and polypropylene are shown in Figure 5. The apparent levelling off of the uptake is typical of a static test, since addition of more bacteria increases the rate again.

By using repetitive transfer to minimal media in which the only carbon source is degraded polymer, it was possible to isolate bacteria which are capable of attacking degraded polyethylene and polypropylene.[11]

The organisms were separated into two groups by their Gram staining characteristics. Each organism in each group was then subjected to a set of standard microbiological identification tests to determine the genera. In certain cases, specific tests were employed where the standard tests could not easily distinguish between two genera. When tentative identification was made, the genera were subjected to any further test which would confirm their identification. The following genera were identified: Pseudomonas,

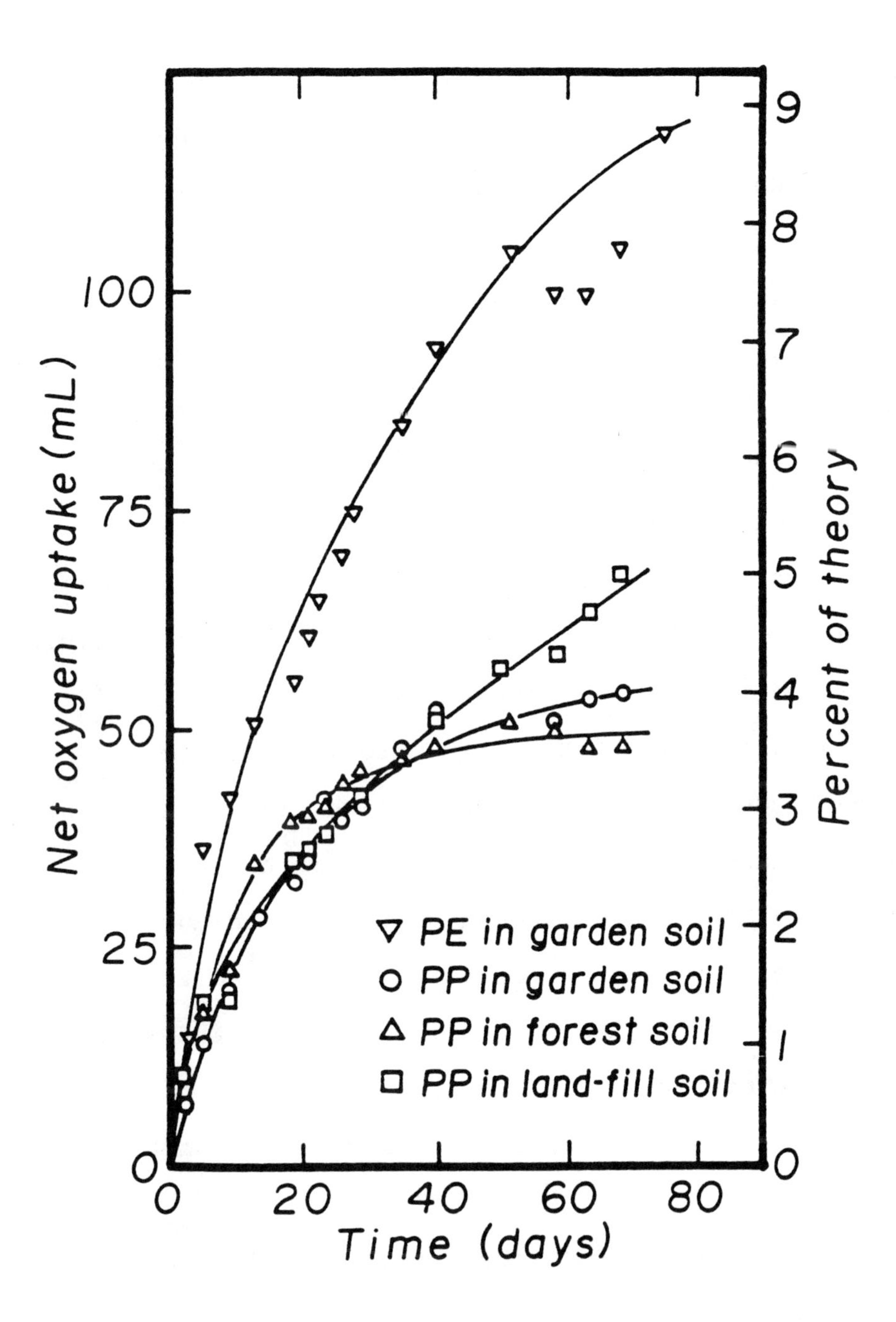


Figure 5. Net oxygen uptake of plastic residues in various soils.

Alcaligenes, Achromobacter, Flavobacterium, and Gamella (all gram-negative), and Arthrobacter, Aerococcus, Cellulomonas, and an Asporogenous bacillus (all gram-positive). Most of these bacteria are common in soils and Pseudomonas and Achromobacter are known to attack both aliphatic and aromatic hydrocarbons.

A further conclusion of this work is that those bacteria which have been demonstrated to attack plastic residues are of relatively common varieties which would be expected to be widely distributed in soils in most terrestrial environments. No fungi capable of utilizing the polymers were isolated by this technique. Furthermore, none of the fungi isolated from the soil were capable of utilizing these polymer as the sole source of carbon.

Studies of the biodegradation of $^{14}C$-labelled Ecolyte polystyrene[12] have confirmed the presence of the radio label in the growing hyphae of bacterial species as well as in proteins, nucleic acid derivatives, and the $CO_2$ developed during the culture. In parallel experiments in our laboratories with a closed terrarium, it was demonstrated that up to 5% of the carbon in a photodegraded $^{14}C$-labelled Ecolyte polystyrene in the soil was converted into $CO_2$ or directly into the growing plants during a six-month period.[13]

Data on the radio carbon tracer detected in various plant species is shown in Table II. These results confirm that polymer fragments produced by photodegradation of certain plastic molecules are indeed attacked and metabolized by soil microorganisms, are assimilated into the natural carbon cycle. Because radio carbon assay of the $CO_2$ produced by plant and animal respiration does not reflect the total carbon assimilated in the bodies of plants and microorganisms, this procedure will tend to underestimate the actual rate of biodegradation of the plastic fragments in a natural environment. In our tests, almost half of the carbon metabolized was found in the bodies of the growing plants.

TABLE II. Total $^{14}C$ Derived from $^{14}C$–Labelled Ecolyte Polystyrene after Five Months in Terrarium

| | d p m | | | |
|---|---|---|---|---|
| | $^{14}C$ from plant leaves | $^{14}C$-labelled $CO_2$ from traps | Total $^{14}C$ | % of initial polymer $^{14}C$ added to soil |
| Mean value | 126,000 | 130,000 | 256,000 | 4.4 |

Because of the living plants in the terrarium test, there is no indication of a slowing of the rate of biodegradation with time. Extrapolation of the data would indicate that substantially all of the polystyrene would be biodegraded in about ten years. This is about the same time as has been estimated for a complete biodegradation of straw,[14] a well known as a biodegradable material.

## Environmental Concerns Relating To Photodegradable Plastics[15]

### Permanent versus Disposable Packaging

Environmentalists seem to believe that all disposable packaging is wasteful of both energy and resources. However, this is not always the case. Permanent packages are often made from materials like steel, aluminum and glass which require a lot of energy to produce and fabricate (see Table III). Disposable plastic and paper packages are much lighter, so the energy cost is even lower. Typical values for beverage containers are shown in Table IV. An analysis of the use of glass milk versus disposable polyethylene is shown in Table V. The energy cost of the bottle is one hundred times that of the plastic pouches. It would have to be re-used many more than 100 times before it would be a lower energy cost system of packaging.

## Resource Conservation

The current world production of plastics used for packaging is about 32.4 billion pounds per annum (see Table IV), of which the three largest are polyethylene, polypropylene and polystyrene, which account for 27.3 billion pounds annually.

The only practical substitute for plastic packaging which could be considered now would either paper or other cellulose-based products. A typical grocery bag from high-density polyethylene weighs approximately 5 grams, while a heavy kraft paper grocery bag weighs 35 grams. Replacement of plastic grocery sacs of this type by paper would this increase the actual weight of garbage from this source by a factor of about seven! Low-density polyethylene grocery and shopping bags are somewhat heavier (ca. 11 g) but are still about one-third that of paper. There are similar advantages to the use of foamed polystyrene fast food containers as compared to paper products.

The resource consequences of a shift from plastic back to paper

TABLE III. Energy Requirements for Production of Material Used in Packaging Applications

| Material, 1 pound | Energy requirement, kilowatt hours (thermal) |
|---|---|
| Aluminum | 33.6 |
| Steel | 6.3 |
| Glass | 3.6 |
| Paper | 3.2 |
| Plastic | 1.4 |

TABLE IV. Energy Requirement per Beverage Container

| Container | Weight, ounces | Energy used per container, kilowatt–hour |
|---|---|---|
| Aluminum can | 1.41 | 3.00 |
| Returnable soft drink bottle | 10.6 | 2.40 |
| Returnable glass beer bottle | 8.83 | 2.00 |
| Steel can | 1.76 | 0.70 |
| Paper milk carton (1 pint) | 0.92 | 0.18 |
| Plastic beverage container | 1.23 | 0.11 |

TABLE V. Comparison of Energy Cost of Manufacture of Disposable and Returnable Milk Containers*

| | Weight ounces | Energy used in manufacture kW/hr | Energy ratio | Heat content kcal** |
|---|---|---|---|---|
| Two–quart glass glass milk bottle | 37.1 | 8.36 | 99.5 | 0 |
| Two–quart plastic pouch — plastic bag (polyethylene) | 0.97 | 0.084 | 1.0 | 317 |

* Complete energy cost estimates of using the system would involve additional costs for the returnable system in transporting the extra weight of the glass bottle (2.3 lb) and in washing and sterilizing it for reuse. An allowance for the energy cost of garbage collection must also be made for the disposable system.
** This amount of heat energy would be obtained by burning the plastic in air; it is sufficient to raise the temperature of about one gallon of water to the boiling point.

TABLE VI. World Production of Plastics (billions of pounds)

| Plastic | Total production | Used for packaging | |
|---|---|---|---|
| Polystyrene | 9.7 | 3.8 | 27.3 billion (Polystyrene, Polypropylene, Polyethylene) |
| Polypropylene | 5.5 | 3.9 | |
| Polyethylene | 39.2 | 19.6 | |
| Poly(vinyl chloride) | 13.0 | 3.1 | |
| Poly(ethylene terephthalate) | 2.9 | 2.0 | |
| | | 32.4 | |

are enormous! Assuming that we take the lower figure (i.e., a factor of one-third the weight of a paper package), the replacement of plastic by paper would require the total weight of garbage by 55 billion pounds per year, and would require the additional production of 82 billion production of timber on all of the land under its jurisdiction is about 550 pounds per acre annum. The U.S. Forest Service estimates that the sustainable production of timber on all of the land under its jurisdiction is about 550 pounds per acre annum. Packaging grade paper requires about 1.1. to 1.2 pounds of wood per pound of paper or about 500 pounds of paper per acre. To produce the required 82 billion pounds of paper would therefore need an additional 162 million acres of forest land devoted to paper production. To put this into perspective, the area of all agricultural land in medium-sized U.S. states such as Michigan, Louisiana or Virginia is about 11 million acres. When we consider the environmental damage already inflicted on the earth by cutting down trees in countries such as Brazil and China, a solid-waste strategy which might result in harvesting another 162 million acres does not appear to be a "giant step for mankind", particularly when we know that forests are important in reducing the $CO_2$ concentration in the atmosphere.

Finally, there is a question of the energy cost of a conversion from plastic to paper. Makhijani and Lichtenberg[16] estimate the energy requirements for the production of paper and plastic (polyethylene) at 3.2 and 1.4 kwh per pound, respectively. A simple calculation indicates that the extra energy required to produce the paper would cost about 224 billion killowatt hours, or 25.5 million kilowatt years. We cannot afford to generate this energy using coal, oil, or gas because of the greenhouse gases produced. A possible solution would be the construction of about fifty 500-megawatt nuclear power reactors. The point is that the issue of materials for packaging is very complex, and simple solutions such as banning plastics can result in very undersheriff ecological consequences.

## Air and Water Pollution

A further consideration is the air and water pollution associated with the manufacture of paper and plastics packages. Table VII shows data produced by the West Germany Federal Office of the Environment (Berlin, 1988) on the air and water pollution associated with the production of 50,000 carrier bags of polyethylene, unbleached kraft paper and "paper combination". The latter is the formulation used for most paper carrier bags approved for use in the FRG. The production of plastic carrier bags causes significantly less air pollution, and as much as 200 times less water pollution compared to that of paper carrier bags.

The report concludes with the statement: "The replacement of polyethylene by paper carrier bags makes no sense ecologically. The production of polyethylene carrier bags requires less energy and in the process results in less burden to the environment. There is no significant difference in the disposal of polyethylene and paper bags at landfill sites or in incineration plants."

Table VII Air and Water Pollution Associated with the Production of 50,000 Carrier Bags

| Environmental burden | Polyethylene | Unbleached kraft paper | Paper combinations |
|---|---|---|---|
| Energy (GJ) for production process | 29 | 67 | 69 |
| Air pollution (kg) | | | |
| $SO_2$ | 9.9 | 19 4 | 28.1 |
| $NO_x$ | 6.8 | 10.2 | 10.8 |
| CH | 3.8 | 1.2 | 1.5 |
| CO | 1.0 | 3.0 | 6.4 |
| Dust | 0.5 | 3.2 | 3.8 |
| Waste water burdens (kg) | | | |
| COD | 0.5 | 16.4 | 107.8 |
| $BOD_5$ | 0.02 | 9.2 | 43.1 |

Comparative Strategies for Litter Abatement

Photodegradable plastics are designed to address the litter problem, not necessarily the problem of garbage disposal or landfill capabilities. Computer simulations are a useful way of assessing the value of various litter abatement strategies. Figure 6 shows the results of such calculations. The assumptions behind this model are as follows. (1) The production of plastics in 1970 in the United States was 19,600 million pounds. (2) The production of plastics increases annually by 6%. (3) The proportion of plastics production which will be used in packaging applications will be constant at 20% per annum. (4) Two percent of plastics packaging will become litter.

Curve (a) shows the accumulation of litter if the average

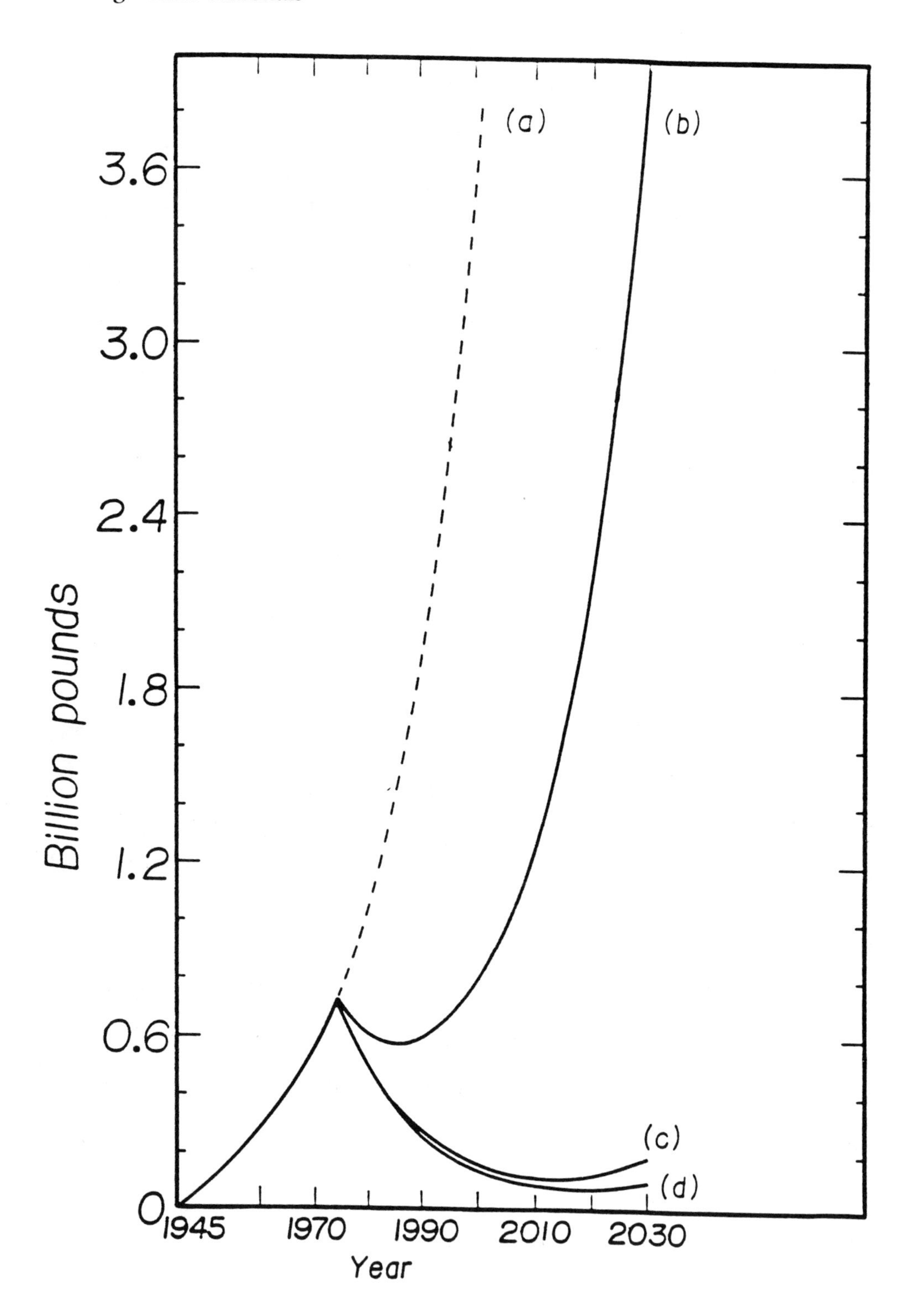


figure 6. Comparative strategies for litter reduction.

lifetime is 10 years. Curve (b) shows the accumulation if, as a result of education of fines, the amount of litter thrown away is reduced by a factor of five. Very little improvement occurs and the exponential increase in accumulated litter starts again 10 to 20 years later. Curve (c) shows the effect of reducing the lifetime of litter to 0.2 years (ca. 2 months), a target well within the range of modern plastics technology. An immediate improvement in litter accumulation occurs and after 20 years, litter has almost disappeared! Curved (d) shows that nearly the same thing occurs, even if people throw away twice as much litter because they think it is degradable. This confirms that the most effective way to deal with the litter problem is by reducing the "lifetime" of the littered object. Photodegradable plastics technology provides the only way we know of to accomplish this objective. These conclusions have been confirmed by actual studies of the accumulation of six-pack plastic rings on Pacific beaches. Those states whose legislation requires the use of photodegradable rings showed a remarkable decrease in the number of such packages accumulated on their beaches.

In conclusion, plastics can be made compatible with nature and the natural environment. In fact, the manufacture of synthetic polymers for fiber and plastics applications should provide for less environmental stress than the use of such natural materials as wool, cotton, and possibly even wood. Moreover, since plastics are based on carbon, a truly renewable source, they will be available even if other materials should become scarce or too energy expensive.

References

1. U.S. Patents 3,753,952, 3,811,931 3,853,814 and 3,860,538.
2. U.S. Patent 3,878,169.
3. M. Heskins, W.J. Reid, D.J. Pinchin and J.E. Guillet, ACS Symp. Ser. no. 25, 272 (1976).
4. L. Alexandru and J.E. Guillet, J Polym. Sci., Polym. Chem. Ed., 13, 483 (1975).
5. G.H. Hartley and J.E. Guillet, Macromolecules, 1, 165 (1968).
6. H.G. Troth and A.N. Ainscough, unpublished results.
7. A. Boberg (to Akerlund & Rausing), German Offen. 2,244,801.
8. M.U. Amin and G. Scott, Eur. Polym. J., 10, 1019 (1974).
9. P.H. Jones, D. Prasad, M. Heskins, M.H. Morgan and J.E. Guillet, Environ. Sci. Technol., 8, 919 (1974).
10. J.E. Guillet, T.W. Regulski and T.B. McAneney, Environ. Sci. Technol., 8,923 (1974).
11. L.R. Spencer, M. Heskins and J.E. Guillet, Proceedings of the Third International Biodegradation Symposium, ed., J.M. Sharpley and A.M. Kaplan, Applied Science Publishers, London, 1976.
12. A.A. Lepidi, M.P. Nuti, G. Bernacchi and A. Seritti, Preprints, International Symposium on Degradation and Stabilization of Polymers, Brussels, Sept. 1974, p.283.
13. J.E. Guillet, M. Heskins and L.R. Spencer, Polymeric Materials Science & Engineering, 58, 80 (1988).
14. N.B. Nykvist, Plastics and Polymers, Oct. 1974, pp.195-199.
15. J.E. Guillet, Plastics Engineering, vol. 30, no. 8, August 1974.

16. A.B. Makhijani and A.J. Lichtenberg, An Assessment of Energy and Materials Utilization in the USA, Electronic Research Laboratory, University of California, Berkeley, Sept. 1971.

QUESTIONS AND ANSWERS

---PRESENTATION BY DR. JIM GUILLET

(University of Toronto)

DR. BAILEY: Bill Bailey, University of Maryland.

Your data sort of indicates that there is a fraction of your photodegraded material that's biodegradable, and this would probably be the very low molecular weight. You have never, I guess, taken the biodegradation to a very high extent to show that it was all biodegradable, is that right?

DR. GUILLET: No. That is correct. The highest degradation we have taken it to is 10 per cent, but it takes time. I mean, it's still degrading. If there was just a low molecular weight material it would stop at about one per cent.

DR. BAILEY: It's a gradual thing.

DR. GUILLET: But there is no change in rate, Bill.

DR. BAILEY: Oh, there is not? That's kind of surprising. One of my comments of course is that I hope that in the discussions we don't segregate photodegradation completely from biodegradation. We have been interested in making photodegradable materials that were biodegradable, and this is by putting in both the ketone and an ester group, and these materials are both photodegradable and biodegradable.

So in a mulch, for instance, where you cover the edges with dirt to keep it from blowing away that part is biodegradable and the part that would be exposed to the sunlight would be photodegradable.

There are a lot of cases where you throw things under the shade that they don't photodegrade very well or if they get in the ocean the water streams out the UV and you can still get biodegradation.

So for many of the things we are talking about in litter and everything else, the ability to have things that are both photodegradable and biodegradable are very important. So I want to keep this ability to talk about degradable things and have a dual mechanism in may of these things, in the discussions that will go on in the next couple of days.

DR. GUILLET: May I just comment? I didn't intend to slight you by not mentioning your processes, Bill, But I restricted my comments to those which had been in industrial production, primarily.

I still think it's important to distinguish between photodegradation and biodegradation because people are asking these properties for different reasons.

There is no evidence that making something biodegradable will solve the litter problem. It may solve the garbage problem. It it does, it's not one that I am aware of. But the only question -- I raise the question of biodegradeability because we want to know that in the long term - five, ten, twenty years - it will be back into a natural form.

But if you have something that is biodegradable but sitting on the top of the soil, as you point out it's not going to be very biodegradable until it breaks up by some mechanism, and so combining the photodegradation which breaks it up into small particles with the biodegradation in the ground is a say Phase 2 solution to our problem, and I think it's very eloquent work.

DR. WOOL: I am Richard Wool from the University of Illinois.

As an expert yourself in free radical chemistry, I wonder if you could compare the two technologies in terms of the propagation phases; namely, the Griffin technology with your technology in terms of how the free radicals propagate to cause disintegration of the material.

As a last question, if you were to add six per cent starch to your material would it be biodegradable?

DR. GUILLET: Well, first of all, I am not aware that the Griffin process catalyzes the oxidation of anything, and secondly... Would you repeat the second question?

DR. WOOL: I was hoping in the first part that you might discuss some of the propagation steps of the free radicals, either from the Griffin technology or from your side. Are you saying that in the Griffin technology there are no free radicals?

DR. GUILLET: I'm sorry, what is the Griffin technology?

DR. WOOL: It is the one introduced ADM today.

DR. GUILLET: So it has a metal salt in it, correct?

DR. WOOL: Yes, correct. Apparently, free radicals are generated and these propagate.

DR. GUILLET: So the propagation step is independent; it's the initiation which makes the difference. So the catalyst primarily causes a more rapid initiation, I believe. But I'm not an expert on that kind of chemistry.

DR. WOOL: In your photochemistry the free radicals have the ability to propagate, don't they, and cause a disintegration process? Or is it just the number built inside that actually gives you the --

DR. GUILLET: It's the peroxy radicals by propagate the process, and presumably the same radicals would evolve regardless of how you initiate the photooxidation.

DR. WOOL: Right.

DR. GUILLET: But what the role of starch would be -- I haven't studied this. I don't know. All I know is if you can show me data and I can repeat them, then I believe you.

DR. WOOL: Thank you.

DR. SHALABY: You noted that the type of soil is quite important to the degradation process. You also mentioned that polyesters and their chemical degradation followed by photo- or biodegradation, and in developing any kind of protocol to test degradation should you recommend that we should be addressing, at least in case of soil as a medium, the count of the microorganism, the pH, the porosity-enchanced percolation of the soil, and also the matter of contents, quality and quantity?

DR. GUILLET: Well, it's my understanding from reading the literature that one of the worst places to bury plastics is in garbage dumps because the soil there is usually very low in bacteria, and there are reports of digging up newspapers that have been buried for 100 years or so and you can still read them.

So that either says the paper is not very biodegradable or garbage dumps are not a very good place to carry out biodegradation. You can have it either way, but it confirms that burying things in the ground is not a good way to dispose of them.

It's clear that if you want to dispose of plastics biologically, if that's your intention, you can get a lot better way of doing it than doing it in a garbage dump. You can put it in a big pot with lots of organisms and heat it up and oxygen or nitrogen, whatever.

The chemistry is well-known. It just has to be done in a better way.

I'm sure you could probably speed it up by a factor of 100 if you did a type of compositely treatment, but that would be very expensive. I would think it would be economically feasible to try to dispose of garbage by composting it in this way.

DR. ENNIS: One of the hypotheses that have been advanced by some environmental groups have been that although a degradable plastic may eventually get to $CO_2$ and water and go back into the carbon cycle that the intermediate breakdown products may themselves have some toxic effect on the environment.

I wonder if any of your studies have addressed that particular aspect or you could give us some more information on that?

DR. GUILLET: The person to ask about that is Dr. Albertsson, who, as I say, has done that.

The breakdown products of polyethylene are relatively simple. As I indicated, they consist of long-chain fatty acids, alcohols, and compounds which are known to themselves be biodegradable and non-toxic.

I see no reason -- and also the other evidence for that is the fact that in many countries, particularly in Israel, this kind of technology has been used for 10 or 15 years.

There is no evidence for any toxic residues in the soil, or in fact, the soil gets better from the point of view of growing things. So there is very little reason to expect any toxic residues.

Furthermore, if you are talking about litter rather than say agriculture where you are getting a high level of continued use of plastic, litter is unlikely to be ever present in a very high concentration, and as the polymer breaks down - say a plastic cup breaks down into small pieces, it blows around, and soon is spread around an area of perhaps several tens of square metres - the actual concentration of plastic in the soil would be in parts per billion, and it would be very unlikely that even if it were highly toxic that it would have any effect on the soil and its microorganisms.

So I don't think it is a serious issue from the point of view of litter.

DR. LOOMIS: I just make one...

You made a comment that rather than changing materials as way to get rid of garbage we should be changing method or something. I am phrasing. It was something like that in the beginning. But you saw litter as a different problem.

I'm not sure why you don't just view litter as a sociological problem rather than changing the material. It's the same thing. If you change the material you tell someone this is going to go away. I mean, it's almost encouraging throwing it away.

DR. GUILLET: Are you suggesting --

DR. LOOMIS: No, I am suggesting that I think the litter problem is sociological and that --

DR. GUILLET: And not technological?

DR. LOOMIS: Absolutely.

DR. GUILLET: And it's not so much volume-wise anyways.

DR. LOOMIS: -- sociology?

DR. GUILLET: Excuse me?

DR. LOOMIS: Do you think that there is no such thing as technology that doesn't change sociology?

DR. GUILLET: No, I am just saying that I am not sure that photodegradable polymers are a cure-all to the litter problem.

DR. LOOMIS: The comment I wanted to ask you is your expertise in photochemistry, why a 50/50 ethylene carbon monoxide copolymer, strictly alternating ethylene carbon monoxide does not apparently photodegrade or it photodegrades at an incredibly slow rate where it, while it cannot nourish Type 2 cleavage, it can certainly undergo a Type 1. Have you ever looked at a 50/50 ECO copolymer?

DR. GUILLET: I'm not aware of that data.

DR. LOOMIS: And there are the copolymers now you make with a nickel catalyst, with a coordination catalyst. They're strictly alternating.

DR. LOOMIS: That's true.

DR. GUILLET: That's the reason, is that there is very little photodegradation takes place in the crystalline regions, so if the polymer was substantially crystalline and had very little amorphous material, than it might not degrade, but I have never worked with this polymer so I would not want to comment on it.

DR. LOOMIS: That is one that's being developed by a couple of different companies as a stable, engineer, plastic-type material, and the question keeps coming up where people say ECO polymers are photodegradable, but there are ECO polymers which are not photodegradable.

DR. GUILLET: If they're trying to sell it, they should be providing you with good data to prove it.

DR. R. WILLIAMS: Rich Williams, Roy F. Weston.

I am curious. Do you know anything about the radiochemical purity of your Carbon-14 preparations?

DR. GUILLET: The what?

DR. R. WILLIAMS: The Carbon-14 preparations?

DR. GUILLET: What do you mean "the radiochemical purity"?

DR. R. WILLIAMS: The chemical purity? My concern is that when you are talking about one per cent to $CO_2$ or two per return to $CO_2$ at very low levels that what you might be seeing is the other compounds that are contaminants within your preparation converted to $CO_2$, you might not be seeing the $CO_2$ from the compound that you are really interested in.

DR. GUILLET: Well, all I can tell you is that we prepare those polymers very carefully and we extract them beforehand to make sure that there is no large, low molecular weight component.

DR. R. WILLIAMS: Typically, it is very hard to get greater than 98 percent chemical purity, and in fact, often --

DR. GUILLET: Okay. So you have got a two per cent error, but...

DR. R. WILLIAMS: When you are talking about one or two per cent return, it's -- you know, that's a concern of mine in that data.

DR. FEIJEN: I just would like to know, what is the relationship between an initial photodegradation of some of your polymers and then a bacterial degradation. Did you do anything in that response?

DR. GUILLET: Did we do anything in between?

DR. FEIJEN: Yes.

DR. GUILLET: We take the powder, that's formed in the first photodegradation. We usually seive it to a certain size, I think about 100 mesh or about something like this, to get a standard kind of sample. Then we bury it in the ground or put in one of these bioxygen determinations.

DR. FEIJEN: And the --

DR. GUILLET: It's not treated in between the two steps.

DR. FEIJEN: Okay. But the amount of oxygen produced is a measure of the bacterial activity, is that so?

DR. GUILLET: Yes. That's right.

DR. FEIJEN: Yes.

DR. GUILLET: It's carbon dioxide that's absorbed actually, but...

DR. FEIJEN: I was a little bit curious because I could not completely understand the picture where you had a faster degradation of polyethylene as compared to polypropylene, and -- so more oxygen production? You had a higher --

DR. GUILLET: It's well-known that linear hydrocarbons biodegrade more rapidly than branch chain hydrocarbons.

DR. FEIJEN: But these were pretreated or something? These were not pre-exposed to light or so?

DR. GUILLET: These were pre-exposed.

DR. FEIJEN: These were pre-exposed because polypropylene, we know from polypropylene it is degraded much faster than polyethylene because it's a tertiary carbon.

DR. GUILLET: Thermally, yes. And with light.

DR. FEIJEN: Also, oxidatively.

DR. GUILLET: Yes, that's correct.

DR. FEIJEN: Yes. So I was wondering, what is the difference in terms of if you have a faster pre-degradation? Is it than obvious that we have more oxygen consumption with polyethylene as compared to the degraded polypropylene?

DR. GUILLET: I don't have the data as to whether the polypropylene had a higher degree of oxidation or not, but I think the very fact that the polyethylene goes faster is strong evidence that it is a biodegradation, because if it were a thermal degradation it would be taking place more rapidly than poly-propylene.

DR. FEIJEN: Okay. Thank you.

DR. IANNOTI: I would like to point out a fact. If you look at land -- Gene Iannoti, University of Missouri.

It doesn't take you very long in looking at a landfill to realize that it is an extremely diverse set of environments. There is about as many ranges of environments from the microbiological point of view that you could ever imagine.

It's one of the most poorly managed waste systems that we have had. We don't try to mix, we don't try to optimize; we just simply bury it. It does not surprise me to find that there is materials in these types of environment that might have been buried underneath something that would block out moisture, that would have toxic compounds, so that if you go back 50 years it's not suprising to find materials that do not degrade.

If we were archeologists we would be very happy to find something and we would base our studies on one, but we are dealing with mass balances here, and what we are talking about is that there are materials left over.

I think the data supports the fact that it is very poorly managed, that degradation is very poor in these environments, but that does not say that there is no degradation. Methane evolves from many different landfills. There is -- 40 percent of the material breaks down aerobically. It doesn't take you very long just looking at your garbage. I would leave it out in the garage waiting for the garbage man. It becomes very odorous. That's breakdown. Okay?

So there is a breakdown in landfills. It's a misstatement to say that there is none, and to use finding something 50, 100, 1,000 years and say that's indicated that there is no breakdown I think is a little bit of a misuse of the scientific fact. Poor management, I go along with.

Some of the other misstatements I enjoy is that some of the environmental people say that plastics are not going to break down in a landfill-type of an environment because it's sealed from the top to the bottom. Their very next statement is that there is going to be release of toxic small particles of polyethylene. I, I don't know how those two facts go together. We will show you some --

I wish I had brought some data. We have data that says it takes very, very little oxygen for this series of free radical formations.

As a matter of fact, we had a very difficult time blocking out this type of reaction. We put them into steel containers, stainless steel containers. We scrubbed out as much oxygen as we could from the air stream by passing over copper columns. We had very short leads of hoses going into it. Still with that, even jacketing the containers, we ended up with measurable amounts of deterioration that we could trace back to events that's very similar to what you attribute to free radical formation. So it takes very little.

I don't see how you at least have an initiation to the breakdown of some of the free radical formation in polymers, even in landfills, when you try to seal that landfill.

DR. CUILLET: I don't think I said that there was no degradation in landfills. I just said that it was very slow. And in fact, that's really one of the real problems of landfills is the fact that you get methane and all sorts of other -- we have a nice big one out by the University. We now find it has been covered up for 30 years and we can't build anything on it because it's -- and probably won't be able to anything on it for another 100 years, I suppose.

So that's another problem relating to landfills as a way to dispose.

I think -- what I feel is that we should be looking at radically new and different ways of disposing things now that we understand more of the chemistry of the disposal process, that things ought to be ground up and made available and put in areas where, if we do want them to degrade, they can degrade rapidly, or put them in a place where they're sealed up completely and forget about them. One or the other.

DR. IANNOTI: I was just emphasizing...

It almost seems to be an assumption that there is no breakdown, emphasizing that there is a significant amount of degradation.

The fact is that we can't stop it. Stopping is more of a problem, you know? That's -- the engineering problem is stopping and reclaiming that land very quickly. If you do promote degradation, you get reduction of toxic materials, it stabilizes a lot faster. I think that's the direction we have to move in the future.

DR. SALAMO: Simon Salamo with International Paper.

I have a couple of comments, one with respect to your first couple of slides. Paper is biodegradable. It's probably the biodegradable material par excellence. What you have in there is plastic - coated paper, both milk cartons and cups, and there the plastic part is the one that's obstructing. We manufacture both, by the way.

The second comment is with respect to litter. I think humans have been littering for thousands of years, ever since our existence, and that's not going to go away. What we are seeing recently is the great increase in environmental impact of our litter, such as dead dolphins or dead pelicans choking on six-pack rings, and I'm not sure that your photodegradable polymers are going to solve that environmental issue.

It's not so much that it's an aesthetic question, it's more of an environmental impact on the wildlife. And photodegradability or biodegradability is not going to solve it until we figure out whether that pelican encountered that six-pack ring six years after it was disposed of, or whatever.

DR. GUILLET: Well, the big advantage of using plastics is you don't have to cut trees down. As we begin to lose all our forests, I think we will no longer have the luxury of using paper as a packaging material.

I might point out, however, that one of the pleasures of being a scientist is to get nice letters from people who think you've done good things, and we got a very nice one form Thor Heyerdahl, the famous explorer, who told us that when he crossed the ocean on his barge he got into the Sargasso Sea and found that, in his words, "The sea was covered with plastic litter."

Now, it's going to be very difficult to send a bunch of Boy Scouts out there to pick it up, and he felt that the invention of degradable material for packages was a major advance, and I happen to agree with him but it's nice to know that somebody else likes it.

Thank you very much. I think that's all our time.

DR. PATEL: My name is Satyen Patel and I am from Pillsbury Burger King.

It's interesting. As I sit out here, and I am here purely as an applications person coming in here to listen and to watch what's going on in the technology area for us to look at carefully for applications downstream.

I do agree that we are very sceptical of using photodegradable stuff for any of our Burger King applications, and part of it has to do with the fact that we might in fact issue a lot of litter. It's an issue of perception.

And I also disagree with the fact that technology has sociological implications, although in this case my fear in that it will have a negative implication on consumer habits.

What I wanted to say more than anything else out here is that I look at different perspectives here and different technologies being thrown out. It's a good start.

I also see some very irresponsible usage and propagation of those same technologies, and I am going to point out Archer Daniels Midland right there. When you were on the David Brinkley Show on prime time television and when you claim that your degradable bags vanish into air, it serves the cause of sensationalizing degradation as an alternative to recycling, but it does not serve the cause for us to propagate degradation in the long run.

In our situation, we are having a lot more problems defending degradation, even though we feel that is the way to go and that's where Burger King's commitment lies versus recycling, and I hope somewhere along in this discussion there is some sort given to ethics.

While we need to propagate degradability and we need to look at solutions from a technological standpoint, things are to be looked at from a degradability standpoint, but they also have to be decoupled from the landfill situation. Maybe you look at it from a different alternative in terms of composting.

But we also have a moral responsibility here to propagate some rational, objectives. Understanding the fact that in the reality of things our Legal Department tells us that they are legislating about 12 different laws per day at the county, city, State levels in every single part of the nation, and it's -- the last thing I want to do out here is to get logical about it because that's not the way the world is operating out there.

I would like to part with just one statement. There is a Japanese proverb that is very apt here. I think knowledge without wisdom is like a load of books on the back of an ass. So I mean, somewhere along the line we need to use it pretty effectively. Thank you. [Applause].

DR. GUILLET: I can only make one comment. As a scientist I have always believed that if you were right, it would survive, if you we're wrong, it would disappear.

So the technology is here now, it's available, it can be used. If it's any good, it will survive; if it isn't, it won't.

And that decision will not be made in this room, but out in the marketplace where the fact that there is so much legislation being proposed is an indication at least of the problem, of the magnitude of the problem, if not the quality of the solution.

Thank you very much.

# ALPHA-HYDROXY ACID BASED BIODEGRADABLE POLYMERS POLY(L-LACTIDE) AND POLYDEPSIPEPTIDES

P.J. Dijkstra[a], M.J.D. Eenink[b] and J. Feijen[a]

a. University of Twente, Department of Chemical Technology, P. O. Box 217, 7500 AE Enschede, The Netherlands

b. Organon International bv, P. O. Box 20, 5340 BH, Oss, The Netherlands

## POLY(L-LACTIDE)

### SYNTHESIS

Poly(L-lactide), PLLA, is a poly-alpha-hydroxy acid which has been investigated extensively for possible use as a biodegradable polyester in medicine (1). The applications studied include bone plates for internal fracture fixation, surgical sutures, urethral grafts and tubular guidance channels for peripheral nerve repair.

PLLA is an optically active, isotactic, semicrystalline polymer of L-Lactic acid, with a glass transition temperature (Tg) of around 58°C and a melting temperature (Tm) around 178°C. Methods for the synthesis of L-Lactic acid polymers include polycondensation of L-Lactic acid (2), ring opening polymerization of L-Lactic acid anhydrosulphite (3) and L-Lactic acid O-carboxyanhydride (4). Using these methods relatively low molecular weights ($3\text{-}10 \times 10^4$) were obtained.

The preferred route for the preparation of high molecular weight L-Lactic acid polymers is based on ring-opening polymerization of the six-membered cyclic diester L-Lactide (5-8). A preferred initiator for this reaction is stannous octoage [tin(II)ethylhexanoate] (5) which has been used in the preparation of the surgical suture Vicryl®, a copolymer prepared from L-Lactide and glycolide.

Although many possible applications of PLLA have been reported, only a few detailed studies on the synthesis and characterization of PLLA have been published. The problems with the reproducibility of L-Lactide, DL-Lactide, or other lactone polymerizations have been recognized by several investigators (9,10,11). However, detailed information on the parameters affecting the reproducibility of the polymerization is lacking to a large extent. Kricheldorf et al. (12) have investigated the effect of the use of different initiators and polymerization temperatures on the optical purity of PLLA obtained. It was concluded that optically pure PLLA can be best prepared using temperatures in between 120° and 150°C, reaction times of at least 24 h, and the use of initiators such as stannous octoate, bismuth octoate, and antimony octoage.

In order to develop a standard polymerization procedure for L-Lactide initiated with stannous octoate we stuided the effect of the polymerization conditions, i.e. reaction temperature and M/I (monomer over initiator) ratio, on the molecular weight and molecular weight and molecular weight distribution of the polymers obtained.

Polymerization reactions were performed in the melt under a dry nitrogen atmosphere at a slight overpressure. The results obtained at four different reaction temperatures (T = 103 - 130°C) and at four different initiator concentrations (M/I = 500 - 50.000) are given in figures 1A and 1B. It is shown that for all polymerization temperatures a maximum in $M_w$ and $M_n$ of PLLA as a function of M/I was found. In general, at lower polymerization temperatures PLLA's with higher molecular weights were obtained. The highest $M_w$ and $M_n$ values were observed at 103°C and M/I = 5.000.

The results of another series of polymerization reactions carried out at T = 103°C varying M/I between 500 and 50.000 have been given in figure 2. PLLA samples obtained had $M_w$ values ranging from 6 x $10^4$ to 6 x $10^5$ and $M_n$ values ranging from 4 x $10^4$ to 5 x $10^5$. Figure 2 shows that an almost inverse relationship between initiator concentration (up to M/I 7.500) and the $M_w$ and $M_n$ of PLLA exists. High monomer conversions (> 95%) were only obtained at M/I ratios <10.000. Increasing the reaction temperature from 103 to 130°C at different M/I values resulted in an increase in $M_w/M_n$ from 1.1 up to 1.6.

KINETICS

In a study on the polymerization kinetics, the crude polymerization products obtained at different reaction times were

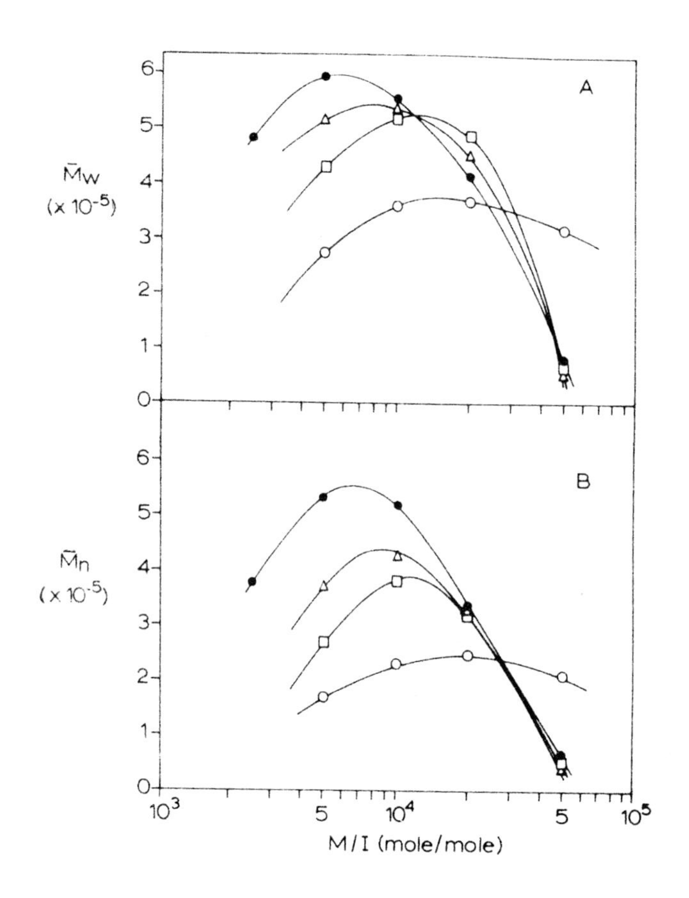


Figure 1A,B. Semilogarithmic plot of PLLA Mw (A) and Mn (B) as a function of M/I (t = 70 h) for T = 103°C (•), 110°C (▵), 120°C (▫), and 130°C (∘).

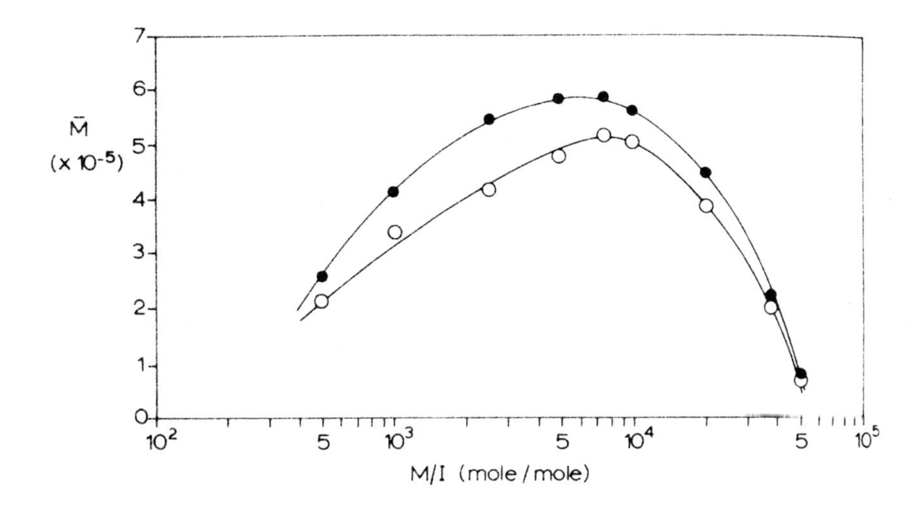


Figure 2. Semilogarithmic plot of PLLA Mw (●) and Mn (o) as a function of M/I for T = 103°C (t = 70 h)

analyzed for their monomer content by 200 Mhz $^1$H NMR spectroscopy, while the purified PLLA samples were characterized by combined GPC-LALLS measurements. Figures 3A and 3B show the $M_w$ and $M_n$ of purified PLLA samples obtained at different reaction temperatures and a constant M/I value of 10.000 as a function of polymerization time. At higher reaction temperatures molecular weights increased faster, but leveled off at an earlier stage as compared to lower reaction temperatures.

To test if the reaction is zero or first order in monomer both 1 - [M]/[M∘] and - In [M]/[M°] ([M] is the monomer concentration at time t, [M∘] is the initial concentration at t = 0, and $k_{eff}$ is the apparent rate constant) were plotted against polymerization time using different temperatures and a M/I ratio of 10.000. The data shows that compared to figure 4A, the linear region in figure 4B extends up to longer polymerization times for the reactions at the lower temperatures (T = 103,110°C). Moreover, higher regression coefficients were found for the lines given in figure 4B compared to those given in figure 4A. The slopes and regression coefficients of the linear regions in figures 4A and 4B appeared to fit best for a first order reaction rate in monomer conversion according to eq.2.

$$- \frac{d[M]}{dt} = k_p[M][P] = k_{eff}[M] \qquad (1)$$

$$- \ln \frac{[M]}{[M_0]} = k_{eff}\, t \qquad (2)$$

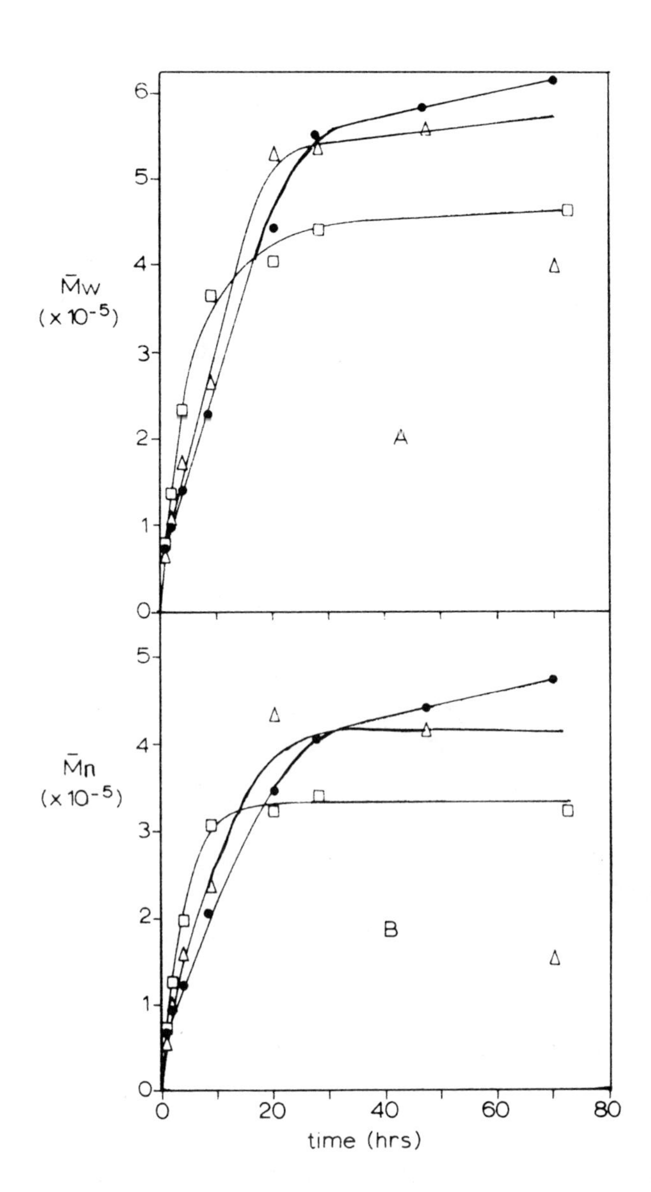


Figure 3A,B. Mw and Mn of PLLA as a function of polymerization time (M/I = 10.000) at T = 103 °C (●), 110 °C (Δ), and 120 °C (□).

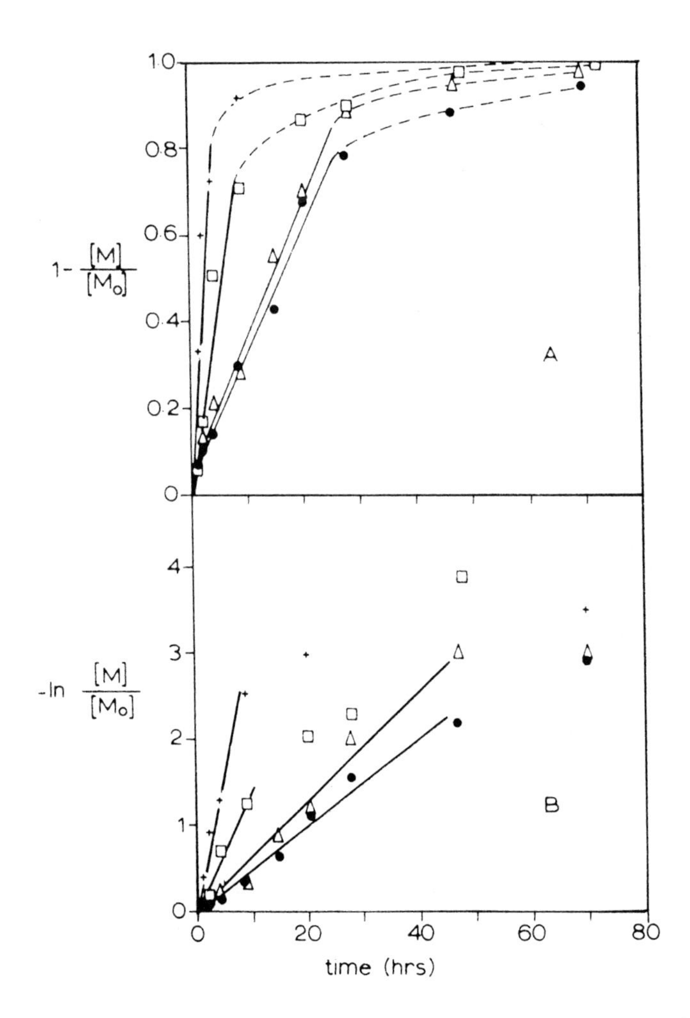


Figure 4A,B. Kinetic curves of stannous octoate initiated L-Lactide bulk polymerizations (M/I = 10.000) at T = 103 °C (●), 110 °C (△), 120 °C (□), and 130 °C (+)

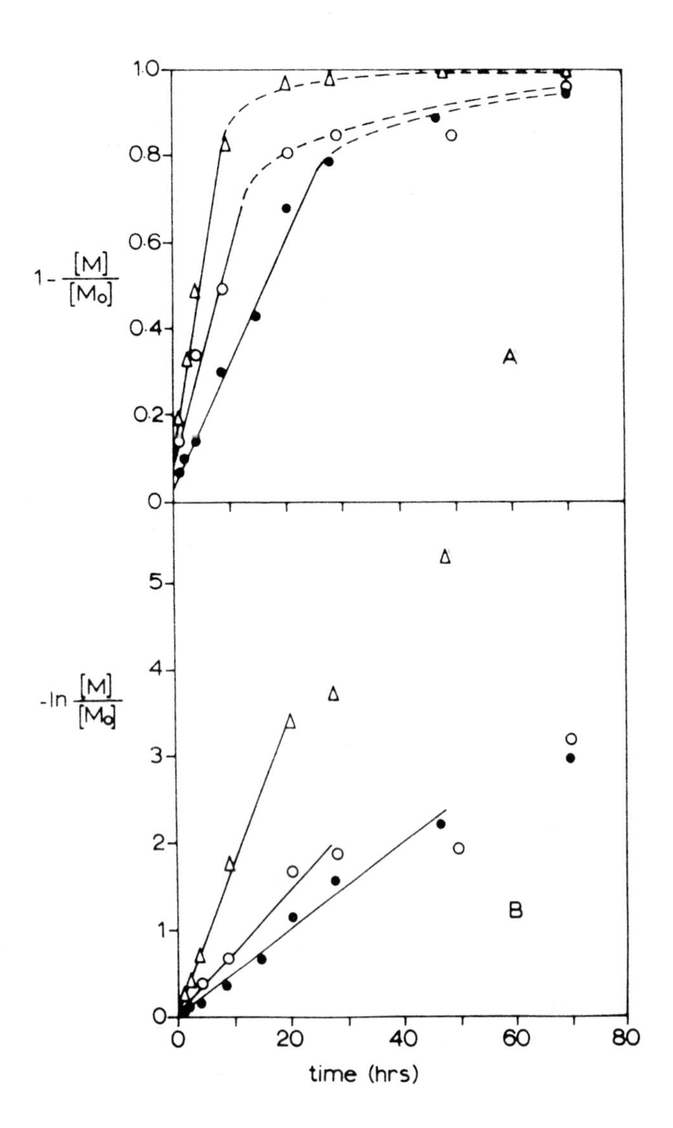


Figure 5A,B Kinetic curves of stannous octoate initiated L-Lactide bulk polymerizations ( T = 103 °C) for M/I = 2.500 (Δ), 5.000 (○), and 10.000 (●)

The linearity of the curves in figure 4B indicates that the concentration of growing species, [P], was constant during the reaction up to high conversions (80 - 85%). The data of figure 4 also show that the polymerization rate increases with the reaction temperature. From this temperature dependence the overall polymerization activation energy was determined. An Arrhenius plot of -In $k_{eff}$ against 1/T gave a value of $E_a$ = 19.2 kcal/mol.

Plots of -In[M]/[$M_O$] versus time for polymerization reactions performed at 103°C for 70 h and M/I = 2.500, 5.000 and 10.000 gave also good regression coefficients indicating that all cases the reaction rate is first order in monomer concentration (figure 5). These results also showed that the reaction rate increased with increasing initiator concentration. A plot of the apparent rate constant $k_{eff}$ against the initial initiator concentration expressed as $I_o/M_o$ (figure 6) shows a linear correlation. This indicates tht the concentration of active species [P], was proportional to the initial initiator concentration, [$I_O$];

$$k_{eff} = k_p [P] = k_r [I_0] \quad (3)$$

$$-\frac{d[M]}{dt} = k_p [M][P] = k_r [M][I_0] \quad (4)$$

$$k_{eff} = k_r [I_0] = k_r [M_0]\frac{[I_0]}{[M_0]} \quad (5)$$

$$E_a = \Delta H^* + 2RT \qquad (6)$$

$$k_r = e^2\left(\frac{kT}{h}\right)\exp\left(\frac{\Delta S^*}{R}\right)\exp\left(\frac{-E_a}{RT}\right) \qquad (7)$$

Substitution of eq. 3 in eq. 2 results in eq. 4, which indicates that the reaction rate is first order in monomer concentration as well as initiator concentration, with a reaction rate constant $k_r$. The rate constants have been calculated with the use of eq. 5 and are given in Table 1. Successively the enthalpy and entropy of activation were calculated from equations 6 and 7. Using the already determined value for $E_a$ (19.2 kcal/mol), a value of 17.7 kcal/mol for $\Delta H^*$ and 6.3 cal/mol.K for $\Delta S^*$ was calculated.

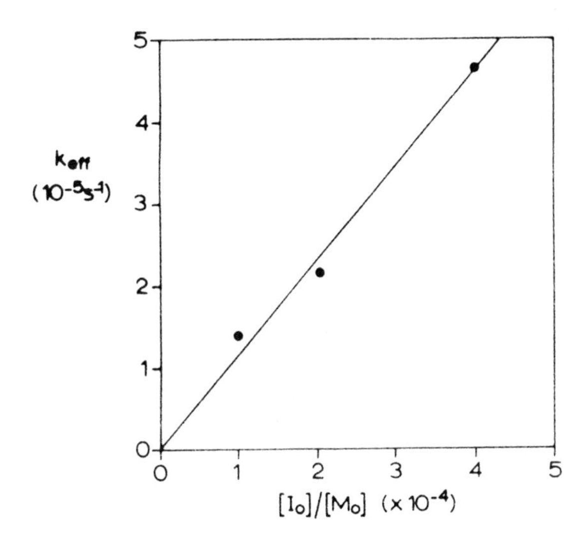


Figure 6. Plot of apparent reaction rate constant keff (T = 103 °C) against initial stannous octoate concentration ([I0] / [M0]).

Table 1. L-Lactide Bulk Polymerization Rate Constants

| T ($^oC$) | M/I (mole/mole) | $k_{eff} \times 10^5$ ($s^{-1}$) | $k_r \times 10^2$ l/mole.s |
|---|---|---|---|
| 103 | 2.500 | 4.68 | 1.47 |
| 103 | 5.000 | 2.18 | 1.37 |
| 103 | 10.000 | 1.38 | 1.74 |
| 110 | 10.000 | 1.81 | 2.28 |
| 110 | 10.000 | 4.04 | 5.09 |
| 130 | 10.000 | 7.50 | 9.45 |

Using a standard procedure for the ring-opening polymerization of L-Lactide, high PLLA molecular weights were obtained at low temperatures (T = 103°C) and optimal M/I ratios. The polymerization reaction of L-Lactide in the bulk appeared to be first order in monomer and first order in initiator.

## POLYDEPSIPEPTIDES

### SYNTHESIS AND CHARACTERIZATION

The properties of homo- and copolymers of PLLA and other poly(alpha-hydroxy acids) have been extensively studied. Other types of biodegradable polymers are poly(alpha-amino acids), poly(cyanoacrylates), polyphosphazenes, polyanhydrides, polyacetals and poly(orthoesters) (13,14). These polymer systems all contain hydrolytically instable bonds in the polymer backbone. By cleavage of labile bonds, the generally water-insoluble macromolecules are converted into water soluble oligomers. The rate of hydrolytic cleavage is dependend on the chemical structure of the main chain, the presence of substitutents, and several physical properties (14,15). Important physical properties are the permeability to water, the crystallinity and glass transition temperature of the polymer (14). These properties can be influenced by plasticization, blending and copolymerization. An interesting class of biodegradable polymers are the polydepsipeptides, polymers composed of alpha-amino acid and alpha-hydroxy acid residues. Because different alpha-amino acids and alpha-hydroxy acids may be incorporated in the polymer backbone a wide variation in polymer structures and properties becomes possible.

The synthesis of polydepsipeptides has been schematically represented in Scheme 1. The first synthesis of sequential polydepsipeptides by solution polymerization in DMF of performed di- or tridepsipeptides was reported by Stewart (16). The synthesis and conformational properties of polydepsipeptides were more extensively studied by Goodman et al. (17 and references cited). Polydepsipeptides, such as poly(L-valine-L-Lactic acid), poly (L-alanine)q -L-Lactic acid) with q = 1, 2 or 3, were synthesized via the solution polymerization of the trifluoro acetate salts of tri- or tetradepsipeptide pentachlorophenyl esters (analogous to Scheme 1, f).

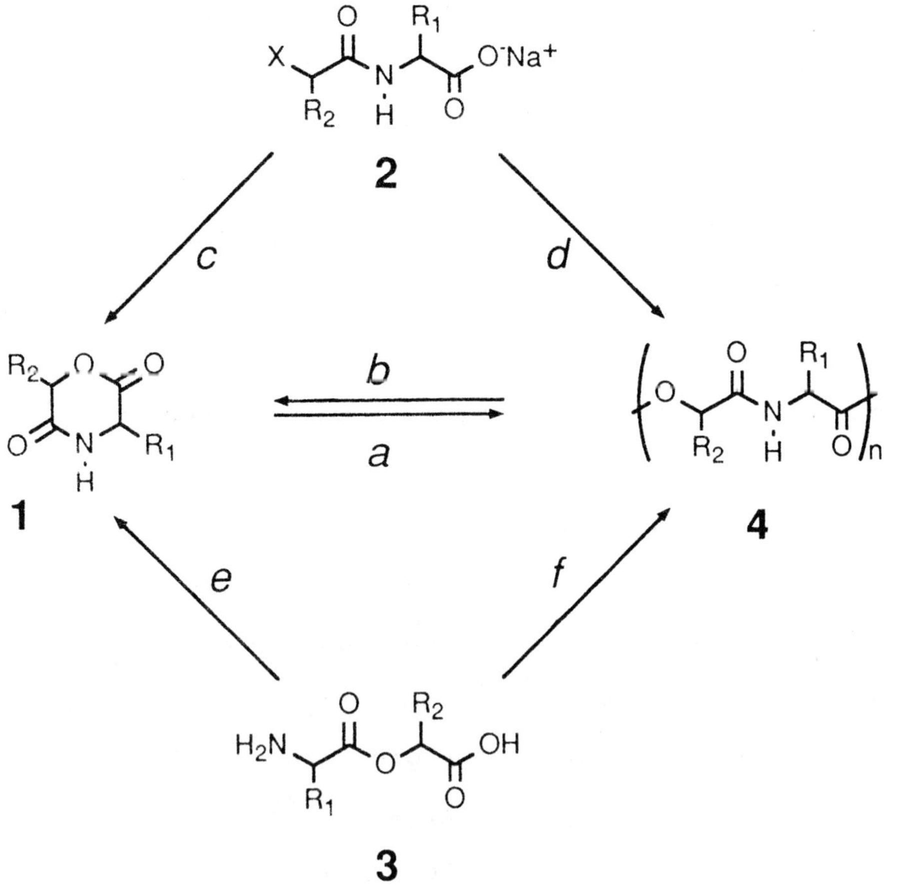

Scheme 1 Pathways for the preparation of polydepsipeptides

Temperature induced helix-to coil transitions of poly(L-alanine)q -L-Lactic acid) solutions in THF or chloroform were observed using $^{1}H$ NMR spectroscopy. Polydepsipeptides containing glutamate esters and epsilon-protected lysine residues all exhibit very stable alpha-helical conformations in organic solvents. This implies that analogous to polylactide, polydepsipeptides incorporating chiral amino acid and/or hydroxy acid residues may be semicrystalline and obtain suitable mechanical properties for application in different biodegradable devices.

The routes to sequential polydepsipeptides based on the multi-step synthetic procedures mentioned above are inadequate for the preparation of large amounts of polymer. The high yields and molecular weights obtained by ring-opening polymerization of NCA's of alpha-amino acid and cyclic mono- or diesters suggest that ring-opening polymerization of cyclo(alpha-amino acid - alpha-hydroxy acid) [1,4-morpholine-2,5-dione] derivatives (Scheme 1, a) could be an attractive route to prepare various polydepsipeptides in a more facile way.

The cyclo(alpha-amino acid-alpha-hydroxy acid) derivatives can be considered as the simplest representatives of the so called cyclodepsipeptides, which are ring systems having regularly or irregularly alternating ester and amide linkages. The synthesis and ring-opening polymerization of the cyclic depsipeptide cyclo(glycine-D,L-Lactic acid) was first reported by our group in 1985 (18). Yonezawa et al. (19) described the ring-opening polymerization of cyclo(DL-valine-DL-lactic acid) and Hwang et al (20) and Kohn (21) mentioned that cyclo(glycine-glycolic acid) could be a useful monomer for the synthesis of polydepsipeptides. The preparation of polydepsipeptides via ring-opening polymerization has only recently become of general interest (22). The strategy used to prepare cyclodidepsipeptides is based on the formation of an ester bond by an intramolecular substitution reaction in bromo- or chloroacyl-alpha-amino acids (Scheme 1,c). Acylation of an amino acid under Schotten-Baumann conditions is a proper methodology to obtain a variety of N-acylated amino acids. Reaction of chloroacetylchloride or (D,L)-2-bromopropionylbromide with either glycine, (L)-alanine or (L)-valine using the procedures described by Fisher et al. (22) afforded the corresponding N-(2-halogenoacyl)-amino acids.

Several attempts to cyclize the N-(2-halogenoacyl)-amino acids in solution using different bases and solvents only resulted in small amounts of oligomers or very low yields of the cyclic depsipeptides. A number of investigators reported that the attempted polymerization of N-(2-halogenoacyl)-amino acid derivatives resulted in the formulation of cyclic didepsipeptides (Scheme 1, c)(24). Goodman and coworkers (25) reported that cyclo(L-alanine-L-Lactic acid) and cyclo(L-valine-L-Lactic acid) were obtained in considerable yields as unwanted by-products in the attempted polymerization of the trifluoroacetate salts of the corresponding didepsipeptide 3 pentachlorophenylesters (analogous to Scheme 1,e).

The tendency of the haloacyl-alpha-amino acids not to cyclize in solution, even at higher temperatures, prompted us to investigate the cyclization reaction in the bulk. The bromo- or chloroacyl-alpha-amino acids were converted into their corresponding sodium salts using a sodium hydroxyde solution. The dried sodium salts were subsequently heated at 15-20°C above their m.p. for at most 1 h. Using this methodology only cyclo(glycine-D,L-Lactic acid and cyclo(L-alanine-D,L-Lactic acid) could be obtained after extended work-up in yields varying from 23-40% (Scheme 1, c). The main side products appeared to be oligomers, as was shown by the $M_n$ values of approximately 300 as determined by GPC measurements. In an attempt to improve the yield of cyclic didepsipeptides the compounds 2 were diluted by the addition of Celite and the cyclized product was directly sublimed out of the reaction mixture. Using this method yields of cyclic depsipeptides varied from 25 to 65% when $R_2$ = $CH_3$ but were very low (~5%) when $R_2$ = H. The low yields of the 6-unsubstituted 2,5-morpholinediones (1, $R_2$ = H) prompted us to a further analysis of the residue remaining after the reaction. $^1H$ NMR analysis revealed that the residues consisted mainly of oligomeric products.

However, further analysis of the products obtained after precipitation by adding DMF solutions of the residues to an excess of water revealed that during the preparation of monomers with $R_2$ - H also the corresponding polydepsipeptides 4 were formed (Scheme 1, d) (Table 2, matrix polymerization). In order to increase the yield of cyclic product we also studies the possibility to depolymerize the

Table 2. Characteristics of alternating polydepsipeptides

| 4 R1 (amino acid residue) | R2 (hydroxy acid residue) | method a) | Mw (x $10^4$) | $[\alpha]_D^{25}$ b) (o) | Tg (oC) | Tm (oC) |
|---|---|---|---|---|---|---|
| H | (D,L)$CH_3$ | RO | 2.3 | -- | 117 | -- |
| (L)$CH_3$ | H | RO | 4.5 | - 29 | 103 | 226 |
| (L)$CH(CH_3)_2$ | H | RO | 7.4 | +44 | 99 | -- |
| (L)$CH(CH_3)_2$ | H | matrix | 3.5 | +43 | 93 | -- |
| (D)$CH(CH_3)_2$ | H | matrix | 4.9 | -40 | 93 | -- |
| (L)$CH(CH_3)_2$ | (D,L)$CH_3$ | RO | 2.5 | -- | 93 | -- |
| (L)$CH(CH_3)_2$ | (D,L)$CH_3$ | matrix | 3.3 | -- | 99 | -- |
| (L)$CH_3$ | (D,L)$CH_3$ | RO | 0.9 | -- | 94 | -- |
| (L)$CH_3$ | (L)$CH_3$ | ref. 17 | -- | | | |

a) RO = obtained via ring-opening polymerization, matrix = obtained by matrix polymerization of the corresponding N-acylated valine (see text)

b) c = 0.002 (g/100 ml) in $CHCl_3$

oligomeric and polymeric products formed during the cyclization reaction using antimony trioxide (Scheme 1, b). This appeared only successful for the compounds containing the glycolic acid residue (6-unsubstituted 2,5-morpholinediones) but not for compounds containing the lactic acid residue. However depolymerization was also accompanied with racemization of the chiral centre of the amino acid residue. Ring-opening polymerization of the 2,5-morpholinedione derivatives (Scheme 1, a) was performed in the melt using stannous octoate as an initiator at temperatures 5°C above the melting temperature of the monomer. The results have been summarized in Table 2. Contrary to the polymerization of L-Lactide relatively high initiator concentrations (M/I = 250 - 2.500) had to be used in the ring-opening polymerizations of the cyclic depsipeptides. it cannot be excluded that complex formation of the initator with the amide group in the monomers causes some inhibilition of the polymerization reaction. From Table 2 it can be seen that polydepsipeptides 4 with the highest molecular weights were obtained when $R_1$ = isopropyl, or when $R_2$ = H. The rather bulky substituent $R_1$ may cause shielding of the amide group for interaction with the initiator resulting in an more facile polymerization. Comparing the optical rotations of the polydepsipeptides 4 ($R_1$=isopropyl and $R_2$ = H) obtained through different polymerization reactions (Scheme 1 routes a and d) indicates that the polymerization is not accompanied by racemization. In this series only the polydepsipeptides 4 {$R_1$=(L)$CH_3$, $R_2$=H or (L)$CH_3$} were semicrystalline.

Using ring-opening copolymerization of 1 ($R_1$ = H, $R_2$ = $CH_3$) and DL-lactide a range of random copolymers containing glycine and DL-lactic acid residues could be synthesized and the degradation behaviour studies as a function of the composition.

DEGRADATION

The extensive literature concerning the degradation of polyesters has been reviewed by Holland et al.(26). After penetration of water into the polymer matrix, degradation starts through a bulk hydrolysis of ester bonds and the molecular weight decreases. Weight loss ensures when the molecular weight of the polymer chains has sufficiently decreased (27-31). In crystalline polymers like poly(glycolic acid) (PGA), poly(L-Lactic acid)(PLLA) and poly epsilon-caprolactone)(PCL) degradation first occurs in the amorphous regions and later in the crystalline domains (29, 33-35). As a result the crystallinity may increase during the initial stages of degradation. The period required for complete degradation of the polymer depends on the type of polymer, the initial molecular weight of the samples, the crystallinity and the glass transition temperature. Although some enzymes were found to affect the degradation of PGA (36) and PLLA (37) in vitro, an enzymatic contribution to the initial stages of degradation of polyesters has not been observed in vivo (38,39). Only a few studies about the degradation of copolymers containing both ester and amide groups were reported (40,41). More recently Kaetsu et al. (42,44) studied the in vivo degradation of melt pressed films of sequential polydepsipeptides. The polydepsipeptides were absorbed after several weeks implantation in rats and were more readily degraded than the corresponding poly(alpha-amino acids).

The mechanism of the hydrolysis reaction can be analyzed using the approach of Pitt and Gu (31) by determining the molecular weight and weight of the different (co)polymers as a function of time. In principle the degradation may occur via an autocatalic or via an uncatalyzed reaction. The reaction rate for an acid catalyzed reaction (autocatalysis) can be determined using equation 9 in which $M_n{}^\circ$ = number average molecular weight, $k_2$ = rate constant, and t = time. The kinetics for the uncatalyzed reaction are described by equation 10 (31).

$$\mathrm{In}(M_n) = \mathrm{In}(M_n{}^\circ) - k.t. \qquad (9)$$

$$1/M_n = 1/M_n{}^\circ + k'.t \qquad (10)$$

We have used equations 9 and 10 to analyze the degradation of the copolydepsipeptides by measuring the changes in $M_n$,app during degradation of the glycine/DL-lactic acid copolymers and PDLA in a phosphate buffer at pH = 7.4 and at 37°C (45-47). The rate of degradation of the (co)polymers was determined during the period before significant weight loss occurred (an example has been given in figure 7). The molecular weight distribution ($M_w$,app/$M_n$,app) and overall composition of the copolymers did not change significantly during this period.

Figure 8 shows plots of In($M_n$,app) versus immersion time for glycine-DL-lactic acid copolymers with different mol% glycine residues according to autocatalytic hydrolysis (eq.9). Regression coefficients calculated for the plots according to equation 9 were much better than those of the plots based on eq.10, and it was concluded that the degradation of the copolymers proceeds via an autocatalytic process. The rate constants obtained for the autocatalytic degradation of solution cast films of the copolymers were compared with those determined for poly-epsilon-caprolactone (PCL) and PLLA. Rate constants of $1.8 \times 10^{-3}$ day$^{-1}$ for PCL and $2.3 \times 10^{-3}$ for PLLA were obtained. The rate constants for the hydrolysis of glycine/DL-lactic acid copolymers were all in the same range ($5\text{-}7 \times 10^{-2}$ day$^{-1}$). The reason for the higher rates of hydrolysis may be due to the absence of crystallinity among other physical and structural parameters for the polydepsipeptides. This is supported by the finding that the hydrolysis constants of PDLA films with molecular weights of $3.2 \times 10^4$ and $14.2 \times 10^4$ were $4 \times 10^{-2}$ day$^{-1}$ and $5 \times 10^{-2}$ day,$^{-1}$, respectively.

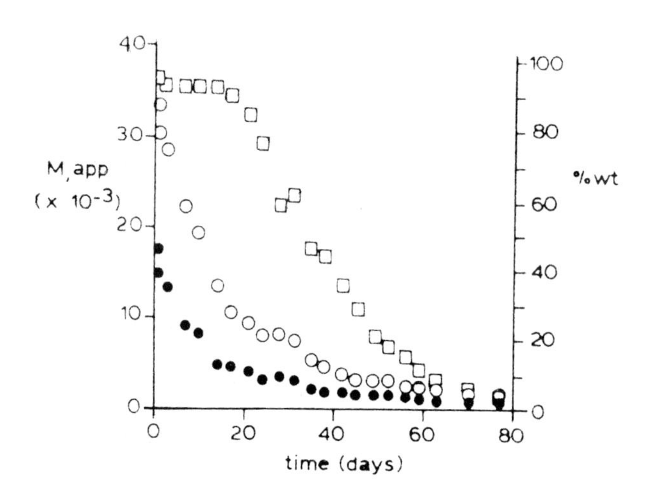


Figure 7. Glycine DL-lactic acid copolymer (20 mol% glycine): Mw,app (o ), Mn,app (● ) and residual weight (□) as a function of time.

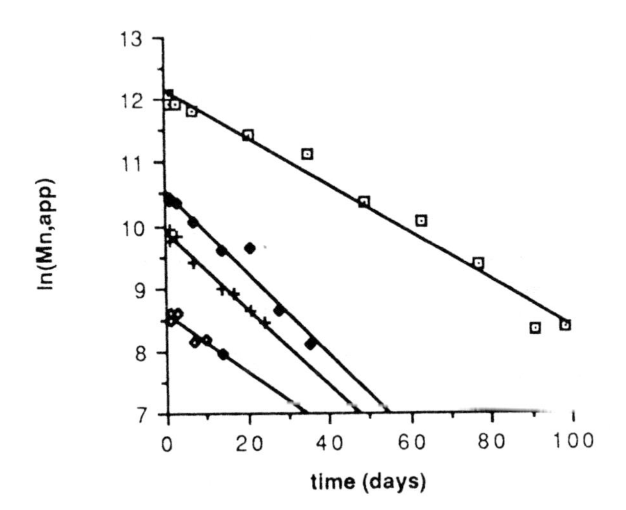


Figure 8. Plots of ln(Mn,app) as a function of immersion time (consistent with eq. 9) for glycine DL-lactic acid copolymers containing (◇), 25 mol% glycine, ( + ), 15 mol% glycine (◆),10 mol% glycine and poly(DL-lactide) (□).

The GPC patterns of all (co)polymers became bimodal just before the start of weight loss (fig. 9). Ming et al. (48) reported that the molecular weight distributors of either glycolic acid/L-Lactic acid copolymers or PLLA processed into polymer plates also became bimodal after in vitro degradation for several weeks. The molecular weight of polymer located in the central part of the plates was much lower than that for polymer situated at the surface. These authors assigned this feature to the fact that the degradation is acid-catalyzed and controlled by ion-exchange at the surface. Acidic polymeric fragments, which are formed inside the polymer mass cannot escape into the aqueous surrounding medium until their molecular weight is sufficiently reduced.

Although the hydrolysis constants and initial moledular weights for the different copolymers are comparable, weight loss was observed at an earlier stage for copolymers with a higher content of glycine units (Fig. 10). this can be explained by the higher solubility of degradation products rich in glycine residues in the buffer solution. Further evidence was obtained from the composition of partly degraded copolymers as determined by $^1$HNMR (Table 3) showing

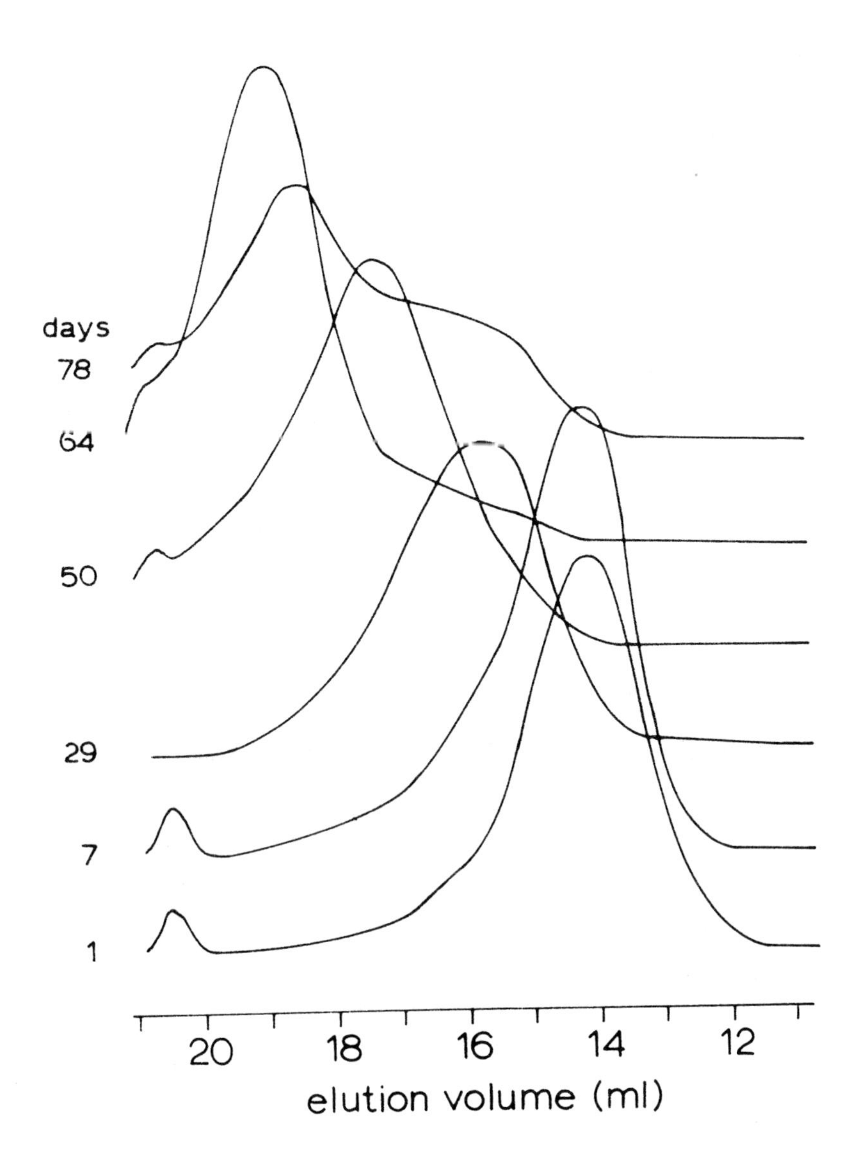


Figure 9. GPC of poly( DL-lactide) after 1, 7, 29, 50, 64 and 78 days of immersion in buffer solution (pH = 7.4).

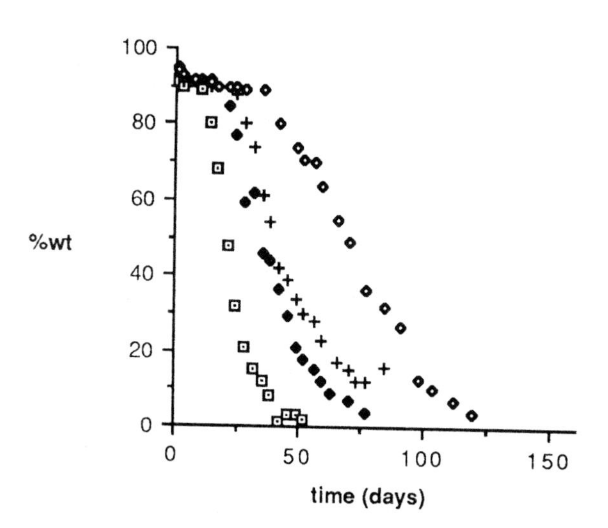


Figure 10. Residual weight of Glycine DL-lactic acid copolymers: (□), 25 mol% glycine, (◆), 20 mol% glycine, ( + ), 15 mol% glycine, and (◇), 12.5 mol% glycine, as a function of time.

Table 3. Composition and resisual weight of the copoly(glycine DL-lactic acid) after different degradation times.

| time (days) | F a) | % wt | time (days) | F a) | % wt | time (days) | F a) | % wt | time (days) | F a) | % wt |
|---|---|---|---|---|---|---|---|---|---|---|---|
| 0 | 77 | 100 | 0 | 83 | 100 | 0 | 87 | 100 | 0 | 89 | 100 |
| 1 | 78 | 92 | 1 | 83 | 95 | 1 | 87 | 94 | 1 | 88 | 94 |
| 7 | 77 | 91 | 7 | 83 | 92 | 7 | 86 | 91 | 7 | 89 | 92 |
| 14 | 81 | 80 | 14 | 83 | 92 | 21 | 87 | 90 | 24 | 90 | 90 |
| 21 | 87 | 48 | 31 | 86 | 62 | 38 | 92 | 54 | 59 | 91 | 64 |
| 31 | 89 | 14 | 45 | 91 | 29 | 59 | 92 | 23 | | | |
| 38 | 90 | 2 | 63 | 94 | 10 | 73 | 95 | 12 | | | |

a) F is the percentage of lactic acid units in the copolymer.

that after longer degradation times residual material contains a higher percentage of DL-lactic acid residues, which is caused by early dissolution of low molecular weight fractions rich in glycine units.

Some of the glycine/DL-lactic acid copolymers were studied with respect to their in vivo and in vitro degradation behaviour (49). For the in vivo examination, discs of the copolymer films were subcutaneously implanted in rats, whereas in vitro studies were simultaneously performed using phosphate buffer at pH=7.4 and 37°C. Figure 11 shows $M_n$,app and the residual weight of the discs as a function of time for both the in vivo and in vitro studies. The decrease in molecular weight, the loss of weight and the tissue reactions of the different copolymers were determined after 2, 5 and 10 weeks.

The in vivo and in vitro degradation behaviour of the polymers was comparable. The decrease of molecular weight of the copolymers and poly(DL-lactic acid) in time appeared similar, and the weight loss for copolymers with a higher mole fraction of glycine units started earlier as was also observed in the more detailed in vitro experiments (see above). The copolymer with the highest content of glycine units disappeared completely within 10 weeks both in vivo and in vitro, whereas the poly(DL-lactic acid) implant lost only 25% weight over the same period.

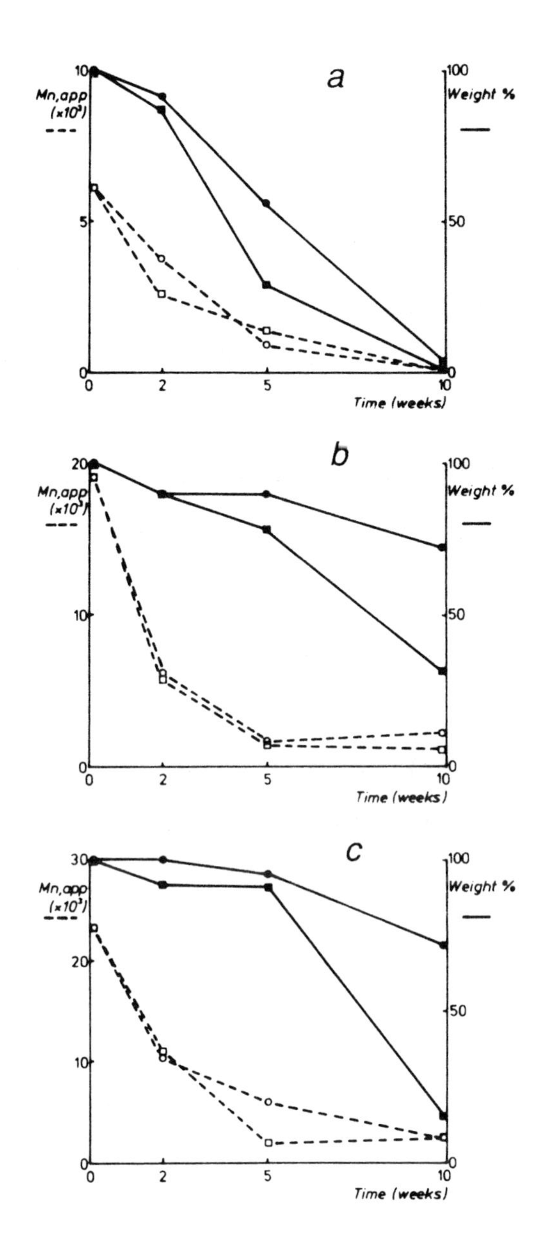


Figure 11. Changes in Mn,app (in vitro (□) and in vivo ( o )) and residual weight (%) (in vitro (■) and in vivo ( ● ))of copolydepsipeptides containing 20 mol% glycine (a), 10 mol% glycine (b), and poly(DL-lactide) ( c ).

Tissue reactions against all materials started with an acute inflammatory reaction caused by the trauma of implantation, followed by wound healing processes, ending in a very mild foreign body reaction for the poly(DL-lactic acid) and a more excessive macrophage mediated foreign body reaction for the glycine/DL-lactic acid copolymers. The tissue reaction was more severe for polymers showing a higher rate of dissolution.

Polydepsipeptides have been synthesized following the pathways given in Scheme 1. Ring-opening polymerization of morpholine-2, 5-diones with different substituents provides a method to synthesize polydepsipeptides with a wide variety in properties. Degradation rates of glycine/DL-lactic acid copolymers were similar but a faster dissolution of the polymer fragments with higher glycine content was observed. The degradation appeared the same in in-vitro and in-vivo experiements.

REFERENCES

1. M. Vert, Angew, Makromol. Chem., 166/167, 15 (1989).

2. a) E. M. Filachioni and C. H. Fisher, Ind. Eng. Chem., 36, 223-228 (1944)

   b) R. S. Nevin, U. S. Patent 4,273,920 (1981)

3. D. G. H. Ballard and B. J. Tighe, J. Chem. Soc. (B), 976-980 (1967).

4. H. R. Kricheldorf and J. M. Jonte', Polym. Bull. 9, 276-283 (1983).

5. D. K. Gilding and A. M. Reed, Polymer 20, 1459-1464 (1979).

6. P. L. Salzberg, U. S. Patent 2,758,987 (1956).

7. R. A. Fouty, Can. Patent, 808,731 (1969).

8. R. G. Sinclair, U. S. Patent, 4,057,537 (1977).

9. F. E. Kohn, J. W. A. van de Berg, G. van de Ridder, and J. Feijen, J. Appl. Polym. Sci. 29, 4265-4277 (1984).

10. A. Schindler, Y. M. Hibionada, C. G. Pitt, J. Polym. Sci, Polym. Chem. Ed. 20, 319-326 (1982).

11. R. D. Lundberg, J. V. Koleske, K. B. Wischmann, J. Polym. Sci, A-1 7, 2915-2930 (1967).

12. a) H. R. Kricheldorf and A. R. Serra, Polym. Bull, 14 497-502 (1985).

    b) R. Dunsing and H. R. Kricheldorf, Polym. Bull, 14, 491-495 (1985).

13. J. Heller, CRC Crit, Rev. Ther. Drug Carrier Syst. 1, 39, (1984).

14. C. G. Pitt, T. A. Marks and A. Schindler, in "Naltrexone Research Monograph", ed. by R. E. Willette and G. Barnett, 28, 232, (1981).

15. C. G. Pitt, T. A. Marks and A. Schindler, in "Progress in Contraceptive Delivery Systems, ed. by E.S.E. Hafex and W. A. A. van Os, MPT Press Ltd., Lancaster, England 1, 17 (1980).
16. F. H. C. Stewart, Aust, J. Chem., 22, 1291 (1968).
17. M. Goodman, Biopolymers, 24, 137, (1985).
18. J. Helder, F. E. Kohn, S. Sato, J. W. van den Berg and J. Feijen, Makromol. Chem. Rapid Commun., 6, 9 (1985).
19. N. Yonezawa, F. Toda and M. Hasegawa, Makromol. Chem. Rapid Commun. 6, 607 (1985).
20. S. S. Hwang, S. I. Hong and N. S. Choi, J. Korean Soc. Text. Eng. Chem, 13, (1981).
21. F. E. Hohn, Ph.D. thesis, University of Twente, Enschede, The Netherlands 81 (1984).
22. F. N. Fung and R. C. Glowaky, Eur. Pat. Appl, EP 322154 (1989).
23. E. Fischer, Chem. Ber., 40, 952 (1907(25).
24. B. Ridge, H. N. Rydon and C. R. Snell, J. Chem. Soc., Perkin I, 2041 (1972).
25. D. Nissen, C. Gilon and M. Goodman, Makromol Chem., Suppl. 1, 23 (1975).
26. S. J. Holland, B. J. Tighe and P. L. Gould, J. Controlled Rel., 4, 155 (1986).
27. J. Heller, J. Controlled Rel., 2, 167 (1985).
28. A. M. Reed and D. K. Gilding, Polymer, 22, 494 (1981).
29. R. M. Ginde and R. K. Gupta, J. Appl. Pol. Sci., 33, 2411 (1987).
30. J. W. Leenslag, A. J. Pennings, R. R. M. Bos, F. R. Rozema and G. Boering, Biomaterials, 8, 311 (1987).
31. C. G. Pitt and Z. Gu, J. Controlled Rel., 4, 283 (1987).

32. R. A. Kenley, M. O. Lee, T. R. Mahoney and L. M. Sanders, Macromolecules, 20, 2398 (1987).

33. C. C. Chu and N. D. Campbell, J. Biomed. Mat. Res., 16, 417 (1982).

34. N. S. Mason, C. S. Miles and R. E. Sparks in "Biomedical and Dental Applications of Polymers', ed. by Gebelein and Koblitz, 279 (1981).

35. C. G. Pitt, M. M. Gratzl, G. L. Kimmel, J. Surles and A. Schindler, Biomaterials, 2, 215 (1981).

36. D. F. Williams, J. Bioeng. 1, 279 (1977).

37. D. F. Williams, Eng. Med. 10, 5 (1981).

38. C. G. Pitt, R. W. Hendren, A. Schindler and S. C. Woodward, J. Controlled Rel. 1, 3 (1984).

39. T. N. Salthouse, J.Biomed. Mat. Res, 10, 197 (1976).

40. Y. Tokiwa, T.Suzuki and T. Ando, J. Appl. Pol. Sci, 24, 1701 (1979).

41. T. H. Barrows, J. D. Johnson, S. J. Gibson and D. M. Grussing in part 34 "Polymers in Medicine 2" of the series "Polymer Science and Technology", Plenum 85 (1986).

42. M. Asano, M. Yoshida, I. Kaetsu, R. Katakai, K. Imai, T. Mashimo, H. Yuasa, H. Yamanaka and K. Suzuki, J. Jpn. Soc. Biomater., 3, 85 (1985).

43. M. Asano, M. Yoshida, I. Kaetsu, R. Katakai, K. Imai, T. Mashimo, H. Yuasa and H. Yamanaka, Seitai Zairyo, 4, 65 (1986).

44. I. Kaetsu, M. Yoshida, M. Asano, H. Yamanaka, K. Imai, H. Yuasa, T. Mashimo, K. Suzuki, R. Katakai and M. Oya, J. Controlled Rel., 6, 249 (1987).

45. J. Helder, J. Feijen, S. J. Lee and S. W. Kim, "Copolymers of D,L-Lactic Acid and Glycine", Makromol. Chem., Rapid Commun., 7, 193-196 (1986).

46. J. Helder, F. E. Kohn, S. Sato, J. W. A. van den Berg and J. Feijen in "Biological and Biomechanical Performance of Biomaterials, ed, by P. Christel, A. Meunier and A. J. C. Lee, Elsevier Science Publishers BV, Amsterdam, The Netherlands 245 (1986).

47. J. Helder, P. J. Dijkstra and J. Feijen, J. Biomed. Mat. Res., accepted for publication.

48. L. S. Ming, H. Garreau and M. Vert, Preprints Kunming International Symposium on Polymeric Biomaterials, May 3-7 (1988).

49. J. M. Schakenraad,. P. Nieuwenhuis, I. Molenaar, J. Helder, P. J. Dijkstra and J. Feijen, J. Biomed. Mat. Res., 23, 1271-1288 (1989).

QUESTIONS AND ANSWERS

PRESENTATION BY DR. P. DIJKSTRA

(Twente University)

DR. LOOMIS: Gary Loomis, DuPont.

A couple of questions. One is on your DSC of your DL block copolymer. With your initial melt you show a double melting point up there around 220 degrees, which then coalesces into a single melting point after reheating. Would you care to comment on that?

DR. DIJKSTRA: Yes, I can. I will. Of course we also were looking at this material for the first time to see those both melting points. It cannot be caused because you have low molecular weight components, because then you may be seeing phase separation or something like that.

Maybe this is due to different crystallite sizes or different types of crystallites in the polymer. But we can only look at this in more detail I think by doing some x-ray studies.

DR. LOOMIS: I can tell you from experience with the stereo complex from the individual polymers that the double melting points are due to different size crystallites. So that's what you should look for here.

DR. DIJKSTRA: Yes.

DR. LOOMIS: And in fact they are so big you can almost see the high melting ones with the naked eye. I mean, you don't need much of a microscope. So I think look for that. And I agree that's probably what it is.

DR. DIJKSTRA: Thank you.

DR. SHALABY: I'm not just Shalaby, I'm Shalaby Shalaby form Johnson & Johnson. And I have two questions.

The first question, were you lucky to make the polydepsipeptide of glycolic and the clycine?

DR. DIJKSTRA: Yes, we made this once, but it is very difficult, and we could only obtain it in very small amounts.

The second problem is that probably this compound has a very high melting point, and ring-opening polymerization then becomes very difficult. Because we generally also see that cyclic didepipeptides with high melting points are very difficult to ring-opening polymerize, and rather decompose than polymerize.

DR. SHALABY: The second question, actually I noticed that you did cast films. Can you melt process these without degradation. The polymers, all didepsipeptides that you talked about. Can you melt process them without degradation?

DR. ANDERSON: Melt process.

DR. DIJKSTRA: Yes, we have compared the data of melt pressed films, which is also done by bit, for instance for caprolactone, and we did not see any difference. We see in all cases the same rates of degradation. The rate constants are always the same, if you use solvent-cast films or melt pressed films.

DR. ALBERTSSON: My name is Ann-Christine Albertsson.

I would like to ask you, because you said something about in vitro-in vivo test. When I do this kind of test, I always get a much higher degradation rate in vivo than in vitro. But did you say you have the same?

DR. DIJKSTRA: Yes, it is the same.

DR. ALBERTSSON: And I have measured different kind of polyesters and polyanhydrides and --

DR. DIJKSTRA: Yes, but what do you expect, why should it be higher?

DR. ALBERTSSON: Because you have a mechanical stress and you have a dynamic system instead of static system.

DR. DIJKSTRA: No, these disks were implanted subcutaneously in the back of the rat. So I think -- and are very readily encapsulated, also. So probably these are the reasons mechanical stress is not so important.

DR. ALBERTSSON: When you had your in vitro, did you have any kind of dynamic in that system?

DR. DIJKSTRA: No, no dynamics.

DR. ANDERSON: I would like to ask a question. If I assume that the break in the weight per cent loss is indicative of loss of mass integrity, and I look at the molecular weights, what your are telling me is that I have to have over or approximately a 90 per cent decrease in molecular weight before I have a loss of mass integrity. Is that true?

DR. DIJKSTRA: Yes. Of course depends on the polymer you have, and so...

DR. ANDERSON: Well, what you showed us.

DR. DIJKSTRA: Yes, but it has to be decreased to maybe molecular weight of around 5,000, before it can dissolute. And this may be different for different polymers. And it also depends on the hydrophilicity of the polymer fragments.

It is well known that caprolactone, for instance, had to be

degraded up to around a molecular weight of 5,000 before dissolution can take place. But we did not, and we see in these systems around the same values.

DR. ANDERSON: The reason I asked my question is because while dissolution is something that we see in vitro, and it certainly happens in vivo, once you lose mass integrity in vivo, you change mechanism, because you change environment. And so one has to be very careful with trying to play off in vitro versus in vivo.

DR. VERT: Michele Vert, University of Rouen, France.

You mentioned that you met the copolymers with the glycine lactic acid and the DL lactide. And you said, if I picked it up right, that you got a random copolymer. Are you sure of this?

DR. DIJKSTRA: It is maybe somewhat exaggerated, but not too much. It is almost, almost random. Although it appears that lactic acid is somewhat easier copolymerized in the system, but it is not a very big difference.

DR. VERT: Did you measure it?

DR. DIJKSTRA: We tried to. But maybe because of the small differences, we could not have very accurate values. And we tried to determine the reactivity ratios in this system. But be it almost, almost the same, but there was maybe some discrepancy because of the accuracy of these determinations.

DR. VERT: You didn't try to copolmerize with a lactide, did you?

DR. DIJKSTRA: No, not at this moment. We only tried to polymerize it with caprolactone.

DR. NARAYAN: Ramani Narayan, Michigan State University.

I'm just curious. What is the rationale in picking up a

combination of a lactide and an amino acid? Were there any specificities which made you pick it up? Or you just picked it up to look at different properties?

DR. DIJKSTRA: From the early work, we knew that these compounds may show good mechanical properties, and there is still a lack of materials with different rates of degradation or -- yes, or diffcront, as most polymers nowadays use, or as polylactic acid, poly-e-caprolactone or copolymers out of. They are very slowly degrading, and there is still some demand for medical applications to have very rapid degraded materials, and these systems cannot afford it.

DR. NARAYAN: Does amino acid give you anything special or --

DR. LOOMIS: Probably, as I see, as we have seen in this glycine, lactic acid copolymers, much higher hydrophilicity, and of that much more rapid degradation of the material. And they are still biocompatible, of course. And they are well known to the human body. I think that's also a very important item to me mentioned.

DR. LOOMIS: One concern that people have with the depsipeptides, the same kind of concern people have in using polyamino acids as bio materials in that when they start to degrade, small fragments will be recognized as foreign bodies and elicit immune response.

I take it from your limited in vivo studies you have not see any --

DR. ANDERSON: Excuse me, foreign bodies do not elicit an immune response. Let's get it straight.

DR. LOOMIS: I'm sorry, okay. Let me rephrase this. I should probably ask Jim to phrase that actually.

Is there an immune response, forget the foreign body, the foreign protein. Does the material elicit immune response? Or have you specifically looked for that?

Is there an immune response, forget the foreign body, the foreign protein. Does the material elicit immune response? Or have you specifically looked for that?

DR. DIJKSTRA: We do not look for this at the moment. The only thing we are doing directly, when we synthesize such a compound, is have it in salt culture with human fibroblasts and see if normal off growth takes place. And as long as that happens, we continue this research.

DR. LOOMIS: And do I also understand it that you have not yet made a polymer that's crystalline, is that true?

DR. DIJKSTRA: Except for the block copolymers, no.

DR. LOOMIS: I guess the question is are there any crystalline depsipeptides, polydepsipeptides in the world yet?

DR. DIJKSTRA: Yes.

DR. LOOMIS: Oh, there are.

DR. DIJKSTRA: I've shown you on the first slide.

DR. LOOMIS: Oh, that was not crystal, I'm sorry.

DR. DIJKSTRA: I said, and as shown by the work of Goutman (phonetic), when you have a stero regular polymer, it shows crystallinity. And we also synthesize this compound by his methods, or somewhat --

DR. LOOMIS: And this was optically? Because I've forgotten which one it was.

DR. DIJKSTRA: Yes.

DR. LOOMIS: I guess the question is, have you looked for the stereocomplex in the depsipeptide? And if so, have you found one?

DR. FEIJEN: I suppose that's the next step.

DR. DIJKSTRA: Maybe I better answer this question, because I'm more directly involved sometimes in this work.

We tried. In the only one special case as I mentioned, with a V line and a glycolic acid we can synthesize an R or an S configuration. And we tried to make stereocomplexes out of it, but we did not succeed yet.

Also, in the homopolymer, you do not see any crystallinity at the moment. So we are looking for other systems, for instance using elanane (phonetic) with lactic acid, so you have two methyl groups and a more symmetrical molecule, which shows crystallinity and may also give rise to stereocomplex formation. But for that we have to look after the synthesis in more detail.

DR. ANDERSON: One comment; you need a couple of amino acids in a string to get a processable antigen site in order to create an antibody to give you an immune response.

DR. LOOMIS: You need a couple of different ones?

DR. ANDERSON: No, a couple in sequence. So these polymers don't have amino acids in sequence.

DR. LOOMIS: They can have two, unless it's --

DR. ANDERSON: That's not enough. That's still not enough. That's still not enough.

DR. FEIJEN: It's purely alternated so they don't have two.

DR. ANDERSON: That's right. It is an alternating copolymer.

The question that I had was, you told me that by introducing the amino acid into the polymer chain, you increase the hydrophilicity. And in the cases you were using, you were putting in veiling. Are you telling me that veiling as an amino acid increases the hydrophilicity of a polymer chain, as opposed to lactic acid?

DR. DIJKSTRA: No, you are correct. I didn't say that.

DR. ANDERSON: Yes, you did.

DR. DIJKSTRA: I mentioned only 40 glycine. Let's not fight here on stage.

DR. ANDERSON: Thank you very, very much.

# PHOTO-BIODEGRADATION OF PLASTICS. A SYSTEMS APPROACH TO PLASTIC WASTE AND LITTER

Gerald Scott
Polymer processing and Performance Group
Department of Chemical Engineering and Applied Chemistry
Aston University, Birmingham B4 7ET

## 1. INTRODUCTION

The problem of plastics litter, at one time considered to be primarily an aesthetic problem in areas of natural beauty (1-8) has more recently been recognized to be a serious threat to many forms of oceanic wildlife due to ingestion and strangulation (9,10). Plastics litter is also a by-product of the agricultural industry and plastics with time controlled degradability are widely used in automated agriculture where there is a positive cost benefit. They have been shown to photodegrade and biodegrade quickly at the termination of the growing season (11-14).

## 2. APPROACHES TO DEGRADABLE PLASTICS

The properties of the ideal packaging material is summarized in figure 1. The essential features are an "induction period" during which the properties of the polymer do not change. This is followed by rapid deterioration of mechanical properties due to some trigger of the environment. It will be argued below that light is the ideal initiator leading to rapid chemical modification with destruction of the integrity of the polymer article but other triggers will be discussed. The third stage involves the incorporation of the polymer into the ecosystem by continued oxidation and assimilation of the oxidation products, which like nature's litter are biologically recycled to carbon dioxide and water.

2.1 Biodegradation of plastics

None of the common packaging plastics are biodegradable. In general they are hydrophobic which largely accounts for their success as barrier materials. To make plastics inherently biodegradable, then, as in the case of the biopolymer PHBV (see Table 1 and scheme 1) is to make them less useful as packaging materials. An alternative strategem is to induce biodegradability. Plasticizers (particularly the aliphatic esters), many antioxidants and unsaturated fats and oils all induce biodegradation (15) either by preferential attack of the microorganisms on the low molecular weight additive, or by accelerating the rate of oxidation of the plastic, thus making it more accessible to microorganisms. In this connection it has been demonstrated by Albertsson (16) that oxidation products are preferentially removed from partially degraded polyethylene leading to purer hydrocarbon polymer.
Starch-filled plastics have been shown to biodegrade under composting conditions. This requires the use of oxidizable coagents (oils and fats) to induce the breakdown of the polymer

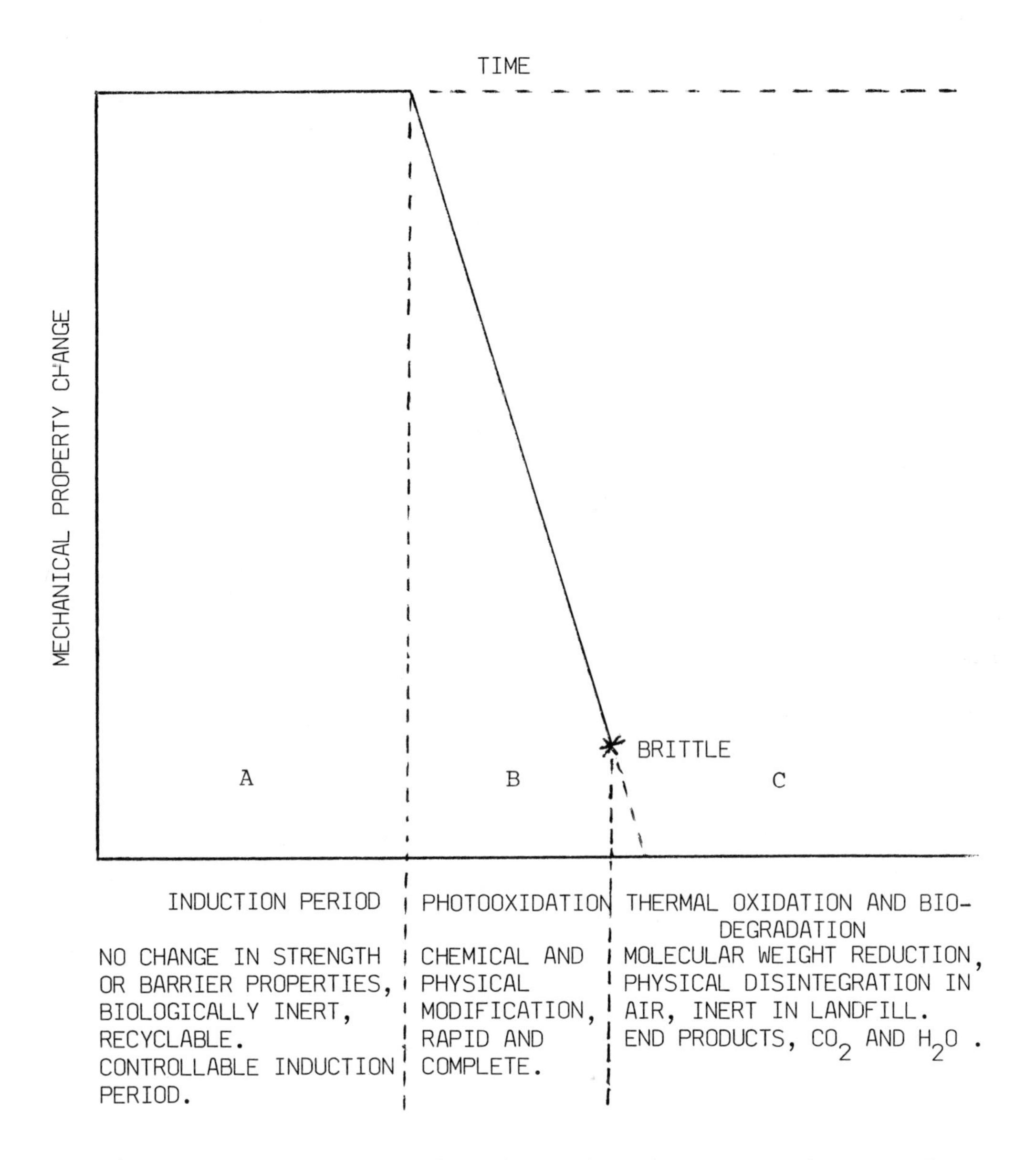


Figure 1. Performance with time of the ideal packaging plastic

**Table 1.** Commercially Available Degradable Plastics

| Common Description | Composition | Trade Name | Manufacturer |
|---|---|---|---|
| Biodegradable Polymers | | | |
| Poly(3-hydroxybutyrate-3-hydroxy valerate) | Biosynthetic copolymer of 3-hydroxy-butyric and 3-hydroxy-valeric acids | Biopol | ICI (UK) |
| Polymers containing a biodegradable filler | | | |
| Starch filled polyethyl-ene (Griffin process) | Physical blend of LDPE and starch | Bioplast<br>Ecostar | Coloroll (UK)<br>St. Lawrence Starch (Canada) |
| Photodegradable Co-polymers | | | |
| Ethylene-carbon mon-oxide co-polymers | | E/CO | DuPont (USA)<br>Union Carbide (USA)<br>Dow (USA) |
| Vinyl Ketone copolymers (Guillet process) | Copolymers of ethylene, propy-lene and styrene with a vinyl ketone | Ecolyte (Canada) | EcoPlastics |

| | | | |
|---|---|---|---|
| Photosensitising/ Photoactivating additives | | | |
| Iron salts | Probably Ferric stearate | PolyGrade | Ampercet (USA) |
| Antioxidant-Photo-activator (Scott-Gilead Process) | Ferric thiolates (sometimes with other metal thiolates) (in poly-ethylene) | Plastor | Plastopil (Israel) Polydress (Germany) |
| | | Greenplast | Enichem Agricultura (Italy) |
| | | Plastigone | Plastigone Technologies Inc. (USA) |
| | | Ecoten (manufacture discontinued in 1975) | Amerplast (Finland) |
| | (in polyropylene baler twine) | Cleanfields | American Brazilian Company (USA) |

Scheme 1

POLYMERS WITH INHERENT BIODEGRADABILITY

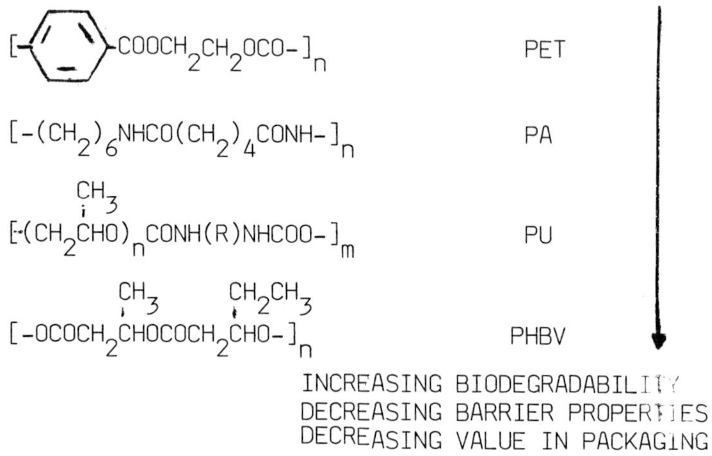

| | |
|---|---|
| $[-C_6H_4-COOCH_2CH_2OCO-]_n$ | PET |
| $[-(CH_2)_6NHCO(CH_2)_4CONH-]_n$ | PA |
| $[-(CH_2CH(CH_3)O)_nCONH(R)NHCOO-]_m$ | PU |
| $[-OCOCH_2CH(CH_3)OCOCH_2CH(CH_2CH_3)O-]_n$ | PHBV |

INCREASING BIODEGRADABILITY
DECREASING BARRIER PROPERTIES
DECREASING VALUE IN PACKAGING

Table 2. MOLECULAR WEIGHT CHANGE IN E/CO PHOTODEGRADABLE POLYMER (2.74% CO) ON UV IRRADIATION

| EXPOSURE TIME, h | $M_n$ | $M_w$ |
|---|---|---|
| 0 | 45,000 | 618,700 |
| 650 | 7,300 | 15,000 |
| 1350 | 11,100 | 39,100 |

matrix (17,18). There is little evidence, however, that this type of material biodegrades under ambient conditions. This is hardly surprising since the starch is securely encapsulated within the polyethylene matrix. A recent report (19) indicates that the permeability of corn starch filled polyethylene can be very much improved by the use of a polyethylene acrylic acid copolymer as a plasticizer, making film products more accessible to microorganisms. For the reasons indicated above, such water permeable materials are probably not very useful in packaging and their use in agricultural mulch is being investigated. Films made from the material are reported to lose 50% of their strength in contact with the soil in a few months, and the attack of microorganisms appears to begin immediately in the environment. Similar claims have been made for other starch-filled polyethylene films by "careful modification of the polymer" (20) but no treatment details have been published. This product biodegrades rapidly with disintegration of the film when subjected to microbiological attack under composting conditions (oxidation at 70°C) (20). It is clear that the products described above are not truly biodegradable polymers. They are primarily polymers with induced thermal oxidizability which subsequently biodegrade by the mechanism described by Albertsson (16). Thermal oxidation is, however, not a very controllable process, and although starch has ecological overtones which are satisfying to the "green" consumer, it seems unlikely that this approach can offer both the packaging manufacturers and the consumer the benefits of rapid biodegradation in the environment coupled with safety in use necessary for the control of plastics litter. In particular it lacks a "trigger" mechanism in which the plastic performs just like normal plastics during manufacture and service but are triggered by some component of the natural environment to degrade rapidly after a controllable induction period (figure 1).

## 2.2 Photodegradation of plastics

It was considerations such as those discussed above that led chemists to consider sunlight as the trigger for chemical breakdown of polymers. Two quite distinct processes which utilize sunlight as the initiator have to be distinguished. The first involved photolysis of polymers containing UV sensitive groups and the second photoinitiators for the oxidation of the polymer substrate.

### 2.2.1 Carbonyl copolymers

It has been recognized for many years that carbonyl chromophores introduced into polymers by oxidation during processing are, at least in part, the reason for the lack of stability of fabricated products to the outdoor environment (21,22). An early patent due to Brubaker (23) claimed that copolymers of ethylene and carbon monoxide were sensitive to light. It was subsequently shown that photolysis occurs, predominantly by the Norrish II process, reaction (1) (24)

$$-CH_2CH_2CH_2\overset{O}{\overset{\|}{C}}CH_2- \xrightarrow[h\nu]{\text{Norrish II}} -CH{=}CH_2 + CH_3\overset{O}{\overset{\|}{C}}CH_2- \quad (1)$$

Degradation of the polymer commences immediately on exposure to UV light with reduction in molecular weight without an apparent induction period and for products that require a period of stability, conventional UV stabilizers have to be added (24). In the later stages of UV exposure the vinyl double bonds appear to become involved in macromolecular enlargement as the molecular weight increases again (see Table 2). Possibly for the reason, photodegraded polyethylene made by this process does not biodegrade (25). By this contrast, carbonyl-containing polymers made by copolymerization of vinyl ketones with olefins (e.g. I) invented by Guillet (26,27), have been reported to biodegrade suggesting a great contribution from the Norrish I process, reaction (2):

$$-CH_2CH_2\underset{\substack{|\\ C=O\\ |\\ R}}{CH}CH_2- \xrightarrow[h\nu]{\text{Norrish I}} R\dot{C}O + -CH_2CH_2\dot{C}HCH_2- \quad (2)$$

This reaction does involve free radical formation and has been shown to lead to photooxidation during the later stages of the UV degradation of polyethylene. The Guillet vinyl ketone co-polymers have been developed commercially and are now the basis of the Ecolyte polymers (29).

These offer considerable potential over conventional plastics in short term uses and as disposable cups and food trays since the photosensitizer is part of the polymer chain and the products should present no problems of toxicity. However, like the E/CO polymers, degradation commences without an induction period and the use of this system in longer-term degradable products has yet to be demonstrated.

2.2.2 Photoinitiated oxidation

In spite of their considerable advantages in foodstuffs applications, the carbonyl copolymers are substantially more expensive than conventional polymers and as indicated above, the packaging industry is very resistant to any increase in cost of their products. Consequently in parallel with the above approach considerable attention has been paid to the use of additives and in particular, non-toxic transition metal compounds to accelerate photo-oxidation. Some of these, notable the organosoluble acetyl acetonates are very powerful photo-prooxidants (30). Unfortunately, like the transition metal carboxylates they are also effective thermal pro-oxidants. By accelerating the rate of hydroperoxide thermolysis during polymer processing (31,32) transition metal compounds of this type interfere unacceptably with the manufacture of plastics due to molecular weight changes during processing (30). However, all metal complexes are not thermal pro-oxidants and some, notably the metal thiolates, II-V, are known thermal antioxidants and processing stabilizers in a wide range of polymers (33).

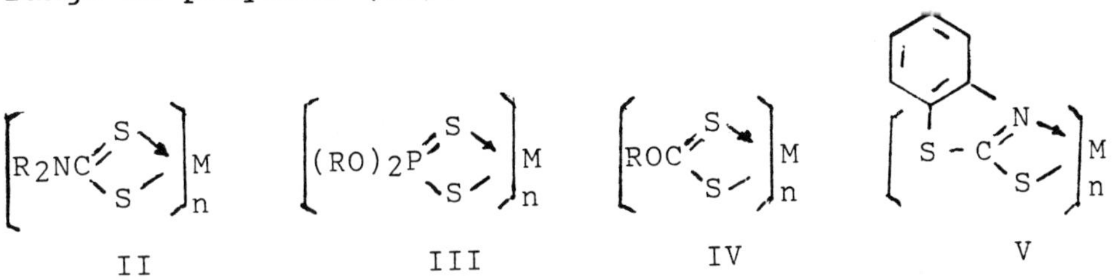

$$\left[R_2NC\begin{smallmatrix}S\\S\end{smallmatrix}\right]_n M \qquad \left[(RO)_2P\begin{smallmatrix}S\\S\end{smallmatrix}\right]_n M \qquad \left[ROC\begin{smallmatrix}S\\S\end{smallmatrix}\right]_n M \qquad \left[S-C\begin{smallmatrix}N\\S\end{smallmatrix}\right]_n M$$

II III IV V

Many of these are also effective photoantioxidants; notably the nickel and cobalt complexes. However, the group II complexes and, particularly the iron and manganese complexes are much less stable and undergo photolysis after an induction period; reaction 3 (34).

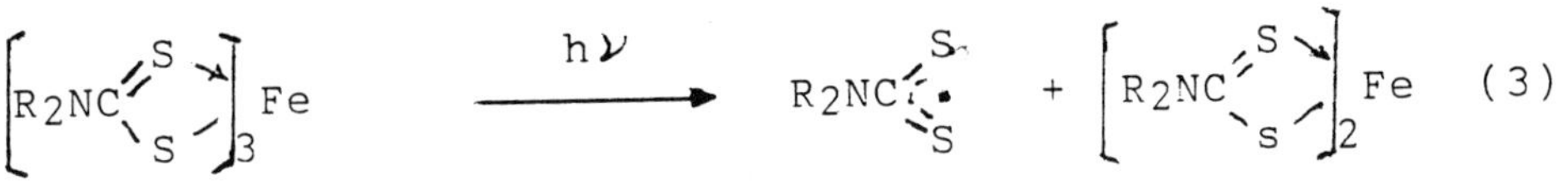

$$\left[R_2NC\begin{smallmatrix}S\\S\end{smallmatrix}\right]_3 Fe \xrightarrow{h\nu} R_2NC\begin{smallmatrix}S\cdot\\S\end{smallmatrix} + \left[R_2NC\begin{smallmatrix}S\\S\end{smallmatrix}\right]_2 Fe \qquad (3)$$

By a sequence of such reactions, the thiocarbamoyl residue which is the effective antioxidant group in the polymer and is responsible for the thermal and photooxidation induction period (33) is slowly destroyed, releasing free ionic iron as macromolecular carboxylates. The latter behave as photo-initiators in the normal way (see figure 2) and have similar activity to the acetyl acetonates. It was the growing understanding of the complex inversion mechanisms which occurs with many metal complex antioxidants (e.g. figure 3) that led to the development of the Scott-Gilead photo-biodegradation processs (35) which is now widely used in packaging because it provides effective antioxidant protection during processing, storage and normal use of the package. The mechanism of the peroxidolylic antioxidants provided the basis for the design of antioxidant-photoactivators with photocatalytic activity similar to the organo-soluble iron compounds but with a controllable induction period. Figures 2 and 3 show that this system in common with other antioxidant-based iron complexes has an inherent advantage over the carbonyl photosensitizers which show the reverse behavior, that is they initially catalyze photooxidation but subsequently cause autoretardation due to competition between the termination reaction (4(b)) and the propagation reaction (4(a)) (28).

$$\mathrm{PhC(=O)Ph} \xrightarrow{h\nu} \mathrm{Ph\dot{C}(\dot{O})Ph} \xrightarrow[(a)]{RH} \mathrm{Ph\dot{C}(OH)Ph} + \mathrm{R\cdot} \quad \begin{cases} \xrightarrow{(b)} \mathrm{PhC(OH)(R)Ph} \\ \xrightarrow[(c)]{O_2} \mathrm{ROO\cdot} \end{cases} \qquad (4)$$

The antioxidant-photoactivation system has proved reproducible and reliable in practice. However, in agricultural technology, particularly in hot climates use of multicomponent systems have been found to provide the increased induction periods required for some crops.

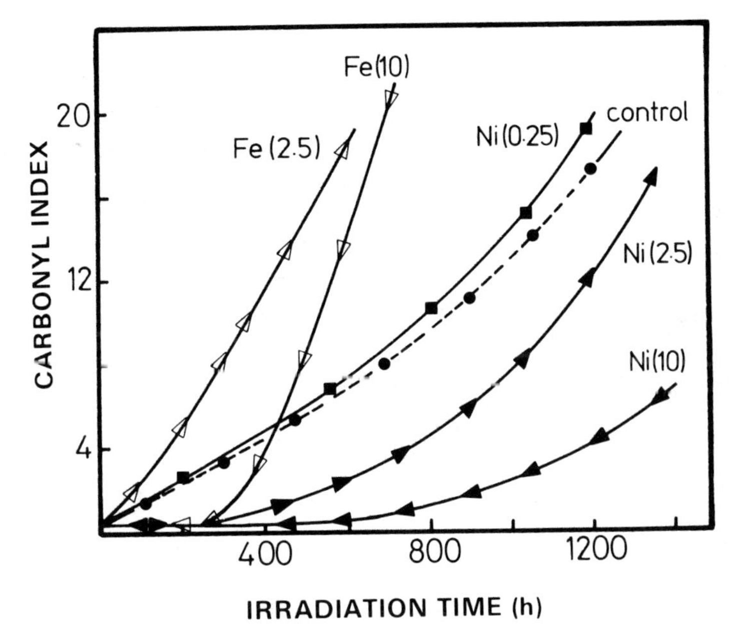


Fig. 2. Photooxidation of PE in the presence of Fe and Ni dithiocarbamates ($10^4$mol/100g)
(After ref. 44 with permission)

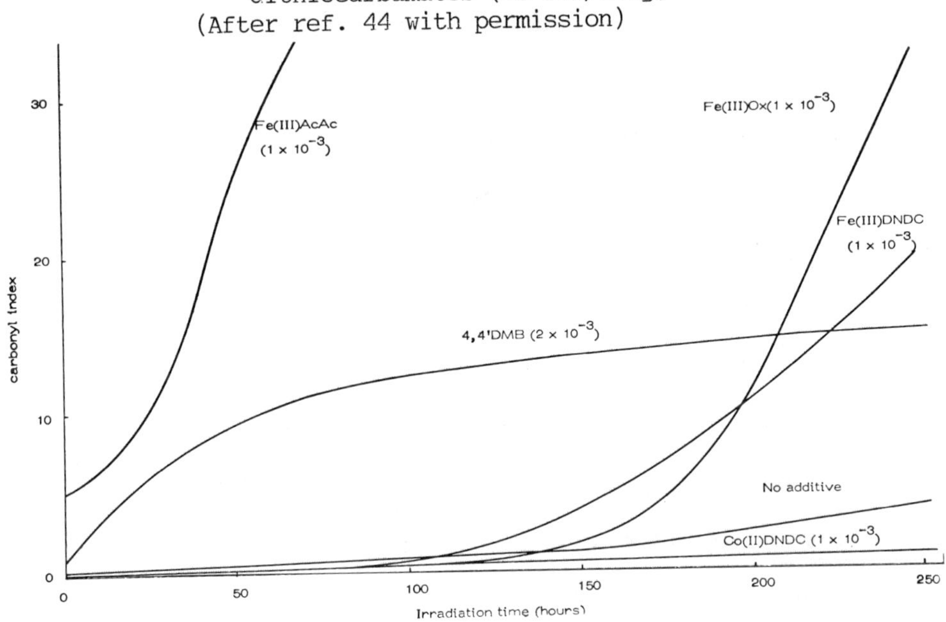


Fig. 3. Photooxidation of PE containing different photoactivators Fe(III)AcAc-ferric acetyl acetonate, Fe(III)Ox-ferric salisaldoxime, Fe(III)DNDC-ferric nonyl dithiocarbamate, Co(II)DNDC-cobalt nonyl dithiocarbamate, 4,4'-DMB -4,4' dimethoxybenzophenone.
(After Ref30 with permission)

Rather surprisingly, a combination of activator thiolate complex with a stabilizing complex actually produces a sharper inversion than the former alone at certain molar ratios (see figure 4). At high molar ratios of stabilizer to activator, however, very long induction periods can be achieved with rapid photooxidation to embrittlement at the end of the induction period. This is shown typically in figure 5 for polypropylene. This system is being used commerically in polypropylene baler twine where a lifetime of one year is required. It has also considerable potential in longer lived items such as tree guards, agricultural packaging and fishing nets where a durability of three years or more out-of-doors may be required.

The success of antioxidant-photoactivator systems lies in the time control offered and ultimate biodegradation of photo-oxidized products. Their potential limitations is in food-contact applications where the additives either have to be approved for toxicity or have to be non-migratory. Some components of the Scott-Gilead system are already approved for dry foodstuffs packaging but obtaining full approval for all systems is an expensive procedure which is not undertaken lightly by any potential user. Recent research has therefore recently been directed toward chemical attachment of the active agents to the polymer backbone. The reasoning behind this development is that if the functional compounds cannot be removed by any physical

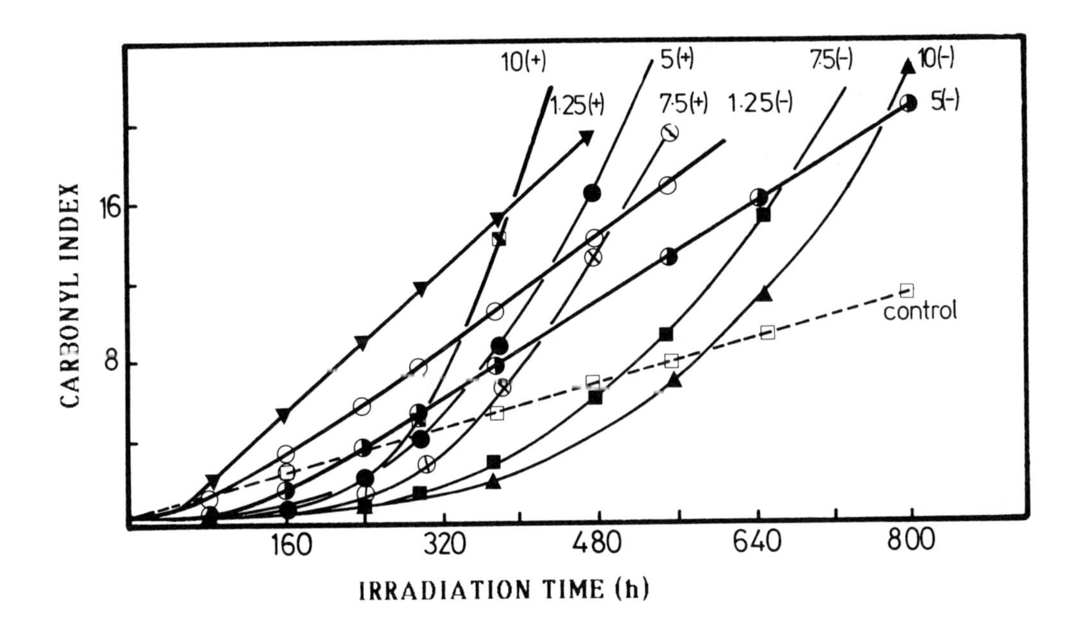


Fig. 4. Photooxidation of PE containing FeDMC in the presence(+) and absence(-) of NiDBC (2.5 x $10^{-5}$mol/100g). Numbers on curves are FeDMC concentrations ($10^4$mol/100g). (After Ref.44 with permission).

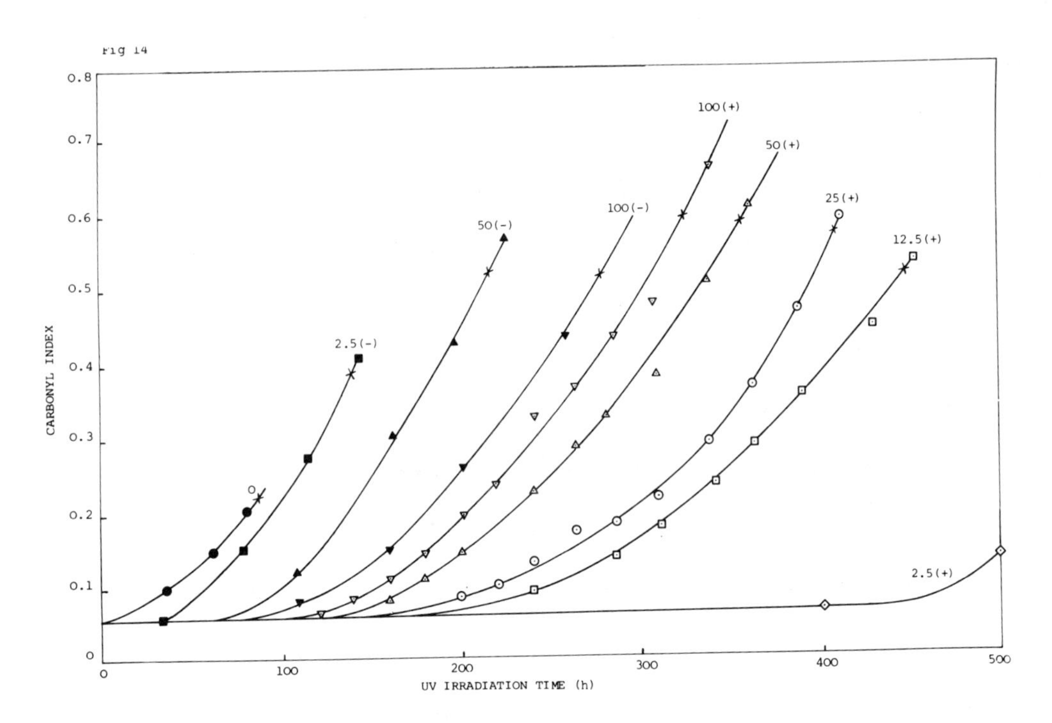


Fig. 5. Photooxidation of PP containing FeDMC in the presence(+) and absence(-) of NiDBC (2.5 x $10^{-4}$mol/100g). Numbers on curves are FeDMC concentrations ($10^5$mol/100g). (After Ref. 13 with permission).

means, then their toxicity has no practical relevance. Several such products are currently under evaluation and activity entirely similar to the additives currently used in agriculture has been observed.

The photograph shows the progress of the degradation of a typical packaging product with "dry foodstuffs" approval over the course of an English summer. Bags exposed out-of-doors in March took one and a half times as long to reach the particulate stage as identical bags exposed in July (~2 months). However, identical products exposed out-of-doors up to the end of August degraded to fine grass-like particles in the same season. Samples exposed in September were not embrittled in the same season but degraded rapidly in the following season.

Extensive studies over ten years on mulching film made by the process have shown that photodegraded particles cannot be found in the soil in the season following use. Similar studies of photobiodegradable packaging litter have confirmed this even although the litter is covered with soil. The evidence is consistent with the continued, albeit slower, thermal oxidation of the particulate material influenced by the free metal ions.

There is also preliminary evidence that weight loss also occurs at a measurable rate even in the absence of

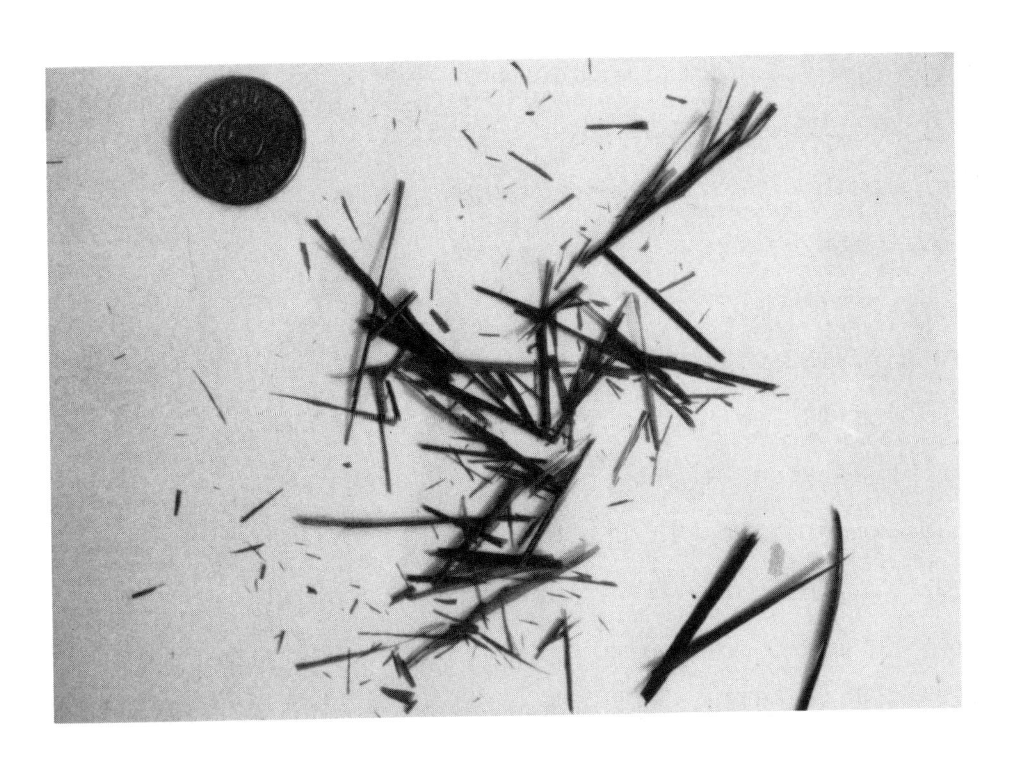

Degraded polyethylene bag after two months exposure out-of-doors in England. Particles right bottom are grass for comparison.

microorganisms due to thermal oxidation alone. On the other hand, photobiodegradable plastics which have not been exposed to the outdoor environment undergo no weight loss in compost over many years.

## 3. A SYSTEMS APPROACH TO THE DISPOSAL OF PLASTICS WASTE AND LITTER.

Many of the reasons advanced by the polymer industries for not using photo-biodegradable plastics in the battle against plastics packaging litter (36) are trivial and not supported by scientific evidence. There are, however, some arguments which do need serious consideration. The first of these is the danger of premature failure of the package during use. It will be evident from what has been said above the antioxidant-based photosensitizers provide a predetermined lifetime for packaging which can be varied from a few weeks to several years. A second concern over possible health hazards has already been discussed but it should be remembered that many packaging uses do not involve direct food contact (e.g. carrier bags, waste bags, etc.) and many of the products already on the market are already suitable from this point of view.

An objection frequently raised by the polymer industries is that degradability is not compatible with recycling (36). This is a convenient argument as it removes the responsibility for action from the producer and transfers it to the user and ultimate disposer of the waste. Recycling is essentially an entrepreneurial activity (see scheme 2) and will be increasingly used for clean and segregated polymer waste if it is available at the right price and

Scheme 2.

SOURCES AND FATE OF PLASTICS WASTE AND LITTER

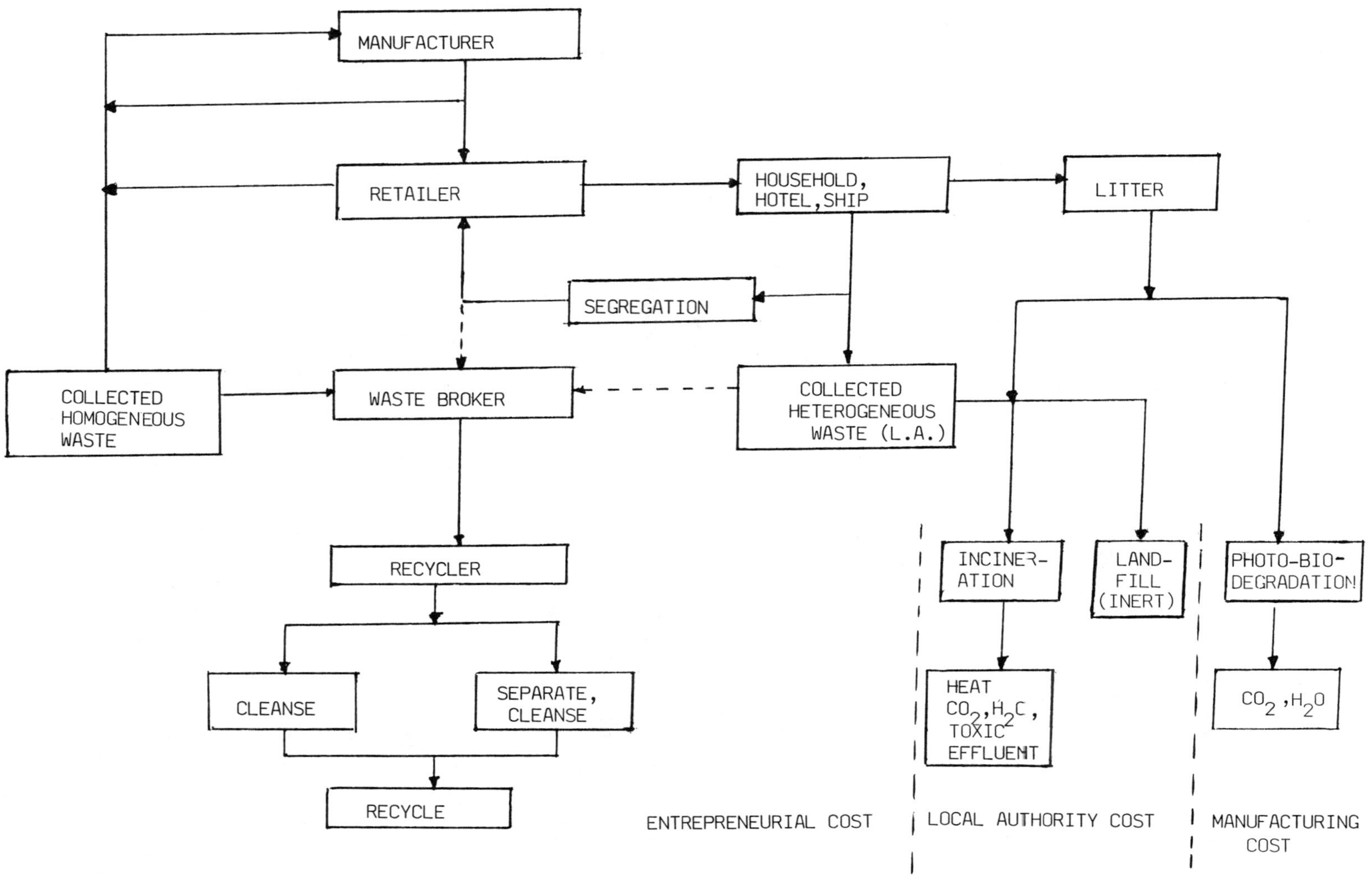

MANUFACTURER

RETAILER

HOUSEHOLD, HOTEL, SHIP

LITTER

SEGREGATION

COLLECTED HOMOGENEOUS WASTE

WASTE BROKER

COLLECTED HETEROGENEOUS WASTE (L.A.)

RECYCLER

CLEANSE

SEPARATE, CLEANSE

RECYCLE

INCINERATION

HEAT $CO_2$, $H_2C$, TOXIC EFFLUENT

LAND-FILL (INERT)

PHOTO-BIO-DEGRADATION

$CO_2$, $H_2O$

ENTREPRENEURIAL COST

LOCAL AUTHORITY COST

MANUFACTURING COST

if there is a stable market for the products.

Examples of useful secondary product that can be made economically are polyester insulation fibers from recovered PET bottles, waste bags from car batteries and discarded crates, etc., see scheme 3. However, by far the major proportion of plastics waste is contaminated mixed collected waste. Even supermarket or industrial waste has to be made suitable for recycling by cleansing and removal of labels, metal tags, etc. Household waste is a mixture of all kinds of metallic, ceramic, biological and plastics waste and only the most intensive separation and cleansing procedures can produce mixed plastics of sufficient quality to recycle. Furthermore even after these energy consuming procedures, the products that can be produced from the polymeric mixtures have much inferior properties to those from virgin polymer (38,39). Thus for example mixtures of the common packaging plastics are incompatible and give products with inferior mechanical performance (e.g. tear strength and impact strength), see figure 6, and although impact resistance can be improved by the use of solid phase dispersants (SPDs or "compatibilizers"), these are generally expensive and increase the cost of the product even more (39), Equally important, the outdoor ageing performance of compatibilized polymer blends are much inferior to homogeneous polymer. The rate of photooxidation of PE-S-blends is shown in figure 7 to increase dramatically with the introduction of SBS (the SPD). It is unfortunate that the major uses that have been proposed for recycled mixed plastics are in outdoor applications (fence posts, park benches, boat docks, cladding etc.). To produce serviceable

Scheme 3

MATERIALS RECYCLING

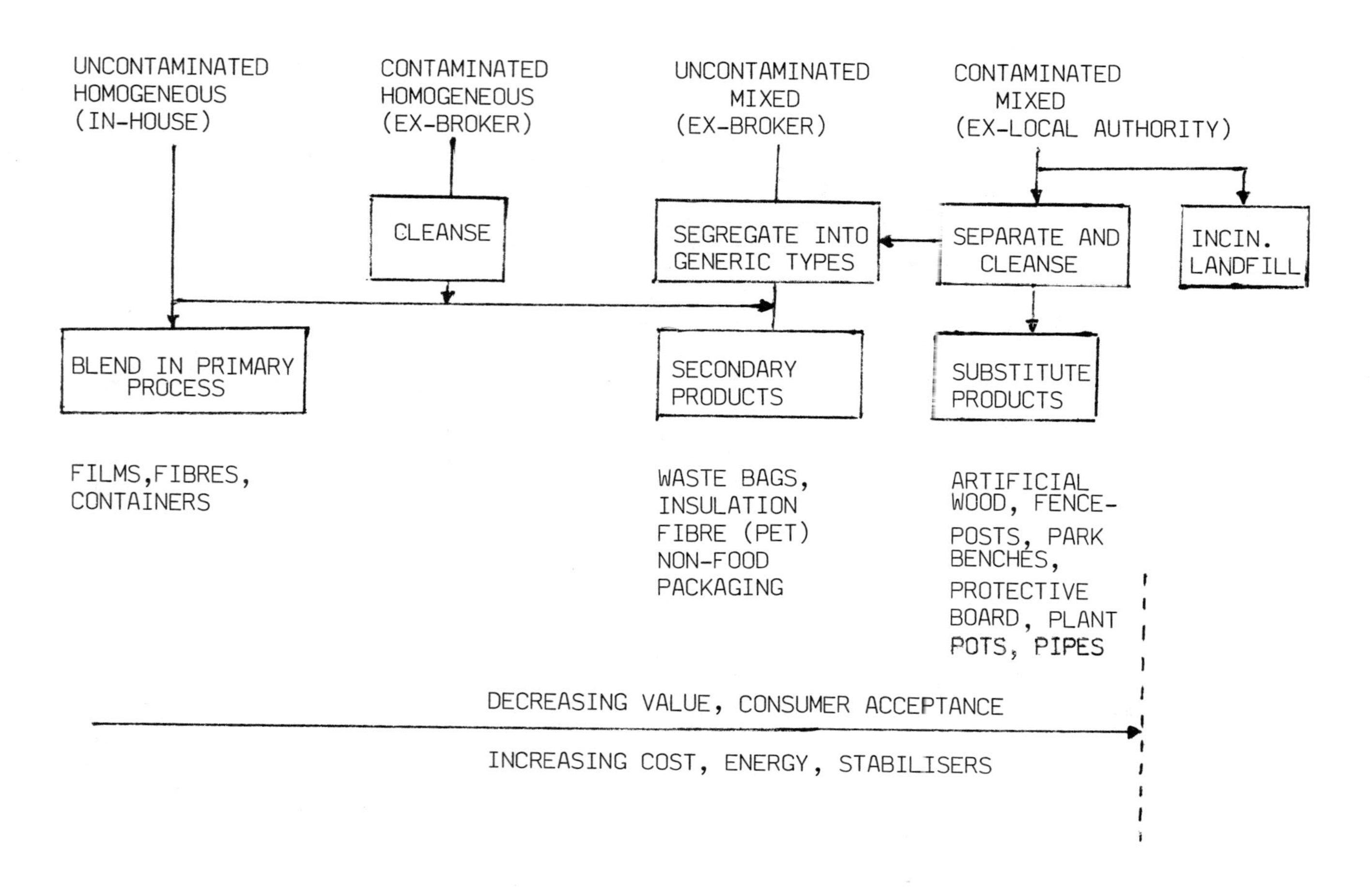

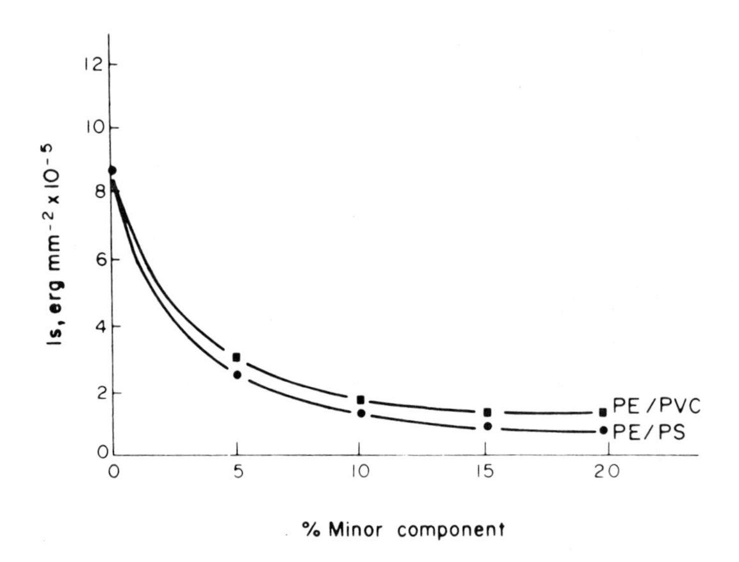


Fig. 6. Effect of minor proportions of PS and PVC on the impact strength (Is) of LDPE.
(After Ref. 39 with permission)

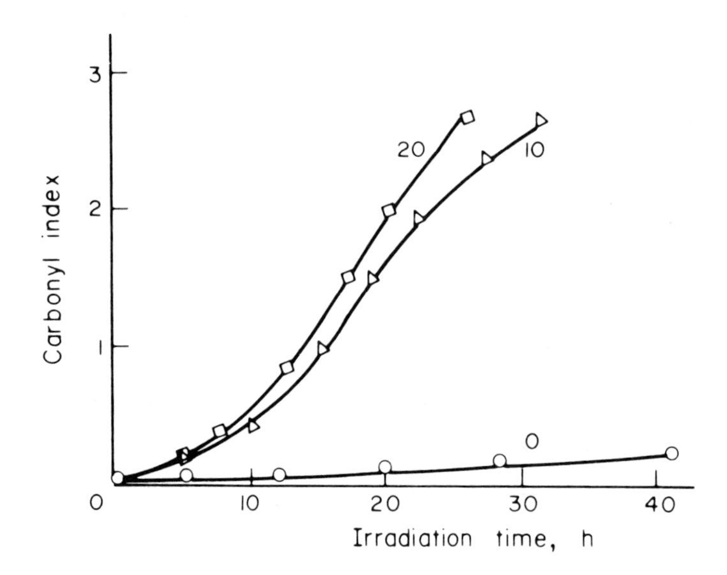


Fig. 7. Effect of an SPD (SBS) on the photooxidation of PE/PS (1:1) blends. Numbers on curves are SBS concentrations (g/100g)
(After Ref. 39 with permission)

products from such waste will clearly require the use of substantial quantities of antioxidants and UV stabilizers. The auxiliary chemical manufacturers could find this profitable but it seems unlikely that for most applications the value of the end application will match the cost of the additives involved. It seems likely that for the foreseeable future most of the contaminated collected mixed plastics waste will continue to make inert landfill. Hopefully, the design of better incinerators will lead to increased energy recovers. (It should not be forgotten that the calorific value of the hydrocarbon plastics is similar to that of oil).

How then do degradable plastics fit into the systems approach to waste disposal. Degradable plastics have to be capable of being disposed of by any of the alternative procedures listed in Table 3. Litter is by definition waste which is located where it cannot be economically collected. Recycling may have a marginal effect in reducing its impact but evidence from all parts of the world suggests that even punitive fines have very little effect.

Recycling, coupled with the public will to collect, return and segregate packaging waste could have a significant effect on litter reduction, but it is unrealistic to think that recycling could eliminate litter. Few would argue that it is not important to conserve materials as a major objective but enthusiasm must be tempered with realism. Human nature being what it is, it is not possible to preduct which items of packaging will end up as litter, in a recycling system, in an incinerator or in a landfill. It follows then that degradable plastics must be compatible with each of these. All degradable plastics burn just like normal plastics. In anaerobic landfill, photodegradable plastics do not biodegrade and indeed, there is some evidence that even paper does not biodegrade under these conditions (36). This is considered to be a positive advantage for plastics in this method of disposal, since it does not lead to the formation of combustible gases with the subsidence associated with the biodegradation of biological materials (36).

## 4. DEFINITION OF TERMINOLOGY AND THE DESIGN OF TESTING PROCEDURES

This problem of terminology is one which clouds the whole issue of environmental degradation of plastics. Claims are repeatedly made by retail chains that their packaging is "biodegradable" because this is the word the "greens" understand. In fact, no packaging plastics (except the small amount of cellophane still used) degrade by a purely biological process. Commercial processes used for degradable packaging are concerned without exception in accelerating the natural degradation processes of plastics in a controlled way to release biodegradable products.

As degradable materials become increasingly part of the armoury in the war on waste and litter, the standards organizations will need to take a lead in defining the terminology used in order to provide the plastics technologist with simple test procedures which distinguish between the processes of thermal oxidation (both during processing and in shelf ageing), photooxidation and biooxidation. This is essential, not only to assist in the design of appropriate legislation to control the litter problem but in order to provide local authorities with a means of evaluating commercial packaging in the context of the Trades Descriptions Act.

Figure 1 describes how an environmentally acceptable package should ideally behave. There are three quite clearly defined stages.

A. The "induction period"

It is essential that packaging plastics should during use have low permeability to water and hence to microorganisms. They should not be susceptible to microbiological or macrobiological (animals and insects) attack for periods in excess of those expected to be used in practice.

The end of the induction period must be accurately controllable by some trigger of the environment. Exposure to microoganisms does not constitute such a trigger since the attack of fungi, bacteria, etc. is unpredictable and, in the case of the hydrophobic polymers, impossible without oxidative modification. Heat and oxygen are also non-discriminating, but sunlight and oxygen are specific to the litter situation and constitute the ideal trigger.

B. Photodegradation

Conventional processing stabilizers and antioxidants which are normally used to protect packaging during service also reduce the rate of breakdown of plastics out-of-doors. However, as discussed above, some processing stabilizers undergo role inversion on exposure to light and if suitable designed give a complete induction period during use followed by a period of very rapid photooxidation and loss of physical integrity.

C. Thermal and Biodegradation

Physical disintegration does not in itself guarantee biodegradation, but photooxidized polymers contain high surface concentrations of carboxylic acids at the broken chain ends which undergo bioassimilation by ß-oxidation leading to complete conversion to carbon dioxide and water via the Krebs cycle. Present evidence suggests (16) that this process accelerates as the surface area of the degraded polymer increases.

Research is in progress in a joint collaborative project between the Universities of Aston and Clairmont-Ferrand under EC sponsorship to evaluate the range of commercial degradable plastics with respect to the three stages outlined above in order to provide a fundamental basis for setting up European standards for degradable plastics. As part of this project, the water extractability, toxicity and biodegradability of any low molecular weight compounds formed will be assessed.

REFERENCES


1. J.J.P.Staudinger, Disposal of Plastics Waste and Litter, SCI Monograph, No. 35, 1970.
2. G.Scott, Waste Disposal, (B.I.M.), 5, 78 (1971).
3. G.Scott, Int. J. Environmental Studies, 3, 35, (1972); 7, 131 (1975).
4. T.R.Dixon and A.J.Cooke, Marine Pollut. Bull, 8 (5), 105 (1977); Dixon, T.R. and Dixon, T.J., Marine Pollut. Bull, 12 (a), 289 (1981).
5. G.Scott, Royal Soc. Arts J., 122, 188 (19794).
6. G.Scott, in Beadle (Ed.), Plastics Forming, McMillan Eng. Eval., 1972, p161.
7. G.Scott, Design Engineering; 77 (Oct. 1972).
8. G.Scott, J. Oil Col. Chem. Assoc. 56, 521 (1972).
9. B.Heneman; Persistant Marine Debris in the North Sea, etc., Report to the U.S. Department of Commerce, 1988.
10. R.Shomura, (Ed.), Proceeedings on the Second International Conference on Marine Debris, Honolulu, April, 1989.
11. D.Gilead, Chemtech., 299 (May 1985); Plasticulture, 43,31 (1979).
12. D.Gilead and R.S.Ennis, Symp. on Degradable Plastics, SPI, Washington DC, 1987, p37.
13. D.Gilead and G.Scott, in G.Scott (Ed.), Developments in Polymer Stabilization - 5, Appl. Sci. Publ, 1982, p71 et seq.
14. D.Gilead, Polym. Deg. and Stab., in press.

15. H.O.W.Eggins, J.Mills, A.Holt and G.Scott, in G.Sykes and F.A.Skinner, (Eds.), Academic Press, 1972, p267.
16. A-C.Albertsson, Europ. Polym. J., 16, 623 (1980).
17. G.J.L.Griffen, IUPAC Microsymposium, Prague, 1975, Symp. on Degradable Plastics, SPI, Washington DC, 1987, p47.
18. W.J.Maddever and G.M.Chapman, Symp. on Degradable Plastics, SPI, Washington, DC, 1987, p4.
19. F.H.Oltey and W.M.Doane, Symp. on Degradable Plastics, SPI, Washington DC, 1987, p39.
20. Reported in Ref. 17 (1987).
21. G.Scott, Atmospheric Oxidation and Antioxidants, Elsevier, 1965, p272 et seq.
22. G.Scott, in N.Grassie, (Ed.), Developments in Polym. Degradation - 1, App. Sci. Pub., 1977, p205.
23. M.M.Brubaker, (DuPont), U.S. Patent 2, 495, 286 (1950).
24. G.M.Harlan and A.Nicholas, Symp. on Degradable Plastics, Washington DC, 1987, p14.
25. R.J.Statz and M.C.Dorris, Symposium on Degradable Plastics, Washington DC, 1987. p51.
26. J.E.Guillet, U.S. Pats., 3, 753, 962; 3, 811, 931; 3, 853, 814; 3, 860 ,533.
27. J.E.Guillet in J.Guillet, (Ed.), Polymers and Ecological Problems, Plenum Press, 1973, p1.
28. C.H.Chew, L.M.Gan and G.Scott, Europ. Polym. J., 15, 791 (1977).
29. A.E.Redpath, Symp. on Degradable Plastics, SPI, Washington DC, 1987, p31.

30. M.U.Amin and G.Scott, Europ. Polym. J., 10, 1019, (1974).

31. D.C.Mellor, A.B.Moir and G.Scott, Europ. Polym. J., 9, 219 (1973).

32. Z.Osawa, in G.Scott, (Ed.), Developments in Polymer Stabilization - 7, Elsevier App. Sci., 1984, p193.

33. S.Al-Malaika, K.B.Chakraborty and G.Scott, in G.Scott, (Ed.), Developments in Polymer Stabilization - 6, App. Sci. Pub., 1983, p73.

34. G.Scott, H.H.Sheena and A.M.Harrriman, Europ. Polym. J., 14, 1071 (1978).

35. G.Scott, Brit. Pat., 1, 356, 107 (1971), D.Gilead and G.Scott, Brit. Pat., 1, 586, 344 (1978), U.S. Pat., 4, 461, 853 (1984), U.S. Pat., 4. 519, 161 (1985).

36. Questions and Answers About Degradable Plastics, SPI Brochure, 1988.

37. Degradability and Plastics Packaging, Industry Council for Packaging and the Environment, Discussion Paper No. 4.

38. G.Scott, Resource Recovery and Conservation, 1, 381 (1976).

39. C.Sadrmohaghegh, G.Scott and E.Setudeh, Polym. Plast. Tech. Eng., 24, 149 (1985).

## QUESTIONS AND ANSWERS

DR. HELLER: I'm Jorge Heller, SRI.

Could you comment on the shelf stability of these materials? Particularly if they are exposed to high temperature and sunlight prior to reaching the consumer. How stable are they?

DR. SCOTT: Well, as I said, they do contain antioxidants, and therefore their thermal stability is very good. I showed some results there on simple thermal oven aging tests, and they are equivalent to conventional plastics, which have to stand up to high temperatures.

There is a combination effect, of course, of heat and light. And what we found is that you certainly get a faster rate of degradation if it is hot, if you get increased temperature. Heat may be in fact more important than light, because of course light doesn't vary that much. Even on a cloudy day, you are still getting 80 percent of the UV through the clouds. So certainly a combination of heat and light is a good thing in the end for getting rid of the plastic.

But don't forget this only happens out of doors. If it was in a supermarket, you would not have that problem.

DR. HELLER: But on shipping, for example, it could be in a truck for long periods of time in hot climate. It could get pretty hot.

DR. SCOTT: Oh, these are very stable under those conditions, as long as there is no light there, yes.

DR. HELLER: Okay.

DR. BAILEY: Bill Bailey, University of Maryland.

You had the photobiodegradation, and I didn't really see much on the bio side of this. And I was wondering if you had data or something that tells how biodegradable it is, and whether microorganisms are there, or whether you take these little grass-like particles, and whether they are biodegradable, or somehow everything just appears and you assume it is biodegradable.

DR. SCOTT: Okay. Well, first of all, Bill, we are just starting the biodegradation work, but it will be based essentially on the work that Ann-Christine has done over the years. And she has shown that the rate controlling step here is the oxidation process.

I should have shown a slide, in fact which I have with me but I didn't think I would have time to show, where she actually exposed oxidized polyethylene to microorganisms and showed that you could actually purify the plastic.

I mean that is the evidence that it is biodegradation which is ultimately responsible for removing the plastic, and not attack on the plastic itself.

DR. BAILEY: Well, there are certain fragments that she is showing that degrade, but -- as I gather, the whole thing didn't degrade. Maybe she will comment.

DR. ALBERTSSON: Yes, my name is Ann-Christine Albertsson.

My comment is that what I have shown is that we have a higher degradation rate, if the material first is oxidized.

And my confusion was the same as Professor Scott already stated, 1975, that the biodegradation is a secondary process, and the oxidation is one of the primary process that happened in degradation of plastics. And I mean, this is a similar degradation that happened with other materials, too. That you have a combined effect.

And in all the experiments, you have an increased biodegradation effect both in the beginning and later on of the materials. And it's not only shown for the first fragments, but since the degradation takes a very long time, I will later show what's happened during the first ten years, and when we see each other next time, I will probably be able to show what has happened during twenty years.

DR. NARAYAN: Ramani Narayan, Michigan State University.

I am a liattle puzzled in listening to biodegradation studies, and I would like to comment on this.

I am left with the distinct impression that whenever any of these equations are shown here and everywhere, we show the total carbon of the biopolymer or the biodegradable polymer converted to $CO_2$ and water, and that's not really true because the microorganisms have to use the carbon to survive, the sustainability of the cell converted into cell mass and biomass.

And in certain microbial systems, doing a lot of work in fermentation, it is a question of carbon balances, where part or even sometimes in certain system, major portions of it is converted into cell mass, and only a part is converted into $CO_2$ and water. For example, if you feed even glucose to yeast, you get ethanol. A part of that carbon goes to cell mass and maybe a part into $CO_2$.

So are we sort of missing the equation? I realize in your

equation you have end product $CO_2$ and water. But are we de-emphasizing something which probably is a more important factor? Because the impression left to me that if you have X units carbon polymer, then we must produce X units $CO_2$. And that equation is never going to be like that.

DR. SCOTT: Well, I would have thought is would be in the long term. Because surely microorganisms also degrade back to $CO_2$ and water. I mean unless you are going to get an accumulation of them, which I think is highly unlikely in view of the biodegradable materials we have around already.

DR. NARAYAN: Yes, but that's over an extended period of time. We are talking about accelerated studies, which is where the correlations are being pulled to the extended period of time. And in that say, well, like the ICI paper this morning showed that they kept it, the polyester for 15 days with microorganisms or incubating it, and then you produce total microbial metabolism to 90 percent carbon with no other carbon source.

DR. SCOTT: Yes, but I mean I don't mind at all how long the microorganisms take to do this, as long as it is no longer a visible plastic waste or a plastic litter. I mean why should we worry about that once it's assimilated into the bio system?

DR. NARAYAN: No, it's not -- yes, I perfectly agree with you. The problem is that in extrapolating these to other, other areas or other comments, maybe even from the generic, the impression which they gather or is left is that X carbon of polymer is totally converted to $CO_2$ and water in that very small period of time.

DR. SHALABY: Shalaby from Johnson & Johnson.

We heard quite a bit today about the use of pro-oxidants, use of photo-inducing agent. If you look at them, they are mostly transition metacomplexes. An the question is should we be addressing the long-term effect of accumulating the transition metacomplexes on the microflora and the soil as well as the water supply?

DR. SCOTT: Okay. Well, so far as I am concerned, the only ones we're interested in here are ion, which I think you would agree is no problem to anybody, because that will never affect anybody. There's so much of it around already.

The other is nickel. An nickel is already around in very large quantities in certain soils. I mean, there are some soils that have very large amounts of this, and they don't take it up into the plants. You can show this. They don't assimilate it into the leaves or the fruit or anything.

Now the EPA has done some legislation at least. I am not sure whether it's officially legislation yet, but they have done some survey of the amount of these metals which can be put onto agricultural land as sludge. And the amount of nickel that you would produce if you covered a whole field of area, one hectare, with degradable plastic, every inch of it, the amount of nickel you would produce from that plastic is only about 1 percent of what the EPA allows, which is 78 kilograms per hectare.

DR. LOOMIS: Can you really make a thick molded piece, like a milk case, go away in a reasonable time? I mean you are stuck with Beer's law. Even if you had a very high quantum efficiency of your photochemical reaction, could you actually have a thick molded piece of polyurethane that would actually degrade in a lifetime?

DR. SCOTT: Oh, yes, no doubt about that. You see --

DR. LOOMIS: How does the light get in through beyond 20 or 30 mills?

DR. SCOTT: It doesn't. It does grittle from the outside. But once it's gone brittle on the outside, then it breaks. And it breaks down into small particles.

So you would get the breakdown. Even although the inside might be protected from the UV light. It's a simple mechanical low, that if you put a flaw into a material and you put a stress on it, it will crack.

DR. LOOMIS: But you are not putting a stress on it. Unless the stresses are molded in.

DR. SCOTT: But you are. This litter is subject to enormouse stresses, because it's being blown around all over the place, up mountains.

DR. ENNIS: Bob Ennis from Plastigone. There has been some confusion around as to -- it's been voiced in a lot of the popular press as to exactly when does the biodegradation process start. Does the plastic have to get down to a certain level of molecular weight before biodegradation begins, or does the biodegradation begin as soon as there are available carboxyl groups or other groups that are made present but the action of the photo process?

DR. SCOTT: Well, this is one of the questions that we hope to answer in the study that we are doing for the EC.

But the prepreliminary evidence suggests that it has nothing to do with the molecular weight of the material. It's to do with the formation of carboxyl end groups. And from that point on, you've got the kind of material which will undergo, by oxidation. Because it will undergo the beater keto-attack followed by being absorbed in the Krebs cycle. And it's the Krebs cycle eventually, of course, which leads to the disappearance of the material as $Co_2$ and water.

DR. ENNIS: So it's really a misnomer --

DR. SCOTT: Exactly.

DR. ENNIS: -- to say that the whole plastic has to break down to small molecular weights before you can start a bio-oxidation process.

DR. ANDERSON: Thank you very much.

# PHOTODEGRADATIVE ADDITIVES

George S. Upton

Ampacet

## OVERVIEW

The use of photodegradables is worldwide in scope. It is in a number of applications from bags to mulch film. The function is to help in a litter problem or to contribute to the economics of agriculture.

The wide disparity in the level of sunlight impacting the world is seen in the world depiction. While this is helpful in gaining a perspective on sunlight it does not address a number of the other issues which impact photodegradation. While these are all interrelated it is difficult to predict without on site testing exactly how a product will perform. Such issues as cloud cover, temperature, wind velocity and the degree of air pollution all have a direct bearing on the effectiveness of a photodegradable product.

## THEORY

Many polymers are impacted by exposure to sunlight. The polyolefins are particularly reactive to sunlight; and, in fact may be said to be made with the seeds of their destruction built in. When polyethylene is made it will have some carbonyl groups present, which are photosensitive. It will also have some residual catalyst which, when it is exposed to another heat history, will continue to react in some way. Every attempt is made to minimize the impact of the reactions. It is interesting to note how much effort is made to produce a polymer which is as clean as possible. Then, depending on the application, the various additives are incorporated to help the polymer withstand the rigors of processing.

The photosensitive reaction is set into motion with the exposure of the product to sufficient sunlight or other UV source. This product will also contain a metal ion initiating complex to insure a more uniform and rapid deterioration. The range that will satisfy this is from 270-370 nanometers. In the case of polyethylene it is most reponsive to 300 nanometers. From this initiation photooxidation can take place. As this continues and direct bond cleavage takes place, a Norrish I and II reaction take place. All of this is the start of a complex oxidative reaction that will ultimately result in the polymer backbone being broken into increasingly smaller lengths. The oxidative reactions lead to the formation of intermediate free radicals which are capable of propagating further reactions leading to polymer degradation. Some of the free radicals that are important are: hydroxy, alkoxy and peroxy. Some of the intermediate products are colored and therefore can absorb light of a longer wavelength than the initiating wavelength. Some by-products that are expected to be thermally unstable and sensitive to catalytic decomposition are peroxide and hydroperoxide. From these degradation reactions thermal decomposition of intermediate products may proceed without light. Thus, a continuing loss of molecular weight will be manifested in the reduction of physical properties. As this deterioration continues the molecular weight gets into the range of 500 MW and at this point become paraffinic in nature. At this level and below the material becomes available for incorporation by microbial action. At this point the material is reduced to its essential elements. It has also been reported that this biologic activity drops off quite rapidly above the 1000-1500 MW range.

The raw materials used in our photodegradable products are all in compliance with the appropriate U.S. regulations for food contact applications. While this is no warranty for the by-products of degradation it tends to mitigate the potential for problems over a period of years.

## FILM PROPERTIES

The major application for photodegradables is currently in film and usually under 4 mils in gauge. The question of impact to film properties was reviewed. The demonstration of film strength and film properties indicate that there are no substantial reduction in film properties. There will be, however, some impact on color. A reddish tinge can be noted in rolls of film, while the single layers do not show this. Similarly, there would be an impact on white and light colors while this would not be noticeable in darker colors. The question of impact on heat seal strength was looked at and no significant impact was seen.

## DETERIORATION

Questions concerning the deterioration and how this occurred over time was looked at. The best data to date indicates that after a minimum of two weeks outdoor exposure the film is substantially degraded. Further examination indicates that when the film is put into envelopes and then tested two weeks later deterioration had continued. Other testing indicated that after exposure and burial and then testing that the degradation had continued. The variety of cities where we had film tested were scattered throughout the U.S. They are listed from south to north without an obvious trend showing. This all gets back to the observation seen in the first slide that there are wide variations in sunlight impact and that other relationships may be the determining force. An example would be where the sunlight impact is less but the air quality is better or the elevation is such that there is less impact on the radiation. The addition of excess material is not helpful in the data we have seen to date. Similarly when reduced levels were looked at the drop off in degradation was noted.

MANUFACTURING - QUALITY CONTROL

Melt index is an excellent tool to monitor how the product is doing both in production and over time. Maintaining uniform MI reassures that athe same levels are being incorporated. It also is a key indicator when age is involved. If a sharp rise is noted it is a prime indicator that degradation has begun and the concentrate should not be used. Some data indicates that the concentrate is quite stable and the film is also. However, the film should be wrapped in black film and the concentrate kept out of the light. Several tests are run prior to shipping. The first one is to blow film and check for uniformity of dispersion. After this step it may be put into a weatherometer to insure that it degrades. The other test is run using an atomic absorption test to check for the level of metals present. This, in conjunction with a mass balance, is a cross check to insure that the proper level of additives is present.

CONCLUSION

The concentrates and the products made from them are predictable within the limitations of the system. It is stable, again, within limits. Tests are available which can show when it is beyond useful function. Degradation condinutes after sufficient exposure has been initiated and then the film buried or removed from the light.

BIBLIOGRAPHY

AMPACET - Internal reports - various.

CHEVRON - Degradable resin technology - 5/5/88 - C.L. George.

CIBA GEIGY - Natural and artifical exposure and weathering of plastics - Kurt Berger.

CIBA GEIGY - Degradable plastics - Peter P. Klemchuk.

UNION CAMP - Effect of photoderadable additives in HMW-Hope - 9/5/89 - A.K. Bose.

ADM - Slide of Illinois soil.

AMERICAN CYANAMID - slide.

BIOGRAPHY

Mr. George S. Upton received a BBA from the University of Oklahoma. After graduation he served over four years in the Army resigning as a Captain. He then re-entered the University of Oklahoma where he received an MBA. After graduation he joined Spencer Chemical, a division of Gulf Oil. There he sold nylon and polyethylene resins for three years before joining Ampacet. Since 1969 he has moved through the marketing side to Regional Sales Manager and now to the Market Manager of Specialty Products. These include both the photo and starch degradable additives as well as Ampacet's product lines for extrusion coating, wire and cable, fibers and specialty antistats. Mr. Upton is an active member in both SPE and TAPPI and has presented papers at Plastics Fairs, Conferences at Clemson on flame retardant fibers and to the API on degradable plastics.

# Photochemical Decomposition of Polymers

## Photooxidation

INIT.:

A) $ROOH \xrightarrow{h\nu} RO\cdot + HO\cdot$

B) $RO\cdot\ (HO\cdot) + RH \longrightarrow ROH\ (H_2O) + R\cdot$

C) $-CH_2CH_2C(O)CH_2CH_2- \xrightarrow{h\nu} -CH_2CH_2\cdot\ [\equiv R\cdot] + -CH_2CH_2C(O)\cdot$

PROP.:

$R\cdot + O_2 \longrightarrow RO_2\cdot$

$RO_2\cdot + RH \longrightarrow RO_2H + R\cdot$

TERM.:

A) $R_{SEC}O_2\cdot + R_{SEC}O_2\cdot \longrightarrow R_{SEC}OH + R'C(O)R'' + O_2$

B) $R_{TERT}O_2\cdot + R_{TERT}O_2\cdot \longrightarrow [ROOOOR]$

$[ROOOOR] \longrightarrow R_{TERT}OOR_{TERT} + O_2$

$[ROOOOR] \longrightarrow 2\ R_{TERT}O\cdot + O_2$

$R_{TERT}O\cdot \longrightarrow R'C(O)R'' + R'''\cdot$

## Ketone Photolysis

NORRISH TYPE I:

$-CH_2CH_2C(O)CH_2CH_2- \longrightarrow -CH_2CH_2C(O)\cdot + -CH_2CH_2\cdot$

NORRISH TYPE II:

$-CH_2CH_2C(O)CH_2CH_2CH_2- \longrightarrow -CH_2CH_2C(O)CH_3 + CH_2{=}CH-$

# Time Required to Degrade LDPE Film

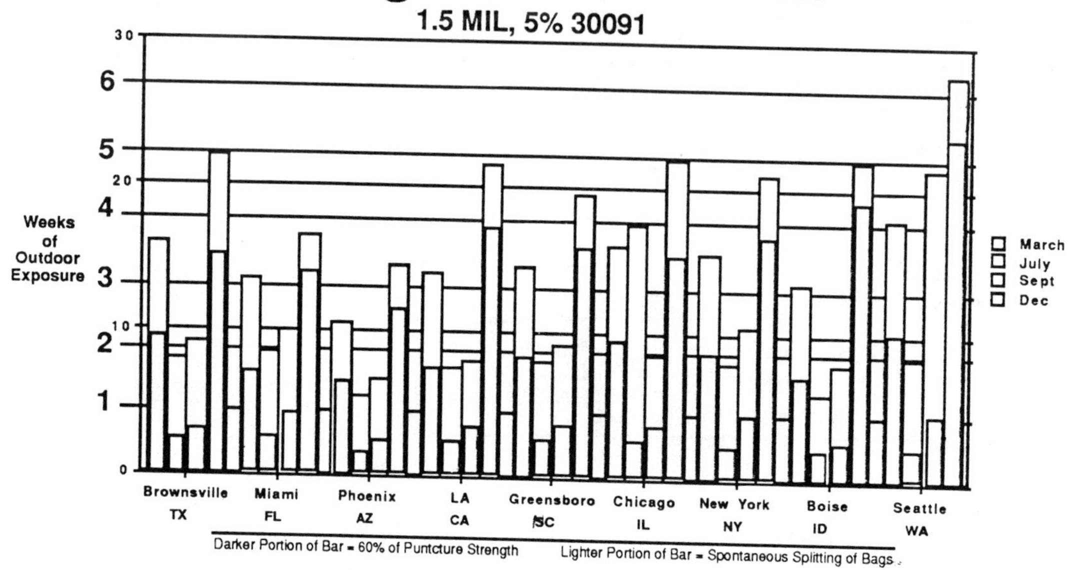


| Percent 30091 MB Used | Film Gauge Tested | Hrs In Weather-Ometer | % Elongation | Melt Index |
|---|---|---|---|---|
| 5% | 2 MIL | 0 | 320 | 1.4 |
| | | 24 | 170 | --- |
| | | 48 | 7 | --- |
| | | 72 | Film Completely Degraded | |
| 5% | 1 MIL | 0 | 170 | 1.5 |
| | | 48 | 100 | --- |
| | | 72 | 90 | --- |
| 8% | 2 MIL | 0 | 390 | 1.4 |
| | | 24 | 350 | --- |
| | | 48 | 140 | --- |
| | | 72 | Film Completely Degraded | |
| 8% | 1 MIL | 0 | 180 | 1.6 |
| | | 24 | 160 | --- |
| | | 48 | 100 | --- |
| | | 72 | Film Completely Degraded | |

Once samples reached 50% of their original Elongation they were buried in soil. The samples were unearthed after 3 months and were completely degraded. Melt Index was run on the remaning film of the 5% usage level and were found to have risen from 1.4 to 7.0. The 8% usage samples were completely gone. No M.I. was attempted.

# Photodegradable Film Trial

Extruded in Princeton on June 15, 1988

| | 100% 5005 Lot # 804087 ControlRoof | Control after 2 Weeks Exposure | 95% 5005 5% Ampacet 30091 | 95% 5005 5% 30091 After 2 Wks Roof Exp |
|---|---|---|---|---|
| Gauge (mil) | 0.62 | 0.62 | 0.63 | 0.63 |
| Variation (+/-%) | 23 | 23 | 17 | 17 |
| Tensile Yield md (psi) | 4568 | 5168 | 4425 | 0 |
| td (psi) | 4097 | 4179 | 3965 | 0 |
| Break md (psi) | 6288 | 7003 | 11044 | 2066 |
| td (psi) | 6005 | 5584 | 8257 | 1995 |
| Elongation md (%) | 268 | 232 | 348 | 0 |
| td (%) | 243 | 302 | 418 | 0 |

# Ampacet Polygrade I (30091) TrialsExtruded at Princeton on June 15, 1988 in Cain 5005, Lot # 604087

(Exposure Period: July)

| Exposure | None Control | 2 Weeks | % Diff. | 2 Wks/2 Wks in Envelope | % Diff. |
|---|---|---|---|---|---|
| Gauge (mil) | 0.63 | 0.63 | | 0.63 | |
| Variation (+/-%) | 17 | 17 | | 17 | |
| Tensile Yield md (psi) | 4425 | 0 | 100.00 | 0 | 0 |
| td (psi) | 3965 | 0 | 100.00 | 0 | 0 |
| Break md (psi) | 11044 | 2066 | -81.29 | 1219 | -40.99 |
| td (psi) | 82571995 | | -75.84 | 1291 | -35.28 |
| Elongation md (%) | 348 | 0 | 100 | 0 | 0 |
| td (%) | 418 | 0 | 100.00 | 0 | 0 |

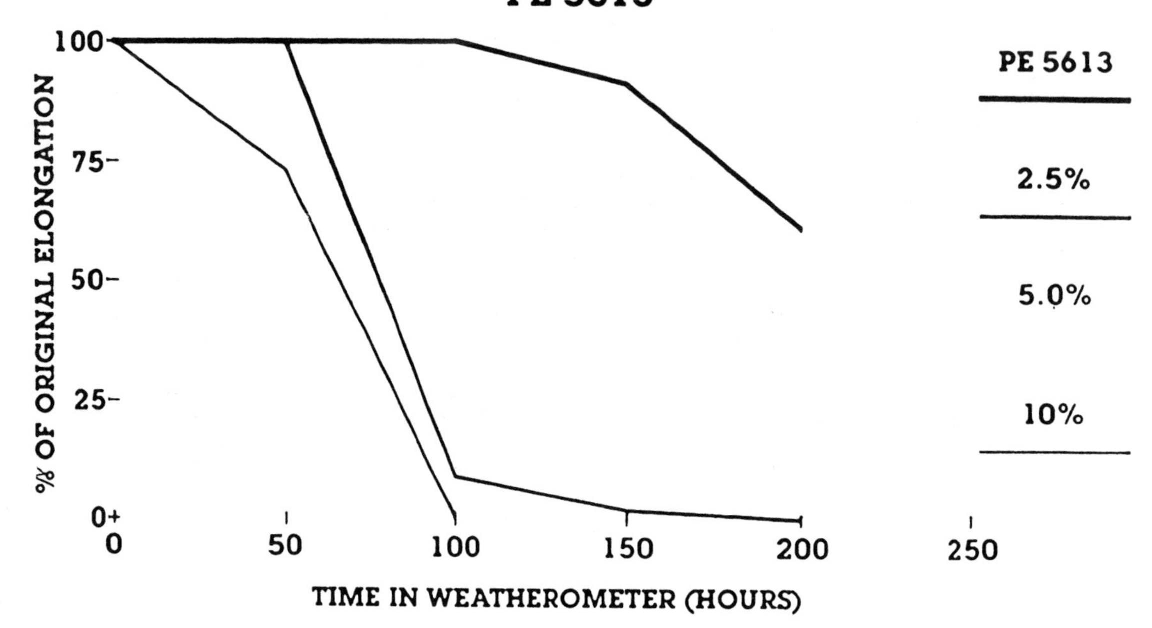


Courtesy of Chevron Chemical Co.

## PHOTODEGRADABLE PLASTIC

A polymeric material that disintegrates under environmental conditions in a reasonable and demonstrable period of time, where the primary mechanism is through the action light

## QUESTIONS AND ANSWERS

DR. ANDERSON: Questions?

DR. NARAYAN: Put it back to what was, based on what you have done so far, if you are to say what would you like to have done, what research needs to be done, or do you think you have done everything and we have got the perfect material? If you were to list 1, 2, 3, could you give us some, based on your experience, what do we need?

DR. AUSTIN: Some of the research is ongoing now that we have actually been delinquent in doing. We know that as you go up in gauge, the product is not very good.

It was tried in the high cone, six-pack container early on and was not fast enough, where the ECO worked extremely well in that application area. So we cannot tell you from drawing a graph, on a density, on a gauge relationship, and that needs to be done. And we are currently in the process of doing that so that we will have a higher comfort level with where that comes from.

We also know if you have nonuniform film, you are in trouble right away. And in terms of nonuniform film, in terms of gauge.

FIRST INTERNATIONAL SCIENTIFIC CONSENSUS WORKSHOP

ON DEGRADABLE MATERIALS

Toronto, Ontario, Canada 2 - 4 Nov. 1989

## Photo-degradable Plastics in Agriculture

by Dan Gilead

Plastopil, Hazorea

Plastic materials have penetrated every sector of Agriculture (Fig.1,2). Many applications such as in machinery and in building are of a permanent nature and do not represent a burden on the environment. Other applications do have a varying impact on the ecology. The packaging of produce in crates, tote-boxes and trays contribute significantly to the solid waste and to the litter problem. Fertilizers and chemicals brought to the farm in plastic packaging materials such as shipping bags or plastic containers when improperly discarded have a damaging impact on the ecology and may be dangerous to man and animals alike. Plastic baling twine and plastic irrigation tubing accumulate in the fields and by the wayside and are causing an eyesore and a nuisance. The greatest problem is caused by agricultural films used for mulching. These films used by themselves or in conjunction with plastic tunnels have created an entirely new agrotechnical method known by the name of Plasticulture. (Fig.3) The main factors affected are soil temperature and humidity. The mulch produces a local greenhouse effect by raising

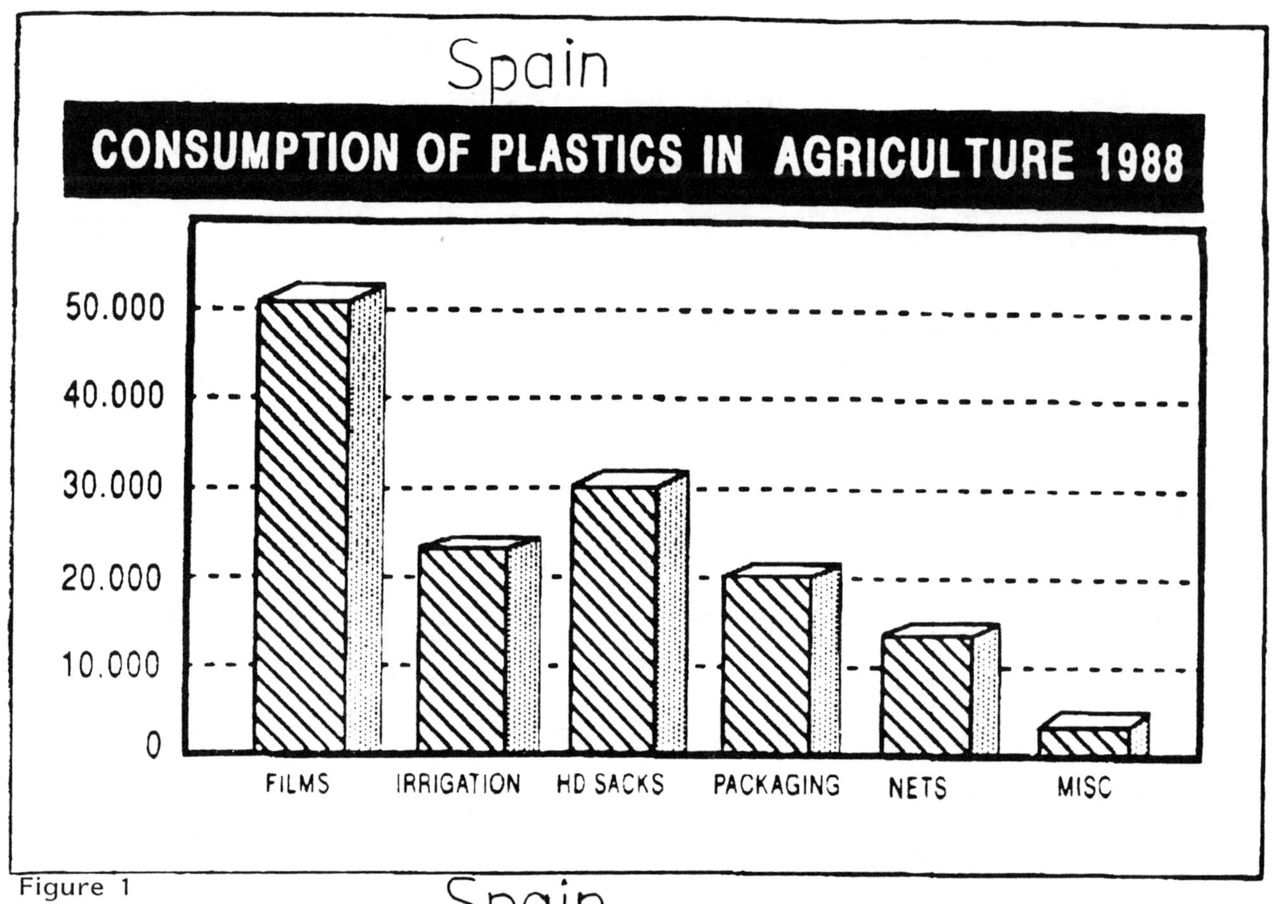


Figure 1

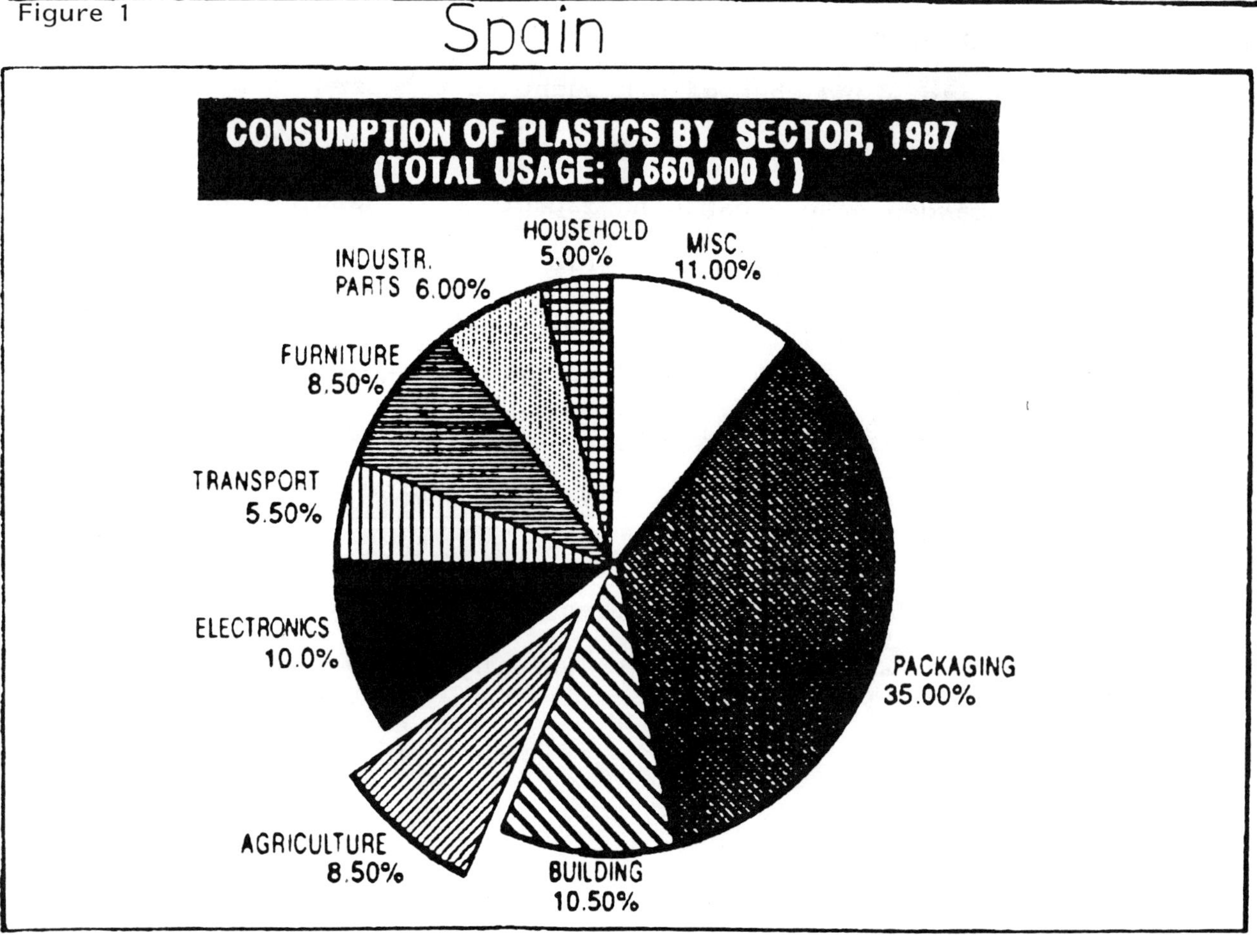


Figure 2

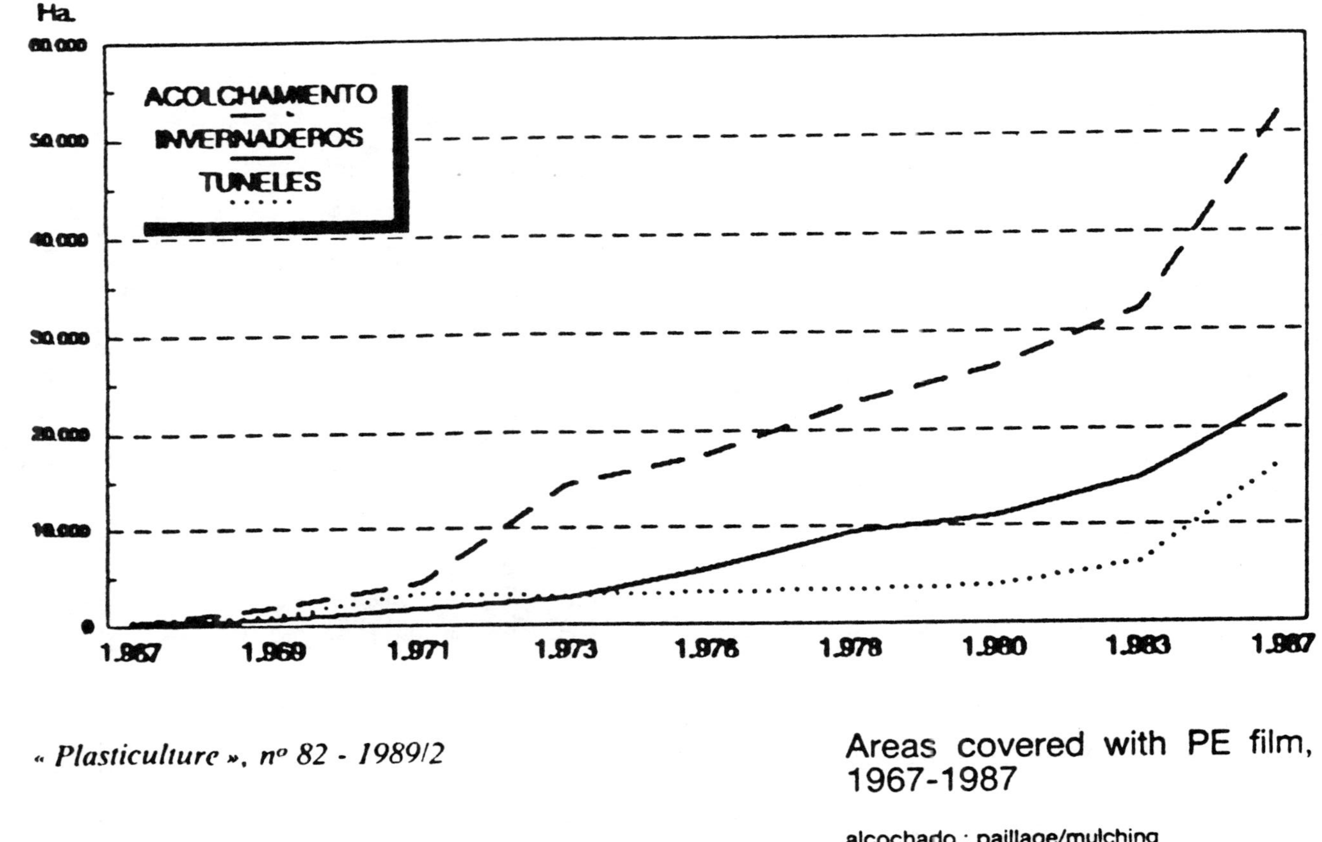


« *Plasticulture* », *n° 82 - 1989/2*

Areas covered with PE film, 1967-1987

alcochado : paillage/mulching
invernaceros : serres/greenhouses
tuneles : (petits) tunnels/(low) tunnels

Figure 3

the soil temperature of the area covered. The humidity of the soil is maintained at such levels so that it is available for seed germination and root growth. The effect on the growing crops is very extensive making it possible to overcome certain climatic limitations for some crops to and to achieve optimum results of bigger and better crops in their actual locations.

These are the benefits of mulching:

It is possible to advance the date of sowing up to one month. Rapidity of germination resulting in earlier maturity and better quality of the crop. Uniformity and improved germination. This makes it possible to save the quantity of seeds sown and even obviates the need for thinning the crops made necessary by overseeding.

A great saving or irrigation water results from the prevention of evaporation by the mulch film, where ever additional water has to be supplied. In unirrigated fields the moisture present in the ground is conserved making it possible to grow crops in arid regions. However and entirely gratitious benefit is obtained by creating ideal conditions of warmth and moisture and oxygen the the root bed directly below the surface. This area is of greater fertility than the lower regions of the soil. It is therefore possible to save on fertilizers, especially on the nitrogenous ones which not only represents a saving in cost, but the reduction in its use has such a salutary effect on the ecology that it has been said that this single factor alone makes the use of mulching film desirable. Chemicals such as pesticides and herbicides are made redundant by the mulch and add another immediate benefit to grower and the the environment. (1)

It is therefore not surprising that the use of mulching films is increasing at a very fast rate. (Table 1) The mulching films however are causing a grave disposal problem. The removal of the films from the field is imperative as the residue will hamper subsequent cultivating operations. This removal represents and arduous and expensive task.

In some crops it is even impossible to remove the residue after the harvest as it has become totally entangled in the plant residue. When mechanical harvesting is practiced the use of mulching films is precluded as the film debris would clog the harvesting machinery. Even after the physical removal of the films from the field one is still left with extremely bulky soiled and partially oxidized residue which has to be dispose of. Burning on the spot is now severely restricted in most localities and even when practised large masses of carbonized insoluble and nonflammable of the polymer are left.

Here the controlled Photo-degradation of Polymers is an elegant and inexpensive solution. However for such a system to be acceptable as an agro-technical method certain stringent criteria have to be applied.

1) Reliable and variable Time Control
2) No Reduction in the mechanical Properties
3) Retention of the mechanical Properties during the Service period

TABLE 1

The Use of Polyethylene Mulching Films in 1986

| Region | Surface area (hectare) | Tonnage (t) |
|---|---|---|
| W.Europe | 200,000 | 50,000 |
| E.Europe | 10 000 | 2,500 |
| Africa and Middle East | 10,000 | 2,500 |
| Americas | 200,000 | 50,000 |
| Asia and Oceania without China | 200,000 | 50,000 |
| China | 1,300,000 | 150,000 |
| World Total | 2,220,000 | 305,000 |

J.C.Garnaud .Plasticulture 1987/2

4) Rapid and Total Destruction of the films

5) No Toxic residue

6) Indefinite Storage Time (if unexposed)

The Scott-Gilead method of controlled photo-degradation does comply will all the above requirements. (2,3)

All methods of induced photo-degradation make use of the ever present energy of solar irradiation. The ultraviolet region of the spectrum and especially the range between 295 - 320 Nm has the necessary energy to break the primary chemical bonds of the polymer (4). It is however necessary to introduce into the polymer additives or intrinsic configuration that can absorb the energy of these wavelengths. The Scott-Gilead system is based on certain metallo-organic complexes which act as chromophores and light controlled catalyst. A protectional-destruction mechanism is obtained which is depending on the light quanta absorbed by the additives, which in turn can be varied in type and concentration as to give reliably reproducible life-spans of a polymer. The action of the catalyst is producing a free radical chain reaction, which is self-accelerating and which will continue even in the dark, in the presence of oxygen. Albertsson and co-workers have shown that the oxidation products of the polymer are continuously removed by micro-organisms (5). The end products of this degradation process are simple compounds such as carbon dioxide and water, leaving no residue differing from the decomposition of vegetation. The amount of irradiation present at a given location will depend on latitude, season and altitude. These irradiations are widely monitored in many countries. (Fig.4) The yearly irradiations are fairly constant and the variations are within manageable limits. (Fig 5 Table 2)

Time Control

It is obvious that degradability must be closely controlled as the films have to perform their agrotechnical tasks before becoming redundant. These time spans will vary greatly. Very short formulations are required when the mulch films are only needed for the acceleration of the germination of the seeds and the first stages of plant development, crops having relatively short growing periods during which they require the presence of the mulch. Crops needing the mulch until the harvest is finished will need longer formulations and finally where double cropping is practiced very much longer life-spans are required. These requirements can be compacted into a few categories: less than 30 days, 30-45 days, 60,90 and 120 days.

The mechanical strength of photo-degradable films

Agricultural films are laid down mainly by tool-bar attachments to tractors. this equipment may be quite sophisticated in doing several actions simultaneously. The film is spread tautly over the prepared seedbed the seed being sown through the films by suitable devices that perforate the films in the exact place of the intermittently spaced seeds. Fertilizers and pesticides may be deposited at the same time. The edges of the films are held down by coulter-wheels and are covered by soil disc-coulters. This is quite an operation and requires apart from the tractor driver on or two men to oversee the whole operation. The speed of the operation is impressive reaching 20 km/hr. The reason for this remarkable speed

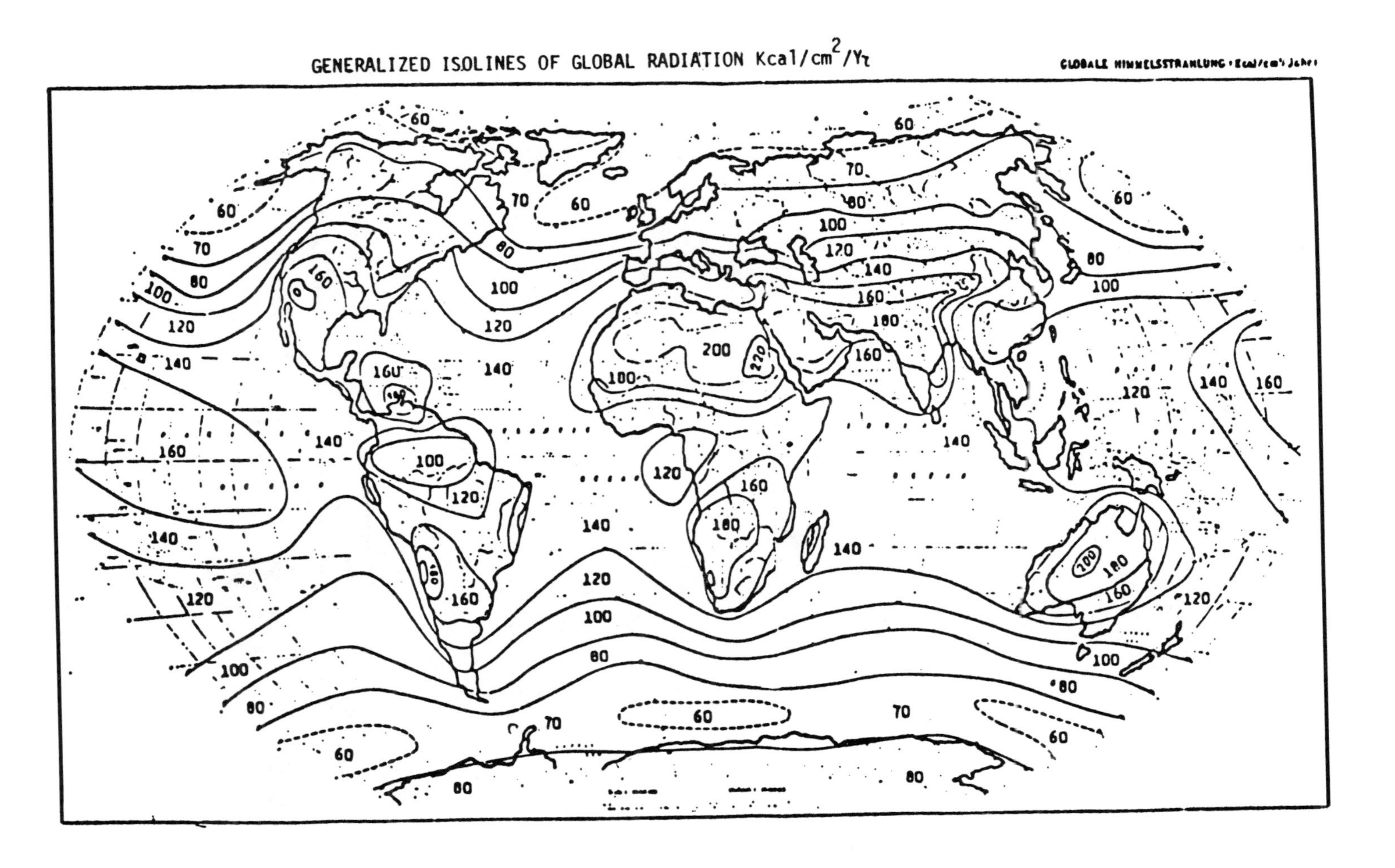
GENERALIZED ISOLINES OF GLOBAL RADIATION Kcal/cm$^2$/Yr

TABLE 2

**Daily U.V. irradiation mouthly averages in mW/min/cm² in Israel**

| | Per diem | per Lunum |
|---|---|---|
| Jan | 6.22 | 193 |
| Feb | 9.89 | 277 |
| March | 17.70 | 549 |
| April | 27.61 | 828 |
| May | 39.12 | 1212 |
| June | 46.00 | 1380 |
| July | 46.40 | 1438 |
| Aug | 41.82 | 1296 |
| Sept | 33.30 | 1000 |
| Oct | 18.70 | 580 |
| Nov | 9.70 | 291 |
| Dec | 4.62 | 143 |

Figure 5

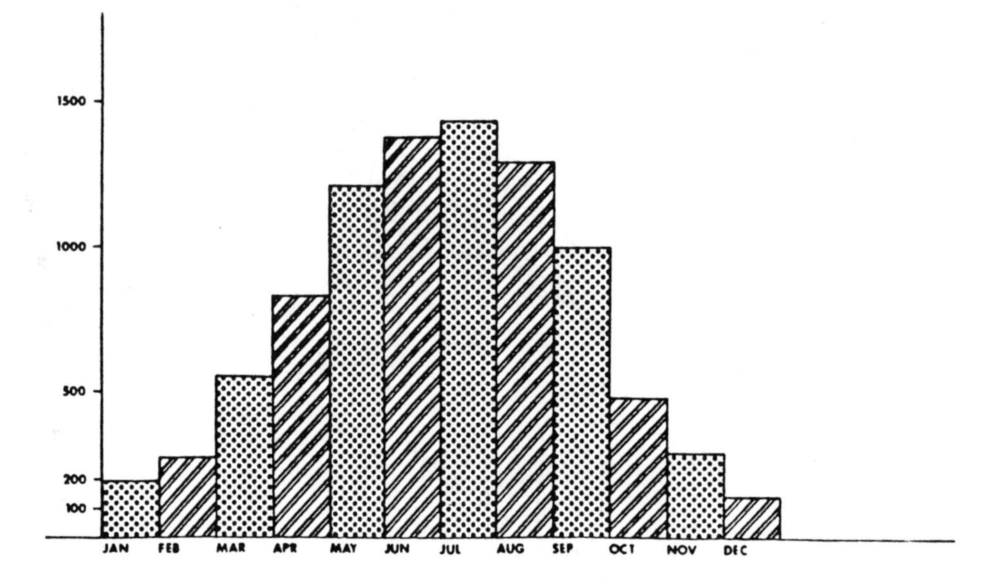

being apart from reducing man-hours per unit area, the fact that very often sowing is a race against time, when the threat rainfall may result in delay in sowing when every day counts. Under these conditions films have to have great mechanical strength, especially tear resistance and impact strength.

Retention of mechanical Properties

The nature of the Scott-Gilead additives is such that they provide stabilization during the induction period and therefore the initial mechanical strength of the films is maintained until the onset of the disintegration process.

Toxicity

The additives are metallo-organic complexes containing iron and may contain a nickel component. The iron which is not considered a heavy metal is physiologically harmless and it is in any case present in any agricultural soil and vast quantities. The Nickel component is per se suspect. Closer inspection however will show that the presence of minute quantities of this element has no effect on the environment whatsoever. Nickel is present in the soil along with most of the other elements of nature in varying quantities ranging from 10 parts per million to 300 p.p.m. (6) Table 3

Simple calculation of the quantities involved will show that in several hundred years of continuous use of photo-degradable films not more than one part per million will be added, ignoring the fact that some quantity will be removed by the crops all of which contain varying amounts of nickel along with other so called trace metals.

TABLE 3

Contents of Nickel and Cobalt in Soils deriving from different origins

| Soil Origin. | Co ppm. | Ni ppm |
|---|---|---|
| Gneis detritus | 14 | 26 |
| Dioritic detritus | 3.3 | 230 |
| Mixed granite & epidiorite detr. | 21 | 22 |
| Various textures | 4 | 13 |
| Mixed granite & schist detr. | 14 | 7 |
| Quartz schist | 1 | 16 |

G. Casalicchio, by courtesy of ENICHEM

Some of which are essential to plant and animal alike. Recent studies by Casaliccio in Italy and Taber in the U.S. A. have proven this preposition by field trials and plant analysis. (6,7) Table 4

Storability

This requirement is of a practical nature as production and marketing factors will engender the need for storage of different periods. The photo-chemical process is initiated solely by exposure to sunlight and this over more or less extended periods. Because of the excellent anti-oxidant properties of the system especially during processessing. Storability will not be reduced but may well be better than that of regular films.

Lastly the Scott-Gilead system by using additives is a very flexible method making it possible to apply it to all polyolefins and other vinyl type polymers. It is therefore unnecessary to keep stock of different polymer resins and grades as simple dry blending during processing will give the desired photo-degradability to the product manufactured.

The photo-degradable films for agriculture have proven to be an important tool in the progress of achieving better and bigger crops. They play an important part in providing food for the expanding population of our planet by bringing more land into the orbit of cultivation. Other plastic products used in agriculture are made to be photo-degradable such as baler twine which is a notorious nuisance on the farm when discarded. Many other applications which will greatly benefit from the principle of photo-degradation are feasible some of which are already in the development stage.

Bibliography

1) Gildead D.; The Use of Photo-degradable Polyethylene Film in the Cultivation of Field Crops in Israel. Plasticulture Vol. 43, 1979

2) Gildead D. and G. Scott; British Patent #22099/78; U.S. Patent #4461853, 4519161.

3) Gilead D. & G. Scott; Time-controlled Stabilization of Polymers Development in Polymer Stabilization-Vol. 5. Ed. G. Scott, Applied Science Publ. London 1982.

4) Ranby B.; Rabek H. R., Photodegradation, Photo-oxidation and Photo-Stabilization of Polymers Wiley Interscience, New York 1975.

5) Albertsson A. C., Europ. Pol. Journal 16, 623, 1980.

6) Casalicchio G. and A. Bertoluzza, Studio su Eventuali Interazioni fra i Residui della Pacciamatura Fotodegradabile e alcune; Colture, 1983. Received by courtesy of Enichem Agricoltura, Milan. (Unpublished)

7) Taber Henry G. et al.; Plant Uptake of Heavy Metals from Decomposition of Plastigone (TM) .21rst N.A.P.A.; Congress (proceedings) 1989 Florida.

QUESTIONS AND ANSWERS

---PRESENTATION BY DR. DAN GILEAD

(Plastopil Corporation)

DR. WOOL: I have a question about rainfall on these films. I am tempted to say the rain in Spain falls maily on the plain, but I won't.

DR. ANDERSON: Thank you.

DR. WOOL: Have you or anyone else studied the problem of a drop of water on these films? Is there an enhancement of the photodegradation underneath the drop, due to a magnification factor? Or is there actual screening?

DR. GILEAD: I wouldn't be able to tell you whether it has, because in the laboratory it is now tested in all these machines.

But as you know, if you go out in the field, the only thing is you have to see whether it meets your predictions. Because you don't know whether it's going to rain or not going to rain. And I have found in no way any difference, not only in Israel by the way, in Germany and Italy, and of course in the United States, especially in Florida and California, there is absolutely no difference in disintegration times when it rains or whether it doesn't rain.

Probably if there is a difference, it's too minute to affect an operation like this, if we are not working on the clock.

DR. ALBERTSSON: I would like to make a comment. When you say about humidity, in all my experiments with different methods, and their effect between -- the effect of metals on polyethylene and together with humidity. Humidity have a very important role in this, and a very high increase of the degradation, if you have high humidity, if you have cycling of temperature and if you have UV light and things like this. Especially humidity is necessary for many of the metals to have a quick effect.

DR. GILEAD: I would quite agree to this. But there is one thing we have to consider. The induction period should be completely immune, or immune from a technological point of view, to the vagueries of the climate. And they should only be controlled by the incidence of light with within reason. It doesn't make any difference if it's two or three days out.

The degradation process, as it is from a physical point of view, is very rapid, and we get these very small bits of film. What goes on from there in vivo we have not tested, and I think it's very difficult to test.

DR. ENNIS: Bob Ennis, Plastigone.

I think that I just wanted to make one comment that I think wasn't emphasized sufficiently by Dan Gilead. And that is that this is a very good example of how a degradable plastic can actually give rise to a new indurstry or a new type of activity that wasn't present before.

With the type of growing of certain plants that couldn't be grown at all without degradable mulch, such as certain types of corn, where you could not take the plastic away from the corn, using a degradable mulch, that corn can now be grown more rapidly and with better economics, using a degradable plastic mulch film.

Similarly, the mid row trench, where the little mini green house is made was something that was impossible to do with regular plastic, and it's only by using degradable plastic can the farmer take advantage of not having to use transplanted plants but can actually use seeds, which are a lot more economical for the farmer to use and gets his product to market earlier.

So it's actually given and advantage and given rise to new industries within this small niche, and I think that perhaps in other areas degradable plastics can do the same thing, if we just keep our minds open to the innovation of the product, we will find a lot of areas the degradables can be used where nothing could be used before.

DR. ANDERSON: Thank you. I think we will take our break now and return in twenty minutes, so that we can begin the next session.

I would ask those that are speaking in the next session to please get your slides in the carrousels during the break. Thank you.

---Recess

# DEGRADATION STUDIES OF POLYOLEFINS

R.G. Austin
Exxon Chemical Company
Baytown Polymers Center
5200 Bayway Drive
Baytown, Texas 77520

## INTRODUCTION

Polyolefins and plastics in general have attracted much attention in recent months mainly due to their visibility and because they are the materials of choice in the packaging industry. Plastics have also gained in the market share in other industry sectors such as electronics, automotive and construction. Their widespread use and growth as well as visibility have made plastics a part of the solid waste disposal issue as municipalities and states in many areas of the country grapple with the solid waste issue. Data show that plastics are about 7% by weight and about 18% by volume of the solid waste stream but nonetheless the public believes plastics are a much greater component [1-3].

Franklin Associates have completed studies that indicate plastics are about 7.2% by weight of municipal solid waste (MSW) for net discards and do not include materials which are recovered or recycled. The same study estimated that 25% of the 24 million tons of plastics produced each year find their way to waste disposal facilities, and most of this plastic is being used in packaging. The plastics packaging materials consisted of polyethylene (PE) estimated at 64 wt%, polystyrene (PS) 11%, polypropylene (PP) 9%, polyethylene terephthlate (PET) 7%, polyvinylchloride (PVC) 5% and others 4%. This accounts for about 11.5% by weight of the total packaging materials discarded with paper (47.8%) and glass (27.1%) accounting for much more. Most of the legislative proposals have focused on PE and PS packaging materials since PE is the plastic used in films for garbage and leaf bags and PS in egg cartons, meat trays and drinking cups to name a few [2-5].

Strategies for MSW management generally include source reduction, recycling, incineration with energy recovery and modern landfilling. The use of plastics in packaging is consistent with these strategies and plastics should contribute to improved management of solid waste. Plastics should be recycled or incinerated to recover heat energy. Today, about 80 wt% of MSW is landfilled and 10% is recycled, while only 10 wt% is incinerated. Recycling of post-consumer plastics products is only about the 1 wt% level with most of this coming from industrial waste and polyethylene terephthalate (PET) soft drink bottles. Other plastics such as high density polyethylene (HDPE), polystyrene (PS) and polyvinyl chloride (PVC) are being recycled to a lesser extent. With an increase in recycling and incineration our dependence on landfills will decrease [2,7-9,14].

The MSW being placed in landfills consists of 70% organics which may degrade to some degree over the active lifetime of a landfill of about twenty years. Plastics, not counted in the organic number, pose no degradable problems in landfills such as leaching of toxic material or generation of methane gas, since they do not biodegrade or decompose in the anaerobic, dark, relatively dry environment of today's landfill. Paper decomposes very slowly if at all, depending on the moisture content. Organic wastes such as food or other wastes like wood, cloth etc. will decay partially to totally within twenty years because biodegradation is a slow process [3,5].

Degradation of plastics is not an effective strategy for managing solid waste. However, it may be useful in reducing the problems with litter and in some agricultural applications such as mulch. Marine and land litter problems are visible issues in terms of aesthetics and the protection of wildlife. Examples of birds trapped by six pack rings and other mammals entangled in plastic fish lines have been cited many times. Plastics cups, packaging films, and discarded bags are among the many materials often found in litter. The very visible plastics litter has received much attention and proposals that plastics be made more degradable have been made to insure that they will disappear from the beaches, roads as well as from landfills [10-13].

RESULTS AND DISCUSSION

Based on the above discussion we set up a research program to evaluate the additives being promoted for the past ten to fifteen years to accelerate degradation in polyolefins, especially polyethylene. Before investigating blends of additives and processibility ivery visible plastics litter has received much attention and proposals that plastics be made more degradable have been made to insure that they will disappear from the beaches, etc. We used the following definitions and designed our experiments to show the effectiveness of these additives in obtaining enhanced degradability.

<u>Degradation:</u> A chemical reaction leading to bond scission in the backbone of a polymer which results in reduction of molecular weight and which is caused by chemical, biological, environmental and/or physical forces.

<u>Environment:</u> A degradation process due to action of combined effects of environmental forces like sunlight, rain, living organisms, temperature, etc.

<u>Photodegradation:</u> A degradation process initiated by light, such as sunlight or a UV source.

<u>Biodegradation:</u> A degradation process in which a living organism, like bacterium, fungus, or enzyme metabolizes or breaks down polymer.

<u>Deterioration:</u> The fragmentation of an article in which the individual fragments retain the intrinsic properties of the original article and which is caused by environmental or physical forces.

Polyolefins are inherently degradable by thermal oxidation or photodegradation but achieving controlled or timely degradation is very difficult. A number of methods to accelerate photodegradation have been discovered and are used commercially. However, polyolefins are not inherently biodegradable and considerable work has been done to accelerate biodegradation of polyolefin packaging materials by using natural additives or fillers. To understand better the potential applications of degradable plastics and the effectiveness of some of these additives in linear low density polyethylene we initiated a research program.

The testing was carried out on blown films of LLDPE containing a minimum of additives for thermal stability. The additives consisted of Irganox 1076 (500 ppm) and Irganox 168 (500 ppm) in all samples. The LLDPE is an ethylene copolymer containing about 6.0 wt% butene or about 1.6 ethyl side chains per 100 carbon atoms. The films evaluated in our degradability studies were prepared the same way as the base LLDPE resin with the same additive packages. Blown films of LLDPE were prepared using a 2.5 inch Egan blown film line with a 12 inch die at about 420$^{o}$F. The LLDPE base film with the above additives, had a 0.918 density and was about 1.5 mil thick.

## ADDITIVES FOR ENHANCED DEGRADABILITY

Two additives, one to enhance photodegradation and a natural additive claimed to enhance biodegradation, will be discussed. The photoadditive, Polygrade I - added as a masterbatch as received from Ampacet at the 5 wt% level - is a metal-containing masterbatch (probably Fe). The second additive evaluated was a corn starch based masterbatch, Ecostar[R], containing about 43 wt% corn starch, silane coated, and an unsaturated oil blended in low density polyethylene and used, after drying, as received from St. Lawrence Starch. The starch-containing masterbatch was dry blended into our LLDPE to produce a 6 wt% starch film sample. The blown film containing Polygrade I was 1.5 mil while that containing the starch was 2.0 mil [15].

The effect the additives have on film properties is an important consideration for determining applications, economics, and processibility. Some typical film properties on LLDPE base resin and LLDPE films containing Polygrade I (5.0 wt%) and/or starch (6.0 wt%) are shown in Table I. It is noteworthy that little loss in initial film properties are found when processing the LLDPE containing Polygrade I under typical conditions. On the other hand, some color (orange-red) in the film is apparent.

Adding starch (6 wt%) to the LLDPE changes the initial properties on the films as well as the processing conditions. Limits on down gauging due to the average particle size of the starch (15-20 microns) are important as well as a processing temperature upper limit (450$^{o}$F). Higher temperatures will start to degrade the starch. The moisture sensitivity of the starch masterbatch is important, therefore requiring drying before use. The properties shown for the 2.0 mil film containing starch (6 wt%) in Table I indicate loss in ultimate tensile, puncture resistance, as well as increase in haze. Similar loss in initial elongation at break (%) for the films are indicated in Figure 1 and 2 for the machine (MD) and transverse (TD) directions, respectively. Other

TABLE I

| Formulation | Base Resin(1) | Polygrade I | Starch Content | Both Polygrade 1/ Starch |
|---|---|---|---|---|
| | | 5wt% | 6wt% | 5wt%/ 6wt% |
| **Property** | | | | |
| Ult Ten (psi)(MD) | ***6530*** | 6000 | ***4270*** | ***4100*** |
| Elm Tear (g/mil) | 66 | 42 | 55 | 41 |
| Puncture (lb/mil) | 4.88 | 5.46 | 1.97 | 2.13 |
| Dart Drop (g/mil) | 49 | 46 | 30 | 37 |
| COF - I/I | .52 | .56 | .35 | .41 |
| Haze (%) | ***23*** | 19 | ***74*** | 62 |

(1)Base Resin Was LLDPE, 0.920 Density, Butene-1 Copolymer

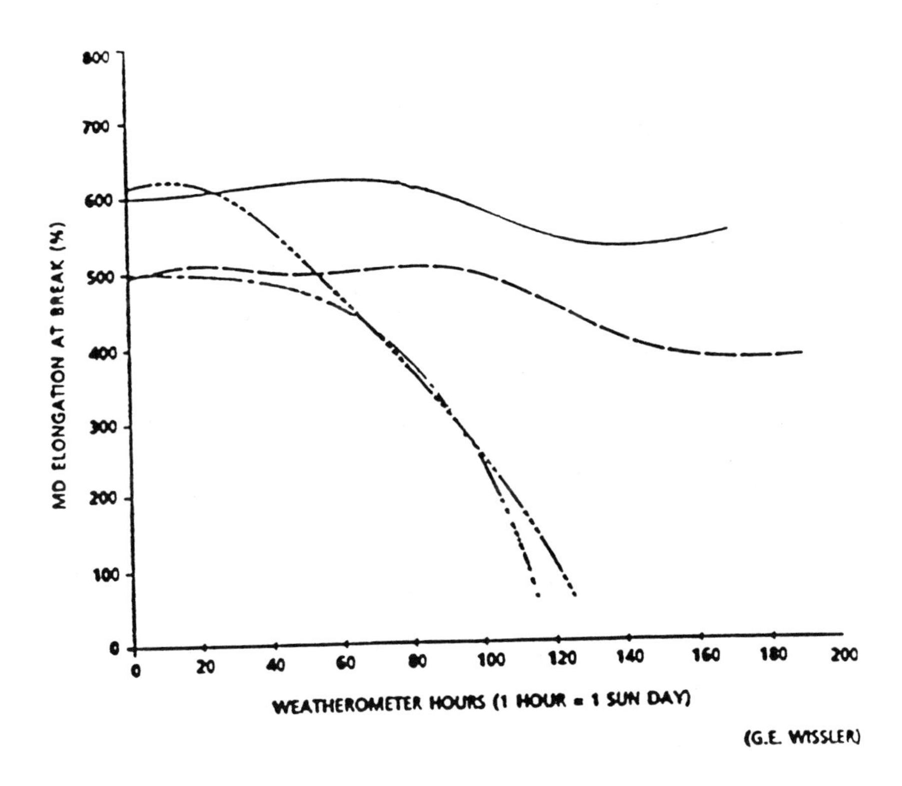


Figure. 1. Elongation at Break (%) Machine Direction (MD) Versus Weatherometer Hours For LLDPE (-), LLDPE/Starch (6wt%), (----), LLDPE/5% Polygrade (-·-·-) and LLDPE with Starch Plus Polygrade (······).

properties such as reduction in gloss and changes in surface were found.

PHOTODEGRADATION STUDY

Testing for comparative photodegradation of these films samples was carried out using a standard weatherometer with a xenon arc lamp with continuous irradiation and inmtermittent "rain" of 18 minutes every 2 hours. The ultraviolet (UV) portion of sunlight, i.e., the shortest wavelengths between 290 and 320 nanometers, causes most of the photochemical reactions to plastics on outdoor exposure. For our comparative study the xenon light source is adequate for simulation of the sunlight's UV spectrum, but actual outdoor testing is used for applications data. Film properties were measured after weathering for a specific time period. Typical data to compare the effectiveness of the photoadditive in the accelerated photodegradation are shown in Figure 1. The control, 0.918 density LLDPE, with the minumum stabilization package shows less than 20% loss in elongation at break in the machine direction (MD) during the 160 hours of irradiation.

The addition of 5.0 wt% photosensitizer dramatically affected film properties during the exposure period. Complete loss of film properties was achieved after about 120 hours in the weatherometer. The LLDPE polymer film containing 6 wt% starch was unaffected as expected. The initial loss in elongation by the addition of starch is noted but the relative photodegradation as evidenced by loss in film properties is similar to the base resin. Combining the starch with the photoadditive produced a similar outcome to adding just the photosensitizers. There was a complete loss of film properties but the degradation rate appears to be slower indicating the starch may be absorbing some of the radiation or at least screening some of the light from the photosensitizer.

The molecular weight of the film containing the photo-degradant decreases by at least an order of magnitude during the exposure, producing in time a very brittle film. No useful physical properties remain in the film. The effectiveness of the photoadditive is dependent on sunlight, UV radiation, and other environmental factors such as rain. However, controlling the photodegradation is difficult and only limited information on the products of degradation are known. Toxicology studies on the products of degrdation should be carried out. The technology could be used for some applications where outdoor litter is the major method of discard. The effectiveness of photodegradation in the marine environment, i.e., under water, will drop off quickly unless the plastic floats for a long period of time.

BIODEGRADATION STUDIES

Biodegradation is the degradation process in which living organisms metabolize or break down substances. Biodegradation is catalyzed by enzymes and its rate greatly depends on environmental conditions such as temperature, oxygen, pH, moisture and a population of microorganisms. The microorganisms secrete extracellular enzymes which are transported for the most part by water. With any polymer or molecule, the enzymes must catalyze the breaking down of the polymer to products low enough in molecular weight to be assimilated by the microorganisms and used as a food source. New strains of microorganisms are continually evolving, but up to now none have been produced which will consume synthetic, hydrophobic, high molecular weight hydrocarbon polymers [3,13,16,17].

The test methods used previously to evaluate biodegradation have included ASTM growth methods using specific fungi or bacteria, or soil burial methods which include weight loss measurements or property studies of various thin polymer films. Labelling experiments using carbon-14 labeled polymers have also been reported for polyolefins in soil burial tests [16,17,27,29,30,31]. These tests are not accelerated and except for the latter C-14 experiment are not quantitative tests for comparison. All these tests show a very slow rate of biodegradation of polyethylene.

In order to establish some comparative data on rates of biodegradation we developed in collaboration with Texas Research Institute (TRI) an accelerated biodegradation test which can be carried out in the laboratory on film samples or granules of polymers. A standardized inoculum was prepared from soil samples from a landfill, a wooded area (deciduous trees) and an activated sewage sludge. It is expected that the microorganisms in these ecosystems would be similar regardless of geographic location and would give a good mixture of species. An enrichment of the extracts of soil and aerobic sludge produced an inoculum and adjustments of pH (5-8) were made to the aqueous medium [28].

On approach to enhancing the biodegradation of polyolefins, inherently resistant to biodegradation, is to add a natural polymer, such as corn starch, which is biodegradable to the polyolefin [3,11,18,20,25]. How effective is this approach in LLDPE blown film in enhancing the film's biodegradation? We have evaluated on a comparative basis the biodegradation of LLDPE and LLDPE containing 6 wt% starch added via the Ecostar[R] masterbatch.

Typically, degradation of PE film is measured by losses in tensile and elongation properties. Most suggested applications of polyethylene with enhanced degradability are in packaging and films and this test was used to follow the film's degradation. Thus, the elongation at break (%) in the tranverse direction (TD) is plotted versus months of exposure of the films to accelerated biodegradable inoculum as shown in Figure 2. Neither the LLDPE nor the 6 wt% starch filled samples show any significant change in elongation properties during the test period. Although as we will show later, some attack on the starch molecules has occurred, this has had little effect on the LLDPE film properties or the polymer itself. Some loss in ultimate tensile strength was found after ten months for LLDPE (less than 20%) but the starch containing film changed even less in our studies.

Other evidence consistent with these film property data can be demonstrated with polymer molecular weight measurements. For instance, the molecular weight of LLDPE was not effected by the biodegradation inoculum after 7 months at pH 7 as shown in Figure 3. The effect of biodegradation on LLDPE molecular weight is shown on a plot of gel permeation chromatographs of pure LLDPE and polymer treated with the biodegradable inoculum for various monthly intervals. The slight difference in GPC's shown is due to the

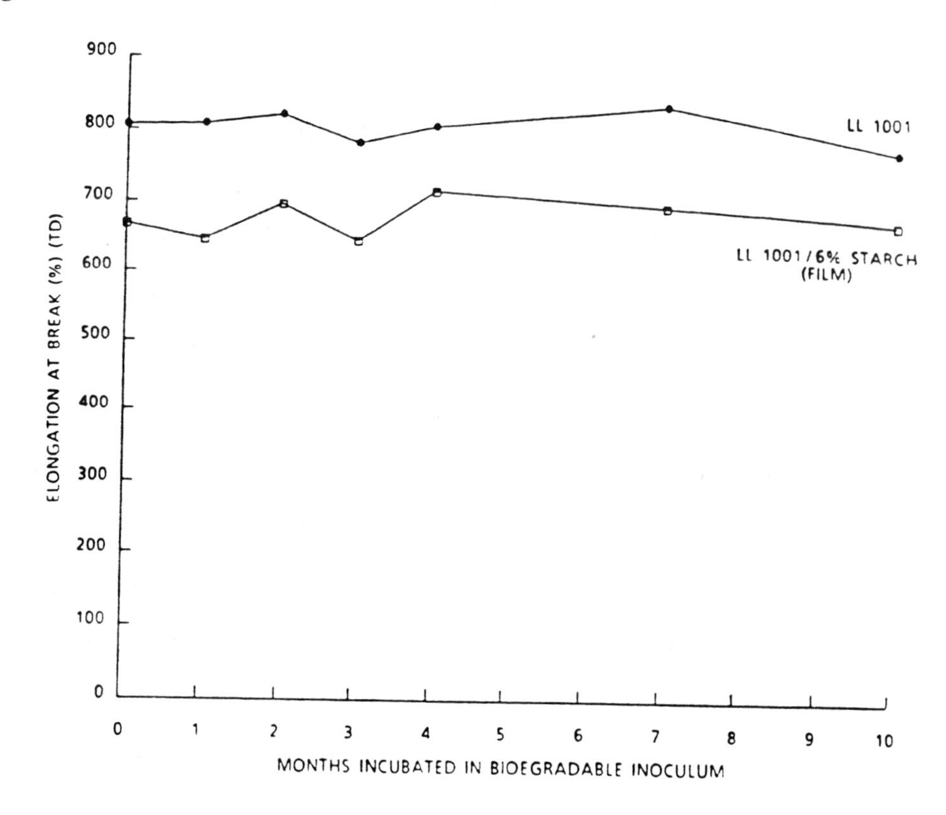


Figure 2. Elongation at Break (%) Tranverse Direction (TD) at Months of Exposure with Biodegradable Inoculum for LLDPE and LLDPE/6wt% Starch Films.

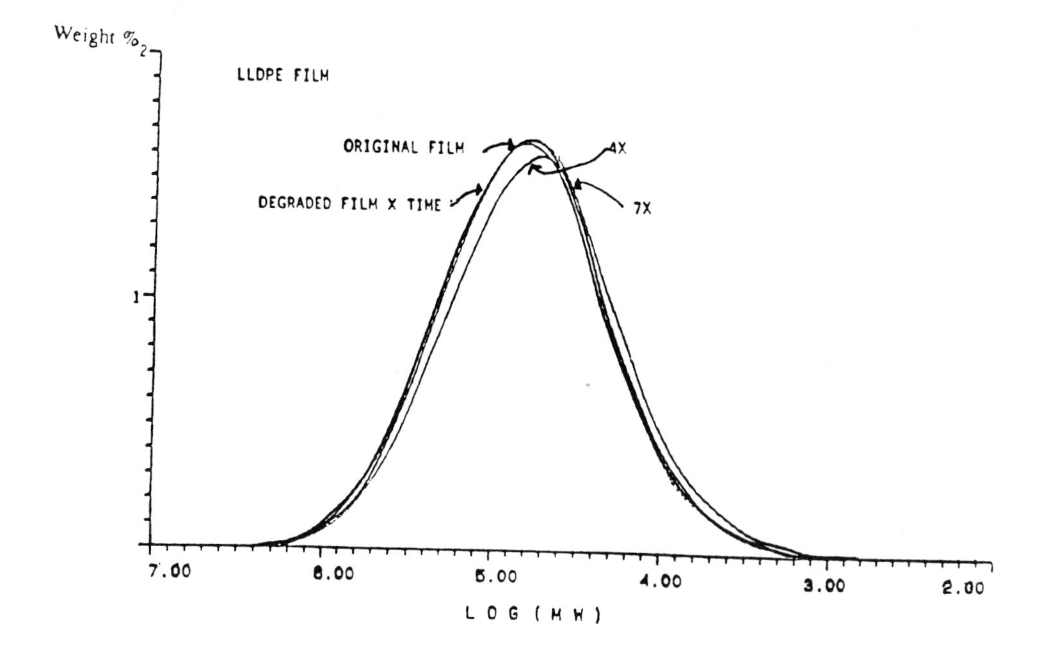


Figure 3. Gel Permeation Chromatograms of Base LLDPE Film Before and After Treatment with Biodegradable Inoculum for X (months) Time.

difference in the columns since chromatograms were run months apart. Similarly, no effect on the LLDPE's molecular weight was found for the starch-containing films after 7 months treatment with inoculum as shown in Figure 4. Based on this study no appreciable effects on the LLDPE are realized by adding the starch. By comparison, the effect biodegradation can have on molecular weight of a biodegradable polyester, polycaprolactone (PCL), is shown in Figure 5 [15,16,26]. The original film had an average molecular weight (Mw) of about 43000 while after seven months its Mw was about 16000 (Figure 5). The number average (Mn) decreases from 24300 to 5900 during the same time.

Another criterion for breakdown of the polymer due to biodegradation, at least aerobically, is the ultimate product of assimilation, carbon dioxide ($CO_2$) [29,30]. In Figure 6, the production of $CO_2$ is plotted against days exposed for LLDPE, LLDPE/Starch, and PCL. The LLDPE generated very little $CO_2$ compared to the control, while PCL generated 140,000 ppm in 100 days. The LLDPE containing starch has generated about 37000 ppm after the same time interval. It is interesting that the starch-containing sample, initially generated $CO_2$ at about the same rate as PCL but after about 30 days leveled off. At this point the total $CO_2$ given off is approximately equivalent to less than 10% of the starch in the film.

Knowing that the starch is being attacked in this inoculum as evidenced by carbon dioxide generation, we investigated the surfaces of the films by scanning electron micrographs (SEM's) of LLDPE before and after 50 days incubation with the biodegradable

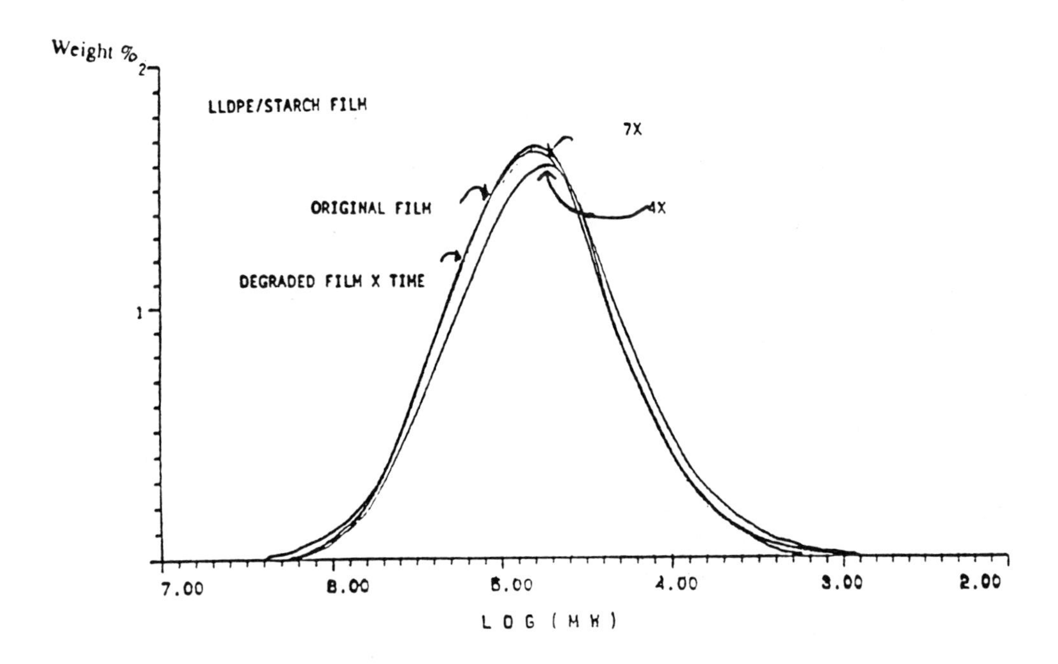


Figure 4. Gel Permeation Chromatograms for LLDPE/6% Starch Films Before and After Treatment With Biodegradable Inoculum for X (months) Time.

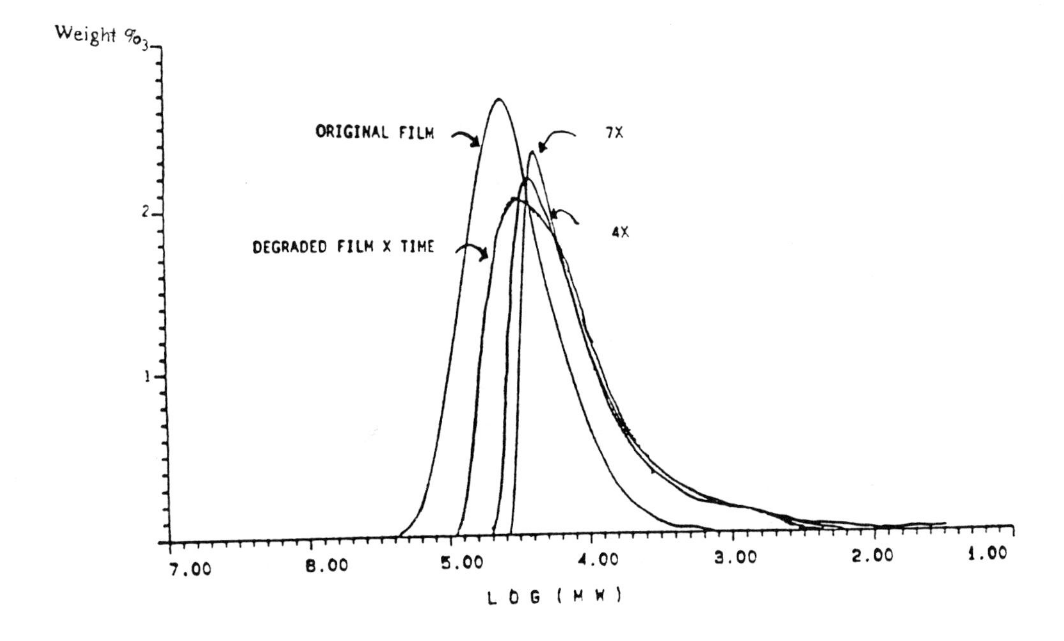


Figure 5. Gel Permeation Chromatograms of Polycaprolactone Before and After Treatment With Biodegradable Inoculum for X (months).

inoculum is shown in Figure 7. There is no apparent differences in the film surfaces. On the other hand, the LLDPE containing 6 wt% starch shows after 50 days preferential attack on the starch molecules by microorganisms. It is clear that there is still a lot of starch left in the film and that the starch at the film surface is attacked (Figure 8). On the other hand, the SEM's of PCL show that the entire film surface as well as the interior is being attacked (Figure 9). In fact, we could have shown micrographs where holes are present in the PCL films. The degradation may indeed effect the surface properties of the films including the starch modified PE. To summarize, in a 6% starch filled LLDPE film, evidence for starch degradation exists but little or no degradation of the PE occurs under our test conditions. The film appearance does change due to the attack on the starch, however.

In summary, photodegradable additives such as Polygrade I are effective in degrading LLDPE in the presence of UV light. The degradation is proved by loss in film mechanical properties and polymer molecular weight breakdown. The additives may be beneficial for degradation of plastics packaging that become litter when exposed to sunlight, but thus far have not been shown to be effective in the marine environment or landfills. The long-term environmental effects of polymer fragments or other products of photodegradation are not well understood. The rate of polymer degradation is dependent on UV radiation and other environmental factors such as oxygen, water, etc. Polymer uses and applications are very diversified and require different levels of performance. Therefore, any application for use of photodegradable polymers must

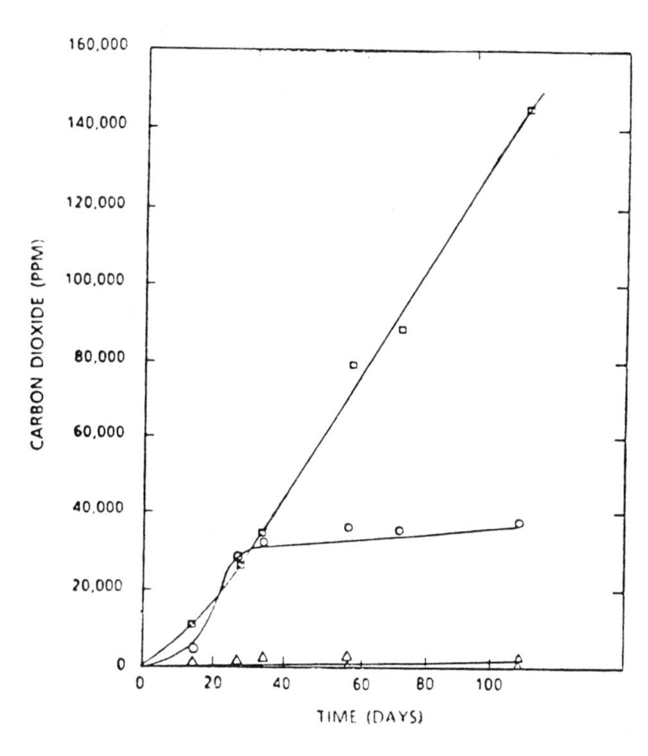


Figure 6. Production of Carbon Dioxide ($CO_2$) During Biodegradation Exposure (days) for Polycaprolactone (□), Polyethylene/Starch (6%)(o), and Polyethylene (Δ).

LLDPE

LL1001
05KV X1100 10U 2280 BPC

AFTER 50 DAYS OF INCUBATION

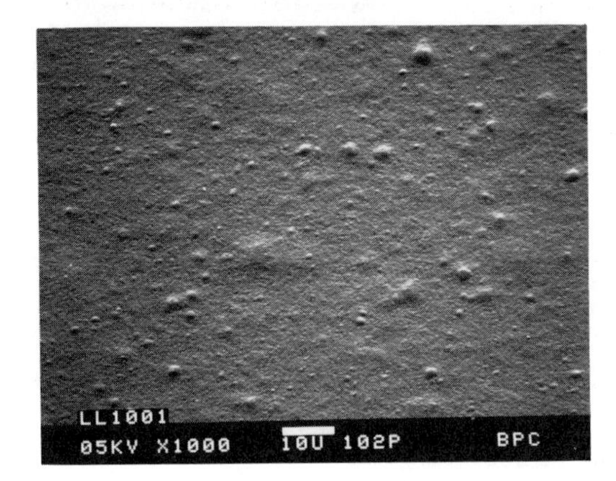


Figure 7. Scanning Electron Micrographs of Neat Film Surfaces of LLDPE Before and After 50 Days Exposure to Biodegradable Inoculum.

LLDPE/6% STARCH

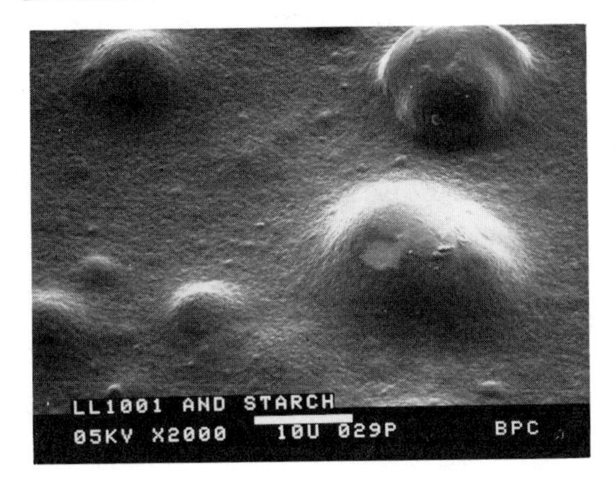


AFTER 50 DAYS OF INCUBATION

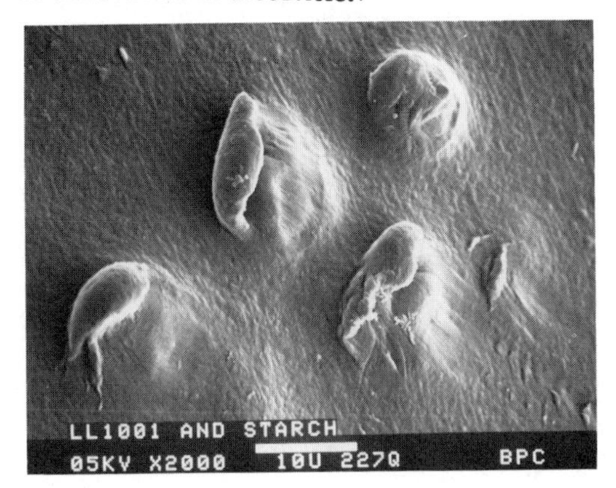


Figure 8. Scanning Electron Micrographs of Neat Film Surfaces of LLDPE Containing 6wt% Starch Before and After 50 Days Exposure to Biodegradable Inoculum.

POLYCAPROLACTONE

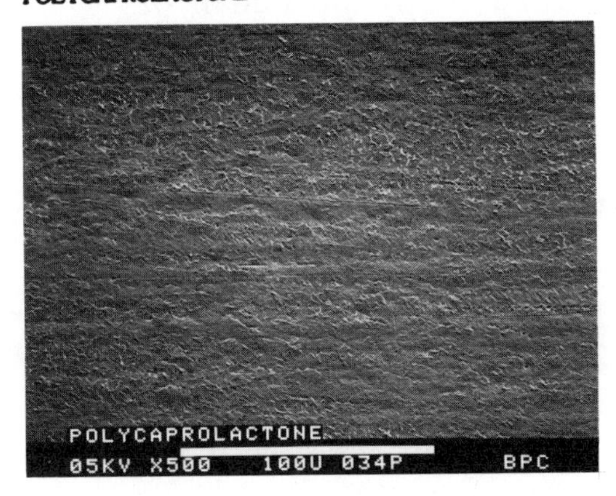


AFTER 50 DAYS OF INCUBATION

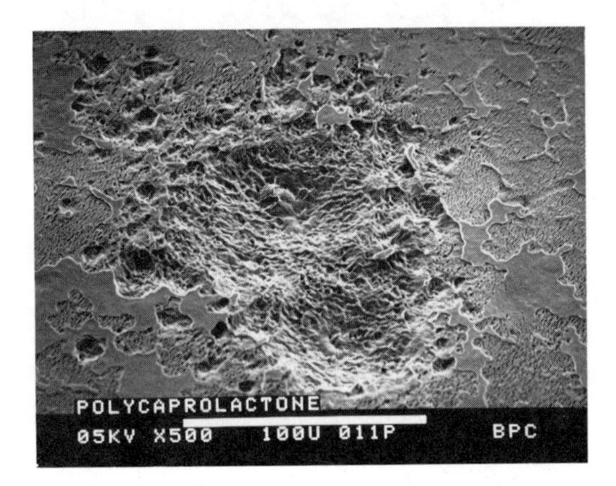


Figure 9. Scanning Electron Micrographs of Neat Film Surfaces of PCL Before and After 50 Days Treatment With Biodegradable Inoculum.

take performance requirements into consideration.

Biodegradation of polymers occurs in certain polymers such as PCL. However, the approach of adding a natural polymer, such as starch at the 6 wt% level, is not effective in enhancing the biodegradation of LLDPE films. The degradation rate is slow with little loss in film properties due to biodegradation although initial physical properties are lost by adding the corn starch. Polyethylene articles made from these blends will not decompose at an accelerated rate either as litter or in a landfill. Based on our accelerated laboratory studies the polymer is not attacked by the microorganisms and penetration of the hydrophobic PE by the organisms to consume all of the starch is slow.

How the plastic product is disposed of is an important criterion in designing a degradable additive. The effectiveness of a biodegradable additive should be well understood and its overall effect on landfills evaluated. If there is high probability that the packaging plastic will be littered on land or in an ocean or lake then these environmental conditions should be tested as well. Before promoting degradable plastics as a solution to our solid waste or litter problems we need more research and development.

# REFERENCES

[1] Plastics World, "How Did Plastics Become the Target?," September, 1989.

[2] "Characterization of Municipal Solid Wastes in the United States, 1960 to 2000," July 11, 1986 and Update, March 30, 1988. Franklin Associates, Ltd. Prairie Village, Kansas.

[3] "Estimates of the Volume of MSW and Selected Components in Trash Cans and Landfills," November, 1989. Franklin Associates, Ltd. Prairie Village, Kansas.

[4] R. Johnson, "An Overview of Degradable Plastics," J. of Plastic Film & Sheeting 4,155(1988).

[5] Modern Plastics, "Industry Weighs Need to Make Polymer Degradable," August, 1987.

[6] Plastics World, "A Way Out," September, 1989.

[7] W. Pearson, "Plastics Recycling in the U.S.," SPE Regional Conference, March 9, 1988, Philadelphia, Pennsylvania.

[8] D.R. Marrow, et. al., "Overview of Plastics Recycling," Converting and Packaging, November, 1987.

[9] D. Brewer, "Recycling Resources: A Plastics-Industry Update," Plastics Packaging, January 1, 1988.

[10] M. Weiskopf, "Plastics Reaps a Grim Harvest in the Oceans of the World," Simthsonian, March, 1988.

[11] Proceedings of the Societiy of the Plastics Industry Symposium on Degradable Plastics, June 10, 1987, Washington, D.C., "Degradable Plastics."

[12] Plastics World, "Are Degradable Plastics the Answer to Litter?," June, 1987.

[13] Taylor, L., "Degradable Plastics, Solution or Illusion," ChemTech, 542(1979).

[14] Plastics World, "Can Plastics be Burned Safely?," September, 1989.

[15] Wissler, G.E., Polyethylene Development, Exxon Chemical Company, private communication.

[16] J.E. Potts, The Encyclopedia of Chemical Technology, John Wiley and sons, 1982.

[17] J.E. Potts, Aspects of Degradation and Stabilization of Polymers, ed. H.H.G. Jellinek, Elsevier, New York, 1978.

[18] G. Griffin, "Biodegradable Fillers in Thermoplastics," American Chemical Society, Adv. in Chem. Ser., 134, Washington, D.C., 1974.

[19] E.J. Reger, "Degradable Plastics For the 90's," Solutions '89, Marketing/Technology Service, Inc., Kalamazoo, Michigan, 1989.

[20] W.J. Maddever, "Starch Based Degradable Films," Solutions '89, Marketing Technology Service, Inc., Kalamazoo, Michigan, 1989 and ref. 11.

[21] D. Gilead and R. Ennis, "A New Time-Controlled, Photodegradable Plastic," ref. 11.

[22] A.W. Carlson and V. Mimeault, "Degradable Concentrates For Polyolefins," ref. 11.

[23] J.E. Guillet, Pure & Appl. Chem., 52,205(1980).

[24] G. Scott, Pure & Appl. Chem., 52,365(1980).

[25] F.H. Otey, A.M. Mark, C.L. Mehltretter, and C.K. Russel, Ind. Chem. Prod. Res. Dev., 13,90(1974).

[26] S.H. Huang, et. al., J. Appl. Polym. Sci., Appl. Polym. Symp., 35,405(1979).

[27] G. Colin, J.D. Cooney, D.J. Carlson, and D.M. Wiles, J. of Appl. Poly. Sci., 26,509(1981).

[28] L.N. Britton and S.A. Lax, Texas Research Institute, private communication.

[29] A.C. Albertsson, S.O. Anderson and S.L. Karlsson, Polymer Degradation and Stability, 18,73(1987).

[30] A.C. Albertsson, Eur. Polym. J., 16,623(1980).

[31] E. Kuster, J. Appl. Polym. Sci., Appl. Polym. Symp., 35,395(1979).

QUESTIONS AND ANSWERS

DR. NARAYAN: Although you didn't bring it up too much, your biodegradation data shows almost no biodegradation or very little with the six percent starch. Is that a correct statement?

DR. AUSTIN: We see some attack of the starch, and we see biodegradation, based on the $CO_2$ generation.

In terms of film properties, we do not see very -- under our test conditions we see very little deterioration of the film properties at that level of starch. Right now we have a test going on using 28 percent masterbatch starch, which is equivalent to 12 percent, and those tests are just starting. The films are a lot different. I can say the initial properties of those films would be degraded as well, but --

DR. NARAYAN: In the tone of the conference, then, obviously you are going to see the next paper which talks about degradation. What could you say would be the issue here in terms of -- what do we need to address? Or what do we need to do to resolve the different data which is going to come out on the six percent starch?

DR. AUSTIN: Well, it's not -- it's not that easy. When you are talking about degradation, you really have to talk about the specific application that you are dealing with. So to try to come up with a test method that is applicable to the variety of applications that plastics are used for, or biomaterials are used for in sutures or whatever, I think you are going to have a very difficult time.

So we may have to separate applications, go through separate, different definitions for each application. I really think we are misleading the public by saying "This material is biodegradable and can be used for some application," where it really may never see a biodegradable situation, versus photodegradable for instance. And I think we have a lot of problems in marine litter. We haven't even found the solution for marine litter, in my opinion.

So I think there's a lot of work to be done. I am open to any suggestions. I don't think there is a universal test or a universal definition. I think in a lot of areas, we've got to look for the application.

Yes.

DR. LOOMIS: You have made a big step here in at least, you know, going to a stable inoculum rather than just burying things in the soil and the cesspool and so forth.

Is that standardized to the point where if you were then to go in and do another test on another series of polymer films, can you reproduce that inoculum so that would you get the same results and pass it around to the rest of us in the --

DR. AUSTIN: We are contemplating handing out a protocol, but big companies move slower than entrepreneurs, so we are addressing that.

We think yes, it will be reproducible to a certain degree, and we have reproduced it ourselves. So each one of these measurements is not done on one film, it's done on multiple. The $CO_2$ measurements are average of 5 measurements. The tensile measurements are an average, because we do the tests on the films already cut for tensile measurements. So there are multiple tensile measurements as well. So we hope it's reproducible and certainly genuine nature.

We think ultimately, though, if you are going to solve a landfill problem, you have got to do tests in a landfill to show that you are going to get a certain degradation rate. But I don't have the ten to twenty years that Ann alluded to study biodegradation. My management just will not allow that. So we had to come up with an accelerated test for comparison.

DR. WOOL: I am Richard Wool from the University of Illinois. I would like to disagree with you on your statement that degradable plastics do not have any part in solid waste management. Let me give you a few important examples.

DR. AUSTIN: I didn't say that.

DR. WOOL: Well --

DR. AUSTIN: I didn't say they didn't have any part. I said at the present time they are not a strategy in solid waste management.

DR. WOOL: I know. I would like to give you a few points that you have overlooked.

DR. AUSTIN: Sure.

DR. WOOL: Point No. 1, the degradable lawn bag for lawn reclamation will give you back about 20 percent of the landfill by taking out the naturally degradable materials.

Point No. 2, the use of degradable plastic films in landfill as a substitute for soils can potentially give you back another 15 or

20 percent of the landfill. And there's several other applications.

If we want to leave the plastic material in the landfill, and in the future we have a way of harvesting the methane from the landfill, then you can make a good argument to have that material be degradable.

Fourthly, in other problem areas, such as in marine pollution, degradable plastics are going to be made. However, only 1 percent of those ends up in either the litter or the marine pollution stream. The other 98 percent or so is going to end up in a landfill. And so whether you like it or not, degradable plastics are going to be part of the landfill continuing problem or solution in the future.

DR. AUSTIN: No comment.

DR. SHALABY: Shalaby Shalaby from J & J.

The first question is pertinent to the development of protocol and for degradation studies. You mentioned that you used a mixture of microorganism from sludge as well as from soil. Do you see any difference between the mixture and the individual source of microorganism?

DR. AUSTIN: We are in the process of isolating enzymes, if you will, and we have no results at the present time.

We have, in this study, which I didn't mention, we did four different sets of tests at four different pHs as well. The results I showed you were at pH 7. We have results now at pH 5, 6, and 8, and we, based on thinking that some microorganisms might be more active in a different pH environment, which we know they will, we have not seen any significant effects on the properties of the films at different pHs.

DR. SHALABY: Second question, which is very quick. We heard earlier that you need to have the appropriate shelf or indefinite shelf life of the filled polymer. In case of a starch, how can you guarantee that the starch-filled polymers will not support microbial growth as you store them?

DR. AUSTIN: I don't support that at all. I do not have the data.

DR. SHALABY: Does anyone do that? Does anyone have that?

DR. AUSTIN: We have not done any shelf life studies on this starch material. Ours are all accelerated tests. I think the starch producers may have that.

DR. SHALABY: If the microorganisms you can see them in the culture, they should be also to see them while you are storing them.

DR. AUSTIN: That's a good point.

DR. FEIJEN: Just one question. Jan Feijen, University of Twente.

Could you explain to me, the 6 percent starch particles, you don't have a continuous pore structure after degradation, I assume. So you have only -- when you expose this to your medium, you only have some removal of material from the surface.

DR. AUSTIN: Yes.

DR. FEIJEN: And if you then look at change in properties, material properties, is that a fair thing to do, if you only modified the surface to a certain extent?

DR. AUSTIN: Well, I am just, based on what I know, polymer science, I am trying to show whether we are getting biodegradation of the polymer by adding starch.

DR. FEIJEN: But I try to --

DR. AUSTIN: Any test you want to recommend that would show that, other than molecular weight --

DR. FEIJEN: But I tried to envisage how you degrade a matrix, if you only get rid of some starch at the surface.

DR. AUSTIN: Well, I think, hopefully -- the starch producers claim that you do penetrate much further in the polyethylene matrix. We contend that it's very difficult to do that because of the hydro- --

DR. FEIJEN: There may be a mechanism, but I would like to know --

DR. ANDERSON: Maybe we'll hear that from the next speaker.

DR. AUSTIN: Yes.

DR. ANDERSON: Could you hold your question to the next speaker?

DR. FEIJEN: Oh, yes, please.

DR. UPTON: Just to address the question that J & J raised, because I can't get the name, I am sorry.

DR. ANDERSON: Excuse me, the name is Dr. Shalaby Shalaby.

DR. UPTON: Dr. Shalaby Shalaby. All right, and I am Dr. George Upton with Ampacet.

An antidotal response to that is if you keep the starch film, contained film cool, dark and dry, it doesn't do too much after a year.

Plus, in an open warehouse down in Orange, Texas, it appeared not to be -- it was not slick, and it was still dry. So it didn't act as an attractant for a host of microorganisms that you could see visually.

DR. AUSTIN: Thank you.

DR. ANDERSON: George, don't leave. I have a question for you and Richard. What's the chemistry of this additive you guys are talking about?

DR. UPTON: The chemistry of which additive are you referring to?

DR. ANDERSON: It's a brand name that you talked about. He used it in his talk. I never heard what it was.

DR. AUSTIN: In the talk that Dick gave, I think you are referring to the ECOSTAR material, is that correct?

DR. AUSTIN: No, Polygrade 1.

DR. ANDERSON: The Polygrade 1, thank you.

DR. UPTON: Is a metal ion, transition metal. It's a photodegradable. We were talking about starch just now.

DR. ANDERSON: But you don't want to reveal its chemistry to us or what? What's the metal ion?

DR. UPTON: There you go, it's a metal ion, and it was in that list of the Illinois soil trace materials.

DR. ANDERSON: Yes, it was probably iron that was there to five significant figures.

Thank you very much, Richard.

# MODIFIED STARCH BASED ENVIRONMENTALLY DEGRADABLE PLASTICS

Dr. W.J. Maddever, Manager - Business Development

Peter D. Campbell, P. Eng., Project Engineer

St. Lawrence Starch Company Limited

Mississauga, Ontario

## INTRODUCTION

The use of natural fillers, such as starch, in polymeric systems has been of interest for many years. In the mid-1970's it was found that by surface modifying the starch acceptable physical properties of polymer starch blends could be obtained.[1] Further it was found that addition of unsaturated hydro-carbons, in the form of fatty acids, would accelerate the reduction of physical properties of these materials when placed in natural environments such as soil burial and municipal compost.[2]

Degradable plastics based on modified granular starch and autoxidants have been available in limited amounts since that time but have only begun to see widespread commercial use in the last three years. This very success has sparked a fierce debate in the public forum over the advantages and disadvantages of use of the products. What has become abundantly clear is the almost complete lack of understanding by the public and by much of the industry of the technical aspects of the degradation process as well as the lack of consensus on the very meaning of the terms degradable, biodegradable and photodegradable. We have come together this week as scientists, not as marketers to try and resolve some of these issues. There is clearly a place for degradable plastics in the marketplace but the success of these materials will depend to a great degree on a clear understanding of what they are, how they work and where they can be applied.

## DEGRADABLE PLASTICS

Degradable plastics are those polymers which contain materials which enhance these already existing but rather slow photo and oxidative degradation processes. Simply removing the stabilizers already incorporated into the polymer is not sufficient as these materials are in most cases included to facilitate processing.

The photodegradation of common polymers is well known and extensive development of UV stabilizers has been carried out over the years.

Biodegradation of polymers is less well understood. Potts et al [3] discussed the biological attack on low molecular weight fragments of polyethylene. More recently Albertsson et al (4,5) by measuring $^{14}CO_2$ evolution from $^{14}C$ labelled polyethylene have shown that over long periods of time oxidative degradation of the polymer occurs and produces low molecular weight carbonyl containing groups which can be metabolized by microorganisms thus producing $^{14}CO_2$ as a product of biological activity.

## BIODEGRADATION

Biodegradation is the breakdown of materials by the action of living organisms. For plastics the most important organisms are bacterial and fungi although larger organisms can play a part.

The breakdown caused by microorganisms can be of three different types:

- a biophysical effect, in which cell growth can cause mechanical damage
- a biochemical effect, in which substances from the microorganisms can act on the polymer
- direct enzymatic action, in which enzymes from the microorganisms attack components of the plastic product leading to splitting or oxidative breakdown.

Degradation can be followed by measuring changes in physical properties of the product, by studying chemical changes in the film or by assessing biological activity.

THE ECOSTAR SYSTEM

Plastics made with the ECOSTAR system incorporate a starch which is modified to make the normally hydrophilic surface of the starch hydrophobic. This starch is then dried to less than 1% moisture (compared to the 10-12% moisture of normal starch). Several different starches can be used. Although corn starch is the most readily available and widely used, rice starch can be used for products which require a fine particle size.

A fatty acid or autoxidant is added to the system to facilitate breakdown of the polymer. In the most recently developed product , known commercially as ECOSTARplus, a combination of organometallic and organic compounds are added to enhance oxidative degradation as well as provide photodegradable characteristics.

BIODEGRADATION MECHANISM

In the ECOSTAR system degradation proceeds by two interactive mechanisms.

Starch is present in the polymer as granules as shown in Figure 1. These granules provide a nutrient source for microorganisms, such as fungi and bacteria, which attack the granules until they are completely removed as illustrated in Figure 2. This weakens the polymer matrix as well as greatly increases the surface area of the plastic.

The holes formed enhance the migration of potential reactants into the matrix. It has also been noted that the surface of the polymer after removal of the starch is more hydrophilic and more easily wetted, leading to further enhanced transport properties.

As the microorganisms remove the starch there is some physical damage to the polymer. In addition extracellular enzymes produced by the organisms likely provide some direct attack on the polymer and may be responsible for the fine cracking of the polymer observed in Figure 2.

The rate of removal of the starch granules is dependant on the nature of the environment to which the material is exposed as shown in Figure 3. [6]

The second mechanism is a result of the formation of peroxides by the autoxidant when it comes into contact with metal salts present in soil, fresh water or sea water. These peroxides then begin to break the polymer chain. This second mechanism is enhanced by the increase in surface area provided by the first

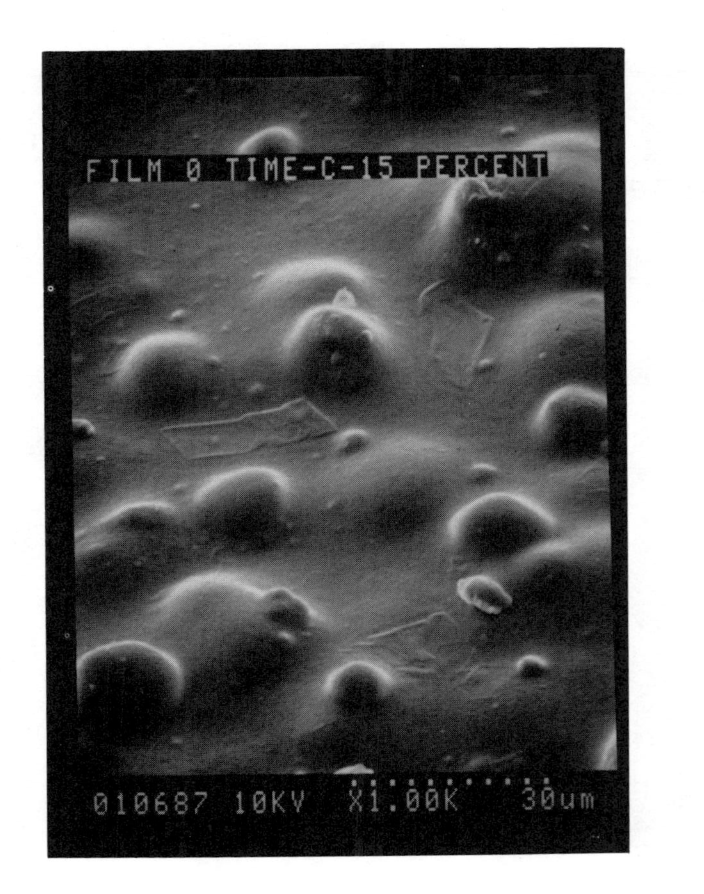


FIGURE 1. Polyethylene film containing 15% starch before exposure to fungi.

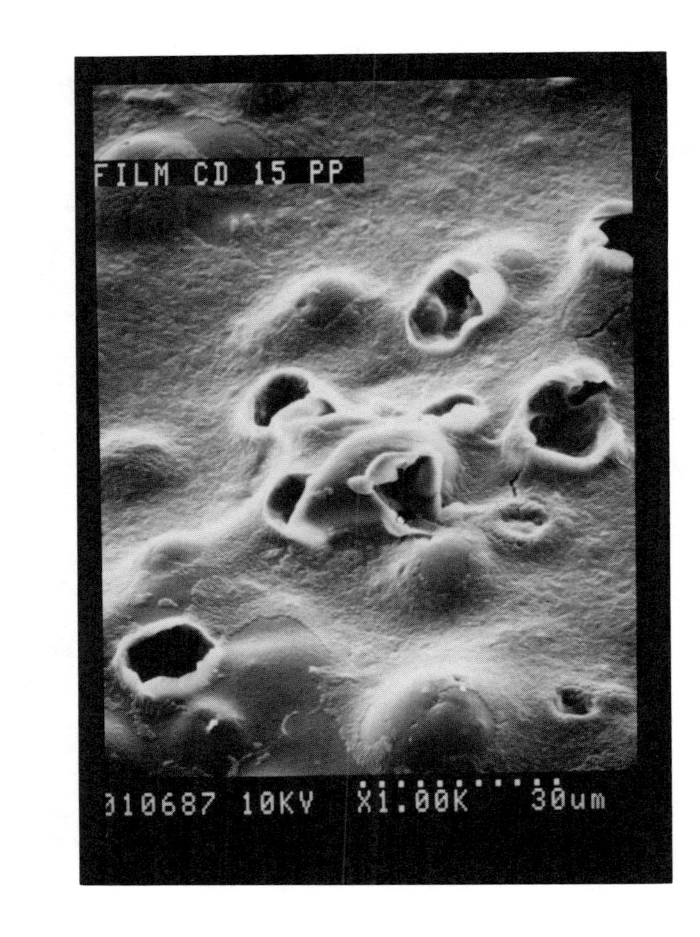


FIGURE 2. Polyethylene film containing 15% starch after 6 weeks exposure to fungi showing removal of starch granules.

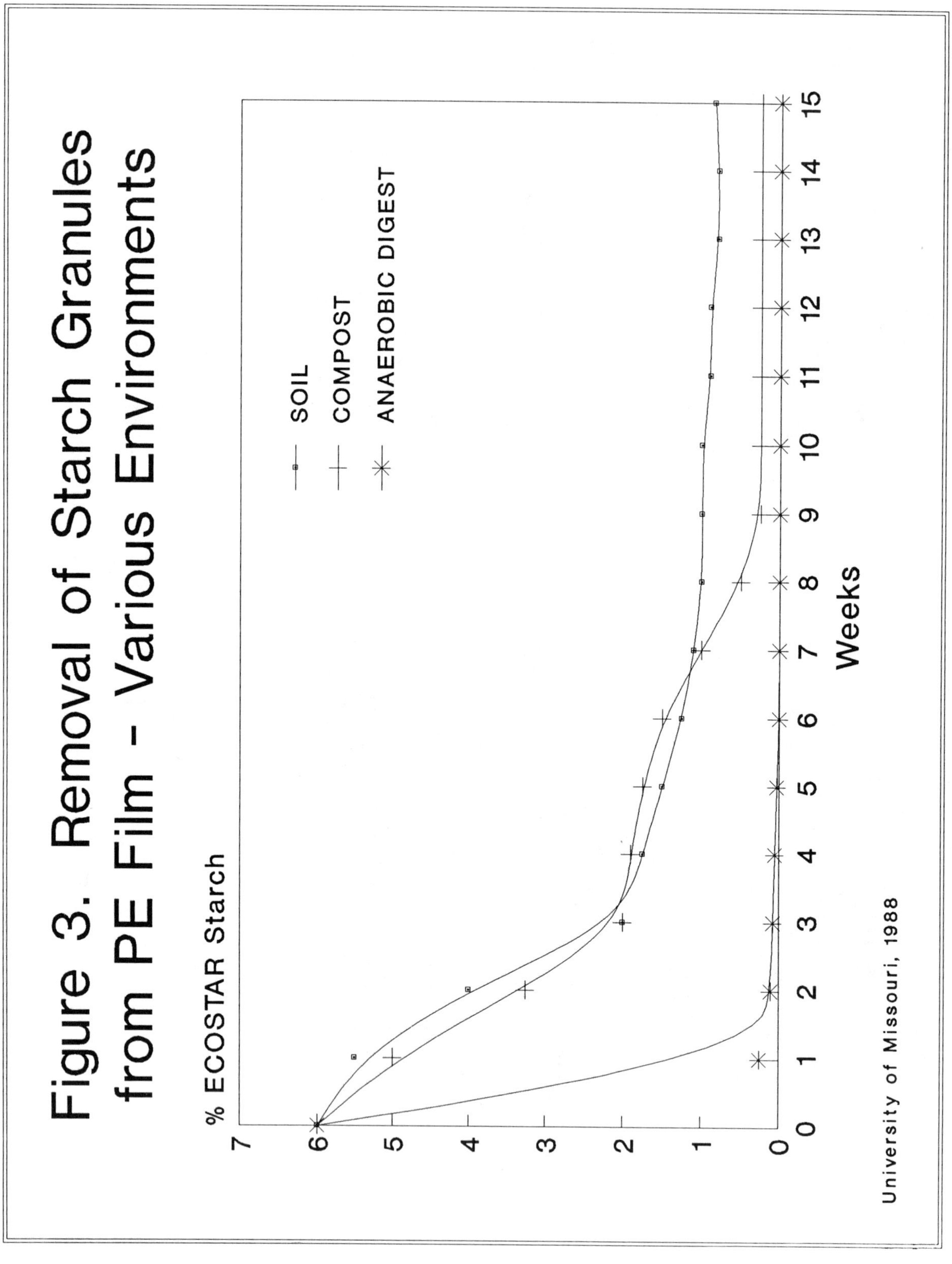

Figure 3. Removal of Starch Granules from PE Film - Various Environments

% ECOSTAR Starch

7
6
5
4
3
2
1
0

0 1 2 3 4 5 6 7 8 9 10 11 12 13 14 15

Weeks

SOIL
COMPOST
ANAEROBIC DIGEST

University of Missouri, 1988

mechanism. In the latest version of the technology the addition of organometallic compounds enhances this autoxidation reaction.

The effect of the oxidative degradation can be observed through oven aging tests followed by physical and chemical testing as shown in Figure 4. Increasing temperature increases the rate of reaction at all addition levels as shown in Figure 4, by what appears to be an Arrehnius relationship. Temperatures ranging from 70°C to 90°C are most commonly used. The lower temperatures have a direct correlation to the performance of the material in an active compost environment which may operate in the range of 50-70°C. The improvement in degradation rate provided by the newest formulation is shown in Figure 5.

Breakdown of the polymer chain weakens the material by reducing the chain length, (molecular weight), eventually to a level that can then be metabolized by microorganisms. As this process is purely biological the products of degradation would be expected to be those of normal biological activity, namely carbon dioxide, water and cell mass and metabolic biproducts.

Measurement of molecular weight distribution of polyethylene containing the additives compared to a control of the same base polymer is shown in Figure 6.

Soil burial tests can demonstrate the results of both the biological and oxidative degradation mechanisms. Simple removal of the starch is not sufficient to result in the magnitude of the loss of physical properties shown in Figures 7[1] and 8.

Figure 8 also indicates the importance of measuring either elongation or toughness (area under the stress strain curve) to

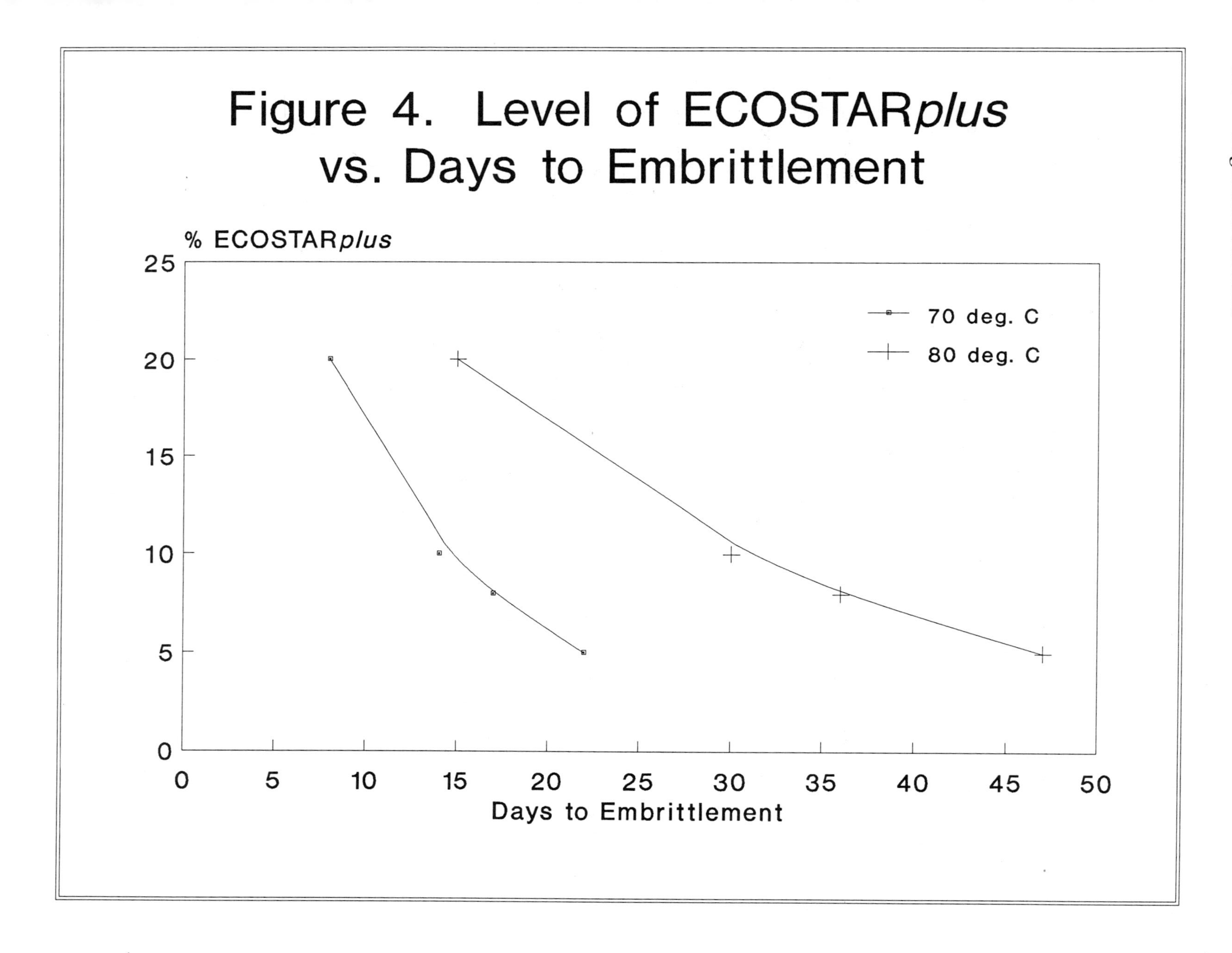


Figure 4. Level of ECOSTAR*plus* vs. Days to Embrittlement

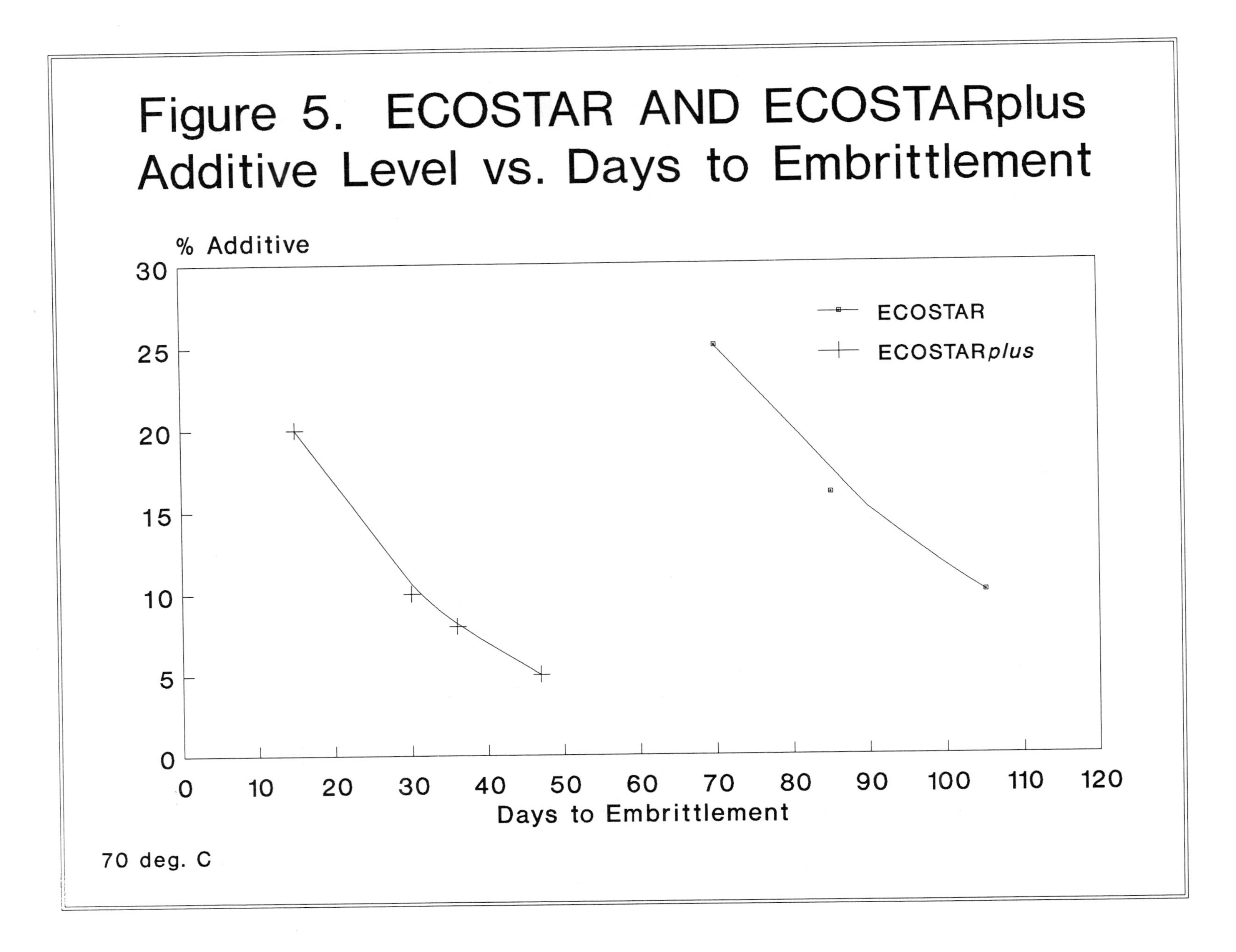
Figure 5. ECOSTAR AND ECOSTARplus
Additive Level vs. Days to Embrittlement
% Additive
30
25
20
15
10
5
0
ECOSTAR
ECOSTARplus
0
10
20
30
40
50
60
70
80
90
100
110
120
Days to Embrittlement
70 deg. C

FIGURE 6

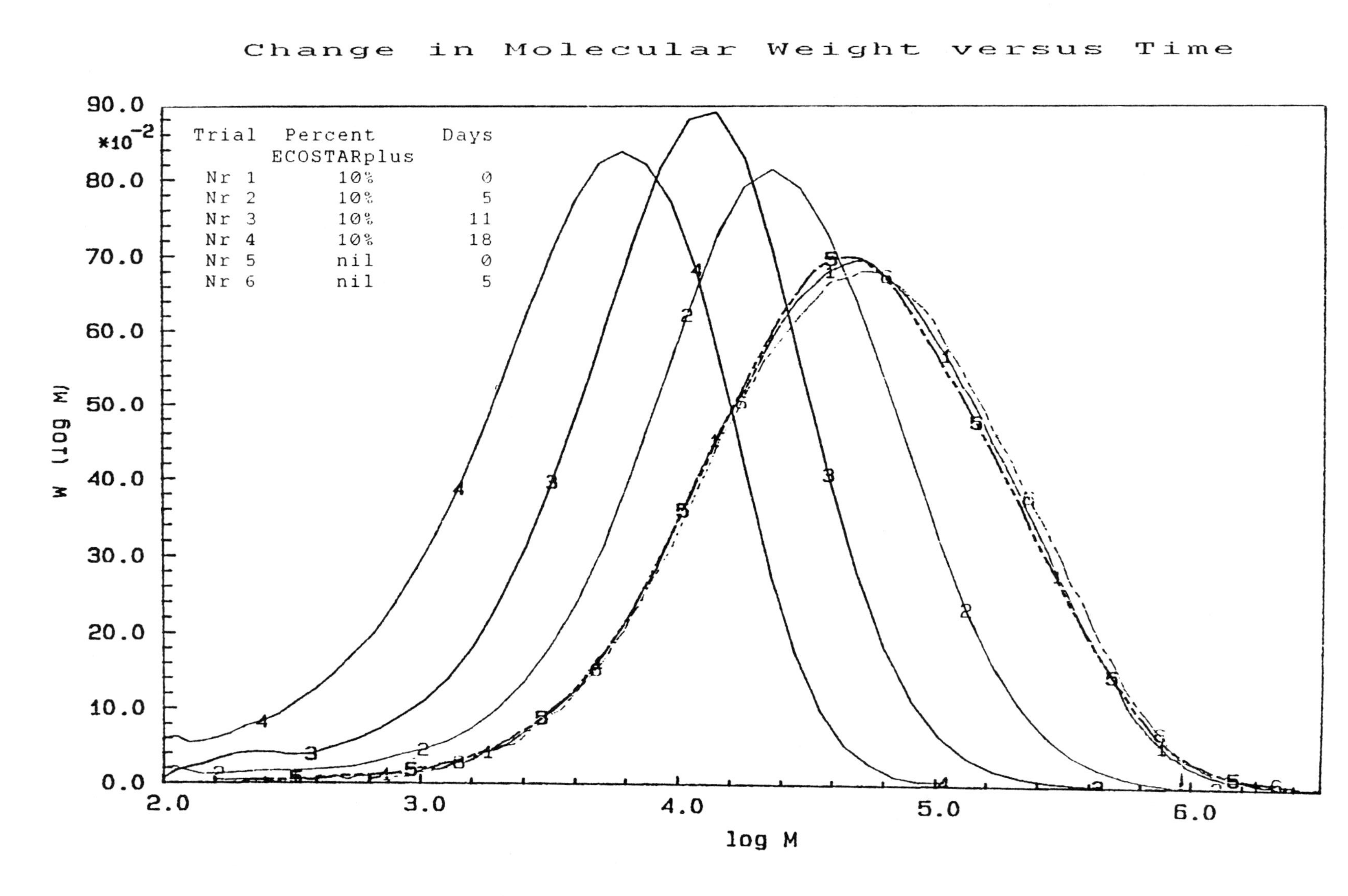
Change in Molecular Weight versus Time
Trial Percent ECOSTARplus Days
Nr 1 10% 0
Nr 2 10% 5
Nr 3 10% 11
Nr 4 10% 18
Nr 5 nil 0
Nr 6 nil 5
W (log M)
*10^-2
90.0
80.0
70.0
60.0
50.0
40.0
30.0
20.0
10.0
0.0
2.0
3.0
4.0
5.0
6.0
log M

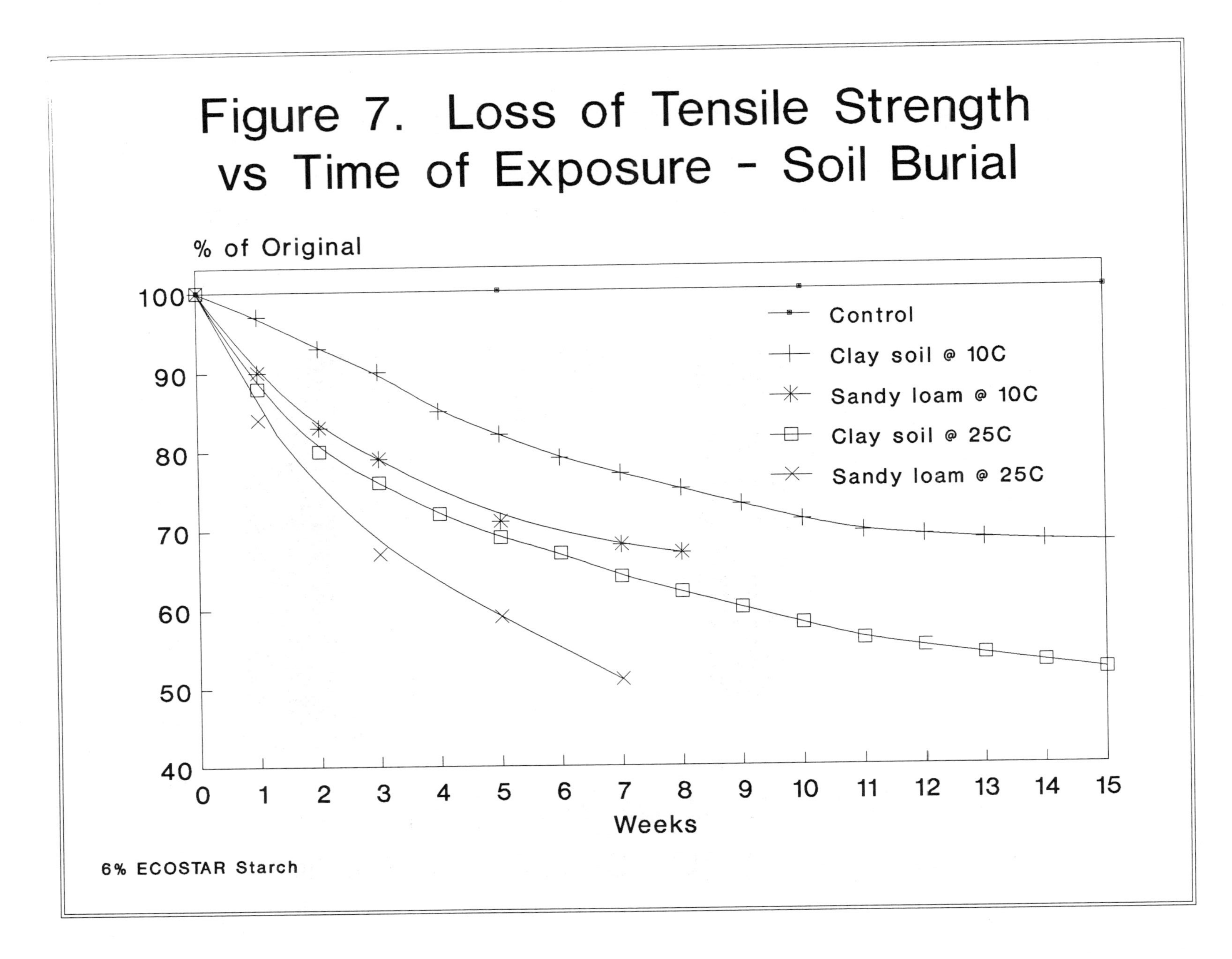


Figure 7. Loss of Tensile Strength vs Time of Exposure - Soil Burial

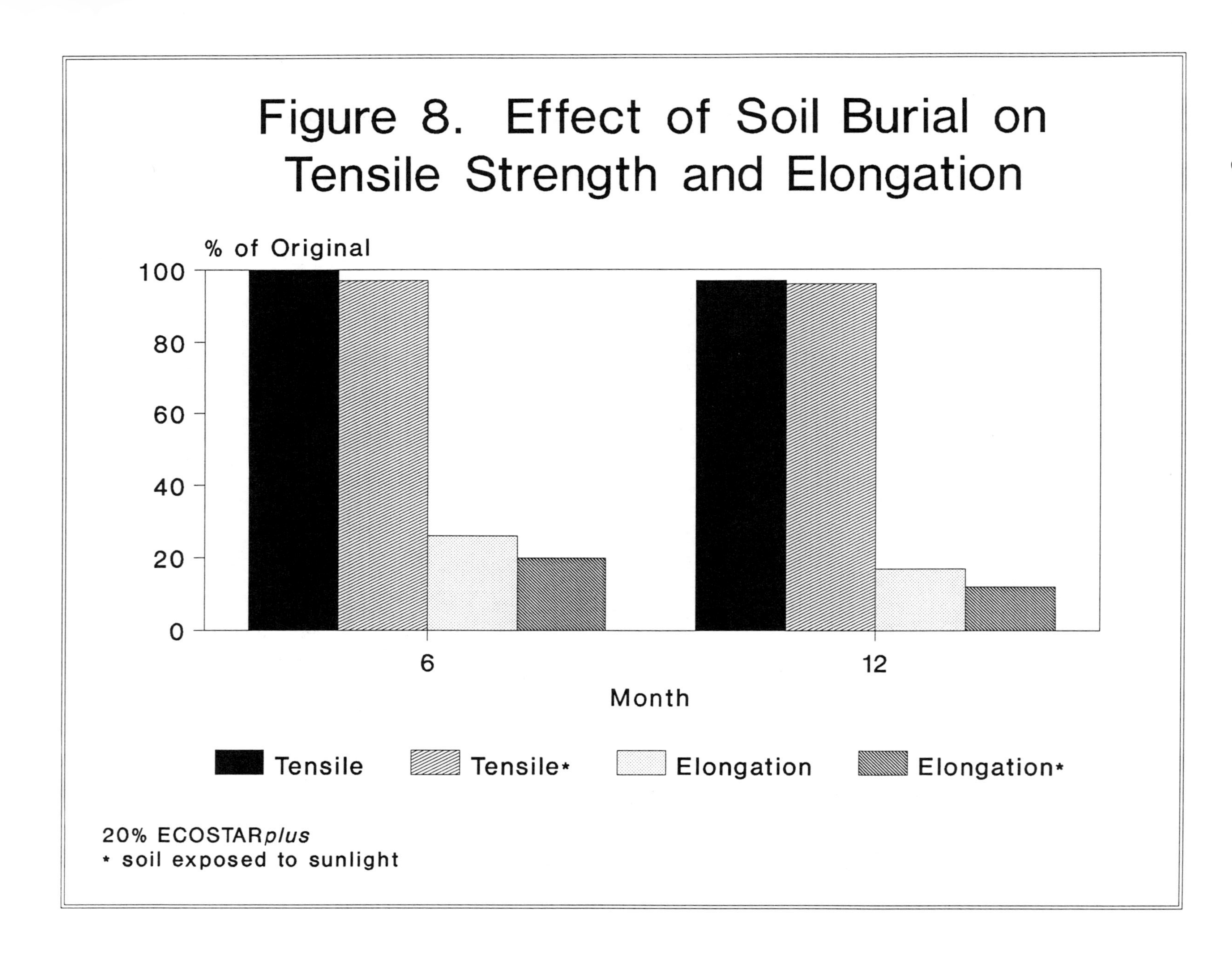


Figure 8. Effect of Soil Burial on Tensile Strength and Elongation

track degradation. Tensile strength remains virtually unchanged until elongation falls to low levels. The slight differences between the burial in shade and in sunlit areas are most likely due to small differences in temperature.

The photodegradation characteristics of the current system are shown in Figure 9 and are compared to those of the original system and a purely photodegradable system. While the photodegradation reaction likely proceeds by the normal mechanisms associated with organometallic systems, Figure 9 illustrates the apparent synergism provided by the presence of the autoxidant.

In addition to the above mechanisms it has been shown that macrodegradation of thin films can take place through attack by certain insects such as woodlice as shown in Figure 10[1].

FACTORS AFFECTING DEGRADATION

The effect of environmental factors such as temperature and biological activity have been demonstrated in Figures 3 to 9. Other factors which will affect the rate of degradation are:

moisture

pH

amount of active ingredients surface area and thickness of article

Other important factors which will affect degradation are the polymer type and the type and amount of antioxidant present in the polymer. It has been reported[7] that linear polymers are more susceptible to direct biological attack than branch polymers. This area has not been investigated to date with the subject systems,

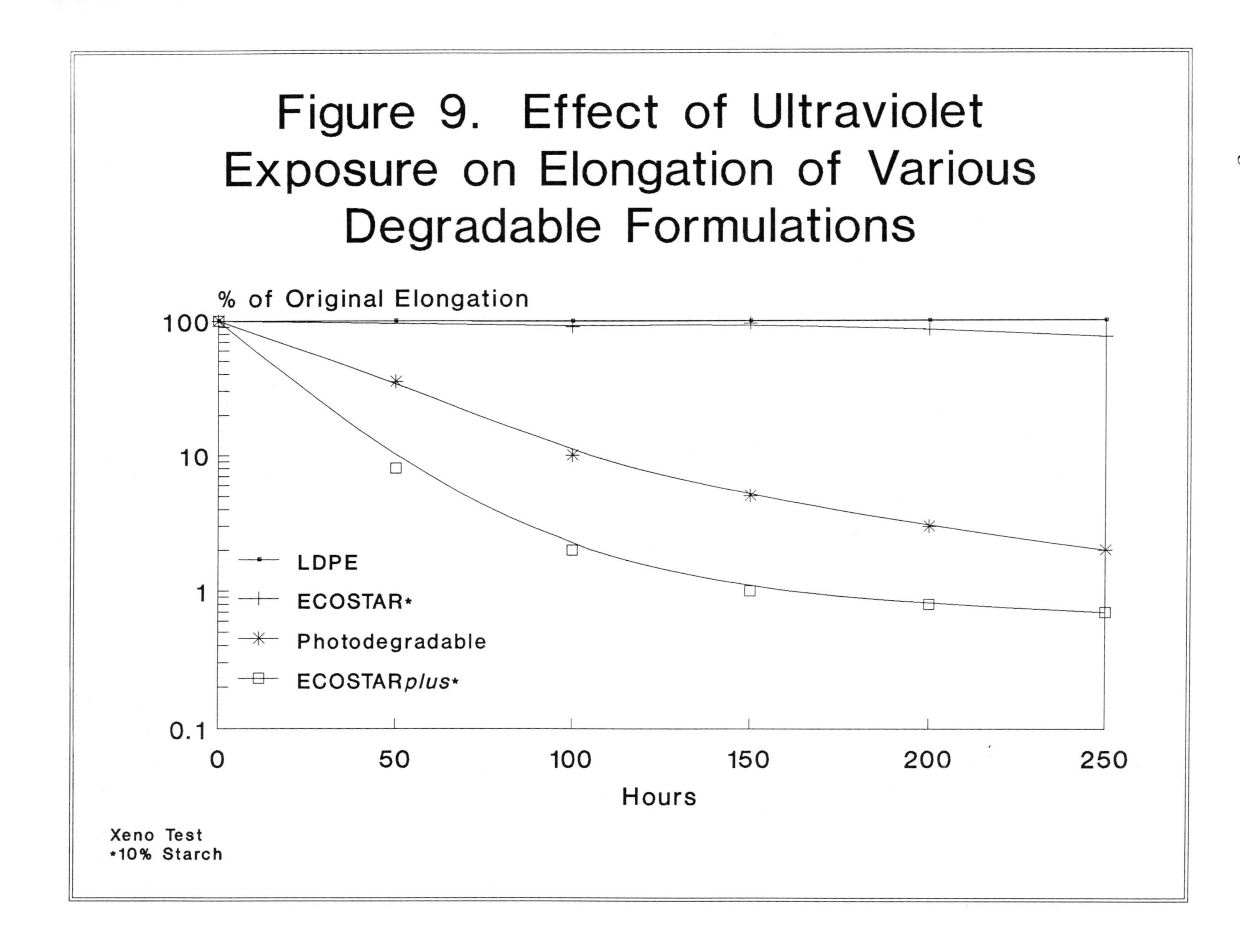

Figure 9. Effect of Ultraviolet
Exposure on Elongation of Various
Degradable Formulations
% of Original Elongation
100
10
1
0.1
0
50
100
150
200
250
Hours
LDPE
ECOSTAR*
Photodegradable
ECOSTARplus*
Xeno Test
*10% Starch

FIGURE 10. Conventional LDPE(left) and LDPE containing 6% starch(right) after 20 weeks exposure to compost.

however the effect of resin selection has been examined with respect to the oxidative degradation mechanism. One complication in the area however is the difficulty in obtaining examples of commercially available resins, particularly LLDPE and HDPE without antioxidants whereas "barefoot" resins are available in LDPE.

Figure II illustrates the difference in degradation rates with an equal amount of the same additive formulation in three resins, one LDPE without antioxidants and two LLDPE resins of different origin. Resins with antioxidants typically display an incubation period as shown in the graph. This indicates the importance of resin selection when designing a product where specific degradation rates are desired.

Figure 12 shows the comparative rates of degradation of a "barefoot" LDPE along with two examples of two different types of antioxidants added to the same base resin. Significant differences between the two families of antioxidants can be observed and lesser differences within the families.

SUMMARY

The mechanisms responsible for environmental degradation of modified starch filled plastics and the factors affecting those mechanisms have been described as currently understood. Oven aging has been shown to be a reproducible, effective accelerated test for predicing relative rates of environmental degradation for various formulations. The importance of resin and antioxidant selections in designing a product for environmental degradation have been

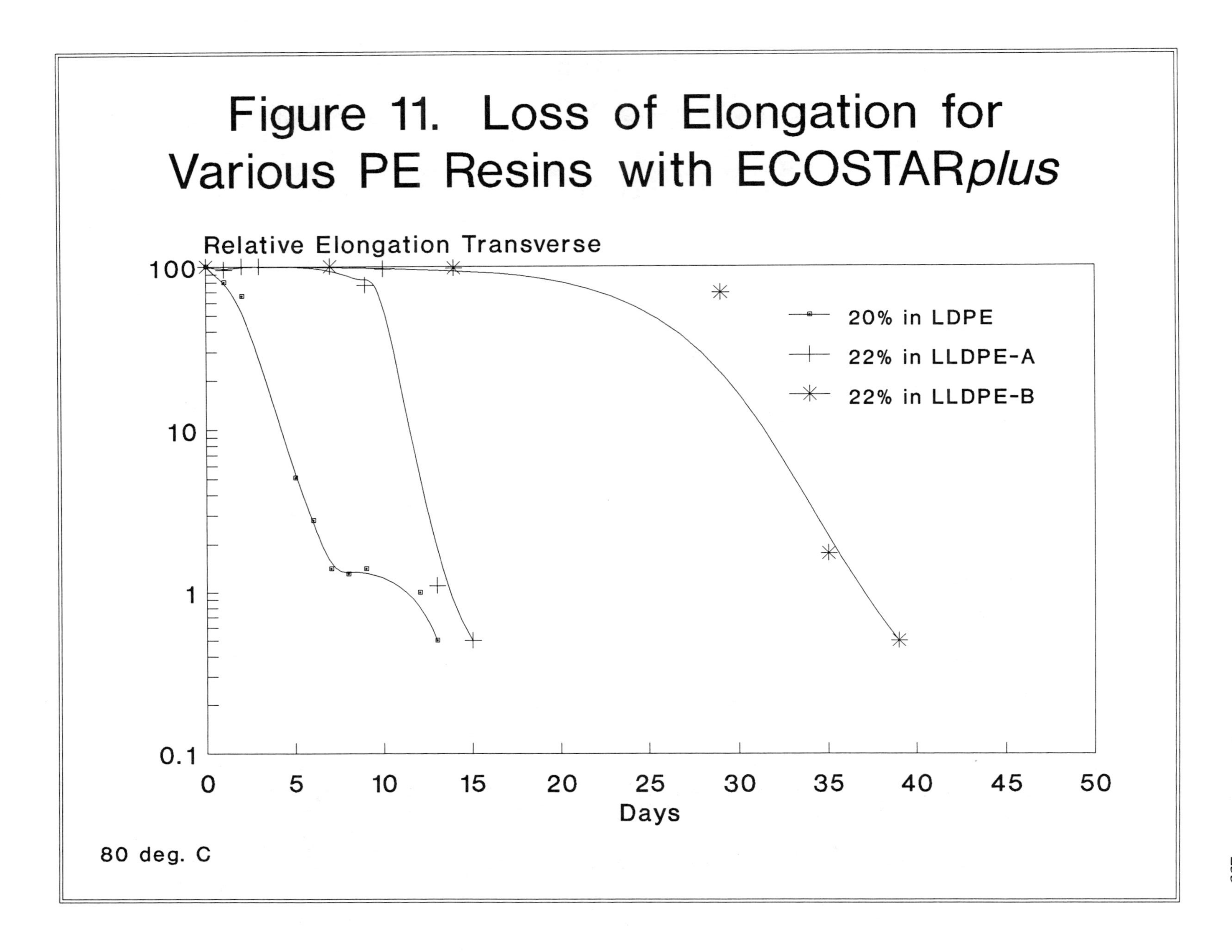


Figure 11. Loss of Elongation for Various PE Resins with ECOSTAR*plus*

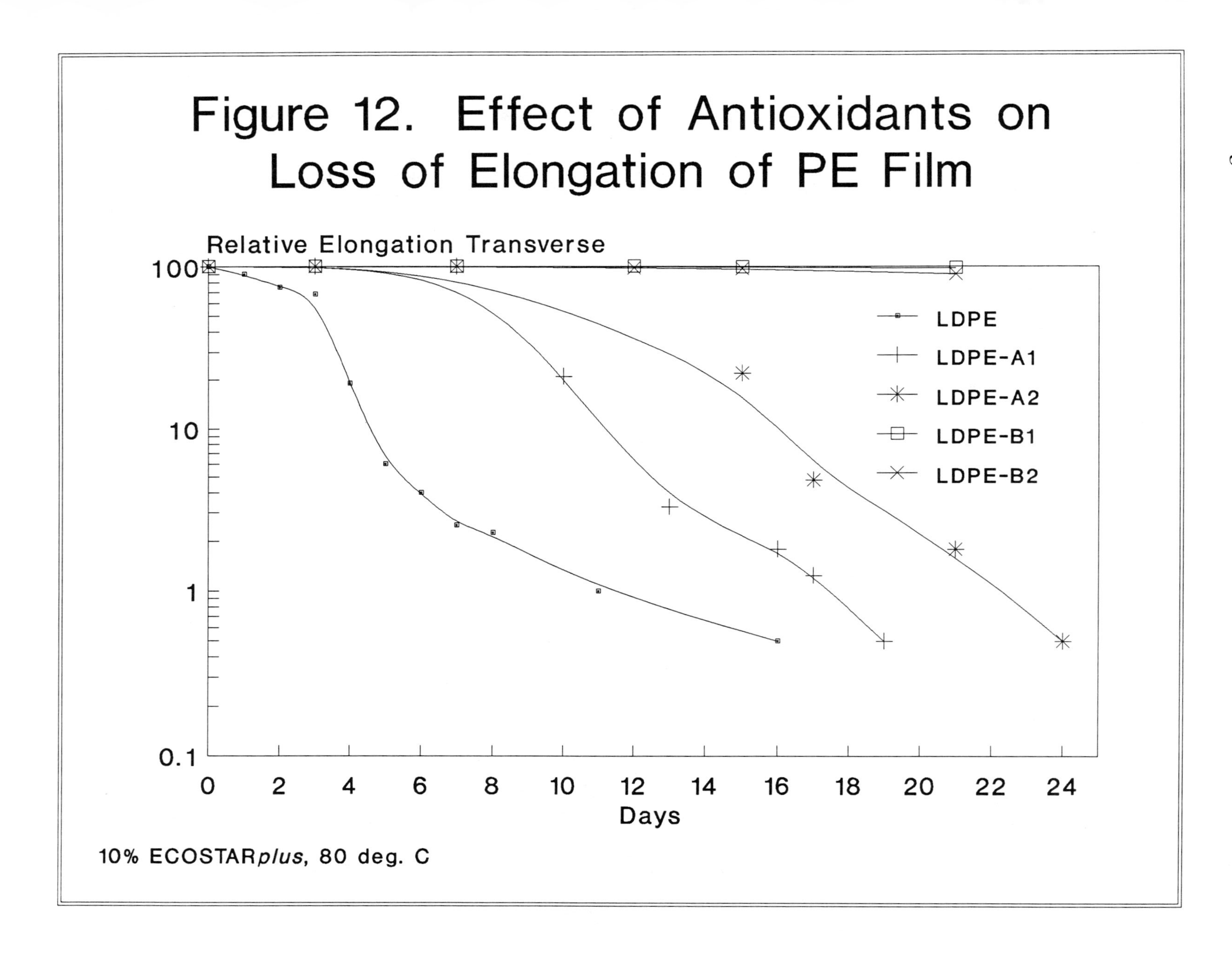
Figure 12. Effect of Antioxidants on
Loss of Elongation of PE Film
Relative Elongation Transverse
100
10
1
0.1
0
2
4
6
8
10
12
14
16
18
20
22
24
Days
LDPE
LDPE-A1
LDPE-A2
LDPE-B1
LDPE-B2
10% ECOSTAR*plus*, 80 deg. C

illustrated. Further investigations using techniques such as carbon 14 labelling in various model environments are necessary to gain a more complete understanding of the chemical/biochemical mechanism involved in the environmental degradation of enhanced degradable plastics.

REFERENCES

(1) U.S. Patent number 4021388

(2) Whitney, P.J. et al., IUPAC 1976

(3) Potts, J.E., et al., Polymer Preprints 13(2) 629,1972

(4) Albertsson, A.C., 8th Annual Conference on Advances in Stabilization and Controlled Degradation of Polymers

(5) Albertsson, A.C., et al., Polymer Degradation and Stability, 18 (1987). 73-87

(6) Tempesta, M. et al., American Chemical Society Meeting, Dallas, 1989

(7) Narayan, R., Degradapak, Chicago 1989.

QUESTIONS & ANSWERS

DR. ENNIS: Bob Ennis, Plastigone.

You showed one slide with a xenon accelerated weatherometer, and you put it up quickly. So correct me if I a wrong, I am a little confused. You showed that the addition of starch material to the plastic made the test results degrade faster, and I am a little bit confused how the addition of starch to your -- in the example you showed of the film could affect an accelerated weatherometer test, where there is no biological deterioration within the plastic.

DR. CAMPBELL: It's not the addition of starch that's important here. The formulation, our ECOSTAR formulation and our ECOSTARplus formulation both contain autooxidants, and those autooxidants can be UV initiated as well.

DR. ENNIS: I understand that. But you showed, on one of your graphs, you showed the plastic with the starch and without the starch had completely different levels of oxidation with the -- or loss of strength, with the one with the starch being accelerated over the one without the starch. And that's why I am a little bit confused as to how that --

DR. CAMPBELL: Well, the one without the starch doesn't have the autooxidant in it, either.

DR. ENNIS: I see. That wasn't clear on the slide.

The second question was, I notice most of your tests were carried out at 80 degrees Centigrade, which is fairly high temperature that one would expose -- would cause a very rapid oxidation of regular plastic under those conditions. I don't see how this compares with the situation under normal conditions in the environment.

DR. CAMPBELL: It's not trying to directly model the environment. It's an accelerated test to select degradable formulations that will perform in the environment. As shown by the comparison between the 80 degrees C and the 70 degrees C slide, there is a correlation that could follow the araneous relationship, so that you could project backwards down to 25 degrees C if you are interested in what's going on in a landfill situation.

DR. REDPATH: Tony Redpath, EcoPlastics.

Peter, a question for you. In general it's a question I think that we may ask everybody here to consider for the consensus sessions tomorrow.

You showed some slides that seemed to clearly indicate via molecular weight distributions that there are changes and molecular weight breakdown. Richard's talk just prior to you showed superimposable GPC curves. What's the difference in the tests? What are we missing? They are nominally -- you know, it's the same measurement, the same test designed to demonstrate the same point, and yet it's black and white in terms of the difference in answers. Can you tell me what the difference is in the tests, the materials? We are missing something.

DR. CAMPBELL: I think what we are missing is what is the rest of the polymer there, and what is in it? There is a great deal of polymer there that isn't the degradable additive, and it can make a very big discrepancy in the test, obviously.

DR. VERT: Michel Vert, University of Rouen.

From the picture you showed us, I got the feeling that starch granules were embedded within the film. They were covered by your film layer. Is that right?

DR. CAMPBELL: Yes.

DR. VERT: Could you tell me what rings the bell to the microorganism, to tell the microorganism that there is a starch granule, there is something to eat inside?

DR. CAMPBELL: Well, Jerry Griffin wrote a paper ten years ago showing that very thin films of polyethylene were -- starch through very thin films of polyethylene was accessible to enzymatic activity. So you might want to refer to that paper.

DR. VERT: But how can the microorganisms know there is something in there?

DR. NARAYAN: It attaches to it.

DR. VERT: Thank God.

Could it be that actually the starch is absorbing moisture through the very thin layer of the polymer --

DR. CAMPBELL: It could be.

DR. VERT: -- swells, cracks the film, and then the microorganism can come?

DR. CAMPBELL: That's quite possible, too. Those mechanisms are not very well understood and --

DR. VERT: Nobody had looked at that so far.

DR. CAMPBELL: That is one place that maybe we can direct some attention.

DR. VERT: Maybe you should do something in that direction.

DR. SHALABY: Shalaby Shalaby from Johnson & Johnson.

I have two quick questions. The first question deals with mechanical degradation. We heard this very briefly this morning, but I notice in all of your slides you have surface damage. And one of the requirements that we heard earlier is to have a stability of the film during storage.

In starch filled films, do you have any information on the mechanical aging of those films as you leave them for a year, not two to three weeks? Do they remain intact? Do they change in property as you leave them alone without any kind of abuse from light or biological environment?

DR. CAMPBELL: Under normal storage conditions we have seen no loss in properties.

With composting formulations, very high loadings of autooxidants and prodegradants, there is a shelf life on those products, because the autooxidation reaction at room temperature is rapid, and it tends to be somewhere around six months for a bag produced for that purpose.

DR. SHALABY: Okay, second question; one of the issues that has arisen during this meeting is selection of reference material. And you showed a slide where litter low density is different from low density. Do you have an idea what the branching, or any of the branching in any of the samples.

DR. CAMPBELL: Of the low density samples?

DR. SHALABY: All of them.

DR. CAMPBELL: No, I am sorry, I don't have any information on that.

DR. SHALABY: I could see a good correlation between the level of branching and the response to degradation.

DR. IANNOTTI: Gene Iannotti from Missouri. A couple of quick statements, a preview for tomorrow.

No. 1, bacteria are chemotactic. They can find and penetrate nonmaterial that they can't degrade to find material that is degradable.

I am going to show you some data tomorrow that same lab, same test --

DR. ANDERSON: Let me just add a comment to that. Bugs follow a concentration gradient which makes them chemotactic. Now are you referring to that, or are you referring to their level of mobility, which is chemokinesis?

There has to be a chemical gradient for those bacteria to move into that area. It's not magic. It's not a magnet.

DR. IANNOTTI: But it doesn't have to be a very large gradient for the microorganisms to find that.

DR. ANDERSON: That's the mechanism of chemotaxis.

DR. IANNOTTI: We are able to find heating plastic, 300 different compounds just in the plastics itself. Some of those that would be metabolized could lead them to the plastic, and then I think you've already seen evidence, somebody mentioned the swelling that goes on with penetration of water. I'll be talking about that more tomorrow.

DR. ANDERSON: That's fine. But the initial -- the initial finding has to be one of a random nature, unless there's something leaching out, which is creating for the initial contact.

DR. IANNOTTI: That could be. It's hard to say.

We have data from some of these plastics that can show that there is not degradation, and we have, I think, much more data that shows there is degradation.

I think we need to know a lot more, and I will point some of that out tomorrow about the mechanisms that are involved. We simply don't know beyond a certain point.

DR. CAMPBELL: The other thing is, you also have an active biological environment where enzymes are being secreted and present, and they can migrate to the starch particles and so on.

DR. ANDERSON: That's the point that he made.

DR. BAILEY: Let's see. Dick Austin was using ECOSTAR in his materials, and he wasn't getting any biodegradation. Do you agree with his results on the ECOSTAR?

DR. CAMPBELL: Well, without seeing any details of exactly what was done...

DR. BAILEY: Well, he wasn't getting any biodegrdation. Do you get any biodegradation at all with your ECOSTAR?

DR. CAMPBELL: I think it's been pretty conclusively shown that you will be biodegradation with anything. It's the relative rates of enhancement of the biodegradation that we are looking at.

DR. BAILEY: Well, but I mean over -- we are talking about over a reasonable length of time. He doesn't show any -- apparently any evidence of biodegradation, and you didn't get any change in molecular weight, you didn't get any change in properties of the polyethylene. I was just wondering if you agree with that, or if you have data that conflicts with that.

Now I realize the ECOSTARplus is a different system than he studies.

DR. CAMPBELL: Right.

DR. BAILEY: But the ECOSTAR is what he studies. I was just wondering if you -- if you had data that would disagree with his that --

DR. CAMPBELL: There is data, and it is in the literature. It has been published over many years from Jerry Griffin, the University of Surrey, University of Brunell, so --

DR. BAILEY: That data, at least you have gone a way, of course, in the ECOSTARplus from that system, and apparently you get a much bigger effect, anyway.

DR. CAMPBELL: Correct.

DR. NARAYAN: I think Mark can address that.

DR. MATLOCK: Mark Matlock with ADM.

I think the difference between the two experiments, and the reason you saw a difference in the molecular weights were, in the original ECOSTAR system, transition metals from the soil migrated in and worked with the unsaturation that was there to catalyze autooxidation and decrease molecular weight. If you have an inoculated biological sludge with a very low concentration of trace metals, you won't see that reaction catalyzed.

In the new formulation, I assume there are some transition metals, and that supplies the catalyst, so that you can get a molecular weight change in that -- in an environment where the transition metals are very low. ANd I think that was the difference between the two experiments.

DR ANDERSON: Thank you very much.

# BIODEGRADATION AND TEST METHODS FOR ENVIRONMENTAL AND BIOMEDICAL APPLICATIONS OF POLYMERS

Ann-Christine Albertsson and Sigbritt Karlsson
Department of Polymer Technology
The Royal Institute of Technology
S-100 44, Stockholm, Sweden

## I. INTRODUCTION

The word biodegradation has become popular these days. It is important to be able to control the life-time of polymers and also to be able to predict what will happen and when. If a material is biodegradable it is often expected to yield non-toxic waste products, but this is not always true. This is not even true for a native polymer. We have e.g. studied the degradation of casein by species of alkali-tolerant Clostridia and found degradation products such as histamine, agmatine, serotonine, tyramine, tryptamine, putrescine and cadaverine (1).

The term biodegradation has often been used without clear definition. In the field of sutures, bone reconstruction and drug delivery the term biodegradation may just be hydrolysis (2). On the other hand, for environmentally degradable plastics, the term biodegradation may mean fragmentation, loss of mechanical properties or sometimes degradation through the action of living organisms.

The general definition must cover all possible situations. The following definition is general and excludes other degradation modes:

> Transformation and Deterioration of Polymers Solely by Living Organisms (including Microorganisms and/or Enzymes Excreted by These)

We have in our work defined the biodegradation as the difference in degradation between a biotic and an abiotic environment (3). There is always a synergism between biodegradation in the biotic environment is observed.

It is possible to distinguish between pure biophysical effects (mechanical damage by swelling and bursting of growing cells), biochemical effects (excretion of different metabolites resulting in pH changes, changing redox conditions, enzyme enhancement, etc.) and true enzymatic effects (both exo- and endoenzymes, leading to enzymatic hydrolytic splitting or to oxidative breakdown of polymeric chains) (4).

Griffin has presented a very detailed differentiation between direct and indirect biodegradation and macrobiological degradation (5).

In natural environments there are a large number of different degradation modes, which synergistically drive to degrade polymers. The term biodegradation should, therefore, be used only when it is possible clearly to distinguish between the action of living organisms and other degradation modes (e.g. photolysis).

Reliable test methods are required to predict whether a polymer is biodegraded or only degraded. Several factors need to be taken into consideration before designing a test method. Table I shows some important factors which will be discussed in this paper. These factors must be determined before setting up biodegradation tests.

TABLE I

| Factor | Comments |
| --- | --- |
| sample | blown film - melted film - powder - film shape |
| properties | polydispersity, crystallinity and additives (GPC, DSC, IR) |
| performance | dynamic or static (stirring), loading |
| definition of biotic environments | fungi or bacteria or both |
| | mixture of strains |
| | inoculation |
| | nutrients |
| | humidity |
| | temperature |
| | petri-dish or nutrient solution |
| | description of growth |
| | description of weight loss |
| | hydrophobic - hydrophilic polymer |
| | microorganisms |

| definition of degradation | weight loss |
|---|---|
| | evolution of $^{14}CO_2$ |
| | changes in tensile strength |
| | changes in carbonyl index |
| degradation products | GC, HPLC, GPC |
| accelerated tests | pH, $O_2$-radicals, enzymes |
| | temperature |

A long tradition of research in the field of biodegradation has given useful information. The use of liquid scintillation countings of evolved carbon-14-dioxide ($^{14}CO_2$) from degraded polyethylene (PE) has been developed and has permitted the measurement of very low degradation rates (<0.1% by weight per year) (5,6).

The shape of the biodegradation curve is another factor discussed in some papers (7,8). In the present work, the biodegradation curve and that of growth of microorganisms are compared.

This paper presents an overview of methods for studying biodegradation. The importance of choosing the well characterized, relevant sample and right biotic environment as well as preparing relevant controls is also discussed. There are several instrumental techniques of importance in work on biodegradation but these are not treated in any depth.

## 2. TEST METHODS

According to <u>ASTM standard</u> a petri-dish test of a polymer's biodegradability is an appropriate initial routine test. Different microorganisms can be inoculated onto the polymer depending on the type of material. The ASTM standard specifies certain standard microorganisms (usually fungi) (9, 19) but it is important that the type of polymer is known so that the correct choice of microorganism can be made. After an appropraite period of time (4 weeks) the polymer shall be inspected for visual growth. It is also customary to calculate changes in polymer mass after the incubation.

Traditional methods for testing the biodegradability of a polymer can use some of the techniques listed below:

1. Visual inspection of mycelium growth on the polymer surface.
2. Quantitative estimation of microbial growth
3. Quantitative estimation of the weight loss of the polymer
4. Measurement of changes in polymer properties, e.g. tensile strength.
5. Measurement of the metabolic activity of the microorganisms by oxygen update or $CO_2$ evolution.

The possibility of evaluating metabolic activities in particularly important. This is also a standard method in microbiology, called respirometry. The decomposition of a polymer can be related to the metabolically evolved gases. Respirometry is one of the most sensitive methods for calculating the degradation of polymers. The term respiration implies a biological oxidation in which molecular oxyten is the ultimate hydrogen acceptor. The product, $CO_2$, is the most highly oxidized form of carbon and in the absence of other carbon sources in a test environment, this $CO_2$ will be a degradation product of the polymer. The major drawback of respirometry is the demand for labelled polymers ($^{14}C$) but if $^{14}C$-labelled monomers are at heand when a new polymer is synthesized the method is reliable and quite simple (5).

## 3. THE SHAPE OF THE SAMPLE

In ASTM standard and similar standards, the shape of the polymer test piece if often specified. For measurements such as tensile strength the standard shape is usually that shown in Fig. 1a and b, for polymer film the parallel-sided strip is recommended in the ASTM standard (Fig. 1c).

In biodegradation tests, however, other requirements must be fulfilled. The most important is to achieve good contact between polymer and microorganisms. The hydrophobicity of most polymers often creates a major obstacle to contact between polymer and microorganism. The addition of surfactants can promote good contact thereby a better biodegradation (3,11). The choice of surfactant needs some consideration. It should not be toxic to the microorganisms nor should it be able to function as a second substrate so that the surfactant is degraded and not the polymer.

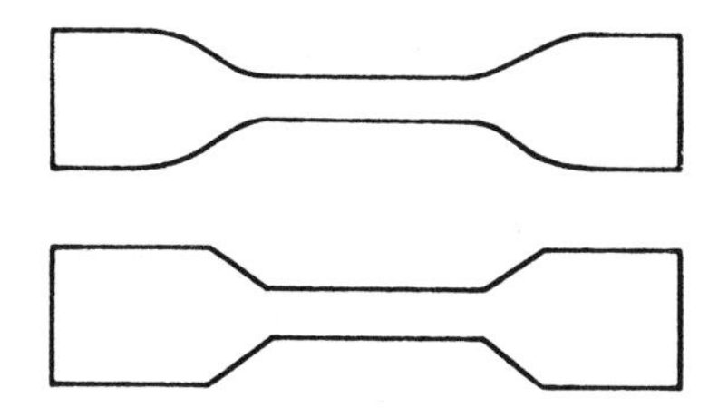

Fig. 1. Example of test piece for tensile strength measurements.
a) and b) broad- and narrow-waisted dumb-bell
c) parallel-sided strip

It is probably favorable to choose powder rather than film as a test sample. A much larger surface area is obtained and more short fragments of the polymer chain will be accessible to biodegradation (12).

There is also a difference between blown films and melted films with regard to surface as well as internal stresses. The properties of the sample are dependent on every detail in the sample preparation and history. Small details like residues or solvents, catalysts or antioxidants may prevent biodegradation and additives like lubricants, plasticizer or the low molecular part of the polymer may increase the rate of biodegradation (5,6,12).

Care must therefore be taken to ensure that the samples will characterize the polymer as well as all additives both before and during the degradation. Important information is the polymer structure, the morphology and the molecular weight. We have shown how the biodegradation is dependent on molecular weight, crystallinity, oxidation, mechanical stress and additives (3,13,14).

## 4. PERFORMANCE

There has been some discussion as to whether a dynamic or a static biodegradation system should be adopted (3). A statis system is of course much simpler and there is no influence of mechanical degradation on the total degradation. On the other hand stirring, would promote contact between polymer and microorganisms and prevents the polymer from floating at the surface in a nutrient solution. The ambition when designing a biodegradation system should also be to imitate nature as closely as possible. In natural surroundings there is always some kind of mechanical degradation due to the wind, the rain or animals. When the polymer is used as implant the system is also dynamic in many respects.

The effect of loading on the total degradation must be overlooked. When low density LDPE samples were mounted on a frame with a load at the lower part as in Fig. 2, several observations were made. The upper parts of the PE films were much more brittle and it was also here that the formation of crazes started (Fig. 3). The films were, however, mounted with two screws giving possibility of extra stress around the screw holes.

When samples were cut from the films (Fig. 4) for infrared spectroscopy (IR) the results showed that the upper sample contained a greater amount of carbonyl groups than the lower part, i.e. the upper part had been degraded more than the lower part.

## 5. DEFINITION OF BIOTIC ENVIRONMENT

The most important factor when designing biodegradation tests is the choice of biotic environment. The proper microorganisms must be chosen and the cultivation (degradation) environment needs special attention (12).

Microorganisms can only be activated with a favorable environment for their growth. The same applies for the biodegradation tests; if the microorganism is to be able to degrade a polymer the necessary nutrients must be available in a suitable form for use as building materials; a proper pH, a suitable temperature and the correct oxyten level must be maintained.

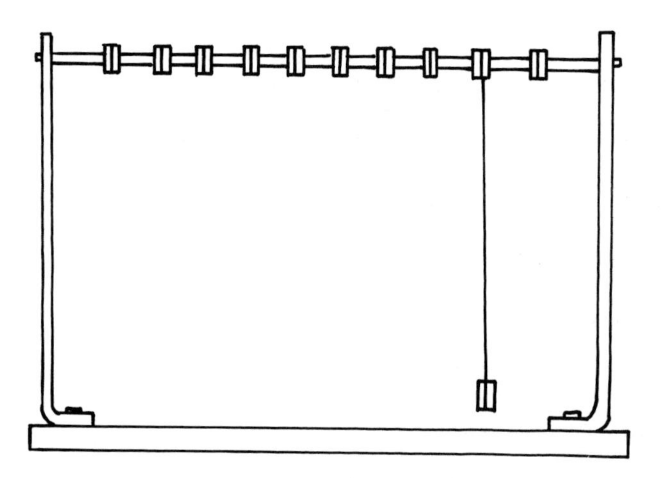

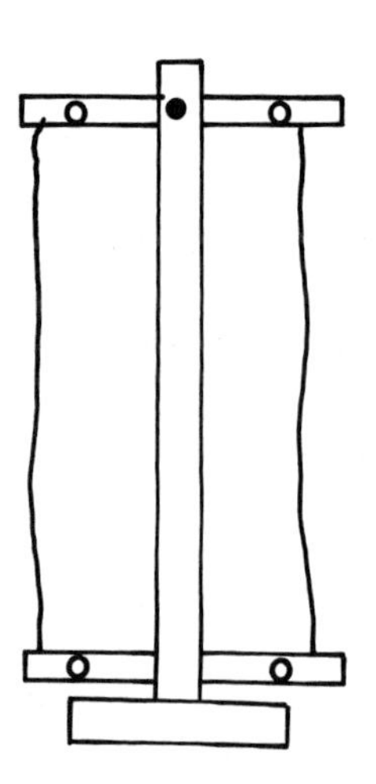

Fig. 2. Equipment for testing of loading effects on LDPE

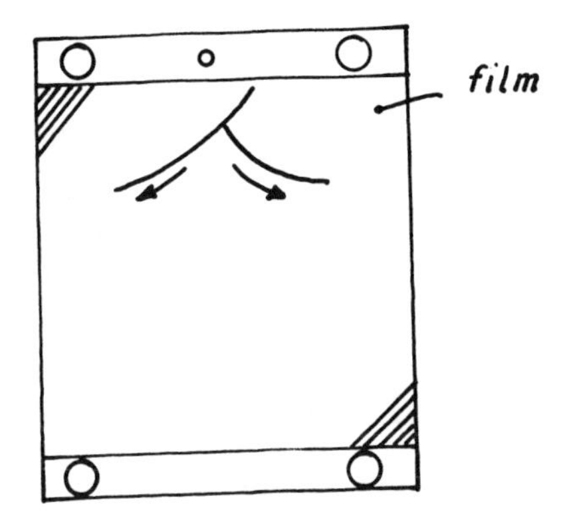


Fig. 3. Crazes in LDPE film after degradation

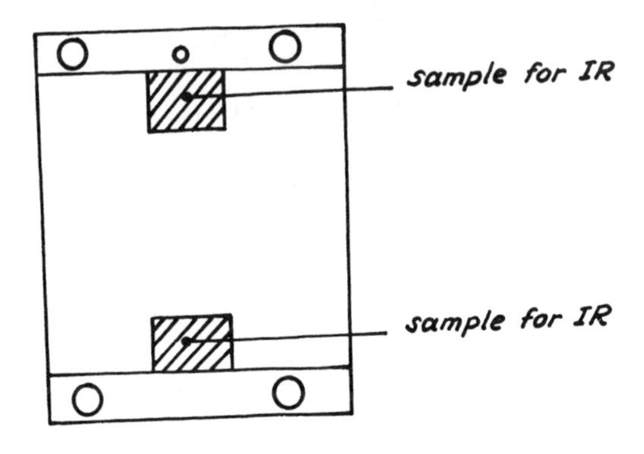


Fig. 4. Samples cut out for IR spectroscopy

On some instances it is more suitable to use pure enzymes than the microorganisms. Many enzymes capable of degrading macromolecules are endoenzymes, i.e. they can only degrade materials which are actually digested by the microorganisms. It is therefore easier to use the pure enzyme and to follow the degradation in a solution containing the polymer and the enzyme. Usually it is more difficult to handle enzymes than intact microorganisms. Enzymes are very susceptible to changes in e.g. temperature, pH and substrate composition. A minor change in, for example, pH is sufficient to render the enzyme inactive.

Pure cultures containing only one kind of microorganism are seldom found outside the laboratory. In natural habitats microorganisms are associated with other kinds of microorganisms. A mixed population is therefore an optimal choice for biodegradation tests. On the other hand, in in initial biodegradative tests on a polymer, a selected number of species should be used in order to properly determine the biodegradability of the polymer.

In general, either bacteria or fungi or both are used for biodegradation tests. The older ASTM standard often recommended a cellulose-degrading organism as reference microorganism for the study of biodegradation. This was not correct. Cellulose is a native polymer with a chemical structure quite different from that of most synthetic polymers. It would be surprising, for example, if a cellulose degrading organism would degrade polyethylene. Nowadays, however, the recommendations are different. Usually a fungus (e.g. Aureobasidium pullulans) and a bacaterium (e.g.

Pseudomonas* aeruginosa) are recommended for testing. The standards specify 4-5 different fungi and the same amount of bacteria for petri-dish testing.

A proper selection of inorganic salts is added as well as some sugar (glucose) to activate the microorganisms and initiate their division. The amount of starting glucose should not, however, be so large that the organisms can multiply and glow solely on this carbon source.

The bacterial growth curve, which describes how the bacteria multiply, can illustrate occurring with time in the biotic environment (Fig. 5).

During phase 1 the organism adapts to the new environment and starts to produce compounds. In stage 2, an exponential multiplication commences, and stage 3 shows a balance between new cells and dying cells. If nutrient and carbon source are limited, stage 4 is eventually reached. With easily accessible nutrients and carbon sources and an active pure culture of bacteria, stages 1 to 4 can last about 24 hours.

A much longer period of time (perhaps years) passes before stage 4 is reached in biodegradation tests where the carbon source is a polymer. Proper humidity and aseptic conditions must be maintained during the very long incubation times. This is

*Pseudomonas is especially noted for its nutritional versatility. Species can be found that are able to degrade starch, cellulose, agar, chitin, phenols, napthalene, hydrocarbons and resins.

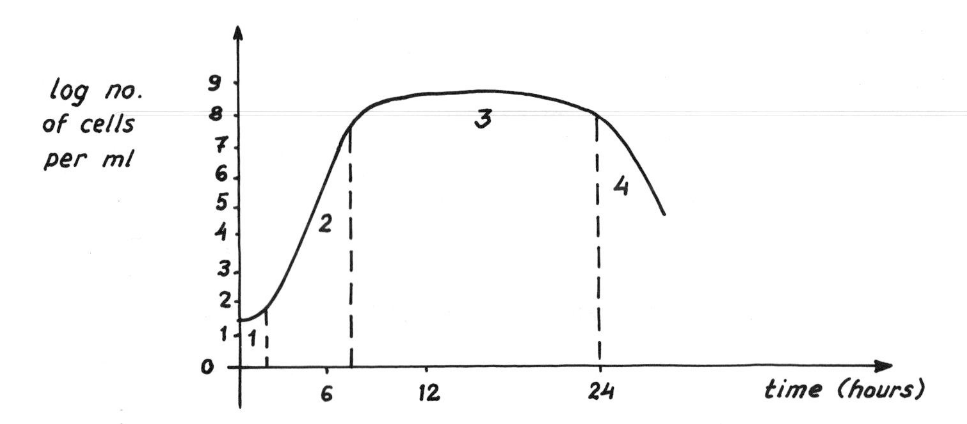


Fig. 5. A bacterial growth curve

1. lag-phase, an enlargement of cells
2. log-phase, exponential multiplication of cells
3. stationary phase
4. decline phase, exponential death of bacteria

especially important for inert polymers such as polyethylene. The microbial growth curve looks very like the biodegradation curve (Fig. 6). The percentage degradation (as evolved $CO_2$) is plotted against incubation time giving a series of curves showing the degradation of athe polymer (in this case LDPE) in abiotic and biotic environments (Fig. 6)(14).

The initial phase (1) cannot be observed, but an increase corresponding to phase 2 in the degradation can be observed. This increase presumably corresponds to the accumulation of weaknesses within the polymer chain which suddenly ruptures, giving a large amount of short-chain fragments accessible to attack by microorganisms. A stationary phase (phase 3 in the microbial growth cure) can also be observed. Phase 4 cannot be observed. Instead a sudden increase in degradation is again visible.

Some photographs of petri-dish incubated PE stribs are shown in Fig. 7.

Fig. 8. Shows examples of arragements for biodegradation tests in nutrient solutions.

In microbiology, growth is usually confirmed with a visual inspection of a petri-dish or nutrient solution shows bacateria or fungi. Absence of growth implies either a non-successful inoculation or that the organisms are not able to grow in the chosen environment (choice of substrate, pH, temperature etc.). For biodegradation tests it is also necessary to show for example weight loss of polymer or to show by microscopic inspection fungi hyphea growing on the polymer surface.

The hydrophobicity of most polymers is often a major obstacle to obtaining good biodegradation tests. In microbiology it is

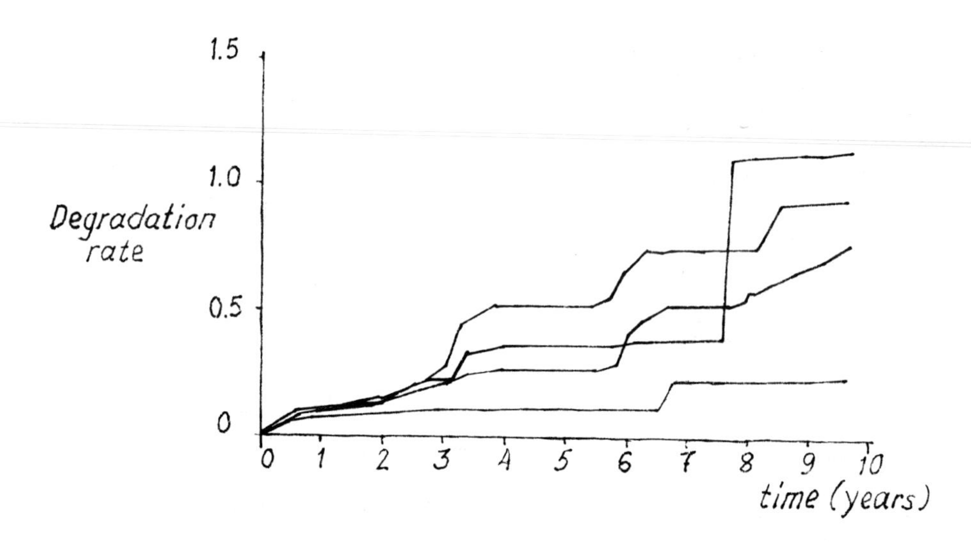


Fig. 6. A degradation curve as obtained by liquid scintillation counting

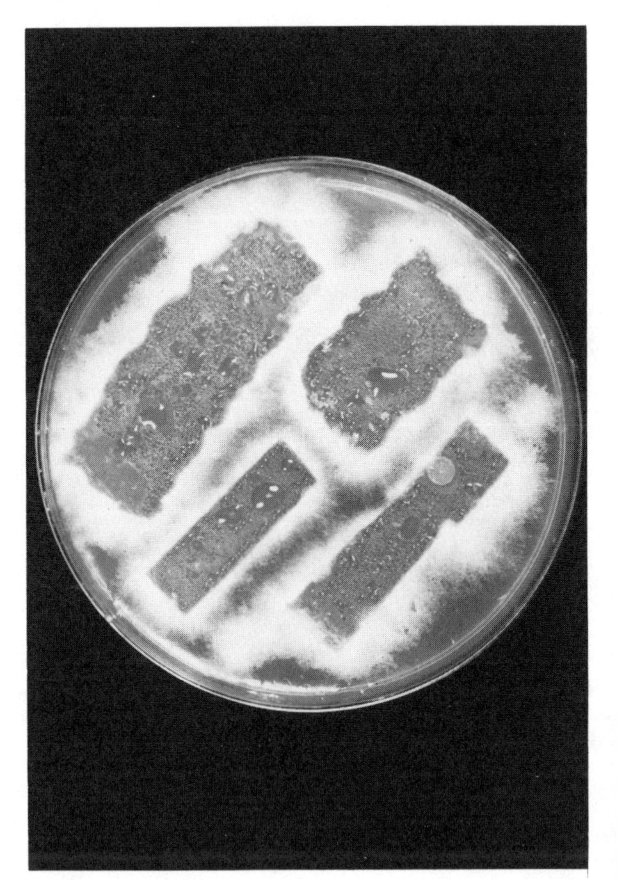

Fig. 7.a) PE incubated with fungi and

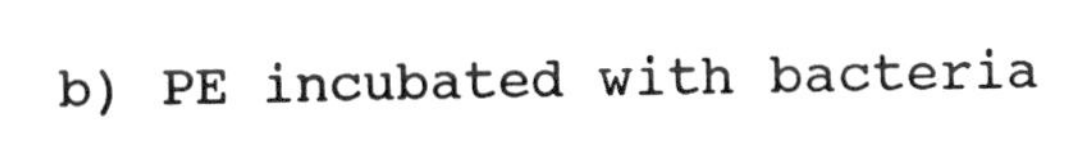

b) PE incubated with bacteria

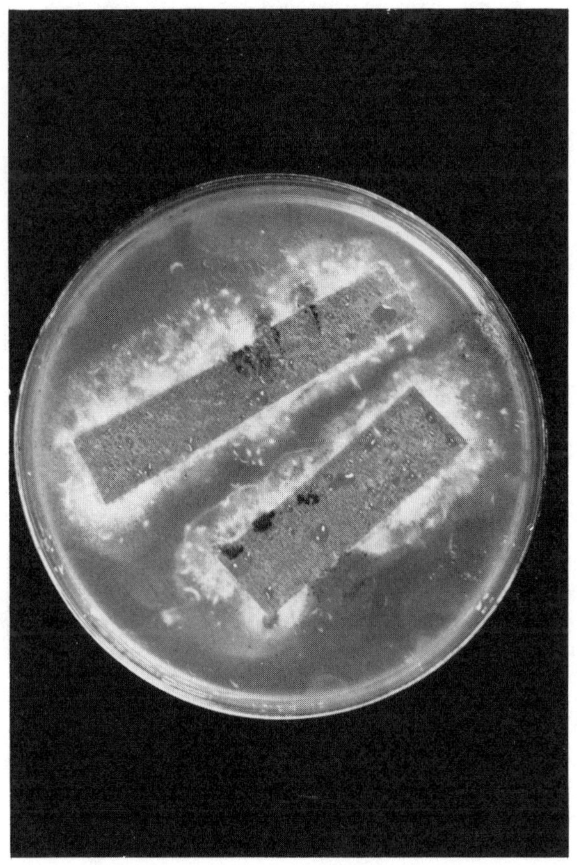

Fig. 8 Arrangements for biodegradation tests in nutrient solutions.

customary to include some surface-tension-diminishing compounds (TWEEN) to obtain good growth conditions. The microorganisms themselves also excrete surface tension diminishing compounds.

## 6. DEFINITION OF DEGRADATION

A series of analyses, among them instrumental ones, can be performed in order to evaluate degradation. Traditional polymer analysis techniques, not further discussed here, include gel permeation chromatography (GPC) for measuring changes in molecular weight, IR spectroscopy for measuring e.g. changes in carbonyl contents, differential scanning calorimetry (DSC) for measuring changes in crystallinity and melting temperature ($T_m$) etc. These techniques cannot, however, distinguish between pure biodegration and other modes of degradation (e.g. photooxidation).

Instead other more sophisticated methods must be used. Among these are liquid scintillation counting showing the evolution of $^{14}CO_2$ originating from the polymer.

In this context the importance of parallel control tests is emphasized. Only a proper set of samples including sterile test material can clearly give an answer to the question of whether or not a polymer is biodegradable.

## 7. ACCELERATED TESTS

Sometimes a quick answer to the question of biodegradabilityi is necessary. An accelerated test is then asked for.

For hydrolysable polymers (e.g. polyesters) it is quite simple. Incubation of polymer in water containing some salts with the pH adjusted to around 7 (20-30°) is usually the quickest way.

For polymers with a backbone structure of carbon, it might be relevant to use oxidation as a primary step and then biodegradation as a second step.

Another means of obtaining quick biodegradation is by using enzymes but, as discussed earlier, this is not always the most effective way and demands more careful performance than using the microorganisms themselves. The chosen test method is, however, dependent on the required properties in a special application.

For both environmental and biomedical use we need degradable polymer with non-toxic degradation products (15,16).

## 8. CONCLUSIONS

A definition of the term biodegradation has been given. The definition rules out all other degradative modes which always (in natural surroundings) synergistically strive to degrade macromolecules. Care must, however, be taken how to use the term biodegradation. In environmental degradation biodegradation is just one factor working synergistically with other environmental degradation modes, e.g. photooxidation, towards mineralization of polymer in nature. In medical applications where often the term biodegradable is used, it is better that there is no enzymatic degradation (i.e. biodegradation) but only hydrolysis. The degradation is often higher *in vivo* than *in vitro* probably because of the dynamics in the *in vivo*-systems.

A series of important factors must be considered before designing a biodegradation test. Among these are the choise of a well characterized sample, a relevant biotic environment and the performance of the tests.

The most important factor in use of biodegradation in environmental and biomedical applications is that the degradation products are non-toxic.

REFERENCES

1. S. Karlsson, Z.G. Banhidi and A-C. Albertsson: Detection by High Performance Liquid Chromatography of Polyamines Formed by Clostridial Putrefaction of Caseins, J. Chrom., 442,267(1988).

2. A-C. Albertsson and O. Ljungquist: Degradable Polymers. IV. Hydrolytic Degradation of Alipyhatic Thermoplastic Block Copolyesters. J. Macromol. Sci. Chem., A25,467(1988).

3. A-C. Albertsson: On the Biodegradation of Synthetic Polymers. II. Limited Microbial Conversion of $^{14}C$ in Polyethylene to $^{14}CO_2$ by Some Soil Fungi, J. Appl. Polym. Sci., 22,3419(1978).

4. A-C. Albertsson: The Synergism Between Biodegradation of Polyethylene and Environmental Factors. In Advances in Stabilization and Degradation of Polymers, Volume 1. A. Patsis Ed., Technomic Publishing Co., Lancaster, Pennsylvania, 1988.

5. G.J.L. Griffin: Synthetic Polymers and the Living Environment. Pure Appl. Chem., 52,399(1980).

6. A-C. Albertsson and Z.G. Banhidi: Microbial and Oxidative Effects in Degradation of Polyethylene. Radiorespirometrical. Measurements. J. Appl. Polym. Sci., 25,1655(1980).

7. A-C. Albertsson: The Shape of the Biodegradation Curve for Low and High Density Polyethylenes in Prolonged Experimental Series. Eur. Polym. J., 16,623(1980).

8. A-C. Albertsson and S. Karlsson: The three stages in the degradation of polymers-polyethylene as a model substance. J.
Appl. Polym. Sci., 35,1289(1988).
9. "Recommended Practice for Determining Resistance of Synthetic Polymeric Materials to Fungi". 1985 Annual Bood of ASTM Standards, Volume 8.03. ASTM-G-21.
10. "Recommended Practice for Determining Resistance of Plastics to Bacteria" 1985 Annual Book of ASTM Standards, Volume 8.03. ASTM-G-22.
11. S. Karlsson, O. Ljunquist and A-C. Albertsson: Biodegradation of Polyethylene and Influence of Surfactants. Polym. Degrad. Stabil., 21,237(1988).
12. A-C. Albertsson. Z.G. Banhidi and L.L. Beyer-Ericsson: On the Biodegradation of Synthetic Polymers. III. The Liberation of $^{14}CO_2$ by Moulds Like Fusarium Redolens from $^{14}C$ Labelled Pulverized High Density Polyethylene. J. Appl. Polym. Sci., 22,3434(1978).
13. A-C. Albertsson and S. Karlsson: Degradation of Polyethylene - studies in abiotic and biotic environment. Prog. Polym. Sci., 14,(1989) in press.
14. A-C. Albertsson and S. Karlsson: The Biodegradation of Light induced Degradation of polyethylene, in "The Controlled Degradation of Plastics", G. Gilead, Ed., in press.
15. A-C. Albertsson and S. Karlsson: Biodegradation of Polyethylene and Degradation Products. Accepted for

publication in "Waste treatment and Polymer Utilization".

16. S. Karlsson and A-C. Albertsson: GPC - A tool for studying the degradation of polymers. Int. GPC Symposium, Boston, Massachussetts, USA(1989).

QUESTIONS & ANSWERS

DR. NARAYAN: I've got a question for you. Could you tell, in all the experiments you have done so far, what is the best percent conversion of carbon from a polymer to $CO_2$ have you got?

DR. ALBERTSSON: You can go -- it is so wide a range, depending on for the first of the polymer and then all the factors that, I mean, if you use my polyanhydrides, they disappear sometimes so quickly, so my student couldn't show it for me. I run down to the laboratory, and it just disappeared.

DR. NARAYAN: Poyethylene I am talking about.

DR. ALBERTSSON: If you are talking about polyethylene, then it is very much dependent on how you make. If you go back to, as I showed you, this powder that you have nothing toxic in, I mean no antioxidant and no solvents and things like that.

And then first the degradation rate is rather slow, and then after while it will start to grow on the surface, and it will sink, and it will increase, depending, of course. But you can play with so many things. So I cannot -- I don't know the highest rate, but I will try to do that kind of experiment later.

You know, from the beginning I was not interested at all to make a degradation rate that was high. I did this just as a study to how does it work, and I used polyethylene as a model for these experiments, and I didn't work to make a fast degradation rate. But I have like 15 percent and things like that, and then I have only measured the part that have gone over to carbon dioxide.

But then some part of it is taken up, and I find in the fungi, like if you take the fungi, you will measure the carbon label, carbon in them. And if you take the nutrient solution, you also have a lot of degradation products in there that is not yet carbon dioxide.

But then some part of it is taken up, and I find in the fungi, like if you take the fungi, you will measure the carbon label, carbon in them. And if you take the nutrient solution, you also have a lot of degradation products in there that is not yet carbon dioxide.

DR. NARAYAN: Would you put this as one of the issues to be researched on, as what percent carbon goes where and what rates? Would that be a good issue for further studies?

DR. ALBERTSSON: I think many of us have these experiments going on. I mean, I have for many years not been doing so much experiments. I had a group that was working with many other things, but now I am increasing this part of the group that is working with this degradation products and how the degradation is going on, because the interest is now decreasing. I mean, it was just a few years ago they told me I was crazy to work with this.

DR. SHALABY: Shalaby Shalaby from Johnson & Johnson.

Among the few goals we'd like to achieve in this workshop is to identify some reference polymers we could use, and also identify media that you could standardize. I was asking you, would it be reasonable to recommend high density polyethylene, poly-e-caprolactone and polyhydroxybutyrate as among the reference to chose?

DR. ALBERTSSON: The problem to use reference material is it's not enough to recognize the reference material. The reference material, you have to recognize how you make it, when you make it, exactly the day, how you use it for every day. Because this is as always with polymer, you have to know all small details, and you can, with a small change --

DR. SHALABY: Actually this is square 1. Square 2, where you go to the -- through the film with such and such specifics. but as far as polymer, generically can you choose this type of material as references?

DR. ALBERTSSON: You can choose it as reference, if you remember all the differences that you have. I mean, you will not be able to repeat and get the same answer, if you make one sample one day and the next sample next day, and you treat them a little different. Because you are working with polymer. That is the most famous molecule we have. I mean, if you are working with small molecules, then you can define it like you did. But here the history is so important.

And that is what I have been working on very much. I mentioned here the factors. That in the beginning I worked with films melted and things like that. Now I try to get, molded films, and I get immediately a big difference.

DR. WOOL: I am Richard Wool, the University of Illinois.

I would like to ask a question about the use of C(14). If you go from C(12) to C(14), it has the potential of changing the chemical nature of that carbon atom.

DR. ALBERTSSON: Yes.

DR. WOOL: And we were quite shocked in the polymer physics community a few years ago, when we did some labeling of polystyrene and found the material became inadmissible.

The origin appears to lie with the enhanced and harmonicity of the bond that is created by going from C(12) to C(14).

In a reaction such as yours, which are occurring very slowly, do you think the additional weakening of the C(14) bond would give you a preferential evolution of C(14)? Do you or anyone else know that C(12) and C(14) have the same reaction rate?

DR. ALBERTSSON: What I have done is I have a -- but this is very long ago, and I don't remember about all the details. We made a lot of calculations, and we also made films with different amount of carbon 14 label on. I mean I had like two millicurie, 4 millicurie, 8 millicurie in films that I had, and I had compared the degradation rate of these films to show that they were the same, even if I had high amount or if I had low amount.

And I also did other kinds of calculations, because this is the first question you get when you do this kind of work. And I have had some discussion with Professor Scott, and also with Dr. Carlson. That is another one of these famous men working in this field, and you know, also his colleague, Dr. Weise, was at this conference 1975 and the rest of us. So I am sure that they can make some comments about this.

Everything I could find out, I tried to show the difference. And, as far as I know, it worked. But there is another thing here, and this is how do you make these films? How do you put in this carbon 14? Because therefore when I made the high density polyethylene film, they were made from the gas, and they were mixed. All the gas were mixed before we made it. So it was randomly in the chain, and hopefully randomly. Because there is next question? Do they react with the same?

So of course there is a difference. And what you can do is all the time make experiments with films that are not labeled and compare. But you have to do this.

DR. ENNIS: Bob Ennis from Plastigone. I think it's been shown that naturally occurring polymers will all degrade, although at quite a slow rate over the years.

I have heard the statement made at a number of meetings now that the addition of these certain additives to the polymers will accelerate the naturally occurring degradation process, but otherwise the process will essentially remain the same. Do you think that is a true statement? Or do you find that there is some fundamental change in the degree of degradation, depending on the additives?

DR. ALBERTSSON: First I have to excuse, it may be that I am becoming a little tired. I travelled 25 hours yesterday.

But what I have done is like if you have a polyethylene with an antioxidant, that will of course degrade slower than if you don't have the antioxidant. But the antioxidant will migrate out from the polymer. And after that you can show how the microorganisms start to grow.

You can't -- if you have other additives, you can increase the degradation rate and so forth, but I have never had any polyethylene without degradation.

DR. ENNIS: My question was whether the basic process of degradation remains the same and is only accelerated by the additives, or whether you think that there is a fundamental difference?

DR. ANDERSON: Let me ask you a question to clarify that question. By process do you mean kinetics and mechanism?

DR. ENNIS: I mean kinetics first, mechanism second.

DR. ALBERTSSON: You can see different kinetics mechanism here. The biodegradation is a little different than this aging effect. If you have two aging effects, they can be like a degration rate in a rather straight line. But the other effect, you will have a different shape in the curve.

And I have tried to discuss that in a paper from '87 about the mechanism, and I was in another paper, but I have changed my mind little. And you will next, or next year have a wider discussion about this.

DR. VERT: Michel Vert, University de Rouen.

Ann-Christine, the experience I have with the polymers, the polyesters which degrades by hydraulic degradation of the chains, is the following: When they are semicrystalline, crystalline, okay, it starts in the amorphous domain, goes rather fast and you get --

DR. ALBERTSSON: Sometimes.

DR. VERT: Usually it's like this. Then you get the crytallites, and the degradation slow down.

DR. ALBERTSSON: Yes.

DR. VERT: And at the level of the crystallites, you have a breakdown which occurs at the folding of the chains within the limallae. And then you would get oligomers, which corresponds to one length, two lengths or lengths of this limallae.

Did you observe the same thing in your poly-e-caprolactone first, and among the degradation products of the polyethylene, do you find some preferential lengths in the paraffin you get or in the fatty acid you get, anything like that?

DR. ALBERTSSON: I dare not answer that question now. Can I take it tomorrow?

DR. VERT: Tomorrow? Oh, fantastic.

# Long Term In Vitro Enzymatic Biodegradation of Pellethane 2363-80A A Mechanical Property Study

B.D. Angeline, A. Hiltner, J.M. Anderson

Department of Macromolecular Science And
Center for Applied Polymer Research
Case Western Reserve University
cleveland, Ohio 44106

Abstract

Tensile deformation and GPC experiments were initially employed to assess the biostability of Pellethane 2363-80A following in vitro exposure to aqueous and enzymic environments; due, however, to the relative insensitivity of these tests to damage accumulation, fatigue tests were also performed. Treatment durations were fourteen and twenty-eight days. The effect of enzyme concentration was investigated by treating samples at two different enzyme levels. For samples treated in the aqueous environments, no degradation was observed. Samples treated in the enzymic environments, however, possessed lower fatigue lifetimes and tensile strengths. GPC experiments indicated that substantial degradation had accurred at the high molecular weight end of the molecular weight distribution. The degradation was found to be dependent upon both treatment duration, and enzyme concentration.

## Introduction

Polyurethanes, particularly the elastomers, have gained widespread use as surgical implants. In the mid 1950's, rigid polyurethane foams were used as bone implants and bone adhesives; soft foams were used in such biomedical applications as valve suturing rings, catheter cuffs, and vascular grafts. Polyurethane elastomers found use in heart assist devices. Most attempts, however, with these early non-segmented polyurethanes failed in the harsh *in vivo* environment. Later work, though, revealed that segmented polyetherurethanes offered a considerable level of stability in the body, and as such proved to be invaluable for use in artificial heart devices, catheters, vascular grafts, and pacemaker lead insulation. (1-3) The advantage of these materials arises from their low in vivo reactivity, ability to be prepared as crosslinked or uncrosslinked elastomers, excellent flex life and long term mechanical properties, and the possibility for extensive variations in their chemistry, processing, and fabrication methods which allow them to be tailor-made to fit a particular need. (4-8)

Segmented polyurethanes are comprised of a hard and soft phase. The soft phase is made up of short elastomeric polyether segments, while the hard phase is a rigid glass-like backbone based upon a diisocyanate chain extended with a diamine, diol or water. Due to the chemical dissimilarity of these two structures, there is a thermodynamic driving force to form a heterophase microstructure in the solid state. The aggregated had segments serve as crosslinks or reinforcing fillers and prevent long range slippage of the elastomeric soft segments. (9) Upon elongation above 150%, the disordered soft segments stress-crystallize parallel to the axis of loading, and the hard segments align themselves in the direction of the strain; at elongations above 500%, hard segment domains are disrupted, forming paracrystalline domains, and the soft segments relax. Such hard segment disruption accompanied by an inability to reorder may lead to irrecoverable stress softening and subsequent hysteresis. (10-12)

The primary focus of this paper is to determine the effect of enzymic environments on the mechanical properties of Pellethane 80A. Upon implantation, surgical trauma and the presence of a polymer initiate and inflammatory response. Here cells release various lysosomal enzymes which may have a deleterious effect on the polymer. (13) These enzymes may catalyze degradation reactions of these polymers, leading to their premature failure. This degradation can be especially catastrophic upon repeated cyclical loading of the polymer, as for instance with artificial heart diaphragms. For this study, papain, a plant thiol endopeptidase, was chosen as the enzyme for the _in vitro_ simulation of the _in vivo_ environment. Papain is very similar in sturcture, homology, sequence, and function to Cathepsin-B, the msot active lysosomal proteinase in the body. Papain can hydrolyze amides, esters, and thiol estersl; it should thus be able to degrade the urethane linkages in Pellethane 80A. (14,15) Zhao and coworkers investigated the mechanical properties of Biomer following aqueous and enzyme exposure and demonstrated that aqueous degradation could be inhibited by ultraviolet stabilizers and anti-oxidants. These additives were not completely effective however against enzymic attack. Degradation was found to occur primarily on the surface; the ether linkages in this material were found to be particularly susceptible to attack. (16,17) Work by Phua et al. compared the effects of papain and urease environments on biomer, and found papain to be more effective in causing degradation; it was concluded that papain degrades the polymer by hydrolysing urethane and urea linkages. (18) Ratner et al. observed changes in the molecular weight and molecular weight distribution for Pellethane 80A following exposure to papain, alpha-chymotrypsin, leucine aminopeptidase, or water. (19) Stokes et al. described two mechanisms which may lead to premature failure of implanted Pellethane 80A pacemaker lead insulation: environmental stress cracking and metal catalyzed oxidation. (20) Glasmacher et al. concluded through tensile testing that Pellethane was biostable following long term immersion in papain environments. (21)

These previous studies suggest that polyetherurethanes are susceptible to enzymic attack and degradation. Such degradation could be especially catastrophic while the polymer is implanted under a load or state of stress. on this basis, we undertook this study to determine the effects of enzymic attack on the mechanical properties of Pellethane 2363-80A.

Materials Experimental

Stabilizer Pellethane 2363 80A pellets were synthesized by Dow Chemical, Granville, Ohio, and fabricated into thin ribbons by Bectin Dickinson Polymer Corp., Dayton, Ohio. Diphenylmethane-4,4' diisocyanate (MDI) was reacted with the prepolymer poly(tetramethylene oxide) and then chain extended with butanediol. To retard degradation, Santowhite (antioxidant) and Tinuvin 328 (UV stabilizer) were added. The Pellethane 80A pellets were then sheet extruded into a ribbon. The structure of Pellethane 80A is shown in figure 1.

The papain enzyme was obtained from the Sigma Chemical Company. The papain solution was twice crystallized, and was contained in a 0.05M sodium acetate suspension with .01% thymol present; its pH was 4.5. Its initial activity and decay rate were determined based upon Arnon's assay of amidase activity, Here, an activated papain solution is allowed to react with N-alpha-benzoyl-DL-arginine-4-nitroanilide hydrochloride (BAPNA). The papain cleaves an amide bond in this substrate liberating a colored species, p-nitroaniline. It is important to note that the papain must be in the presence of the activators EDTA and cysteine to bring about a reaction. The former acts as a heavy-metal binding agent and the latter as a mold reducing agent, and when present, these compounds free the sulfhydryl group necessary for catalytic activity. (15) For this assay method, one unit of enzyme activity is defined as the amount of enzyme that will hydrolyze 1 micromole of substrate per minute at 25°C ad pH 7.5. The specific activity is expressed as the unmber of units of activity per

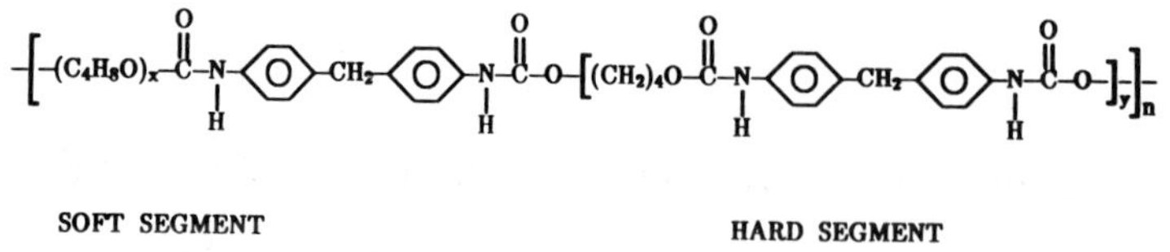

$$-\!\left[-(C_4H_8O)_x-\overset{\overset{O}{\|}}{C}-\underset{\underset{H}{|}}{N}-C_6H_4-CH_2-C_6H_4-\underset{\underset{H}{|}}{N}-\overset{\overset{O}{\|}}{C}-O-\left[(CH_2)_4O-\overset{\overset{O}{\|}}{C}-\underset{\underset{H}{|}}{N}-C_6H_4-CH_2-C_6H_4-\underset{\underset{H}{|}}{N}-\overset{\overset{O}{\|}}{C}-O-\right]_y\right]_n-$$

SOFT SEGMENT HARD SEGMENT

Figure 1. Chemical Structure of Poly(etherurethane)

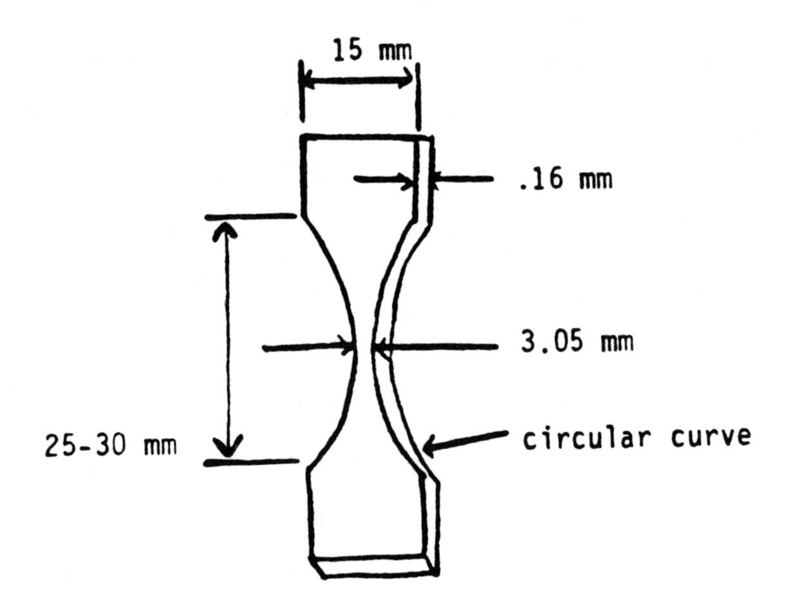


Figure 2. Sample Geometry of Polymer Samples for Mechanical Tests

microgram of protein. Using this method, the activity of papain was found to decrease linearly at a rate of 8.5% per day. The enzyme treatment solutions wre changed once every 48 hours to provide a relatively constant enzyme activity. The average specific activity of papain was 11.96+ 1.02 units (U) per microgram. The enzyme solutions used in the treatments contained 75,000 U/ml and 42,000 U/ml and were designated high and low concentration solutions respectively.

There were three sample treatments in all. The first treatment solution was made up of 0.05M cysteine and 0.02M EDTA in a pH 7.5 buffer solution. These solutions were also changed once every 48 hours. The next two treatment solutions contained the same concertrations of cysteine and EDTA in a pH 7.5 buffer solution, but also had papain added to yield the aforementioned high and low enzyme activities. To retard bacterial growth, sodium azide was added at a concentration of 0.02% w/v.

After the treatment solutions were prepared, they were transferred to covered glass petri dishes containing a strip of Pellethane 80A ribbon. The dimensions of the ribbon were approximately 60mm x 130mm x .2mm. The Pellethane 80A was then completely immersed in 60 ml of solution, and the dishes were placed in a 37°C oven until the solutions were changed again. Prior to changing the solutions, the Pellethane 80A was rinsed with distilled water for 1 minute. Two treatment durations were investigated, 2 weeks and 4 weeks.

After completion of the treatments, the Pellethane 80A ribbons were removed and cleaned prior to mechanical testing and GPC analysis. The samples were first sprayed for 1 minute with distilled water and then sonicated in a distilled water solution. Following this second sonication, the samples were rinsed in double distilled water for 3 minutes, placed on aluminum foil, and then left to dry in the dark for 2 days. The specimens were then further dried in an oven at 60°C for 20 minutes. The samples wre then labelled and wrapped in aluminum foil in a refrigerated desiccator until tested. SEM studies showed that these cleaning treatments did not alter the apparent surface topology or morphology of the specimens.

Mechanical Testing and GPC Analysis

Samples were cut out following treatments according to the geometry in figure 2. The sample geometry was narrowest in the center to prevent rupture at the grips.

For tensile testing, an Instron model 1123 was used. Samples were run in air at room temperature. The crosshead speed was 254 mm/min (1016%/min.). A 1000 lbs. load cell was used on the 0-20 lbs. full-scale setting. A minimum of five samples was tested for each sample type.

For fatigue testing, an Instron model TTC was used. Here, the specimens under went a zero-tension fatigue test; the maximum load was 20.0 MPa (based upon the initial cross-sectional area at the narrowest point). The crosshead speed was 508mm/min. (2032%/min.) yielding a frequency between 3 and 5 cycles per minute. Failure generally occurred between 2 and 4 hours. For fatigue tests, a 2000 g laod cell was employed. An average of 3 specimens was run per treatment type. A schematic of the loading scheme is shown in figure 3. The permanent set experienced by the polymer corresponds to the initial region on the strain axis where the load returns to zero at an higher strain than when it was loaded on the first half of the cycle.

Molecular weights were determined using a Waters GPC equipped with a model 510 pump, model 410 differential refractometer, and a Spectra-Physics model SP270 integrator. The ultrastyrene gel columns were maintained at 40°C, at a pressure of 700 psi and a flow rate of 1 ml/min. Pellethane 80A samples were dissolved in tetrahydrofuran to give a concentration by weight of 5%. The samples were allowed to dissolve for 48 hours at room temperature with mold agitation and were then filtered to remove any impurities.

For all mechanical tests performed, the Student t test for unpaired samples was used. A 95% level of confidence ($p<.05$) was used to determine the significance of the results. (22)

## Results and discussion

Table 1 shows the ultimate rupture stresses and displacements for each sample treatment. The only samples whose rupture stresses and displacements were statistically different from the untreated control were those treated in enzyme environments. Rupture stresses among the four enzyme treated samples did not differ significantly when compared to one another, suggesting that a single loading experiment

COMPARISON OF ULTIMATE TENSILE PROPERTIES AND FATIGUE LIFETIMES FOR TREATED AND UNTREATED PELLETHANE 80-A SAMPLES

TABLE I.

| SAMPLE | RUPTURE STRESS (MPA) N=6 | ELONGATION % N=6 | CYCLES TO FAILURE (N) AT 20 MPA, N=3 |
|---|---|---|---|
| 2-WEEK TREATMENTS | | | |
| UNTREATED | 68.9 ± 6.7 | 697 ± 47 | 495 ± 26 |
| ACTIVATORS | 61.0 ± 3.3 | 731 ± 17 | ——— |
| LOW ENZYME (42,000 U/ML) | 45.8 ± 4.3* | 600 ± 53* | 430 ± 11* |
| HIGH ENZYME (75,000 U/ML) | 47.9 ± 7.0* | 650 ± 72 | 375 ± 29* |
| 4-WEEK TREATMENTS | | | |
| DIST. WATER | 72.4 ± 3.9 | 644 ± 72 | 511 ± 17 |
| ACTIVATORS | 66.2 ± 9.8 | 661 ± 53 | 456 ± 21 |
| LOW ENZYME (42,000 U/ML) | 48.5 ± 9.0* | 610 ± 61* | 314 ± 12* |
| HIGH ENZYME (75,000 U/ML) | 46.6 ± 6.7* | 602 ± 40* | 265 ± 8* |

* DENOTES SIGNIFICANT DIFFERENCE FROM UNTREATED CONTROL AT THE 95% LEVEL OF CONFIDENCE ($p<0.05$) USING THE STUDENT T TEST FOR UNPAIRED SAMPLES.

was not sensitive enough to detect the varying levels of degradation. Fatigue tests were therefore employed since they are considered to be more sensitive to damage accumulation and would also better approximate in-use service conditions.

The samples treated in the aqueous environments showed no degradation, and in fact, had higher fatigue lifetimes and rupture stresses. Similar results were reported by Zhao in his Biomer studies, (16), and have been observed in our current studies with segmented polyurethanes (23). In previous studies, degradation was noted for unstabilized Biomer films. It has been proposed that such degradation is free radical in nature, and the incorporation of anti-oxidants and ultraviolet stabilizers act as free radical scavengers which retard degradation.(16) The papain treatments, however, may not have been influenced by stabilizers. All enzyme treated samples showed similar rupture elongations, though a slight decrease was noted for all samples exposed to enzyme. A decrease in rupture extension may be indicative of degradation of disruption of the soft elastomeric polyether segments.(24,25)

Upon inspection of the fatigue lifetimes in Table 1, the extent of chemical degradation and subsequent loss of biostability becomes more evident. As the amount of damage experienced by the polymer increases, its fatigue lifetime decreases as stress concentration arises at the sites of damage; such damage may be chemical in nature (i.e. broken chemical bonds), or physical (i.e. holes or pits). (26-28) The untreated and non-enzymic treatments showed comparable fatigue lifetimes, indicating an insignificant degree of degradation. The enzymic environments, however, showed a statistically significant level of degradation. The four week high enzyme treated sample showed about a 46% decrease in fatigue lifetime when compared to the untreated control. Similarly, the four week low enzyme treated sample showed approximately a 27% decrease in fatigue lifetime when compared to the untreated control. It also appears that the extnet of degradation was also dependent upon the treatment duration as well as concentration. All four week treatments at the same enzyme concentration. It is also worth noting that the sample experiencing the two week high enzyme treatment showed a 20% higher fatigue lifetime than the sample undergoing the four week low enzyme treatment. All samples failed by a parabolic crack front, characteristic of rubber experiencing fatigue failure. An attempt was made to measure crack propagation, but was unsuccessful due to the inability to accurately ascentain the time of crack initiation. Our current studies indicate that crack initiation time measurements may be more useful in assessing biostability than propagation time measurements.

Plots of the stress strain curves obtained for the various treatments are provided in Figures 4 and 5. Treatment types are abbreviated in these graphs (i.e. 2L corresponds to 2 weeks low enzyme). In Figure 4, the stress-displacement curves for the untreated and four week treatments are presented. In Figure 5, the stress-strain curves for all the enzymically treated sample are plotted. In comparing the stress-strain curves for the enzyme treated samples, it becomes apparent that the curves are almost identical. These curves re-emphasize the need for fatigue tests to differentiate the effects of treatment time and concentration since

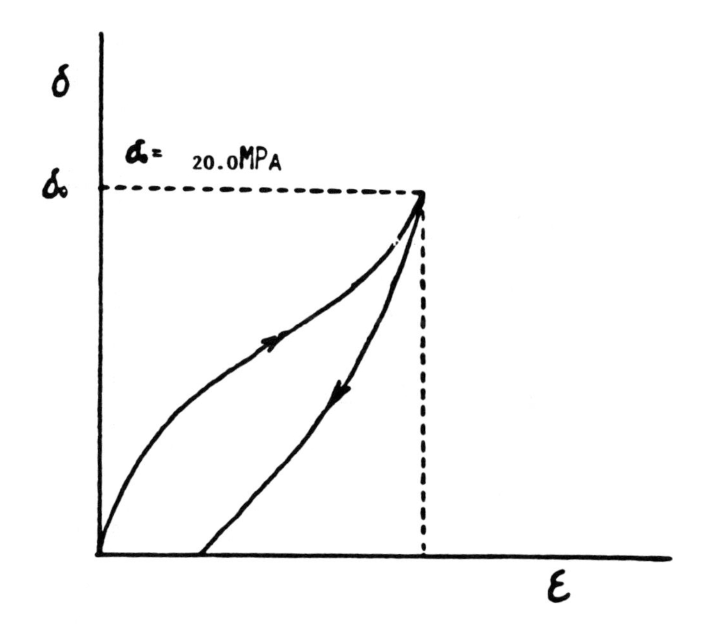


Figure 3. Schematic of Fatigue Loading Program

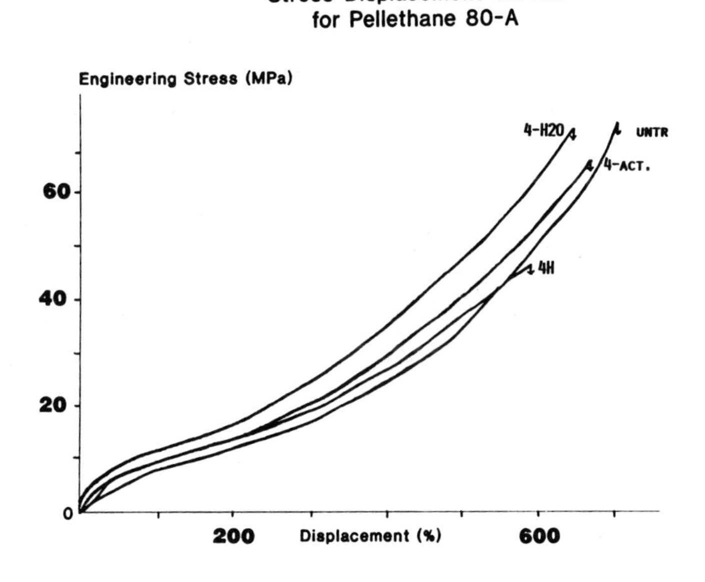


Figure 4. Stress-strain curves for Treated and Untreated Pellethane 80A

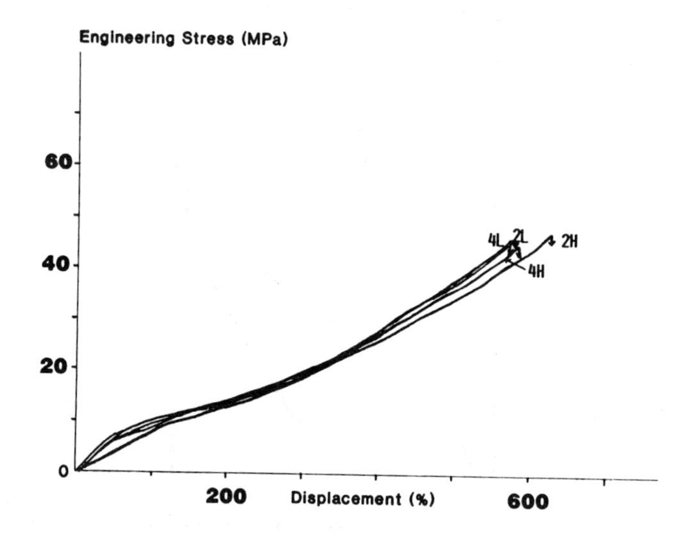


Figure 5. Stress-Strain Curves for Enzyme Treated Pellethane 80A

## GPC Results for Treated and Untreated Pellathane 80-A

Table II.

| Specimen | $M_N$ | $M_W$ | $M_Z$ | Polydispersity |
|---|---|---|---|---|
| Untreated | 30,000 | 93,000 | 2,180,000 | 3.15 |
| 4 wks. Activated (0.05 M cysteine, 0.02 M EDTA) | 31,000 | 97,000 | 2,190,000 | 3.18 |
| 2 wks. High Enzyme (75,000 U/ml) | 29,000 | 84,000 | 820,000 | 2.88 |
| 4 wks. High Enzyme (75,000 U/ml) | 33,000 | 77,000 | 140,000 | 2.34 |

these tensile tests are apparently for less sensitive to damage.

Additional evidence of chemical damage was provided by GPC experiments. In comparing the GPC results in Table 2, a very slight increase in molecular weight and polydispersity is observed for the activating agent treatment. The enzyme treated samples showed comparable values for the number-average molecular weight when compared to the untreated control, but a substantial decrease in the weight average and Z-average molecular weights were observed; a decrease in polydispersity was also obtained. As treatment duration increased, the molecular weights and polydispersities decreased. Since the weight and Z-average molecular weights are more sensitive to the higher end of the molecular weight distribution, and since these values are substantially lower than the control, the damage done to the polymer occurred primarily at the higher end of the molecular weight distribution. Apparently, as the papain attacked the Pellethane 80A, long chains were degraded into shorter chains, causing a corresponding decrease in polydispersity. The GPC curves, in fact, did show a diminution of the high end of the molecular weight distribution. The peak in the GPC curves, which corresponds to the number average molecular weight stayed approximately the same for each sample. The decreased mechanical performance for enzyme treated samples could be explained in part by this chemical degradation, since mechanical properties are especially dependent upon the higher molecular weight chains in the molecular weight distribution.

In conclusion, Pellethane 80A was biostable in aqueous environments at 37°C for the two and four week treatment times investigated. Pellethane 80A was found to be unstable in enzymic environments at 37°C. Single loading tensile tests did not reveal significant difference in mechanical properties following in vitro enzyme treatment, while cyclic stress fatigue tests and GPC showed a decrease in properties following enzyme treatments. The level of degradation was found to be dependent upon both enzyme concertration and treatment duration.

Acknowledgement: This work was supported by the National Institute of Health, Grant HL-25239.

References

1. Williams, D.F., Biocompatibility of Clinical Implants, CRC Press, Boca Raton, Fla., 1981, 2, 128.
2. Lee, H. and Neville, K., Handbook of Biomedical Plastics, Pasadena Technology Press, Pasadena, Calif., 1971, 14-21.
3. Lelah, D.L., Cooper, S.L., Polyurethanes in Medicine, CRC Press, Boca Raton, Fla., 1986, 159-181.
4. Wilkes, G.L., Polymers in Science and Technolgy, vol. 8,1975, "Polymers in Medicine and Surger," Plenum Press, N.Y., 45-75.
5. Wilkes, G.L., Dziemianowica, T.S., Opher, Z.H., and Artiz, E., "Thermally Induced Time Dependence of Mechanical Properties in Biomedical Grade Polyurethanes," J. Biomed. Mater. Res., 1979, 13, 189.
6. Boretos, J.W. Detmer, D.E., Donachy, J.H., "Segmented Polyurethane: A Polyether Polymer II. Two Years Experience," J. Biomed. Mater. Res., 1971, 5, 373.

7. Wilkes, G.L., and Wildnauer, R., "Kinetic Behavior of the Thermal and Mechanical Properties of Segmented Polyurethanes, "J. Appl. Phys., 1975, 46(10), 4148.

8. Takara, A., Tashita, J., Kajiyama, T., and Takayanagi, M., "Effect of the Aggregation State of Hard Segmented Poly(urethaneurea) on Their Fatigue Behavior After Interaction with Blood Components," J. Biomed. Mater. Res., 1985, 19, 13.

9. Coury, A.J., "Factors and Interactions Affecting the Performance of Polyurethane Elastomers in Medical Devices," J. Biomat. Applications, 3(2), Oct. 1988, 130.

10 Estes, G.M., Seymour, R.W., Huh, D.S., and Cooper, S.L., "Mechanical and Optical Properties of Block Polymers, I. Polyester urethane," Poly. Eng. Sci., 1969, 9(6), 383.

11. Fischer, E., and Henderson, J.F., "Effect of Temperature on Stress-Optical Properties of Styrene Butadiene Block Copolymers," Rubber Chem. Technol., 1967, 40(5), 1373.

12. Bonart, R., "X-Ray Investigations Concerning the Physical Structure of Cross-linking in Segmented Urethane Elastomers," J. Macromol. Sci. Phys., B2, 1968, 115.

13. Robbins, S.L., Cortan R.S., Pathological Basis of Disease, Saunders Pub., Philadelphia, Pa., 1979, 55.

14. Gray, C.J., Enzyme Catalyzed Reactions, Van Nostran Reinhold Co., London, 1971, ch. 4.

15. Arnon, R., "Papain," Methods in Enzymology, XIX, 1977, 226.

16. Zhao, Q., Marchant, R.E., Anderson, J.M., Hiltner, A., "Long Term Biodegradation in Vitro of Poly(ether urethane urea): A Mechanical Property Study," Polymer, 28, 1987, 2040.

17. Marchant, R.e., Zhao, Q., Anderson, J.M., Hiltner, A., "Degradation of a Poly(ether urethane urea) Elastomer: Infra-red and XPS Studies," Polymer, 28, 1987, 2032.

18. Phua, K., Castillo, E., Anderson, J.M., Hiltner, A., "Biodegradation of a Polyurethane Elastomer in Vitro," J. Biomed. Mater. Res., 21, 1987, 231.

19. Ratner, B.D., Gladhill, K.W., Horbett, T.A., "Analysis of in Vitro Enzymatic and Oxidative Degradation of Polyurethanes," J. Biomed. Mater. Res., 22, 1988, 509.

20. Stokes, K., Urbanski, P., Cobian, K., Polyurethanes in Biomedical Engineering II, Elsevier Pub., Amsterdam, 1987, 109.

21. Glasmacher-Seiler, B., Erens, B., Doersch, M., Reul, G., Wick, G., Pietsch, H., Abstracts of the Eighth European Conference on Biomaterials, Society of Biomaterials, Heidelberg, FRG, Sept. 7-9, 1989, Abstract A5-1, 54.

22. Fraser, D.A.S., Probability and Statistics, Dunbury Press, North Scituate, Mass., 1976, 247.

23. Upublished results

24. Caspary, R., and Schnecko, H.W., "Craze Formation in Highly Stretched Polyurethane Elastomers," Makromol. Chem., 182, 2109.

25. Priss, L.S., Tyulenev, A.I., and Vishniakov, I.I., "Determination of the Degree of Crystallinity of Elastomers Under Stretching," Polym. Sci. U.S.S.R., 1982, 24(1), 59.

26. Higdon, A., Ohlsen, E.H., and Stiles, W.B., Mechanics of Materials, John Wiley and Sons, New York, N.Y., 1960, 378.

27. Brown, L., Arnold, R., Yahtzee, Y.G., Strength of Materials, Piskin and Boemker Publ., New York, N.Y., 1947, 264.

28. Singer, F.L., Strength of Materials, Harper and Broth, New York, N.Y., 1951, 378.

QUESTIONS AND ANSWERS

DR. FEIJEN: I ahve one question, Jim. I agree completely that I think we show test molecular weight, molecular weight distribution, which is an excellent method to see whether something is going on.

Do you have an explanation for the preferred attack of your enzyme on the higher molecular weight molecules?

DR. ANDERSON: We don't. We don't

DR. FEIJEN: I cannot explain it. Why should they cut more in a high molecular weight than in a lower up to a certain level?

DR. ANDERSON: I don't have an explanation for it.

DR. FEIJEN: In this type of polyurethane, do you have any treatment before you have used it?

DR. ANDERSON: Oh, these are commercial polyurethanes, which do have additive packages. We know that they have Saniwhite and Tinuven in them. We know that they have been extruded with amino extruding aids.

they are the stuff that's going out there. These -- we call them model materials. We can appreciate some of the conditions under which they have been prepared, formulated but we certainly don't understand all of them. And I don't have an explanation for the preferreds high molecular weight cut, but that needs a look at.

DR. SHALABY: Shalaby, Johnson & Johnson.

You give us a good example how you can disturb the biological environment and don't get what you expect.

DR. ANDERSON: And don't what?

DR. SHALABY: Don't get what you expect. You did not expect to see that the tissue reaction would be parrallel to increased stress cracking.

This will take us actually to some of the discussion we had earlier about using the microorganism and measuring or monitoring biodegradation. We had no control on how viable or how stable the system we are using was.

DR. ANDERSON: Yes.

DR. SHALABY: You could start with certain microflora count, and end up with half of it or perhaps none at all. This is one of the things that we should be talking about tomorrow, that you don't have a constant bath, not a variable bath.

DR. ANDERSON: Let me say that in our in vivo system, we are able to sample the exudate, so we know what the cell concentration is in a temporal sequence with time.

With the in vitro system, we change our papain solutions every two days. We know that they drop, per 24 hours, by 11 per cent. So we are about at 80 per cent when we pour out the solution and put new in. We don't want to drop any lower than that. But we do monitor enzyme activity with time as well as -- in the in vivo situation, cellular concentration.

DR. SHALABY: My point is when you go to the biodegradation of plastics, using sludge or soil, we should really be careful how constant the bath is when you work with it.

DR. GREISLER: Howard Greisler, Loyola University in Chicago. That first study was very interesting, Jim, and I certainly agree with you that the mere presence of a macrophage population does not necessarily relate to biodegradation, unless the macrophages have been activated. So I have two questions about the question of cause and effect as related to the macrophage in terms of the pelathane cracking.

The first is do you have evidence that the macrophage is actually facsitizing and then becoming activated by the pelathane?

And the second question --

DR. ANDERSON: Let me answer the first question.

DR. GREISLER: Okay.

DR. ANDERSON: These pelathane specimens are so large that the process of the cell eating them never occurs. The cell sits on the surface. The surface is infinite in size, compared to the cell.

So the cell sits on the surface, and then undergoes some sort of activation process, which I presume is intrinsic to the chemistry at the surface of the material and the proteins that may be on it. But we don't totally understand that.

We do know that the cell does release the constituents once it's adhered.

DR. GREISLER: Which constituents have you measured?

DR. ANDERSON: We have looked at, in general, we have looked at alkaline phosphatase, acid phosphatase, peptidase, superoxide anion, and we are continuing -- Interluken (phonetic), and we are continuing to look at a wide variety of agents that can control other biologic processes.

But we are also interested in those agents which can effectively cause surface bioerosion, biodegradation, whatever you want to call it. That is super oxide anion and hydrogen ion for us.

DR. GREISLER: If you are saying products of macrophage release, such as Interluken I, then presumably the activation would also lead to their release of basic fibroblast growth factor and PDGF.

DR. ANDERSON: That's right.

DR. GREISLER: And not only would the macrophage be inducing the cracking potentially, but might induce a myofibroblast reaction with sort of a filling in of those cracks with scar reaction, which might not be such a clinical problem in that case. Do you see that histologically?

DR. ANDERSON: Well, the case system is not a system that allows tissue to come up to it. So I can't probe that with that system.

I think conceptually that might occur, but that's an experiment that has to be looked at. Our goal is to try to come up with in vivo systems and in vitro systems which are appropriate model systems for looking at the potential for degradation.

DR. GIBBONS: Don Gibbons, 3M.

It was interesting, Jim, the point you are making. And I only want to add a comment. That we are interested in the crack propagation, which is really the factor that is important in your fatigue measurements.

And we think that this is an important -- the controlling factor probably, in the lifetime of an elastomer, since we are interested in the, as many other people are, for ventricular assist pumps. And I think that we are learning, both ourselves and there is some literature, to show the crack propagation in face separated elastomers is actually quite -- the mechanisms properly different than that you see in the normal cross link. And that probably the loss of the high molecular weight component, which is the point that you are making which none of us can explain, would correlate directly, because you've got a large number of molecules going from one domain to the other. When the large ones have gone, you are only left with the short ones, crack propagation.

DR. ANDERSON: That's right. And the system is so sensitive to loss of the high end because of crack propagation.

Now the problem with our materials is, in the in vivo one, Don, they were 400 per cent extended. We have got to cut back on that, because we think, and I raised the question earlier this morning with the fungi, the first steps that we start to see in terms of macrophage interaction on surfaces, and Shalaby knows this very well, is pitting that occurs on the surface.

And we think that the pitting is the antecedent to crack development. That maybe these are stress concentrators created by cellular interaction. But if you have a prestress material, that crack will grow like a shot. You rarely see pits.

We have to cut back and start taking a look at less modified materials, in order to try and tie up the mechanism.

DR VERT: Michel Vert, University of Rouen.

Jim, you pointed out that how much it is difficult to show whether enzymes are active in vitro or not.

The experience I have with the polyesters is that it's very

difficult to show it, and in some example, we even -- we show the stabilization of the polymer, when monitoring the degradation by molecular weight measurement and size exclusion chromotography. So that's very uncomfortable, you feel uncomfortable with it, when you say this.

As a physical chemist, I worry about the effect of ionics trend, and the amount of dissolve species you have in the solution on the change you have between the solid state material, the water and so on as I mentioned this morning, and with respect to the ion -- did you fix the -- a similarity of your solutions, when you changed the amount of enzymes?

DR. ANDERSON: Yes, we think we did.

DR. VERT: Did you use the saline as a --

DR. ANDERSON: Well, we use a buffer.

DR. VERT: A buffer. Is the -- or I should say is the ionic strain fixed?

DR. ANDERSON: I think it is, but I don't remember. It's in the paper.

DR. VERT: Because, I have no evidence for that, but when you -- I don't know how much, in milligram per millilitre is there, 45,000 units per millilitre? Is it a large amount of material you put it?

DR. ANDERSON: Yes, we use -- yes, we use a larger amount.

You see, these specimens are tensile specimens, but they are small tensile specimens. Ann-Christine showed that in terms of methodologies, that, you know, you've got to fudge it a little bit. But we have got small dog bones, and they require -- they require a fair amount of material, a fair amount of solution. Our enzyme solutions are anywhere from 50 to 100 millilitres. That's a hell of a lot of solution.

But you need that amount of solution, because your specimen is large, and you need a large speciment, so that you can carry out appropriate mechanical testing.

DR. VERT: Let me go on with my thinking, okay?

I was wondering whether the difference of concentration in the enzyme, in the solution, could modify the uptake or water or something like that, so...

DR. ANDERSON: I don't think that that's --

DR VERT: Let me finish.

DR. ANDERSON: Come on, that's a stretch. Right now that's too far of a stretch in order to get the fatigue differences we're getting in this polymer; absolutely.

DR. VERT: The fatigue testing were made in dry state or humid state?

DR. ANDERSON: They were made in a wet state.

DR.VERT: Wet state.

DR. ANDERSON: So the others were made in a wet -- and the controls were made in a wet state, too.

DR. VERT: Think about it. If you have a slight difference in water uptake with the samples, with a fatigue machine you couldn't get a slight difference between the two. But who knows.

DR. ANDERSON: Well, we have a limited amount of resources.

Our efforts are now being directed towards the in vivo specimens. And what we are doing here is to extract those materials and to look at the polymer extracts that come out and attempt to characterize them chemically and understand them.

So with the idea that if we can understand that, we can then start to look at chemical species that have been modified in the polymer and then get a better idea of mechanism.

DR. VERT: Yes, that's right.

DR. D. WILLIAMS: David Williams, Liverpool, England.

Could I just make a comment about or in relation to the macrophage on the surface of the pelathane? You are absolutely right. That this sort of pelathane samples you are working with are very large in comparison to the microphage.

We have some experience in Liverpool on using a microfibrous polyurethane. Not perithane, but a polyethyurethane, where we have a microstructure, which involves one or two micron fibers. If that microstructure is available for macrophages to get into that, then you can see very, very clearly, very early, macrophages wrapping themselves around those one or two microfibers. They are activated, and they are eating up that polyurethane in a matter of weeks.

If you have a situation where the macrophage just can't get in there and do that, then it's reasonably stable. Not totally but reasonably stable. Macrophage will get activated on the surface just so they can get around it.

DR. ANDERSON: Oh, yes. Thank you.

DR. FEIJEN: Jim, I have one last question for you. Just -- I still don't get it completely.

If you have a polyurethane, you have a surface, and in case of the enzyme. So the enzyme is absorbed and the surface starts to do something, how can we then see that in the molecular weight distribution? You should see differences in the distribution in terms of that the bulk is still not attacked.

DR. ANDERSON: That the what is not attacked?

DR. FEIJEN: the bulk material, the bulk of the material. And the surface, you start doing something at the surface.

DR. ANDERSON: Right.

DR. FEIJEN: Do you have any evidence of that, that you have a process like that? I still cannot comprehend --

DR. ANDERSON: From our extractions we are starting to see that. That it's obviously surface dependent. And that's what starts to go first. But that was what we're talking about here. I mean this symposium is directed towards surface degradation of materials. I mean that's where these things start.

DR. FEIJEN: That's how these things start. But I cannot see, from your data, in time that you -- how this is developing. You see? You say we have a preferential attack at the higher molecular weight situation. But at the same time you get changes in distritution. So how can you draw that conclusion actually?

DR. ANDERSON: You are looking at the in vitro operative system.

DR. FEIJEN: you have almost 100 per cent conversion to draw that particular conclusion. So I agree we are looking at a surface, but from the surface you are coming into the bulk. From the bulk you are coming into the final distribution. And how to explain these data then?

DR. ANDERSON: Well, we are attempting to do it. But we are concentrating on the in vivo situation, because the in vitro situation we appreciate is artificial. One of the questions you can raise is why should the enzyme attack the material?

DR. FEIJEN: Let's ask the enzyme.

DR. ANDERSON: Well, why shouldn't it?

# The Role of Active Species Within Tissue in Degradation Processes

D.F. Williams
Institute of Medical and Dental Bioengineering

University of Liverpool

## Introduction

The conditions under which polymers are especially susceptible to degradation include those of elevated temperature, particularly in the presence of oxygen, electromagnetic radiation (i.e. X-rays, gamma-rays, UV), mechanical stress at elevated temperatures and ultrasonic vibration. It is not often that biological environments are considered seriously in the discussion of polymer degradation since they do not usually offer these specific conditions to polymers; there are some circumstances, however, where this possibility becomes of significance. There is probably no adequate definition of a biological environment and so it is difficult to be comprehensive in this discussion. There is one particular environment which is of extreme importance here, however, and which may be used as an example. This is the environment of the human body, and the subject of polymer degradation within biological environments is discussed from this perspective.

The most important and obvious difference between this environment and any other is the unique responsiveness of it to the presence of a material. Obviously, the material itself has to reside within the tissues of the body for as long as its function is required, without suffering undersirable degradation, whilst at the same time not exerting any underirable influence on the tissues. These tissues are both exquisitely sensitive to, but strongly unforgiving of, the intrusion of foreign bodies into them, so that this mutual co-existence is no trivial matter.

The phenomena relating to this mutual co-existence are collectively referred to by the term biocompatibility. This is defined[1] as the 'ability of a material to perform with an appropriate host response in a specific application'. The host response refers to the effect of the material on the tissue. The appropriate host response may be minimal where the material is designed to be ignored by the tissues, or specifically functional if the material is designed to play a biological or pharmaceutical role rather than a simple mechanical or physical part. In either case, the host response is largely determined by the chemical reactivity of the material. In the case of polmer-based materials, this reactivity will be concerned with phenomena such as the leaching of additives, catalysts and other residues as well as the degradation of the polymer.

From the above it can be seen that polymer degradation is one of the most important factors leading to the acceptability or otherwise of pollymers for medical use. Indeed, experience has shown that high molecular weight polymers themselves are extraordinarily well accepted by the body, being non-toxic by most criteria, and that it is only additives and breakdown products that cause problems. The details of these tissue responses can be found in other reviews.[2-5]

There is a further reason why polymer degradation is so important in the context of medical applications. Whilst it is clearly a fundamental requirement for most materials used to replace tissues that they are perfectly stable in the body, there are many applications where the function of the material is only transient. Under such circumstances, there is sound logic in assuming that the materials should be removed once the function has been performed. It is frequently difficult to surgically remove materials from tissues and such removal is associated with the risk of morbidity. The alternative approach is to make the materials intentionally degradable so that they are spontaneously eliminated once the function has been performed. Polymers provide a very suitable class of materials for this purpose, although of course the kinetics and toxicology of breakdown products have to be considered very carefully.

It is necessary here to comment on the terms degradation, biodegradation and deterioration. The term biodegradation is often used to denote degradation occurring in biological environments. Whilst this may seem to be a logical use, it is necessary to be more precise. A better definition of biodegradation[6] stipulates that it is 'the gradual breakdown of a material mediated by specific biological activity'. The significance of this is that it is not sufficient for this degradation to occur within a biological environment for it to be defined as biodegradation. It is necessary for there to be specific mechanisms associated with that environment involved in the degradation, whether that be mediated by cells, enzymes, bacteria or other biological species. In the context of implantable biomaterials, a polymer undergoing hydrolysis which can be reproduced both qualitatively and quantitatively by a simple inorganic solution cannot be described as biodegradable; it is only when effects over and above this simple hydrolysis are seen that the degradation may be classed as biodegradation.

The term biodeterioration is also often used in similar circumstances and, in spite of several efforts to specify its meaning, there is no universal agreement. Usually biodeterioration is used to denote changes which are both destructive and undesirable. It is important to note that the use of the term degradation implies structural changes in the polymer but not necessarily undesirable structural changes. Moreover, phenomena of 'polymer degradation' in biological environments which are associated with changes in properties are not necessarily associated with depolymerization, chain scission or other classical structural changes. They may, on the other hand, be due to water absorption, leaching of extractables and so on and these phenomena must clearly be considered.

## MECHANISMS OF POLYMER DEGRADATION *IN VIVO*

### General Comments

In general, polymeric-based materials have one significant

advantage over metals in the context of medical applications, for although the isotonic saline solution that comprises the extracellular fluid is extremely hostile to metals, it is not normally associated with the degradation of synthetic high molecular weight polymers. Most polymers are susceptible to degradation under certain conditions, but these conditions and the kinetics of the reactions are extremely variable. In the context of the present discussion, the degradation processes may be divided into two types. First are those which involve the absorption of some kind of energy which then causes the propagation of molecular degradation by secondary reactions. Secondly, there are hydrolytic mechanisms which can result in molecular fragmentation, usually being seen in hetero-chain polymers.

The conditions under which the first of these general processes takes place include elevated temperatures, especially in the presence of oxygen to give thermal oxidation, electromagnetic radiation, mechanical stress at elevated temperatures and ultrasonic vibration. Clearly, the physiological environment within the human body does not offer any of these conditions to an implanted polymer.

Hyrdolysis, on the other hand, is quite feasible in the aqueous extracellular fluid. A number of conditions have to be met. Firstly, the polymer has to contain hydrolytically unstable bonds. Secondly, for any significant degradation to occur in a reasonasble time, the polymer should be hydrophilic. Thirdly, the hydrolysis

Has to take place at a physiological pH, which is around 7.4. Thus, polymers can be placed in a ranking order of predicted susceptibility to _in vivo_ degradation.[7,8] (a) hydrophobic, no hydrolyzable bonds (most stable); (b) hydrophilic, no hydrolyzable bonds (may swell, but little or no degradation); (c) hydrophobic, hydrolyzable (surface activity only or very slow degradation); and (d) hydrophilic, hydrolyzable (bulk degradation).

## Hydrolysis

Hetero chain polymers, particularly those containing oxygen and/or nitrogen are generally susceptible to hydrolysis. Depending on the structure, this hydrolysis may be favoured by either acid or alkaline conditions. Among the polymers which have been shown to degrade by hydrolysis _in vivo_ are certain polyamides and poly(amino acid)s, some polyurethanes and both aliphatic and aromatic polyesters.

## Enzyme-mediated Hydrolysis

According to the above definitions, if hydrolysis of a polymer takes place _in vivo_ by a mechanism and with the same kinetics as that achieved in a simple _in vitro_ aqueous environment, then it cannot strictly be considered as biodegradation. It is necessary for there to be some active role of the biological environment for biodegradation to take place and here we may consider the role of enzymes. The possibility exists that enzymes may be responsible for other degradation processes (e.g. oxidation) with far more stable polymers, but the question has been discussed on several occasions as to whether enzymes are able to influence degradation rates in hydrolyzable polymers. There is some logic in this hypothesis of enzyme-accelerated polymer degradation since enzymes have the characteristic ability to catalyze chemical reactions. There are some difficulties in confirming this hypothesis, however, since enzymes are normally so substrate specific and we do not usually associate their catalytic effect with synthetic high molecular weight polymers.

It now seems clear from the literature that enzymes are able to influence hydrolytic degradation of certain polymers. This information is laragely derived from in vitro experiments in which polymers have been incubated in enzyme solutions. Williams, for example, has demonstrated the effects of hydrolytic enzyme activity on groups of degradable polyesters, including poly(glycolic acid) and poly(lactic acid).[9,10] A commercially available preparation of poly(glycolic acid), in the form of a surgical suture Dexon(R), was incubated with a variety of enzymes, in solution, including acid phosphatase, bromelain, carboxypeptidase, chymotrypsin, esterase, papain, leucine aminopeptidase and others. Compared to controls containing the appropriate buffer without enzymes, many of these preparations had no effect as measured by changes in mechanical properties over a three week degradation period, but those enzymes with esterase activity increased the rate of degradation. Leucine aminopeptidase in particular had a dramatic effect on the degradation.

Other studies have reported the effects of enzymes on these

hydrolytically unstable polyesters in vitro.[11-12] It is far more difficult to demonstrate that in vitro degradation is associated with enzyme activity, however, and the situation is far from clear. Williams [13] has described experiments from which a certain amount of circumstantial evidence was obtained concerning this phenomenon, especially demonstrating that the kinetics of degradation are influenced by the nature of the polymer's immediate environment in the tissue. A host response that is more active cellularly and in which there is a greater release of enzymes into the extracellular fluid, tends to produce greater amounts of degradation in some materials. This is not, of course, conclusive evidence of the direct involvement of enzymes, but does suggest that they can play a role. It must be remembered, however, that there are other components of the extracellular fluid, for example lipids, which could also be involved.[14]

The potential role of enzymes in the degradation of polymers which are more stable than the aliphatic polyesters has also been studied by Williams [15] and others.[16-17] It has been shown that poly (ethylene terephthalate), and aromatic polyester, is degraded by enzymes with esterase activity and that nylon 6,6 and certain poly (etherurethan)s are degraded by a variety of enzymes under in vitro conditions. These studies were carried out using specially synthesized radiolabelled polymers. The only type of polymer not to suffer any degradation at all in the experiments of Williams was poly (methyl methacrylate).

It should be noted that susceptibility to this type of degradation will vary with the structural characteristic of the polymers, just as it will with other degradation mechanisms. Crystallinity, molecular weight and polydispersity are among the features which are important here. With the aliphatic polyesters, for example, it has been shown[11,18] that gamma-ray irradiation can alter the susceptibility to attack by enzymes, presumably by altering one or more of these features.

## Degradation of Homo-Chain Polymers

As noted earlier, it is not anticipated that hydrolytically stable homo-chain polymers would be susceptible to degradation under the ambient conditions found in the body. Activation engergies for the degradation of the high molecular weight polymers used in surgery vary from 30 to 90 kcal $mol^{-1}$ and such reactions generally require heat, UV light or high energy radiation, preferably in the presence of oxygen, to proceed.

Whilst this prediction of <u>in vivo</u> stability is largely borne out in practice, there is some evidence that other factors are involved and that unexpected degradation mechanisms operate within the body. One of the first observations of degradation of these polymers was made by Oppenheimer[19] who was working on the carcinogenic properties of plastics. In attempting to elucidate the mechanisms by which plastic films induced tumours after subcutaneous implantation in rodents, certain radiolabelled polymers were employed. $^{14}C$-labelled polystyrene, polyethylene and poly (methyl methacrylate) were implanted and urine, faeces and respiratory $CO_2$

were mointored for periods over a year. With the polystyrene, nothing radioactive was excreted in the urine until 21 weeks, but some radioactivity was detected after this time. With polyethylene, radioactive species were excreted after 26 weeks and with poly (methyl methacrylate), this occurred after 54 weeks.

Liebert el al used techniques of dynamic mechanical testing and chemical analysis to study the degradation of polypropylene, in particular comparing the fate, in vivo, of material with and without antioxidants. Neither hydroxyl nor carboxyl levels changed after implantation of antioxidant-containing material in animals, but in the antioxidant-free material, hydroxyl concentration started to increase immediately and then linearly with time. Carboxyl groups were detected after 90 d. A slight shift in molecular weight distribution and a decrease in tan delta were similarly observed. The authors attempted to compare this degradation with oxidative degradation at elevated temperatures and concluded that the rates of degradation in vivo were much higher than the extrapolation of high temperature oxidation rates to 37°C would predict.

The suggestion again arises, therefore, that enzymes or some other biologically active species are involved with degradation. Since several enzymes have oxidative activity and since certain cells release peroxides, this does seem to be a reasonable hypothesis.

## Cellular Degradation

It has been shown above that degradation rates may be affected by various components of the tissue environment. Obviously, there are many different types of cell within this tissue and since those cells with phagocytic ability are normally able to remove debris from the tissue by engulfment and digestion, the possibility of cellular digestion of polymers must be considered. In general terms, it seems unlikely that cells could do much damage to monolithic smooth-surfaced polymers because they do not have any specific mechanism for their recognition. The situation with particles may be a little different since a macrophage should readily digest fragments up to 5 $\mu$m in size, thereby exposing the particle to concentrated biochemical attack. However, such cells only have a limited lifetime, measured in weeks and it may be impossible for sufficient degradation to have ocurred before the cell itself dies. There is much evidence to show that cells are able to phagocytose this type of foreign paticle, eliminate them from that site and deposit them, on completion of their life cycle, else where in the body. The lymph nodes or liver are common sites for deposition.

There is, on the other hand, very little evidence of polymer degradation taking place directly under the influence of cells. Kossovsky et al.[21] have claimed that macrophages have been observed to cause pitting on the surface of silicone polymers by virtue of the peroxides released onto the surface and Williams et al.[22] have observed some degradation of labelled polyester by macrophages in vitro.

Bacterial Degradation of Polymers

If enzymes are able to cause polymer degradation, it might be expected that bacteria could also be involved in this process, especially as much of their activity is mediated by enzymes. It is certainly known that bacteria are capable of degrading some non-proteinaceous macromolecular structures through the action of intracellular enzymes. The question therefore arises as to whether bacteria in contaminated wounds might affect the rate of polymer degradation, this clearly being important with wound-closure materials but also in other situations. It is quite probable that the degradation of the collagenous material catgut, used as a suture, is modified when it is used in infected wounds, where enzymes of both hydrolase and peptidase activity are present, and it is a clinical observation that the absorption of catgut takes place more rapidly under these contaminated conditions.[23]

The situation regarding synthetic absorbable sutures is not so clear. Seberseri _et al._[24] have demonstrated that poly (glycolic acid) sutures are degraded faster in infected urine compared to sterile urine. However, Williams[25] showed that the same material degraded slower in the presence of _Strep.mites_, _E.coli_ and _Staph.albus_ compared to sterile broth conditions. Experiments with _Staph. albus_ _in vivo_ confirmed this finding and suggested that these synthetic absorbable sutures possessed a significant advantage over catgut. No confirmation of these observations has been published, but is interesting to note recent experiments in which bacteria, both aerobic and anaerobic, have been shown to display growth patterns, and especially far greater survival times, when they are in the presence of polyester substrates, possible indicating utilization of carbon derived from the polymer in their metabolism.

## Biomedical Polymers

In this section, the evidence concerning the degradation of selected biomedical polymers is reviewed.

### Poly (lactic acid) and poly (glygolic acid)

These are the best known examples of the poly (alpha-hydroxy acid)s, of general formul $\text{+ OCHRCO +}_n$, which are degradable in the body, with R = H in poly (glycolic acid) (PGA) and R - Me in poly (lactic acid)(PLA). For well over 20 years these materials, either as homopolymers or copolymers, have been used in clinical situations by virtue of their degradability.

It is difficult to determine from the literature what is the precise mechanism of degradation of these polymers. It is widely believed that chain scission occurs through simple hydrolytic reactions, although species preseul ( OCHRCO $)_n$, which are degradable in the body, with R = H in poly (glycolic acid) (PGA) and R - Me in poly (lactic acid)(PLA). For well over 20 years these materials, either as homopolymers or copolymers, have been used in clinical situations by virtue of their degradability.

It is difficult to determine from the literature what is the precise mechanism of degradation of these polymers. It is widely believed that chain scission occurs through simple hydrolytic reactions, although species present other than water, such as the various anions and cations as well as enzymes, may influence the kinetics.

When PGA fibers are implanted (in the form of surgical sutures) they may lose strength completely within a short time, for example 20 days, but suffer insignificant mass loss during this time. Complete mass loss may not occur for periods of two to three times longer than this, during which time crystallinity changes occur. Tyupical commercial preparations will initially have a crystallinity of 80% with 70-80 Angstrom ellipsoidal crystallites. Water adsorption takes place with characteristics consistent with hydrophilic polymers. Zaikov[26] claims that the limiting water absorption by such fibers initially is around 0.14 g water per g polymer. The water preferentially seeks the accessible amorphous sectors and hydrolyzes the ester bonds within these areas. This initiates the reduction in mechanical properties but mass loss does not occur yet. Following degradation of the amorphous regions, the crystallites are affected and the rate of mass loss becomes much greater. Chu[27] has also reviewed the mechanisms of degradation of PGA and PLA in physiological environments. Also referring to PGA commercial sutures, he describes two distinctive phases of degradation, the first, taking place over a 20 day period, being associated with attack on amorphous regions and tie-chain segments, free chain ends and chain folds. Some glycolic acid is released in this time, but it is only when all or most of the amorphous regions have been removed by hydrolysis that the water can begin to attack the crystalline regions. As the last of the amorphous regions are lost, the rate of glycolic acid production temporarily decreases (the degree of crystallinity is now at a maximum). The second phase of degrtadation starts more slowly than the first because of the greater difficulty of hydrolyzing the crystalline regions but this rate increases and glycolic acid is then released rapidly.

Poly (ß-hydroxybutyrate)

A further interesting example of an aliphatic polyester that has been considered in the context of biodegradation is poly (ß-hydroxybutyrate) (PHB). These polymers $\{ OCH (Me) CH_2 \}_n$, prepared either as homopolymers or copolymerzed with hydroxyvalerate (HV), $\{ OCH(Et)CH_2 \}_n$, have been introduced as potentially useful degradable materials in recent years because of their putative degradability in various biological environments.

PHB is a naturally occurring polymer, being a principal energy and carbon storage compound that is synthesized by a range of prokaryotic cells. These energy reserve substances are produced within cells in the presence of excess nutrient, but are broken down during periods of starvation to provide energy. PHB in particular is synthesized by bacteria.[28] It is possible to isolate the PHB and prepare useful material from the polymer. Since it is known that bacteria other than those which initially produced the PHB are able to affect this material under some circumstances (for example, wh;en the polymer is buried in the soil), it has been assumed that degradation would also occur in other biological environments.

However, the degradability of PHB and its copolymers in animal or human tissues or related environments is questionable. As described by Miller and Williams[29] high molecular weight fibers of PHB and PHB/PHV copolymers (up to 27% PHV) do not degrade in tissues or simulated environments over periods up to six months. This stability is severely influenced if the polymer is predegraded by gamma-ray irradiation. Material fabricated in different way may also show different behaviour with respect to degradation. Holland <u>et al.</u> examined the hydrolysis of cold-pressed tablets of PHB and PHB/PHV and demonstrated degradation, the kinetics of which varied with fabrication route, molecular weight and HV content.

Poly(ethylene terrephthalate)

Poly(ethylene terephthalate) is the major aromatic polyester used medically, with extensive application in vascular prostheses. Although apparently clinically successful, failures do occur because of degradation and extensive studies have recently been reported. The basic degradation mechanisms of PET in acid and alkaline media have been known for some time, with the hydrolytic route usually being observed

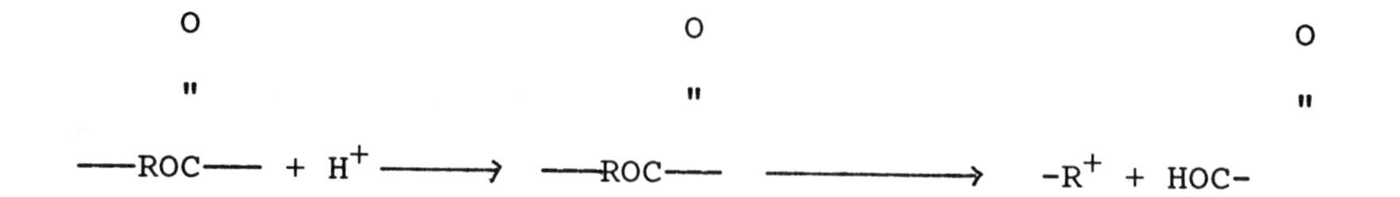

Water diffusivity into PET is very slow, however, with a diffusivity rate of $10^{-9}$ $cm^2s^{-1}$ so that degradation is slow and, depending on the configuration, is a surface-dominated phenomemon. Work of Rudakuva and colleagues[30] has demonstrated the slow loss of strength of PET mesh following implantation in various species including dogs, rabbits and humans. After eight years, the strength was reduced to nearly one-half of the original value and the elongation to failure considerably lowered. However, there was little loss of substance and no detectable change in molecular weight ($M_V$= 2 x $10^4$). The rate and mechanism of degradation does appear to have been influenced by infection, the normally surface-dominated effect changing to one in which internal changes are seen, especially in amorphous regions. It was claimed that this was due to the change in pH and the release of lysosomal enzymes.

Vinard _et al._[31] have reported on a series of failures of Dacron arterial prostheses, mostly arising because of false aneurysms or infection. Fiber breakage and thinning were obvious in scanning electron microscopy of these prostheses removed from patients. The strength of these explanted materials varied from 10 to 98% of the original strengths.

## Polyamides and Poly(amino acid)s

Hetero-chain polymers involving the amino group might be expected to be unstable in aqueous environments and especially biological environments. Polymers of the nylon variety and poly(amino acid)s are briefly mentioned here.

Many years ago, Harrison[32] showed that nylon fabrics lost about 80% of their tensile strength during a three year implantation period. Some nylons are both hydrophilic and hydrolyzable although the extent of the water absorption is variable. Nylon 6 with a water content at saturation of 11% hydrolyzes fastrer than nylon 11, which has a value of 1.5%.

The hydrolysis of these polyamides has been discussed by Gilding[33] and others, the primary attack being on the oxygen atom of the carboxyl group.

$$\mathrm{RC(=O)NHR'} + \mathrm{H^+} \longrightarrow \mathrm{RC(OH)=\overset{+}{N}HR'} \longrightarrow \mathrm{RC(=O)OH} + \mathrm{H_2NR'}$$

Both acid and amine end groups are formed. The protonation reactions will vary depending on the pH of the environment and Gumargalieva _et al._[34] have discussed the effects of various constituents of the physiological environment, for example phosphate, carbonate and bicarbonate ions, on kinetics and reaction products.

It is also interesting to note the influence of enzymes on the rate of degradation. Smith _et al._[15] have shown that a radiolabelled nylon 6,6 is degraded _in vitro_ by the enzymes papain, trypsin and chymotrypsin. Esterase had no effect. Williams[13] has also shown this polymer to degrade faster in tissues that are inflamed (that is, containing a significant infiltration of enzyme-releasing cells) than in quiescent tissue, whilst several authors have reported on the rapid rate of degradation of nylon sutures in the enzyme-rich region of the eye.[34]

Synthetic poly(amino acid)s and polypeptides are also susceptible to hydrolysis. It is now well known that these materials may be degraded by enzymes in vitro, where the enzyme substrate specificity in relation to the bonds broken is unchanged. Much data is available showing the solubility and digestibility of poly(amino acid)s by proteolytic enzymes.[35] For example, synthetic polylysine is degraded by trypsin[36] the chief products being dilysine and trilysine. No lysine appears to be produced, indicating that the terminal peptide bond is not attacked. This is in contrast to the acid hydrolysis of polylysine, where random scission degradation occurs, with lysine as one of the products.

The biodegradation of poly(amino acid)s and the role of enzymes in particular has been discussed by Dickinson et al.[37] Poly(2-hydroxyethyl-L-glutamine), for example, was found to be degraded in vitro by pronase and papain, but trypsin and collagenase had no effect. Oligomers, of degree of polymerization 4 to 9, were the main digestion products. Degradation in vivo was variable, most effects being seen during the time of active cellular infiltration.

The hydrolytic instability of the amide bond in synthetic amino acid polymers has also been used in the formation of intentionally degradable soluble polymers for drug delivery, as described by Duncan and Kopacek.[38] These polymers are designed to give controlled biodegradation via the introduction of segments suceptible to effects by specific enzymes. For example, polymers of N-(2-hydroxypropyl)methacrylamide and p-nitrophenyl esters of N-methacryloylated amino acids are reacted with compounds containing an aliphatic amino group with the formation of an amide bond. If this bond originates in an amino acid specific for a certain enzyme, an enxymatically cleavable bond is formed.

## Polyurethanes

A wide variety of polyurethanes exist and many of these have been used or contemplated for use in medical applications. Urethane, urea, ester and ether groups may be present and different classifications based on the dominant groupings are available. In biomedical applications, polyurethanes are usually described as either poly(ester urethane)s or poly(ether urethane)s. The former group found the earliest use but *in vivo* degradation and disintegration became apparent in applications such as bone adhesives and arterial prostheses[39] and generally little use has been made of them in recent years.

On the other hand poly(ether urethane)s should be more stable and applications have been seen in the artificial heart, pacemaker leads, flexible leaflet heart valves and catheters. Whilst success appears to have been achieved in several of these situations, it is clear that stability cannot be guaranteed and several problems have arisen.

The most significant observations here have been with pacemaker lead encapsulation. High failure rates of these leads have been reported and two degradation mechanisms identified which appear to be responsible for fissuring or cracking of the poly(ether urethane) (PEU). First is the possibiltiy of metal ion catalyzed oxidation of the PEU, the metal being derived from the lead.

Secondly, the PEU is reported to be susceptible to environmental stress cracking. In respect of the oxidation, Thoma and Phillips studied the effects of metal ions and did not find any significant oxidation under in vitro conditions and so the role of this mechanism is unclear, even though the literature demonstrates metal-ion complexation and stiffening of the soft segments or PEU and related effects.[40]

Stokes et al.[41] have reported on the phenomenon of enironmental stress cracking and demonstrated the significance of residual stress. Material affected by this phenomemon is characterized by crazes with little or no reduction in strength or molecular weight. The lower the ether content of the PEU, the greater the resistance to this cracking; annealing of the materials to reduce residual stresses is also helpful.

Several recent detailed studies have defined some of the parameters concerning the biodegradation of PEUs. Smith et al.[42] prepared a series of radiolabelled PEUs and studied the effects of various hydrolytic and oxidative enzymes on these preparations. The PEU synthesized with a polyether of molecular weight 1000 was affected by esterase, papain, bromelain, ficin, chymotrypsin, trypsin and cathepsin, but not by collagenase, xanthine oxidase or cytochrome oxidase. The characteristics of the degradation were very variable, however, and while it was suggested that ethylenediamine was amongst the degradation products, the mechanism of degradation was difficult to define. Anderson and colleagues[43] have made observations on similar PEUs in animal experiments, and found specimens to become pitted, with the pit dimensions of 1-2$\mu$m being compatible with the cytoplasmic structures of cells, especially macrophages, that can attach to polymer surfaces.

Ratner _et al._[16] have also studied the enzymatic and oxidative degradation of polyurethanes. The PEUs and one poly(ester urethane) were studied _in vitro_ and degradation was assessed by changes in molecular weight and polydispersity by GPC methods. Varying degrees of both enzymatic degradation (leucine aminopeptidase, papain and chymotrypsin) and oxidative degradation (hydrogen peroxide) were seen, although the amounts involved, as with the work of Smith _et al._, were low.

One of the potential applications of PEUs is in the area of heart valve and arterial reconstruction. Mitrathane, a poly(ether urethane urea), is an elastomer, synthesized from poly(tetramethylene glycol), methylene diphenyl diisocyanate and a diamine, that has been used in these situations, but has proved to be unreliable because of degradation.[44,45] Laboratory experiments have shown that some degradation occurs in simple aqueous solutions but that this is insufficient to explain the considerable destruction _in vivo_. Again enzymatic degradation was considered as a possible mehcanism.

## CONCLUSIONS

The discussions and data presented here suggest that the environment of the human body can be surprisingly aggressive to polymers. Many suffer hydrolytic degradation but the kinetics and mechanisms of the processes may be significantly influenced by various species present, including especially enzymes, but also anions, cations, cells, lipids and so on.

References

1. D.F. Williams, 'Definitions in Biomaterials', Elsevier, Amsterdam, 1989, p.49.
2. D.F. Williams, J.Mater.Sci. 1987, 22, +3421
3. G. Meachim and R.B. Pedley, in 'Fundamental Aspects of Biocompatibility', ed. D.F. Williams, CRC Press, Boca Raton, FL, 1981, vol I, p.107.
4. M. Spector, C. Cease and X. Tong-Li, CRC Crit.Rev.Biocompat., 1989, in press.
5. BN. Matlaga, L. yasenchak and T. Salthouse, J.Biomed.Mater.Res., 1976, 10, 391.
6. Ref. 1, p.60.
7. D.F. Williams, Annu.Rev.Mater.Sci. 1976, 6,237.
8. F.F. Williams, J.Mater.Sci., 1982, 17,1233.
9. D.F. Williams, and E. Mort. J.Bioeng. 1977, 1, 231.
10. D.F. Williams, Eng.Med. (Berlin), 1981, 10 8.
11. C.C. Chu and D.F. Williams, J. Biomed. Mater.Res. 1983, 17, 1029.
12. C.C. Chu, D.F. Williams and J. Dwyer, J. Appl. Polym. Sci., 1984, 29, 1865.
13. D.F. Williams, ASTM Spec.Tech.Publ. 1979, 684, 61.
14. C.P. Sharma and D.F. Williams. Eng.Med. Berlin, 1981, 10,8.
15. R. Smith, D.F. Williams and C. Oliver, J. Biomed. Mater.Res., 1987, 21, 995.
16. B.D. Ratner, K.W. Gladhill and T.A. Horbett, J.Biomed.Mater.Res., 1988, 22, 509.

17. S.K. Phua, E. Castillo, J.M. Anderson and A. Hiltner, J.Biomed.Mater.Res., 1987, 21, 231.
18. N.D. Miller and D.F. Williams, Biomaterials), 1987, 8,129.
19. B.S. Oppenheimer, E.T. Oppenheimer, J. Danishefsky, A.P. Stout and E.F. Eirich, Cancer Res., 1955, 15, 333.
20. T.C. Leibert, R.P. Chartuff, S.C. Cosgrove and R.S. McCuskey, J.Biomed.Mater.Res., 1976, 10, 939.
21. N. Kossovsky, J.P. Heggers and M.C. Robson, CRC Crit.Rev.Biocompat., 1987, 3, 53.
22. D.F. Williams, R. Smith and C. Oliver, in 'Biological and Biomechanical Performance of Biomaterials', ed. P. Christel, A. Meunier and A.J.C. Lee, Elsevier, Amsterdam, 1986, p.239.
23. T.E. Bucknall, J.R. Soc.Med., 1981, 74, 580.
24. O. Sebeseri, V. Keller, P. Sprang, R.T. Scholl and F. Zingg, Invest. Urol., 1975, 12, 490.
25. D.F. Williams, J.Biomed.Mater.Res., 1980, 14, 329.
26. G.E. Zaikov, J.Macromol.Sci.Rev.Macromol.Chem., 1985, C25, 551.
27. C.C. Chu, in 'Biocompatibility of Degradable Polymers', ed. D.F. Williams, CRC Press, Boca Raton, FL, 1989, in press.
28. D.H. Williamson and J.F. Williamson, J.Gen.Microbiol., 1958, 19, 198.
29. N.D. Miller, in 'Biocompatibility of Degradable Polymers', ed. D.F. Williams, CRC Press, Boca Raton, FL, 1989, in press.
30. T.T. Daurova, O.V. Voronkova, S.D. Andreev, T.E. Rudakova, V. Moiseev, L.L. Razumova and G.E. Zaikov, Dokl. Akad. Nauk., SSSR, 1976, 231, 919.

31. E. Vinard, R. Eloy, J. Descotes, H. Guidicelli, P. Patra, R. Berruet, A. Huc and J. Chaauchard, J.Biomed.Mater.Res., 1989, in press.

32. J.H. Harrison, AmJ. Surg., 1958, 95, 3.

33. D.K. Gilding, in Fundamental Aspects of Biocompatibility', ed. D.F. Williams, CRC Press, Boca Raton, FL, 1981. Vol I. chap. 3.

34. B.E. Cohan, Am.J. Ophthalmol., 1979, 88, 982.

35. H.A. Sober, 'Handbook of Biochemistry', CRC Press, Boca Raton, FL, 1968.

36. S.G. Waley and J. Watson, Biochem. J., 1953, 55, 328.

37. H.R. Dickinson and A. Hiltner, J. Biomed. Mater. Res., 1981, 15, 591.

38. J. Kopecek, in 'Systemic Aspects of Biocompatibility', ed. D.F. Williams, CRC Press, Boca Raton, FL, 1981, p.159.

39. J.B. Walter and C.G. Chiaramonte, Br.J. Surg., 1965, 52, 49.

40. R. Benson, S. Yoshikawa, K. Knutson and D. Lyman, Adv. Chem. Ser., 1982, 199.

41. K. Stokes, P. Urbanski and K. Cobian, in 'Polyurethanes in Biomedical Engineering II', ed. H. Planck, I. Syre, M. Dauner and G. Egbers, Elsevier, Amsterdam, 1987, p.109.

42. R. Smith, C. Oliver and D.F. Williams, J.Biomed.Mater.Res., 1987, 21. 995.

43. R.E. Marchant, J.M. Anderson, E. Castillo and A. Hiltner, J.Biomed.Mater.Res., 1986, 20, 153.

44. R.W. Paynter, H. Martz and R. Guidoin, Biomaterials (Giildford, Engl.), 1987. 8, 94.

45. R.W. Paynter, I.N. Askill, S.H. Glick and R. Guidoin, J.Biomed.Mater.Res., 1988, 22, 687.

QUESTIONS AND ANSWERS

DR. JAMES ANDERSON: Jim Anderson, Case Western Reserve. Very nice talk, David.

I Think you gave a very good overview. I just wanted to make one comment, and it doesn't pertain to you Carbon 14 work because yours is in vitro and you are using an already fully differentiated cell line macrophage but I want to urge caution to those people who are doing in vivo work whether in biomaterials or in environmental polymers.

And that is a lesson that some of us learned some 20 years ago when they did radio labelling experiments with the cyano acrylates and they implanted them in an effort to carry out a similar study as you did. What they saw was variations in cell differentiation in the genotypic and phenotypic expression of those cells that was read as pre-cancerous changes and attributed to the polymer, when in fact the radio label loading was so high, it was the radiation from the polymer that caused these cellular changes.

And I think we have to be careful when we go to polymers which will degrade very slowly because the only way we can study them is to have high radio label loadings which in fact may change the cells that are in contact with them.

DR. WILLIAMS: I agree entirely, Jim.

DR. ANDERSON: Thank you.

DR. SHALABY: Shalaby Shalaby from Johnson & Johnson.

In the use of those esterase to contribute to the degradation of the polyesters you mention is non-specific. Can you comment on the net charge of these esterase at pH 7.2, are they basic, they have a positive charge on the surface or asymmetric?

DR. WILLIAMS: I couldn't comment offhand, no.

DR. SHALABY: Because just the mere presence of a base could actually catalyze the degradation of the chain regardless of whether it is an enzyme or not.

Another thing I also would like to mention, could you get the same effect of esterase regardless of the type of buffer you use?

DR. WILLIAMS: I haven't done that experiment because we, having looked at a wide variety of enzymes, chose very specific conditions for each enzyme. I haven't varied as far as I can recall, we haven't varied the nature of the buffer within the individual enzyme.

DR. SHALABY: Okay, the last question I have. When you mention the effect of inflammation on different sutures, one of the important things that you look at is the change in pH. Have you looked at the effect of pH and the different environments?

DR. WILLIAMS: Yes, we have, yes.

DR. SHALABY: Is it acidic?

DR. WILLIAMS: You can get varying change of pH. We have looked at the effect of pH on the degradation of these materials and seen within the physiological range that we would expect to see, we have seen very little effect of pH on the degradation that could not influence any of the data I am presenting here. I agree that's important.

I should add that we have tried to study the effect of bacteria on the degradation of these materials and that becomes a very important factor then.

DR. ADEWOLE: Akin Adewole, Himont.

What essentially guides your selection of implant sites; I'm referring specifically to your recent article published from Paris in connection with shot bowel syndrome whereby transplants are being made now for the human bowel. Do you see any application in this area?

DR. WILLIAMS: That's a difficult question, what chooses our implantation site. Well, there are two answers to that. One is if you are looking at it clinically, then you choose the appropriate site where you have a problem.

He's from Scotland. I'm flattered. The trouble is I'm Welsh.

The second point, actually it's a very important question is to which implantation site you choose to study, both the host response and the degradation is extremely important. You have to choose one which -- well, there are two sorts of experiments.

The first one you choose, those which are - shall we say - experimentally reproduceable and surgically convenient and so we tend to choose subctaneous intramusclar or intraperitoneal sites. That's the first level of experimentation.

You then have to go to what we would refer to as a type of end use test in which you would put material in the more appropriate site. That might be within the bladder or within the urethra or the small bowel but then that's what you do in the second stage. I'm not quite sure how more I can answer that.

DR. ADEWOLE: Thank you.

DR. WILLIAMS: Don?

DR. GIBBONS: David, I would just like to emphasize again the point you made. And that is that once the material has become particulate, the whole name of the game can completely change so that because you are exposing in many cases, one has to assume a different set of enzymes inside the cell in the sense that the phagocytic responsibility there is not necessarily the same as that on an external environment and enzymes that are secreted and one can see that very much in a whole set of implants related to ligaments whereas if you take the polyethylene teraphthalates then they tend to lead to a mcuh stronger response when they cause particles due to wear.

So again I just want to emphasize that bulk, your point about morphology bulk versus size and therefore whether it can be phagocytized makes a very important difference.

DR. BRUCK: Stephen Bruck.

I would like to make a comment on biodegradation polymers that you get into. Take polyurethane which was implanted subcutaneously for two years and no degradation was seen. Having extruded the material, however, you can take additionally six weeks cracking tests close to it.

In our case the material was implanted in the air which blood flow, not the subcutaneous tissue, was in the environment and you get another type of degradation. So what I'm saying in effect is that the word "biodegradation" as applied to implants have to be specific not only to the materials but the manufacturing procedures which are basically a step which you can't control always and foresee, and therefore, what you get is fabrication standards which should have been involved where you can compare the material at a certain stage and how it degrades to be able to then predict the environmental impact upon them.

DR. WILLIAMS: I agree entirely. That was perhaps implicit if not explicit in what I was saying about characterizing the conditions. No material scientist would dream of doing any study unless he fully characterized the material both intrinsically and in relation to any present technology.

DR. SCOTT: Scott, Aston.

I wanted to reinforce the statement you made about the definition of degradation. It certainly doesn't always involve chain scission, in fact, if you look at any textbook on polymer degradation, you find that there is always a large section on elimination of side chains and of course the classic case is PVC where you can lose the whole of the hydrogen chloride from the colecule without reducing the molecular weight at all.

I think the problem has arisen in that we are talking about oxidative conditions and under oxidative conditions, then you do generally get chain scission, but as a croad definition, I would agree with you.

DR. LOOMIS: I am going to ask these last couple of questions be fast because we are way off schedule.

DR. VERT: Michele Vert.

David, I guss you definition on biodegradation mentioned the work "specific" and I guess we didn't get a consensus in Chester probably because of that word.

I get confused by what you mean by this "specific". You mentioned several times that enzyme, for example, can "contribute" and then you correct, "influence". Contribute is a specific activity. Influence can be for many things which have nothing to do with a biological activity. It could be a physical, chemical effect and so on.

Could you clear up your mind on this label for the people who are not very familiar with this area?

DR. WILLIAMS: Well, in the interests of Anglo-French collaboration here, Michele, I think I would agree with you that I was perhaps not quite specific enough with my definition of specific.

I put that word in, perhaps I really was trying to talk of a definite biological activity rather than there being no biologic activity at all, rather than there being rather loose. Perhaps by using the word "specific" I have gone too far in terms of the preciseness. Perhaps we can discuss that in general.

I don't think we are too far apart on that but I agree that perhaps specific is too strong a word there.

DR. VERT: Yes, maybe biological activity is enough.

DR. WILLIAMS: Well, yes, it may be, yes.

DR. VERT: The second point is involving the comparison between in vivo and in vitro data. I guess it is quite difficult to base the demonstration of a biological activity on the in vitro conditions when we don't know exactly which factors are contributing.

And the fourth point is about the audio activity of the P.E.T. I have got involved in a research program dealing with P.E.T. for bottling, for bottles, to make bottles. And we have got a problem because there is a lot of residual, not residual but degradation products in the P.E.T. which are connected with the processing and one of these degradation products is an acetaldehyde which is regularly produced as soon as you heat up the P.E.T.

How have you treated your P.E.T. samples, did you heat them up until they're melting or up to melt?

DR. WILLIAMS: No, we didn't.

DR. VERT: Okay. I guess you should present the data in a sense that in order to stress the ratio of release activity with respect to the initial activity and I guess your slide was in DPM, right?

DR. WILLIAMS: DPM, yes.

DR. VERT: You should put it because what's the amount of what your activity with respect from the...

DR. WILLIAMS: Yes, I agree with you. I was asked to present an overview here and there was a lot more data which I wasn't actually putting on there, yes.

DR. VERT: Because it is very small.

DR. WILLIAMS: It is small, yes. I haven't got the figure with me.

DR. VERT: One per cent, less?

DR. WILLIAMS: It is not less than one per cent.

DR. VERT: Because you can have up to 100 DPM acetaldehyde in there.

DR. WILLIAMS: Okay.

DR. LOOMIS: A quick question from Dr. Albertsson.

DR. ALBERTSSON: I will just give a short comment. You say that polyethylene is enough but then we have to remember that you can measure part of the polyethylene off-site very quickly and you can also measure how water goes into the polyethylene.

DR. WILLIAMS: I didn't say it was inert. I have never used that term, Ann-Christine.

DR. LOOMIS: Thank you, Dave.

# Evaluating Biodegradable Plastics with *in vitro* Enzyme Assays: Additives Which Accelerate the Rate of Biodegradation

Paul Allenza, Julie Schollmeyer and Ronald P. Rohrbach
Allied-Signal Research and Technology
Des Plaines, IL

## SUMMARY

A common element in the developing technology for producing biodegradable plastics is the use of biodegradable additives such as starch or cellulose. Methods to determine the degradation of such materials have proven inadequate either due to the length of time necessary to perform the tests, or due to problems in their reproducibility and quantitative. A method based on the use of specific enzymes in an *in vitro* testing method has been developed which allows rapid testing and quantitative of degradation products without interference due to microbial growth and metabolic products. Crucial to the use of enzymes for such assays was the finding that certain additives, most notable surfactants, dramatically increase the rate and extent of enzymatic degradation of the incorporated starch or cellulose.

## INTRODUCTION

The biodegradation of plastics and/or the additives which are incorporated into plastics can be viewed as taking place in two steps or phases. As shown in Figure 1, the initial phase results in a deteriorated plastic article by virtue of the biodegradation of one of the components of the plastic, or through the action of other environmental stresses, such as oxidation or ultraviolet light. The deteriorated plastic can then be biodegraded only if the deterioration was sufficient to have caused a decrease in molecular weight to the point where microorganism are capable of transporting the fragments and metabolizing them (1,7). This second phase or series of events results in the actual biodegradation, or decomposition of the material. For most of the commercialized biodegradable plastic materials it is the first phase of deterioration which contributes to the degradability of the article. For example, in the case of corn starch as an additive in polyethylene, the starch is hydrolyzed by extracellular microbial enzymes commonly found in soil. This loss of the starch component weakens the article. Likewise, in those cases in which a pro-oxidant is added to polyethylene to promote degradation, the pro-oxidant reacts with oxygen to cause breaks in the polyethylene chain (2,3). Again, in this case the primary degradation events cause the deterioration of the material and loss of its desirable physical properties. Whether the products of biological, chemical or physical deterioration can be metabolized and completely degraded is often uncertain (4,5,6,7).

Figure 1.

Process of Deterioration and Degradation

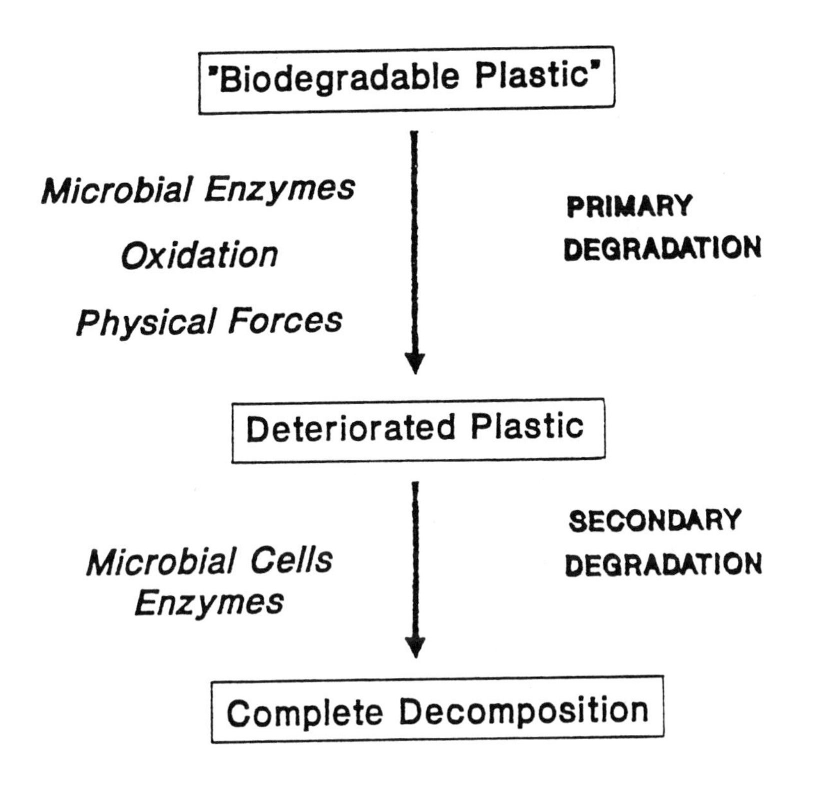

It is the objective of most degradation tests to measure the rate and extent of this primary degradation. To evaluate the actual decomposition of the article is almost inevitably a much slower process. Few technologies are available which can be expected to cause demonstrable decomposition of the plastic polymer in a reasonable period of time. It is the purpose of this study to contribute to the understanding of this initial mechanism of deterioration, specifically the role of microbial enzymes in the hydrolysis of biodegradable additives, and to make possible more efficient and standardized methods for determining "biodegradability". It has been found that the rate and extent of the enzymatic degradation of biodegradable plastics can be significantly improved with the use of additives such as surfactants in the enzyme mixture.

## MATERIALS AND METHODS

Preparation of Enzymes Solutions. In the standard assay method the biodegradation of starch in plastic films was based on the measurement of glucose produced as the starch was hydrolyzed by a mixture of amylases. The enzyme mixture was prepared as a ten-fold concentrate containing 305 units/ml each of amyloglucosidase (*Aspergillus niger*), beta-amylase (*Aspergillus oryzae*), beta-amylase (*Bacillus subtilus*) and 0.2 units/ml pullulanase (*Enterobacter aerogenes*) in 0.1 M acetate buffer, pH 4.8. All enzymes were purchased from Sigma Chemical Co., St. Louis, MO and reference to enzyme activity (units) is according to the supplier's specifications. The enzyme solution was immediately dialyzed at 4°C against the same buffer until the concentration of glucose (present as a contaminant in the commercial enzyme preparations) was below 1 mg/ml. The concentrated enzyme solution was then reconstituted to contain 1mM $CaCl_2$ and 3 mM sodium azide in 0.01 M acetate buffer, pH 4.8. The final preparation was stored at 4°C after sterilization by filtration through a 0.2 - filter.

For the enzymatic degradation of cellulose, an enzyme solution was prepared as a five-fold concentrate with 0.2 g/ml cellulase (Cellulase TV Concentrate, Miles Laboratory, Elkhart, IN) in 0.1 M acetate buffer, pH 4.8. The enzyme solution was dialyzed against the same buffer at 4°C until the glucose concentration was below 2mg/ml. Due to the rapid degradation of dialysis membranes by the cellulase solution it was necessary to change the membrane approximately every 30 minutes. In some cases the glucose concentration was reduced by repeated filtration and dilution of the enzyme solution using a 10,000 molecular weight cutoff Amicon ultrafiltration membrane and filter apparatus (Amicon, Danvers, MA). In either case, the concentrated enzyme solution was reconstituted to contain 0.2 mM sodium azide and 0.1 M acetate buffer, pH 4.8 and then filter sterilized.

Degradation Assays. For the determination of the rate and extent of starch or cellulose degradation, plastic samples containing either material were immersed in 5 ml of the appropriate enzyme solution in sterile screw capped tubes. The tubes were incubated at 45°C and agitated at 150 rpm. At appropriate time intervals 10 -1 samples were removed and glucose concentrations determined using a glucose analyzer (Beckman Instruments, Fullerton, CA). In all cases appropriate controls were run simultaneously and included samples containing the enzyme only (no plastic), the plastic only (no enzyme), and enzyme with beta-cellulose or pearl starch to assure full enzyme activity with a known substrate. In some cases, where it was necessary to run the assay for several weeks, fresh enzyme solution was added after removal of the original solution. When this was done, new initial glucose readings were recorded and the total degradation was based on the cumulative glucose production.

Additions to the Standard Degradation Assay. For the determination of enzyme degradation in the presence of surfactants or other additives, the addition was made to the enzyme solution immediately before adding the plastic sample. Appropriate glucose standards were also used as controls to assure no interference of the additive with the measurement of glucose concentration by the glucose analyzer. The surfactants used included Igepal CO-630 (GAF Chemicals Corp., Cincinnati, OH), cetylpyridinium chloride, Triton X100, and sodium dodecyl sulfate (Sigma Chemical Corp., St. Louis, MO).

Plastic Samples. Plastic films were either commercially available materials containing 6% corn starch, or were provided by the Northern Regional Research Center, USDA, Peoria, IL (40% starch, 45% ethylene acrylic acid, 15% urea), or were produced for the purpose of this investigation and consisted of 40% pearl starch or beta-cellulose in high density polyethylene.

## RESULTS

For the determination of the rate and extent of enzymatic hydrolysis of the starch component in starch-containing "biodegradable" plastics, a useful enzyme based testing method was developed by modifying standard assay procedures. The standard assay consisted of an enzyme mixture containing alpha amylase, glucoamylase, and pullulanase, with the addition of sodium azide to prevent microbial growth. The degradation of starch itself could be readily measured in such an assay system by monitoring glucose production as the combined action of the glucose. The same assay method used for the purpose of measuring the degradation of plastics. Samples were placed in the enzyme solution and incubated at 45$^{o}$C with moderate agitation. As shown in Figure 2, the amount of starch which was degraded in these plastics varied significantly with the type of plastic used. However, in all cases the amount of starch degraded was quite low, less than 20% of the total starch content, and with some plastics as little as 5% of the starch was hydrolyzed to glucose. This was consistent with the findings of Gould et. al. (8) who concluded that enzymes did not significantly degrade starch in some plastics. This observation appears to be especially true with regard to the degradation of starch in the EAA based polymers perhaps because the starch is incorporated into the polymer in a manner which allows the formation of a more intimate mixture of the two components (9,10).

Several studies have demonstrated that microorganisms are capable of complete or near complete removal of starch from such plastics

**Figure 2.** Enzymatic Degradation of Starch in Plastics

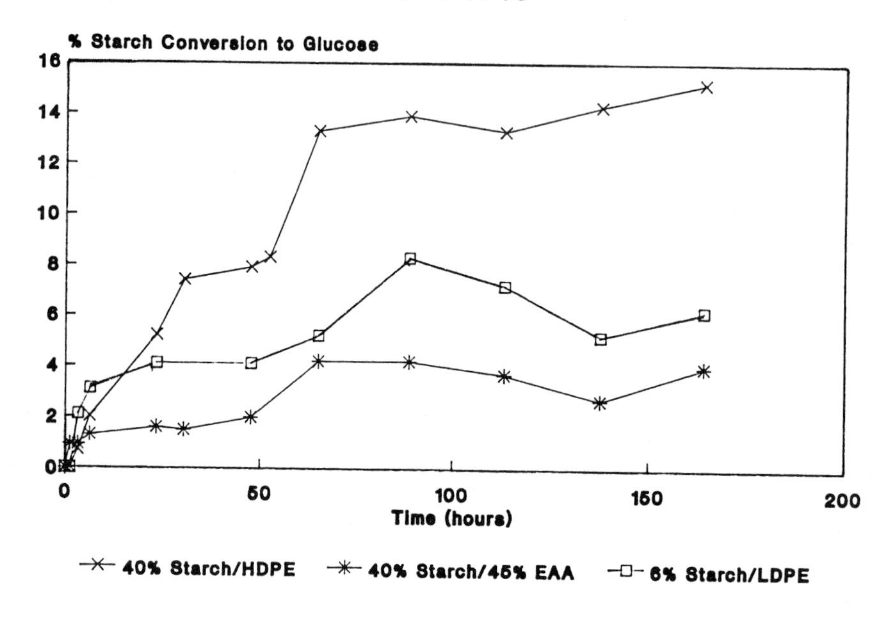


**Figure 3.** Starch Degradation in Plastics
Effect of Additions to Assay Mix

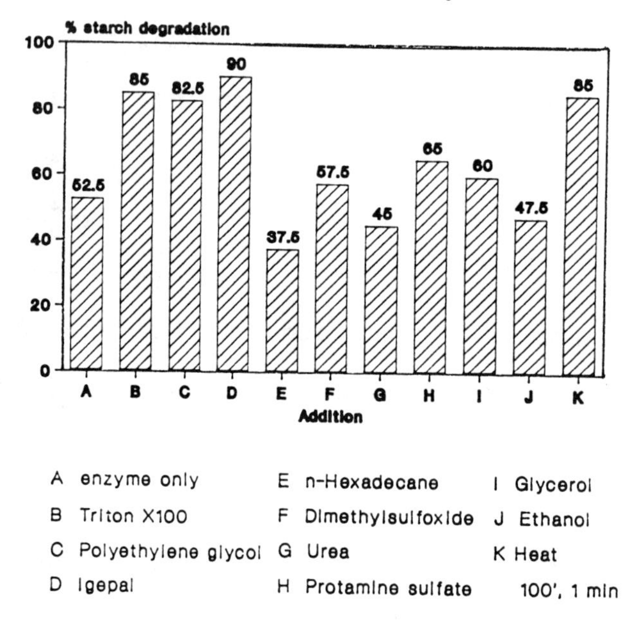

(8,11,12). Regardless of the type of microorganism used, it is the extracellular microbial amylases which are responsible for the initial degradation of the starch. Therefore, such enzymes must be able to hydrolyze the starch within these plastics. The fact that some microorganisms have been found to be more efficient in this process than others (8,13) suggests that there may be particular characteristics of enzymes which can promote this activity. In an effort to better understand the limiting aspect of the degradation of starch by microbial amylases, a variety of additions were made to the standard enzyme assay mixture. Figure 3 summarizes the results of these experiments and the significant effect that some of the assay additives had on the amount of starch which was degraded. Most notable was the effect of the surfactants (Igepal and Triton X100) and polyethylene glycol on the enzyme degradation. The treatment of the polymer by boiling also resulted in an improvement of the degradation of the starch; however, increasing degradation by this physical disruption of the material was not further examined. The effect of the surfactants on the degradation of the plastic was examined more carefully using three different starch containing plastics, one of 40% starch in HDPE, one of 40% starch with 45% EAA, and one of 6% starch in low density polyethylene. In all cases (Figures 4, 5 and 6) the surfactants greatly improved the rate of degradation and the extent to which the starch was degraded by the standard mixed amylase assay mixture. Figure 7 is a scanning electron micrograph of the starch/HDPE plastic before and after enzyme degradation (with surfactant) and clearly shows near complete removal of the starch in the absence of microorganisms.

Figure 4. Degradation of 6% Starch Film
Effect of Additives

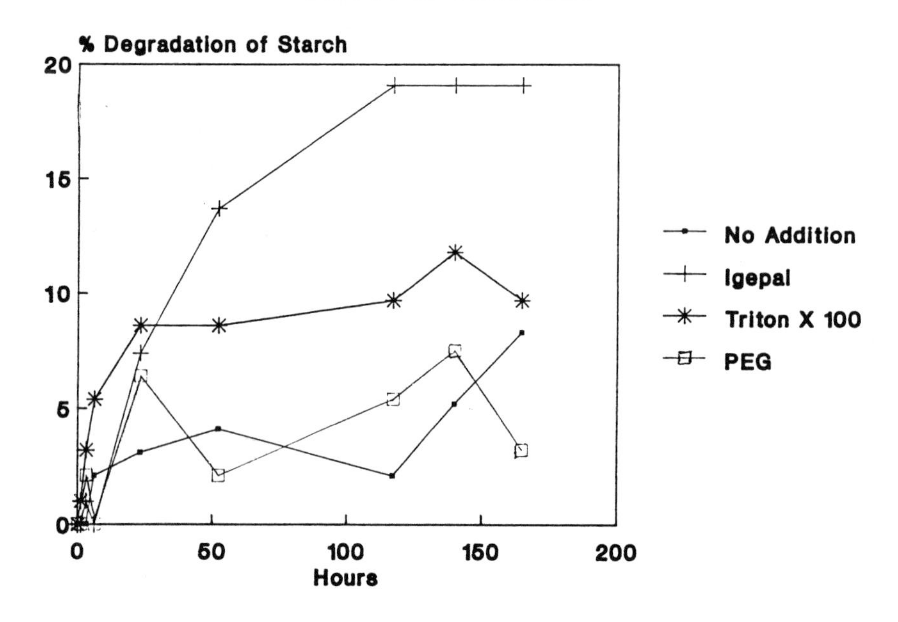


Figure 5. Degradation of 40% Starch/
EAA Film: Effect of Additives

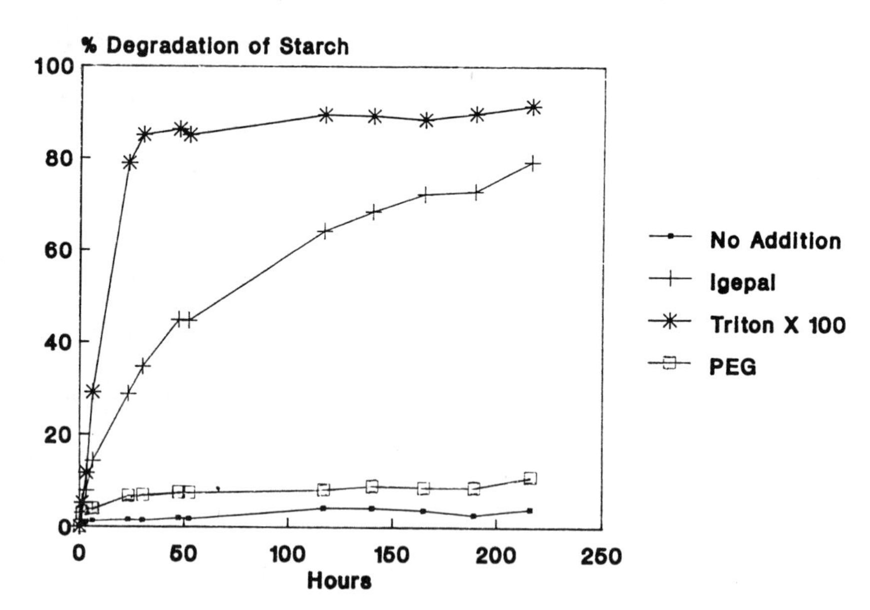

Figure 6. Degradation of 40% Starch/ HDPE Film: Effect of Additives

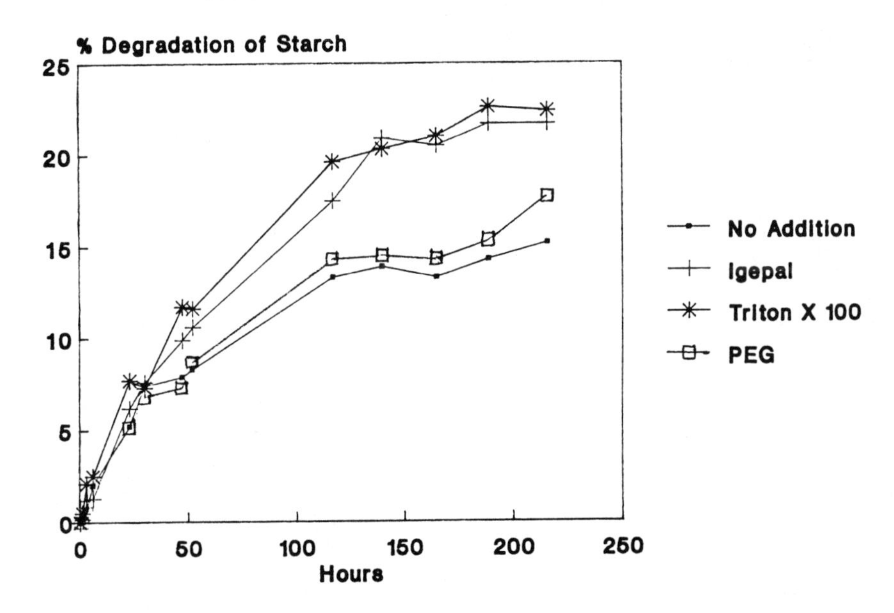


**Figure 8.** Biodegradation of Cellulose/HDPE Effect of Surfactant

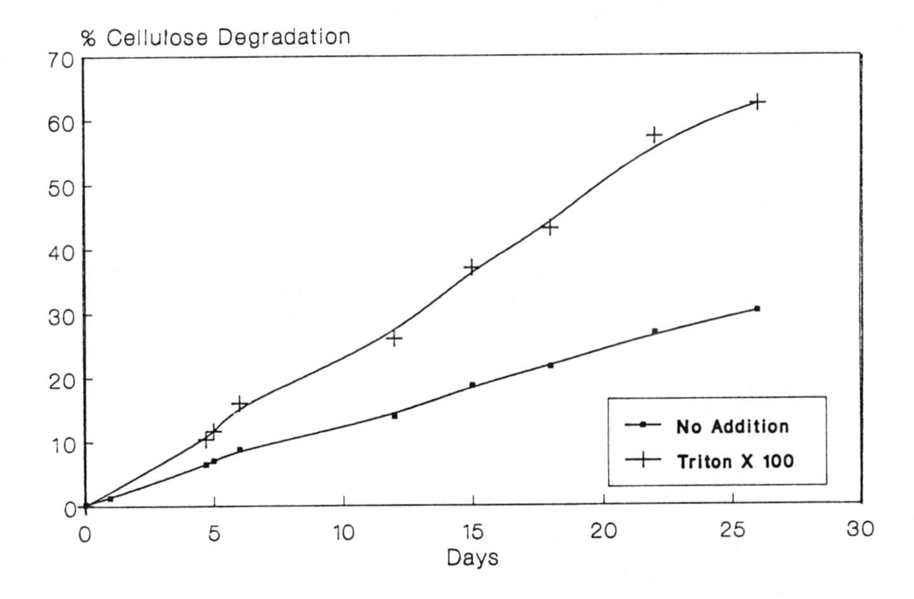

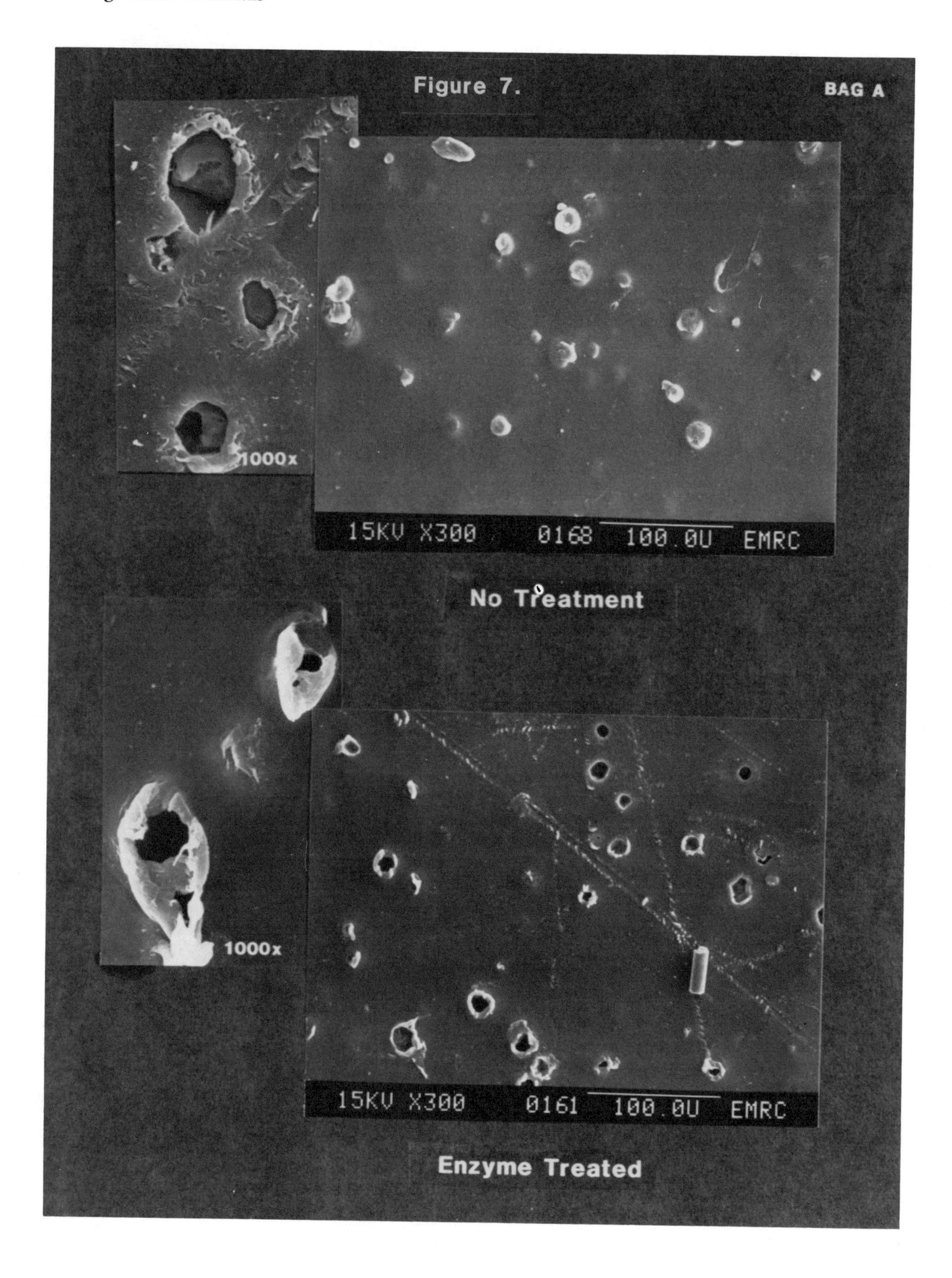
Figure 7.
BAG A
1000x
15KV X300 0168 100.0U EMRC
No Treatment
1000x
15KV X300 0161 100.0U EMRC
Enzyme Treated

To determine if this effect was unique to starch, for example due to the surface properties of starch granules or an interaction between starch and EAA, surfactants were added to an assay mixture designed to measure the degradation of cellulosic materials. Figure 8 shows the degradation of a sample of HDPE containing 40% beta-cellulose. The increased degradation in the presence of the surfactant demonstrates the usefulness of the surfactant for the biodegradation of cellulosic materials as well as starch containing materials.

To determine if this improved degradation was due to the efficiency of the enzymatic hydrolysis of the carbohydrate alone and unrelated to the incorporation of the starch or cellulose in a plastic film, the degradation of both substrates (not incorporated into plastic) was measured with and without the surfactant additives. As illustrated in Figure 9 and 10, the surfactant had little or not effect on the breakdown of these carbohydrates. Therefore, it is clear that the effect of the surfactant is related to the incorporation of the starch or cellulose in the plastic, but the mechanism by which these surfactants interact with the composites and/or the enzyme is not obvious. The type of surfactant was shown to influence the degradation enhancement. As shown in Table 1, the nonionic surfactants Igepal and Triton X100 worked much better than the anionic or cationic surfactants. The effect of the surfactant is at least in part due to its effect on the activity and stability of the amylases. For example, enzyme activity measured by pearl starch hydrolysis was significantly reduced in the presence of the charged surfactants. To further characterize the role of the surfactants the

**Figure 9.** Pearl Starch Degradation
Effect of Additives

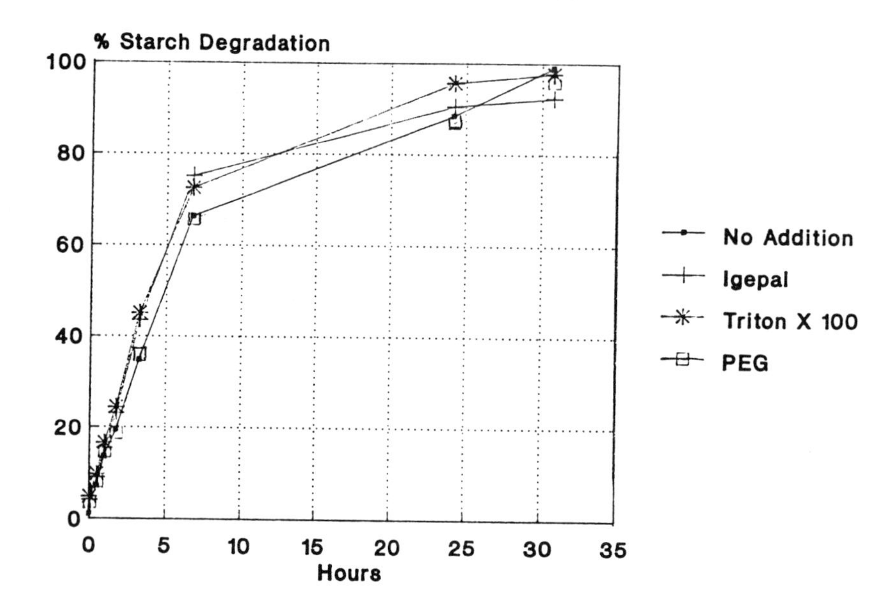


**Figure 10.** Cellulose Degradation
Effect of Additives

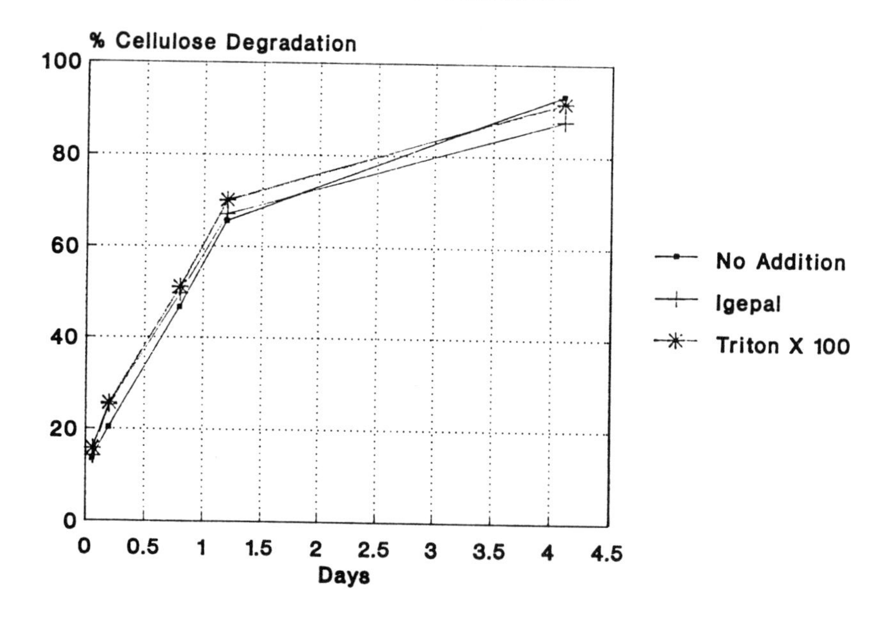

**Table 1. Effect of Surfactant Type on Degradation of Starch in Plastic Films**

| Addition to Assay | % Degradation of Starch | | |
|---|---|---|---|
| | Pearl Starch | 40% Starch + EAA | 40% Starch + HDPE |
| no addition | 95.3 | 5.0 | 5.7 |
| Igepal | 103.4 | 100.2 | 7.9 |
| Triton X100 | 99.0 | 102.3 | 8.0 |
| SDS | 35.0 | 10.5 | .8 |
| CPC | 59.9 | 7.0 | 1.2 |

degradation enhancement the surfactant treatment was separated from the enzyme degradation process. A plastic sample containing 40% starch or cellulose in HDPE was pretreated with Igepal or SDS for seven days at 60°C. The sample was then extensively washed in running water for several hours to remove residual surfactant. After this treatment degradation of the samples was measured with an enzyme solution containing no additives and compared to the degradation of an identical sample washed and treated in the same manner but without the surfactant. Figures 11 shows that the pretreatment did improve the degradation of the starch in the plastic sample; however, the degradation was not improved to the extent that it was when the enzyme and surfactant were added together (see Figure 6). The degradation of cellulose in a similar film was increased approximately to the same extent as was observed when the enzyme and surfactant were added together (data not shown).

## DISCUSSION

The results show that enzyme based assays can be useful for the determination of degradability of plastics provided appropriate additions are made to the assay mixture. Of the reagents tested, surfactants clearly had the most dramatic effects. The type of surfactant which is used also effects the enhancement of degradation, with the nonionic surfactants giving the best results. This difference among surfactants can be attributed to several factors including the effect that the surfactants have on the activity and stability of the enzymes, the effect on the interaction of the enzyme

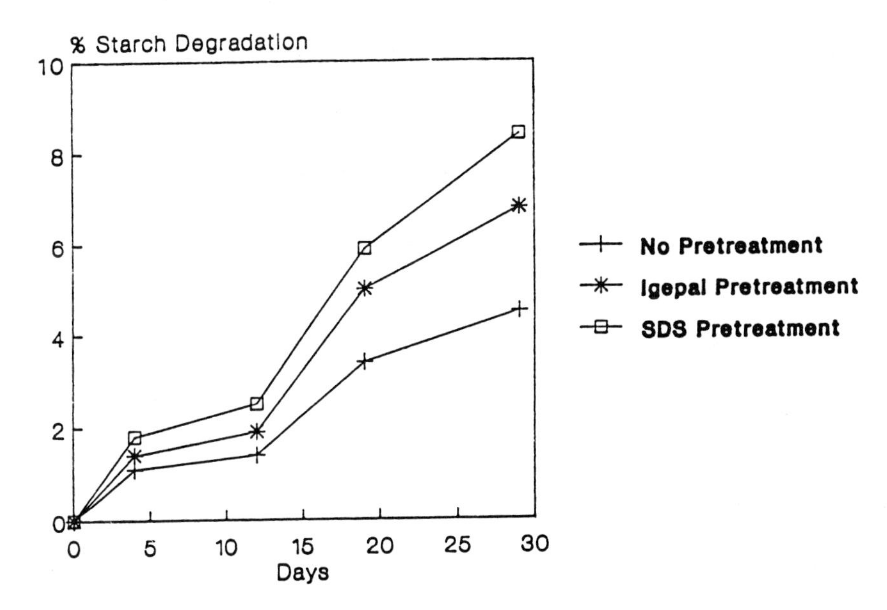
Figure 11. Surfactant Pretreatment:
Starch/HDPE Film
% Starch Degradation
10
8
6
4
2
0
0
5
10
15
20
25
30
Days
No Pretreatment
Igepal Pretreatment
SDS Pretreatment

with the plastic, and the differences in the critical micelle concentration for the surfactants. For the purposes of this study it was not necessary to determine the optimal surfactant / enzyme / plastic combinations.

The mechanism by which surfactants accelerate the enzymatic attack appears to be either through promoting the interaction of enzymes in the aqueous phase with the relatively hydrophobic plastic, or through the direct action of the surfactant on the plastic resulting in crazing and stress cracking (14,15) and thus increasing the availability of biodegradable domains. The fact that pretreatment of the plastic with a surfactant caused an increase in the rate of degradation suggests that it is the latter mechanism which is most important since in this example the enzyme and surfactant were not present together. It should be noted that residual surfactant may have been present, perhaps enough to promote an increased level of degradation. In addition the extent of degradation was not equal to that observed when the enzyme and surfactant were added together. It may therefore be that both mechanisms play a role in the acceleration of enzymatic biodegradation by amylases and cellulases.

The use of such surfactants in enzyme assays makes it possible to use such systems to accurately measure biodegradation of plastics containing starch or cellulose, and likely other degradable materials as well. Enzyme based assay systems can be easily standardized, are quantitative, and when performed with the appropriate additions, can be quite rapid. Enzyme assays should prove quite useful in determining the relative degradability, or susceptibility to biological mechanisms of deterioration, for a variety of "biodegradable" plastic products. It has the added advantage of providing direct, measurable information regarding the mechanism of degradation processes. Such assay methods can be criticized because they do not accurately represent conditions commonly found in natural environments, composts, or landfills. However, extracellular enzymes are the agents of starch and cellulose hydrolysis in any of these environments and no standardized assay system will ever accurately represent natural environments or degradation conditions. Enzyme based assays allow an acceleration of natural processes for comparison of plastic formulations. When microorganism based assay methods are used, and they are certainly necessary for monitoring degradation in may cases, surfactants could still be expected to accelerate the degradation process. Addition of surfactants should accelerate enzymatic degradation regardless of whether the enzymes are added to the system (as in an assay) or generated in situ during microbial growth.

In addition to their utility in biodegradation assay methods it is clear that surfactants could be used to greatly increase the rate of biodegradation of starch and cellulose in plastics after disposal in a landfill or other environments. It may be possible to accomplish this acceleration by adding the surfactant to the plastic at the point of disposal or incorporating the surfactants into the plastic material for release at an appropriate time. An understanding of the role that surfactants play in the acceleration of biodegradation by enzymes will be instrumental in determining the best way to bring about this acceleration.

## Literature Cited


1. Potts, J.E., in Encyclopedia of Polymer Science and Engineering. Vol 2. 1985. J. Wiley and Sons, Inc.; p 632.
2. Jones, P.H., D. Prasad, N. Heskins, N.H. Morgan and J.E. Guillet. 1974. Env. Sci. Technol., 8: 919.
3. Osawa, Z., N. Kurisu, K. Nagashima and K. Nadano. 1979. J. Appl. Polym. Sci., 23: 3583.
4. Albertson, A.C., and S. Karlsson. 1988. Polym. Mat. Sci. Eng. 58: 65.
5. Mungai, V.W., J.A. Cameron, J.F. Johnson and S.J. Huang. 1989 Polymer Preprints. 30: 1289.
6. Albertson, A.C., and S. Karlsson. 1988. J. Appl. Polym. Sci. 35: 1289.
7. Klemchuk, P.P. 1989. Modern Plastics 66: 48.
8. Gould, J.M., S.H. Gordon, L.B. Dexter and C.L. Swanson. 1988. Proc. Second Nat. Corn Util. Conf. Nov 17-18, 1988.
9. Otey, F.H., R.P. Westhoff and W.M. Doane. 1980. Ind. Eng. Chem. Prod. Res. Dev. 19: 592.
10. Otey, F.H., R.P. Westhoff and W.M. Doane. 1987. Ind. Eng. Chem. Res. 26: 1659.
11. Cole, M.A. 1989. Abst. Ann. Meet. Am. Soc. Microbiol. 89:Q204.
12. Proc. of the SPI Symp. on Deg. Plastics 1987. Washington, D.C.
13. Veda, S. and J.J. Marshall. 1980. Starch/Starke 32: 122.
14. Hoffman, R.D. 1982. J. Appl. Polym. Sci. 27: 2119.
15. Karlsson, S., O. Ljungquist and A.C. Albertson. 1988. Polym. Deg. and Stability. 21: 237.

QUESTIONS AND ANSWERS

DR. WOOL: I'm Richard Wool from the University of Illinois.

We have done extensive studies on the accessing of starch in plastics and our basic finding, and I will discuss in detail in my talk this afternoon, is that it occurs through a percolation process, such that if you -- in digestion experiments of this type, you need 31 per cent by volume starch.

Now, when you correct for densities, that turns out to be around 40 per cent by weight of starch, but it depends on the density of the plastic that you use in making this correction.

If you are a point or two below 40 per cent, then you would not get extensive degradation such as you observed with a high density polyethylene. If you are a point or two above 40 per cent, then you would have almost complete degradation. I think what you are seeing here with the variability of your additives is changes in the connectivity. That's one point. And so you are just shifting to the left or to the right of the percolation threshold.

Secondly, the percolation effect is very, very sensitive to skin effects. Okay, if you have a plastic rich skin, then you can completely kill off the enzyme accessibility to the starch.

DR. ALLENZA: I agree. I just want to point out that in these studies, we are not trying to make a point about how much of the starch is degradable in these materials.

DR. WOOL: It has very little to do with the other additives.

DR. ALLENZA: Well, I think in the pre-treatment some of this that shows out there is a relative effect.

DR. LOOMIS: With a given plastic.

DR. FEIJEN: Jan Feijen, University of Twente. I was wondering, do you use this starch particle the same, you use the same starch particles as you use in the control experiments because yesterday I heard that for embedding the starch particles in the matrix, sometimes they are Islamized at the outside and if you are considering what the excess of enzymes at the interface between the starch and the polymer, and I can't imagine that you increase the surface but if you use Islamized starch, then this theory becomes a little bit less straightforward.

DR. ALLENZA: Okay. This was not Islamized starch. It was untreated and this was just made for the purpose of this study and in fact, we had looked at surface acetaldehyded starch granules and found much poorer degradation so this is just a model film that was made for this and it was surface treated.

DR. FEIJEN: Okay, thank you.

DR. SHALABY: You attributed the effect of the surfactants to this stress cracking. On the other hand you also mentioned that the charged surfactant are different from the non-ionic surfactant and I found also that your acrylic co-polymer is much more responsive to surfactant then polyethylene alone. Would you exclude the effect of routing and the increased transport to the load of surfactant?

DR. ALLENZA: Transport in what way?

DR. SHALABY: Of water and enzymes into the bulk of the system. It is just a matter of mere routing, increase routing.

DR. ALLENZA: I agree that would be part of it. And that's why we did the one study to try to pretreat the plastic.

DR. SHALABY: Especially the Triton and the acrylic acid.

DR. ALLENZA: Right. That would go along with the stress cracking and crazing making it more accessible.

DR. SHALABY: Thank you, Paul.

# PLASTIC DEGRADATION AND RECYCLING THROUGH SELECTIVE SOLUBILITY

Melissa Farrah Bouzianis,
Marketing Consultant to Belland Inc.

When one thinks of degradable plastics, bio- and/or photodegradable plastics come to mind. However, there is another category of degradable materials that is neither bio- nor photodegradable and yet will play an important role in helping to solve the plastics pollution issue. These unique plastics have what is called controlled degradation. That is, they degrade under predetermined conditions and in a specified time frame.

Commercially, the controlled degradable materials are known as "selectively soluble plastics". These materials are engineered to degrade in a specified pH range or under desired conditions. The majority of the selectively soluble plastics are water insoluble, yet are readily degraded in mild alkaline solutions. These same plastics have the ability to degrade in the ocean, forming stable soluble salts. In the municipal waste water systems, they do not interfere with the normal flocculating or dispersing operations. They have also been found, through extensive ecological and toxicological tests to be non-toxic. When left on the roadside, they will break down slowly in the presence of soil. Lastly, complete combustion produced only non-hazardous by-products.

Degradable plastics are very desirable in helping to answer the plastics waste issue. Another path that is equally as desirable is recycling. A question that must be addressed by every degradable material is how they will impact recycling efforts. Selectively soluble without plastics can be handled by existing recycling equipmetn without any addition of equipment or investment. Since they are solubilized in the presence of milk alkaline solution, they will dissolve during the alkaline wash of most recycling process and are therefore, easily separable from other materials, including other plastics.

Another benefit of the selectiely soluble materials is their ability to enhance or enable the recyclability of other materials. For example, the wash times for PET bottles is currently three times longer than is needed to clean the bottles, but it is neceessary in order to remove the paper/plastic label and adhesive. If the adhesive and label were made from a selectively soluble plastic, they would wash off easily and completely during the first few minutes of the recycling alkaline wash.

In a similar application, selectively soluble labels can be used to promote reuse of containers of toxic materials, such as pesticides. After the container reaches the customer, it can be emptied and shipped back to the supplier. The supplier can wash it thoroughly and remove the label, then refill the container and ship it to another customer. This would save not only the cost of landfilling a toxic container, but would also be a form of source reduction by reducing the number of containers going to landfill.

## I. SELECTIVELY SOLUBLE PLASTICS

### A. Chemistry and Manufacturing

Selectively soluble plastics are based on acrylic chemistry. The commercial grades of selectively soluble plastics are predominantly copolymers of (meth)acrylic acids with (meth)acrylic esters such as ethyl acrylate, methyl acrylate, butyl acrylate, etc. A typical example of a selectively soluble polymer is a copolymer made of ethyl acrylate and acrylic acid:

```
-(CH2 -CH)   CH2  -CH-
       |   x        |
     COOEt        COOH
```

Faster solubility is incurred by using a higher ratio of acid to esters.

The molecular weight of most commercial grades is in the range of 150,000 to 200,000. In addition to the acrylic grades, a selectively soluble polystyrene has been produced that is approximately 70% styrene and 30% acrylic acid.

A reactive polymerization extrusion process is used to produce the selectively soluble plastics. During polymerization, reactive pendant groups are incorporated into the backbone of the main polymer chain. These reacative groups are, in most cases, carboxyl (-COOH) groups, which react in the presence of alkali to form a stable salt form of the polymer.

B. Types of Plastics

Over 1600 different plastics have been produced by varying the acid and ester types and concentrations. These plastics differ in their performance characteristics, including physical properties, time to degradation and in the medium under which degradation will take place. By varying the concentration of acids, different physical properties in the final plastic form can be attained. Twenty-five products have been commericalized.

In addition to the pure acrylic materials, other monomers have been polmerized, either by themselves, or as a copolymer with another monomer. The most noted one is the styrene acrylic copolymer. The benefit of the styrene is the reduction in manufacturing cost, due to the lower monomer costs, and the performance attained. This grade can be foam molded as well as injection molded. Fast food plastics is the most noted application area for this material. And, in fact, a major research program has been undertaken to prepare fast food plastics using the selectively soluble polystyrene for McDonalds of West Germany.

C. Foams

Selectively soluble plastics are commercially available as films, adhesives, coatings and injection molding and extrusion resins. Films have been manufactured for use as label stock material, laundry bags, garbage bags, overwrap films, agricultural fimls, and a host of other packaging applications. They are available in clear and pigmented versions and with a wide range of physical properties. The films can be purchased directly, or the resin can be purchased for forming by a converter.

Adhesives can take the form of hot melt or pressure sensitive adhesives. They are used with the label stock film to form a selectively soluble, pressure sensitive label for a variety of uses. Other applications include adhesives for non-woven fabrics and interlayers for multilayer bottles.

Temporary protective coatings are the most well known use for the selectively soluble coatings. The ability to remove the coating using a mild caustic solution has tremendous environmental implications over other temporary coatings. Some of the applications for thoese coatings include automobiles, boats, antigraffiti protection, tarnish protection for silver and silverplated products and camouflage coatings.

Resins for both extrusion and injection molding are available. The resins can be filled, reinforced, pigmented, or natural. The extrusion resin can be case or blown into film. The injection molding resin can also be foam molded.

All the forms of selectively soluble plastics can be engineered to degrade in a predetermined timeframe and under desired conditions. The main trade-off is in physical property performance. The performance characteristics of these plastics ranges from the elastic to high modulus materials.

## III. DEGRADATION

### A. Degradation Method

The degradation method of these plastics is one of controlled degradation through selective solubility. That is, these plastics are water insoluble until they come in contact with alkali solution where they will undergo an ionic transfer to produce a stable salt form of the polymer.

This formation of the salt is the means by which the polymer dissolves into solution. Shown in Figure I and II are the degradation mechanisms under two different alkali solutions, Figure I in the presence of sodium hydroxide, and Figure II in the presence of ammonium hydroxide.

B. Solutions Which Promote Degradation

There are a variety of alkaline solutions which initiate the solubilizing mechanism. They include sodium hydroxide, ammonium hydroxide, potassium hydroxide, caustic soda solutions and other caustics with pH values in the desired range. Many of the commercial grades are designed to respond to the pH range of 10 to 12.

This pH range can usually be achieved with a 2% concentration. Stronger solutions will also degrade the plastics, but are not recommended due to special handling requirements. There is an optimum of pH range for the fastest degradation. A solution with a pH of 13 will dissolve the plastic slower than a solution with a pH of 11 as shown in Figure III. Weaker bases will also work, but take longer.

C. Degradation Rate

The rate of degradation can be influenced by a number of factors including heat, agitation, pH, and the addition of alcohol to the degradation solution. Figures IV and V illustrate dissolution rate as a function of condition. In Figure IV, it can be noted that rate of degradation is directly proportional to an

# FIGURE I

## Dissolving Reaction in Sodium Hydroxide

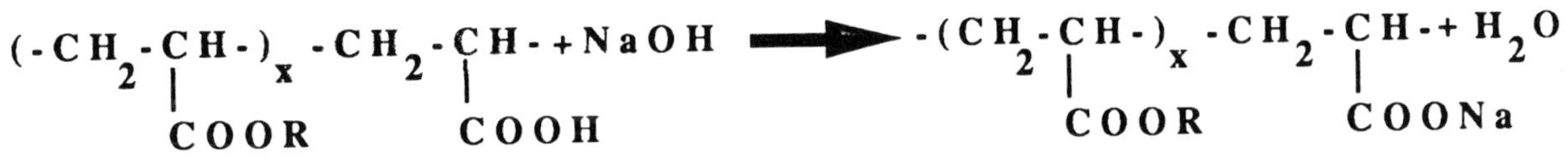

$$(\text{-}CH_2\text{-}\underset{\displaystyle COOR}{\underset{|}{C}}H\text{-})_x\text{-}CH_2\text{-}\underset{\displaystyle COOH}{\underset{|}{C}}H\text{-} + NaOH \longrightarrow \text{-}(CH_2\text{-}\underset{\displaystyle COOR}{\underset{|}{C}}H\text{-})_x\text{-}CH_2\text{-}\underset{\displaystyle COONa}{\underset{|}{C}}H\text{-} + H_2O$$

# FIGURE II

## Dissolving Reaction in Ammonium Hydroxide

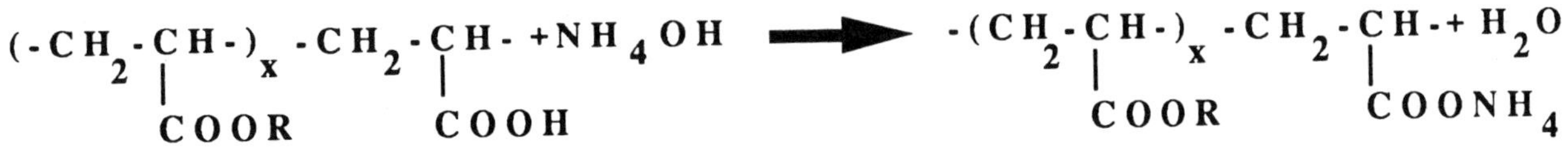

$$(\text{-}CH_2\text{-}\underset{\displaystyle COOR}{\underset{|}{C}}H\text{-})_x\text{-}CH_2\text{-}\underset{\displaystyle COOH}{\underset{|}{C}}H\text{-} + NH_4OH \longrightarrow \text{-}(CH_2\text{-}\underset{\displaystyle COOR}{\underset{|}{C}}H\text{-})_x\text{-}CH_2\text{-}\underset{\displaystyle COONH_4}{\underset{|}{C}}H\text{-} + H_2O$$

**FIGURE III**
**Dissolution Rate as a Function of pH**

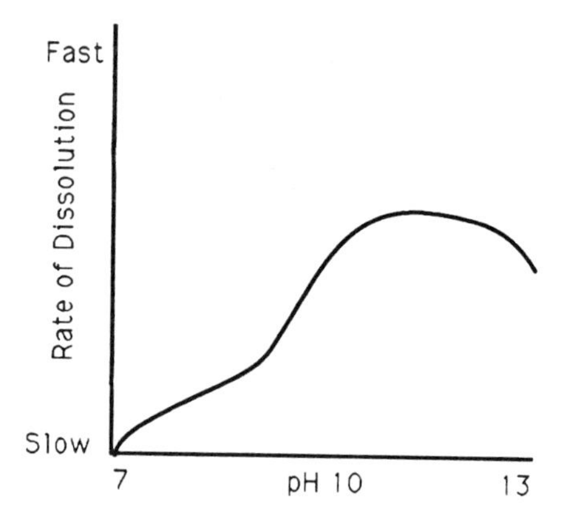


**FIGURE IV**
**Dissolution Rate as a Function of Temperature or Agitation**

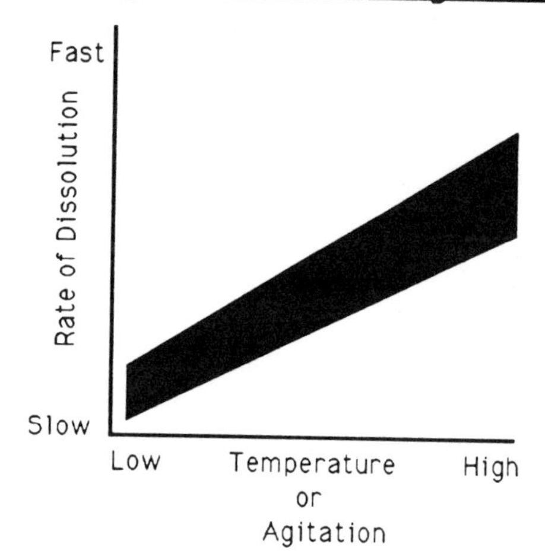


**FIGURE V**
**Dissolution as a Function of Solutions Mediums**

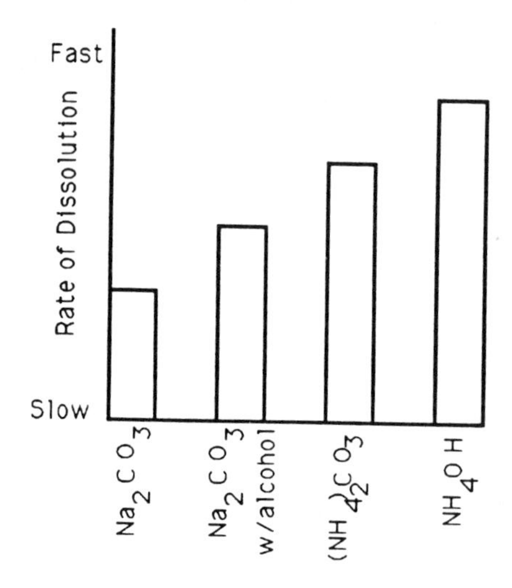

increase in both temperature and agitation. Figure V indicates the effect of various solutions of the rate of degradation, with ammonium hydroxide achieving the fastest dissolution rate.

## IV. ENVIRONMENTAL DEGRADATION

For many applications such as those used in fast food restaurants and school cafeterias, it will be possible to collect the majority of the used plastic items and expose them to the appropriate degradation medium. However, in other applications, this will not only be improbable, it most likely will be impossible. For these types of applications it is necessary to review how the selectively soluble plastics will behave in the environment.

### A. Ocean Degradation

Should the selectively soluble plastics be discarded in the ocean, they will dissolve in the ocean water. The degradation mechanism in this environment is thought to be one similar to that which occurs in the alkali medium. That is, the ocean water contains soluble salts which react with the pendant carboxyl groups to solubilize the polymer, rendering it a stable salt. Both field tests and laboratory tests have shown total degradation in ocean water within a short period of time.

The time to degradation is thought to be a combination of factors including the number of carboxyl groups on the main polymer chain, the thickness of the plastic, turbulence of the water, alkalinity of the water and to a lesser degree, temperature of the water. Table I lists approximate time to degradation in the ocean water for both actual field tests and laboratory tests. There is no standard test methodology to examine degradation in ocean water. Therefore, a procedure was developed and followed for both field tests and laboratory tests.

B. Composting

Composting of materials is another good method of reducing the volume of waste. The selectively soluble plastics breakdown in active composts is a period of approximately four to six weeks. It appears to be a combination of the moist environment and slight alkalinity in pH, resulting from the normal addition of lime to the compost pile, that brings about the degradation process.

C. Landfill

The selectively soluble plastics are not degrade in a landfill. There is normally not enough moisture and/or alkaline environment to cause any degradation.

D. Roadside

Should these plastics be discarded along the roadside, there is evidence that athey will degrade when they come in contact with moist soil. This is shown in Figure VI, by the results of a field test conducted on two soft drink cups. The red, white, and blue cup is coated on the inside with polyethylene (PE) and on the outside with a selectively soluble plastic. The other is only wax coated. The purpose of the selectively soluble coating is to allow

TABLE I

Ocean Degradation Studies
Field and Laboratory Tests

| Test | Sample | Time to Degradation |
|---|---|---|
| Field | Films - FBC 2620 | <17 days* |
| Laboratory | Films - FBC 2600 | <24 hours |
| Laboratory | Tensile Bars GBC 1000 | 24 hours |
| Laboratory | Molded tub GBC 1040 | 48 hours |

*Note: Due to inclement weather conditions, samples could not be viewed before 17 days in the ocean.

**FIGURE VI**

**Recycling**

$$-(CH_2-\underset{\displaystyle COOR}{\underset{|}{CH}}-)_x-CH_2-\underset{\displaystyle COONa}{\underset{|}{CH}}- + HCl \longrightarrow (-CH_2-\underset{\displaystyle COOR}{\underset{|}{CH}}-)_x-CH_2-\underset{\displaystyle COOH}{\underset{|}{CH}}- + NaCl$$

**FIGURE VII**

**Recycling**

$$-(CH_2-\underset{\displaystyle COOR}{\underset{|}{CH}}-)_x-CH_2-\underset{\displaystyle COONH_4}{\underset{|}{CH}}- + HCl \longrightarrow (-CH_2-\underset{\displaystyle COOR}{\underset{|}{CH}}-)_x-CH_2-\underset{\displaystyle COOH}{\underset{|}{CH}}- + NH_4Cl$$

the potential degradation of the paper, should the cup be discarded in the environment. After three months, it can be seen that the selectively soluble coated cup showed a greater degree of degradation, not only of the coating itself, but also of the paper, leaving the PE inner liner intact.

## E. Materials Recycling

Since recycling of materials is an efficient way of recovering and reusing resources, it is important to determine what effect new materials will have on existing recycling systems. The selectively soluble plastics are solubilized during the wash step in the recycling process. This makes them easily separable from other materials as well as from other plastics. Should the volumes be large enough, they can also be recovered from the wash solution and reused. If however, they are discharged into the municipal waste system, they will not adversely affect the system.

## F. Municipal Waste System

Many flushable applications are evaluating the use of selectively soluble plastics to reduce the incidence of clogged systems by flushed, non-selectively soluble materials. The solubilized plastics take the form of low molecular weight polyacrylates, similar to those used in municipal waste water treatment facilities as flocculants and dispersants. The net effect of these plastics on the water treatment facilities is zero.

Test results on these materials in the waste water treatment facilities have a BOD be 68, and a COD of 2,000.

### G. Incineration

In many instances, waste to energy is an efficient way of recovering the resources of plastics, particularly since their starting materials are all petroleum based. In the case of selectively soluble plastics, the components of incineration are carbon dioxide and water.

## V. RECYCLING THE SELECTIVELY SOLUBLE PLASTICS

One of the greatest virtues of the selectively soluble plastics is their ability to be recovered from solution and recycled. This is accomplished by first filtering out non-selectively soluble materials followed by a pH shift to precipitate the material and then a drying and repelletizing stage.

Recycling is most easily accomplished in closed systems where only one type of selectively soluble plastic is used, or where it is easy to collect the plastics, such as a fast food restaurant, school cafeteria or car coating operations.

### A. Recovery Method

As discussed earlier, the solubilized polymer is a stable salt, formed by an ionic transfer. This ionic reaction can be reversed in the presence of a weak acid to reprecipitate the polymer and form a salt as a by-product. The reprecipitation reaction is indicated in the reactions. Shown in Figure VI and VII. In Figure IV the sodium form of polymer is reacted with hydrochloric acid to form the polymer and sodium chloride salt. In Figure VII, the same principle is used, resulting in the formation of ammonium chloride salt.

### B. Properties

In closed systems, the properties of the recovered plastic is constant because materials entering the recycling stream are well known and may or may not be all made from one material. In other instances, the concentration is different and unpredictable. In this instance, the properties of the recyclate would be less predictable.

### C. Number of Recyclings

Tests to date indicate the physical properties of the plastic are not destroyed by the recycling procedure. The limitation to the number of recycling is dictated by the number of heat histories the material sees.

## VI. TOXICOLOGICAL INFORMATION

A significant number of tests have been completed to ascertain the ecological and toxicological properties of the selectively soluble plastics. The tests chosen are a myriad of standard procedures used to determine the safety of the products in a wide variety of conditions. The test results, summarized in Table II show the safety of these materials.

### A. Acute Toxicity

The acute toxicity of 4 different selectively soluble polymers

Table II

TOXICOLOGICAL TEST RESULTS

SUMMARY

| | |
|---|---|
| Acute toxicity | Non-Toxic |
| Aquatic toxicity | Non-Toxic |
| Bacteria Toxicity | No influence on soil or sludge respiration |
| Water treatment | BOD = 68 mg/g sample<br>COD = 2000 mg/g sample |
| Compost | Degrades 3-4 weeks |
| Ocean | Degrades <2 weeks |

was determined through tests on rats. The lowest toxic dose exceeded 5,000mg polymer per kg of body weight. Due to the extremely low toxicological effects, it was not possible to designate an LD50. the selectively soluble polymers were characterized as "non-toxic" according to the guidelines of the laws on chemicals.

B. Aquatic Toxicity

The acute toxicity of 4 selectively soluble polymers was determined through tests on water fleas (Daphina magna) [OECD-Immobilization Test].

The results also proved that the selectively soluble polymers are non-toxic. No effects were noted on water fleas, which are known to be extremely sensitive to very small quantities of environmental materials. Additional tests on trout, carp and green algae also showed the materials to be non-toxic.

C. Bacteriai Toxicity - Degradability

Various tests (ready biodegradation, OECD screening test, respiration in the Hach apparatus) performed on various systems (soil, sludge) have shown distinctly that the tested polymers did not have a negative influence on soil and sludge respiration. Low rates of decomposition were observed.

D. Water Treatment Plant

Should the dissolved selectively soluble polymer (polyelectrolytes) enter a water treatment plant, it may result in absorption in the sludge. Tests have shown that there is no significant influence of the sediment action process, or flocculation behavior of the sludge.

Table III gives both theoretical and as measured BOD and COD test results. The difference between the theoretical BOD of 1877 and the measured results of 68 indicates that there is no biodegradation occurring.

E. Compost

Studies have shown that selectively soluble plastic films, buried in the soil for 3 to 4 weeks will lose their structure and disintegrate.

F. Ocean

Selectively soluble plastics, films and molded parts, degrade in the ocean water in a matter of a day to a few weeks. The rate of degradation is dependent on the type of selectively soluble polymer, the thickness of the sample as well as the conditions in the ocean.

VII. COMMERCIAL AVAILABILITY

Selectively soluble plastics have been commercially available in Switzerland since 1983, where manufacturing for worldwide distribution takes place. Formal North American market introduction took place in May of 1989.

From the 1600 plastics, 25 were commercialized as standard products. They represent the range of materials available and should, in most cases, satisfy the requirements of the majority of applications.

For those applications where a specialized grade is necessary, the main frame computer, which contains formulation and property information on every selectively soluble plastic that has been manufactured, is searched to determine if a product exists, or if not, what formulations come to the closest to the desired product. The candidate materials are then produced and tested. Minor product modifications are then made to reach the desired performance requirements. Currently there are 7 to 8 million pounds of annual capacity available for the production of the selectively soluble plastics.

## VIII. CONCLUSIONS

Selectively soluble plastics have a place in helping to reduce the plastics pollution issue. Their main benefits include controlled degradation, solubility in non-toxic solutions, non-harmful to the environment, the ability to be recovered from solution for reuse and the ability to enable the reuse and improve the recyclability of other materials, such as other plastics and glass.

As the world continues to search for answers to the plastics pollution issue, one important place to look for the answer is in the solution!

Mrs. Bouzianis is an independent plastics and chemicals marketing consultant, currently working for Belland Inc. She is involved in the creation and implementation of the North American marketing plans for Belland's unique "selectively soluble" plastics and reactive extrusion polymerization systems. She participated in the preparation of degradable plastics and municipal solid waste reports for the U.S. Environmental Protection Agency and the Office of Technology Assessment of the U.S. Congress on behalf of Belland.

Prior to her work with Belland, she was employed as a senior consultant with the Chemicals and Plastics unit of Arthur D. Little, Inc. She was responsible for a number of acquisition analysis and the development of marketing plans for both domestic and international companies. Her area of expertise was in high performance plastics and environmental issues related to plastics.

Her earlier experiences were with the Plastics Division of the General Electric Company. There, she held several technical and marketing management positions.

Mrs. Bouzianis is the author of two recent articles dealing with the chlorofluorocarbon issue and its impact on the plastics industry. She has also authorized several technical papers on structural foam molding.

A graduate of the University of Lowell in Lowell, Massachusetts, she holds a B.S. degree in Biological Sciences and an M.S. in Plastics Engineering. She is a member of the Society of Plastics Engineers.

QUESTIONS AND ANSWERS

---PRESENTATION BY MS. MELISSA BOUZIANIS

(Belland Plastics)

DR. HELLER: First, I would like Farr & Associates Reporting, Inc. to point out that these materials are well-known as enteric coatings in the pharmaceutical industry.

MS. BOUZIANIS: Very true. Yes, very true. They're very similar to those materials, that's correct.

DR. HELLER: Yes, you mentioned that. They are enteric coatings, yes? The name is Jorge Heller.

MS. BOUZIANIS: Yes.

DR. FEIJEN: Jan Feijen, University of Twente. What happens in your regeneration of reprecipitation process if you leave these polymers too long in the basic solution?

MS. BOUZIANIS: As far as the reprecipitation?

DR. FEIJEN: Yes, so you dissolve them in basic solution.

MS. BOUZIANIS: They are in a basic solution. If you leave them in the basic solution for extended periods, nothing happens. They are very stable salt forms of the polymer.

DR. FEIJEN: Are you certain? I think those ester groups you have in the polymer will hydrolyze in time and change the properties of your polymer in time. So if you re-collect it and you put it in a basic solution, you will get back another polymer.

MS. BOUZIANIS: I'm not sure if that's the case. Our tests indicate to the extent that we have left them in the solution that that is not the case, but perhaps we haven't left them long enough.

DR. LOOMIS: How long have you actually left them in the solutions?

MS. BOUZIANIS: From my understanding, I believe less that 24 hours. Beyond that I don't believe. For the applications that we are targeting, it doesn't make any sense to leave them in any longer, frankly.

DR. ANDRADY: I'm Tony Andrady from Research Triangle Institute.

MS. BOUZIANIS: Hello.

DR. ANDRADY: Hi. I thought because I have been named in your presentation, I would reply your queries up front rather than waiting for question after my speech later on in the afternoon.

MS. BOUZIANIS: Thank you.

DR. ANDRADY: As far as the sea water degradability is concerned, sea water has a pH of around 8, and not surprisingly materials designed to dissolve at that temperature, at that pH does dissolve and I have sent the data to the Department of the Navy and I thought they would have already communicated that to you.

MS. BOUZIANIS: I haven't received it.

DR. ANDRADY: And as far as the laboratory testing is concerned, as I mentioned, we are still in the process of developing, but rather than going to a weight loss to define the degree of the dissolution, I tend to favour a test which looks at the deterioration of strength of a film as the dissolution progresses because ultimately film materials are used in load bearing applications, whether it is in a plastic bag or some other container, and the test that I am experimenting with now is to place a certain stress on this material, on a strip of material, like a stress relaxation for the polymer people and immersing it in a constant pH, constant temperature bath and monitoring the degree of stress as a function of time and these curves provide valuable information on the rates of decay.

And now I have a question. Have you looked into resin or monomer or oligomer content in your films?

MS. BOUZIANIS: Yes, we have.

At this point in time with the process the way it is, we do not have a Lewis System or a film true to basically to remove the residual monomers. It is our intent to do that in large production quantities so that there is a fairly high residual solvent content, not monomer content, more from a solvent standpoint, the alcohols that are used to solubolize the monomers and the initiators for that matter. So there is a high alcohol content, however, that we expect that with the addition of the Lewis System that we will be able to reduce the residual solvents and monomers down to about three to five parts per million.

DR. ANDRADY: Thank You.

MS. BOUZIANIS: You're welcome.

DR. HAMMER: Jim Hammer from 3M Company.
Your polymer may indeed be biodegradable or degradable but the process of going into solution, I would not define as degradable. If it is to be classified as degradable, then the polyacrylic acid that you get or the polymer that you get going into solution must degrade at that point.

So merely the dissolution process is not part of degradation or biodegradation. If something happens to the acrylic acid after the dissolution, then biodegradation is part of that process.

MS. BOUZIANIS: That's a good point and I think that's a good point and I think that's one of the reasons that we are here and interested to find out where we do fit.

DR. GILMOR: David Gilmor, University of Massachusetts. Just a quick comment.

For agricultural uses, you are going to have to look at this product very closely. One of the reasons being a lot of soils, particularly in the New England states tend to be very acidic, so you might not only have to worry about the ultimate fate of the dissolved polymer, but the possibility of reprecipitation and the fate of re-formed polymer as well.

MS. BOUZIANIS: That's a good point. As a matter of fact, one company that we have spoken to about it has a concern because rain has taken on such an acidity that there may be a problem with that. So we are obviously looking at that, yes.

DR. RUSE: Dorin Ruse, Centre for Bio Materials, Toronto.

I see that you are introducing actually a time bomb here. Your polymer doesn't degrade at all. It's an acid salt, an equilibrium we are talking about. You introduce that in the water, in the ocean, you go swimming, you swallow that stuff and it precipitates in your stomach. What's the pH that precipitator takes place, I think this is...

MS. BOUZIANIS: That is a valid concern. It is our contention that with the International Marpol Treaty against ocean dumping of plastics that there will be a very few and very selective applications where it is necessary to use a plastic to perhaps discharge other materials, biodegradable materials such as food stuffs into the ocean.

It is not our contention to say that everything that is being made not is going to be made into our plastics for the discharge of these materials into the ocean. That is not our intention. But there will be very specific, very small applications where it may be a factor and I don't believe that that will affect the ocean.

DR. LOOMIS: I would just like to make one comment. While it is obvious that just dissolving the polymer in the ocean is actually not degrading it, you have to admit that it has to be a heck of a
lot easier to degrade something in solution that a solid and we don't have all the surface science to work about.

So I think that the system should obviously be studied further but one would think on priority that dissolving it in the ocean is going to make it a good food for the microorganisms.

---Recess at 9:30 a.m.

# BIODEGRADATION OF A POLY(GLYCOLIC ACID) EPINEURAL TUBULIZATION DEVICE

D.F. Gibbons and T.W. Lewis

3M Center, St. Paul, Minnesota 55144

## ABSTRACT

Resorbable nerve repair devices made from a mat of chopped polyglycolic acid fibers were placed around the dissected, but intact, sciatic nerve of Sprague Dawley rats. Non-resorbable silicone tubing and a sham operation were controls. The resorbable device did not produce a fibrous capsule at either the luminal or outer surface. The chronic inflammatory response to the resorbing fibers disappeared when all of the fibers had resorbed and left only loose, vascularized connective tissue around the nerve; this result is quite similar to that produced by the sham. The silicone tubing was surrounded by a fibrous capsule on both its luminal and outer surface. Neither the resorbable nor silicone implants induced any changes in the architecture of the nerve.

## INTRODUCTION

The principal objective of peripheral nerve repair is to align and maintain the proximal and distal stumps of the severed nerve in the pre-injury alignment so that axonal regeneration across the repair site may occur. Clinically accepted methods of repair involve microsurgical suture techniques in which either epineural or fascicular (or group fascicular) sutures are utilized for the direct coaptation of the nerve stumps. Although these methods are now common practice, there are a number of factors related to these repair techniques which may limit the degree of reinnervation, thereby inhibiting the return of normal motor or sensory function. Included in these factors are the invasion of scar tissue and disorganized axonal regeneration due, at least in part, to the presence of the sutures.

As an alternative to placing sutures at the repair site, tubulization has been suggested as a method for nerve repair for almost 100 years (1). Reports of biodegradable tubulization procedures have included the use of collagen (2), polyglycolide-co-lactide (PGL) mesh (3), and polyglycolic acid (PGA) extruded tubes (4). In each case acceptable nerve regeneration was reported as a result of the procedure. We have also reported successful nerve regeneration in a non-human primate model utilizing a non-woven tubulization device made from PGA (5). All of these studies have reported rather detailed histologic evaluation of the axonal regeneration, but none has focused primary attention on the tissue response to the body absorable device and the sequence of events leading to complete device biodegradation.

This paper describes the histologic evaluation of a PGA non-woven tubulization device implanted around an intact rat sciatic nerve. The intact nerve model was utilized in order to demonstrate any potential deleterious effects of the products of the degradable device on normal nerve tissues. Our study includes not only the response of the neural elements but also the surrounding tissue to the degrading device. Since a non-absorbable silicone tubulization device has also been reported (6), the histologic response to a device of this composition was also evaluated and served as a control.

METHODS AND MATERIALS

Figure 1 illustrates the form of the PGA resorbable tubilization device. The non-woven tubulization devices used in this study were fabricated from Dexon® Beige braided #2 suture; the individual resultant fibers were 10-20 micra diameter. The braided suture was cut into 7 mm tow and debraided in dichloromethane under high shear in a blender. A non-woven mat was formed by an air-lay process, and the mat was formed into the U-shaped tube using a Teflon form and hexafluoroisopropanol (HFIP) as the bonding solvent. A flat rectangular piece which fits into the top of the U-shaped tube was fabricated using similar techniques. Excess processing solvent was removed under high vacuum. Combustion analysis indicated residual combined fluorine and chlorine content of approximately 1000 ppm in the devices implanted. Gas chromatography analysis of the devices sterilized by exposure to ethylene oxide gas revealed only trace amounts of ethylene oxide (1 ppm), ethylene glycol (57 ppm), and ethylene chlorohydrin (1 ppm) as residuals. The devices sterilized by gamma irradiation (1.5 Mrad) were completely hydrolyzed and analyzed by C-13 NMR and High Pressure Liquid Chromatography (HPLC). No effects of irradiation were detected by NMR methods. Small amounts of the following organic acids formed by gamma induced free radical processes were detected by HPLC: oxalic acid (520 ppm); tartaric

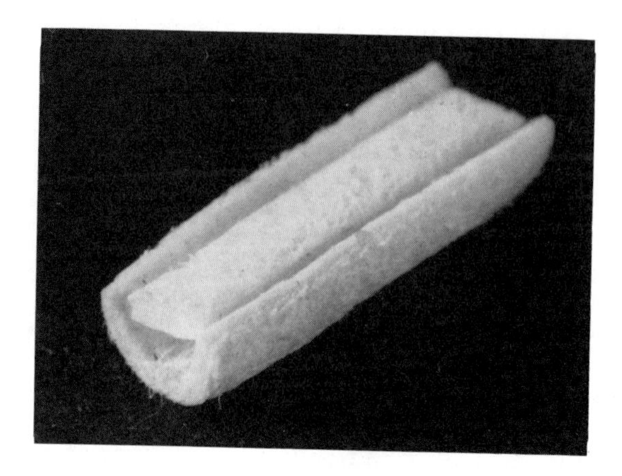

Figure 1. Resorbable Tubulization Device

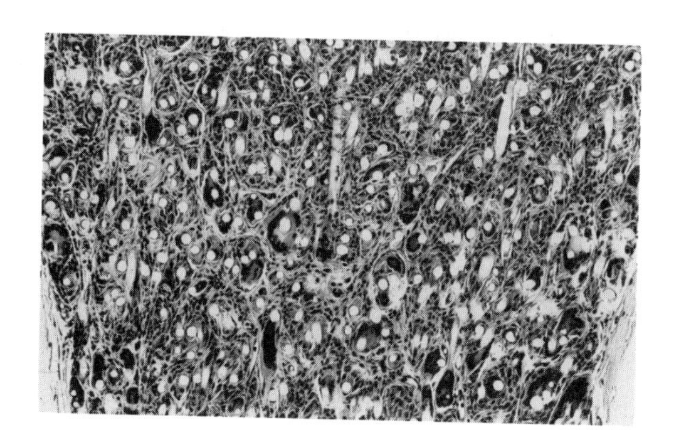

Figure 2(a). Tissue Response to Device at 1 Month X100

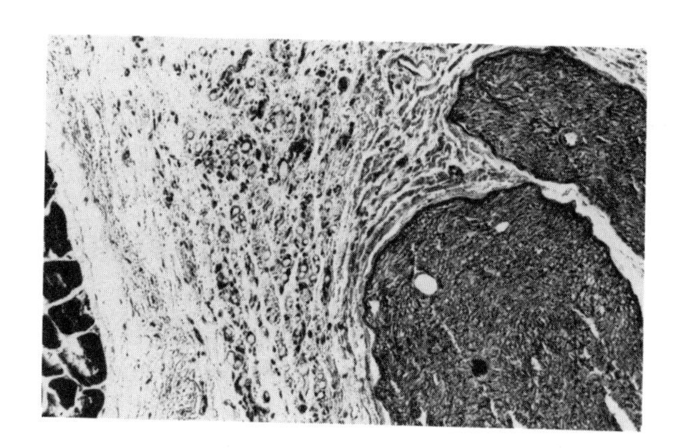

Figure 2(b). Tissue Response to Device at 3 Months X100

acid (210 ppm); acetic acid (370 ppm). Based on these analyses, we concluded that the processing and sterilization method did not substantially alter the chemical composition of the PGA suture.

The PGA tubulization devices used in this study were 1 cm long and had an inside diameter of 1.35-1.5 mm. The average pore size as measured by Coulter porosimetry* was 39.5 micron and the average porosity 90%. The wall thickness for both the tube and the rectangular lid was approximately 0.5 mm. The silicon tubing (Dow Corning Silastic cat. no. 602-265) had a 1.57 mm inside diameter and a wall thickness of 0.84 mm. The lumen size of the tubing was chosen according to Mackinnon (7) in order to avoid any damage caused by sciatic nerve compression.

Thirty-two male Sprague Dawley rats weighing 300-350 grams each were used in this study. The rats were randomly divided into four study groups defined by the time between device implantation and necropsy (one, three, six, and nine months). The sciatic nerves were exposed by an incision on the lateral side of each femur. Under the operating microscopy, the nerves were mobilized approximately 1 cm. Extreme care was taken not to traumatize the nerve during the dissection. Each rat had both sciatic nerves isolated and treated in one of four ways: mobilization and elevation only (sham); PGA tubulization device (gamma sterilized); PGA tubulization device (ethylene oxide sterilized); longitudinally split Dow Corning Silastic tubing. After the nerve had received

*Coulter Electronic Inc., Hileah, Florida

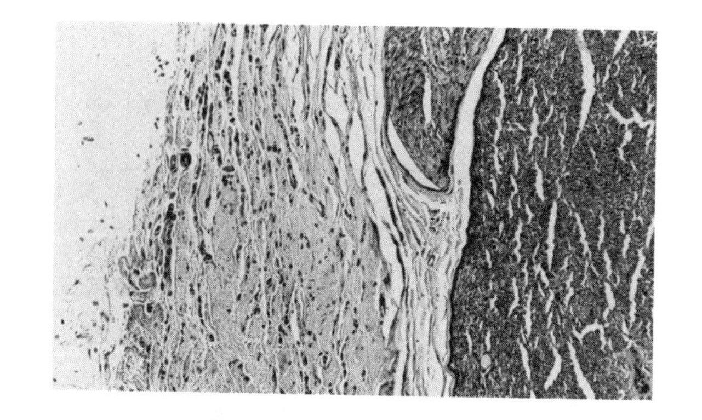

Figure 2(c). Tissue Response to Device at 6 Months X100

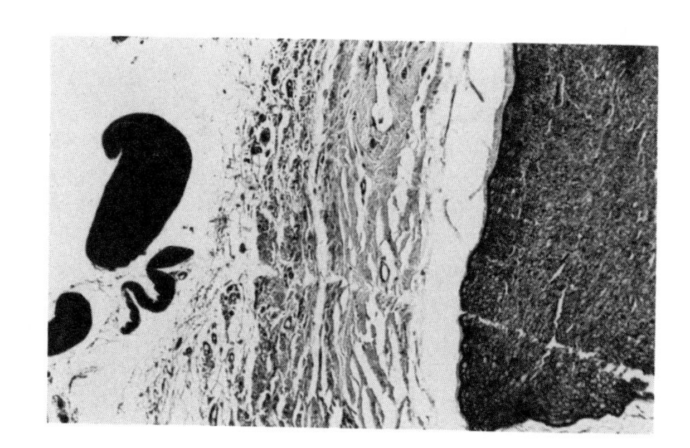

Figure 2(c). Tissue Response to Device at 9 Months X100

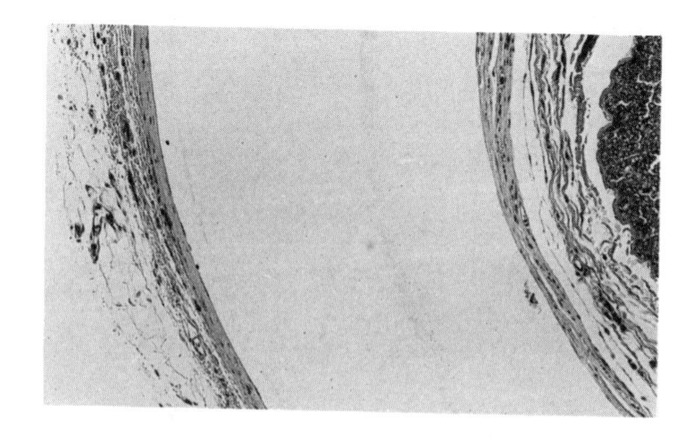

Figure 3(a). Tissue Response to Silastic Tubing at 3 Months X100

the designated treatment, the incisions were closed with skin staples*.

Animals were sacrificed at times corresponding to their study group designation. The animals were sacrificed in a $CO_2$ gas chamber and the sciatic nerves were exposed as in the initial surgery. The implantation sites and sham sites were removed en bloc, pinned on paraffin blocks at their normal length, and immersed in 10% neutral buffered saline for at least 48 hours prior to processing. The tissues were processed for standard histology. Transverse sections approximately 2 mm thick were taken from the proximal, distal and mid sections of the device. After dehydration the sections were processed for paraffin embedding. Sections were cut at 4.0 micra thick and stained with Hemoloxylin and Eosin, Masson's Trichrome, Giemsa, Severe-Munger (8).

## RESULTS

We were unable to identify any differences between the cellular response elicited by the sterilization proceedures, namely ethylene oxide and gamma irradiation. Therefore, the histology presented in these results will be that from the implants sterilized with ethylene oxide gas.

*The animals used in this study were held in compliance with the Laboratory Animal Welfare Act of 1966 (P.L. 89-544), as amended by the Animal Welfare Act of 1970 (P.L. 91-579) and amended by the Animal Welfare Act of 1976 (P.L. 94-279).

Figure 2 a-d show the tissue response associated with the resorbable polyglycolic acid tubulization device at 1,3,6, and 9 months. Figure 3 (a) and (b) show the tissue responses to the non-degradable, silastic, tubing at 3 and 9 months. Figure 3 (c) shows the result of the sham operation at 9 months.

At the end of 3 months, Figure 2 (a) and (b), the principal response is the ingrowth of capillaries, deposition of collagen, and the Foreign Body Giant Cell (FBGC) response around each of the PGA fibers. By 6 months the majority of the fibers have resorbed and the response is characterized by capillaries, collagen, and only a few FBGC's surrounding the remaining fibers, Figure 2 (c). It was also apparent that the effective thickness of the tissue response to the device decreased rapidly after the first month, Figure 2 (a-d). This corresponds to the resorbtion of PGA and consequent "collapse" of the thickness of the wall in the device. At 9 months all the PGA material is resorbed. The remaining tissue ia areolar collagen and capillaries, the inflammatory cells having disappeared, Figure 2 (d). The collagen organization is quite similar to that from the sham.

A major difference between the response to the fibrous network of the resorbable device and the silastic tubing is that there is no fibrous capsule formed around the PGA non-woven device, Figure 2 (a). The silastic tube elicits a fibrous capsule on the luminal and outer surfaces, Figure 3 (a).

No changes in the silastic nerve architecture were observed at any time with either the resorbable or silastic implants, Figure 4 (a). The perineurium and proportion of myelinated fibers remained

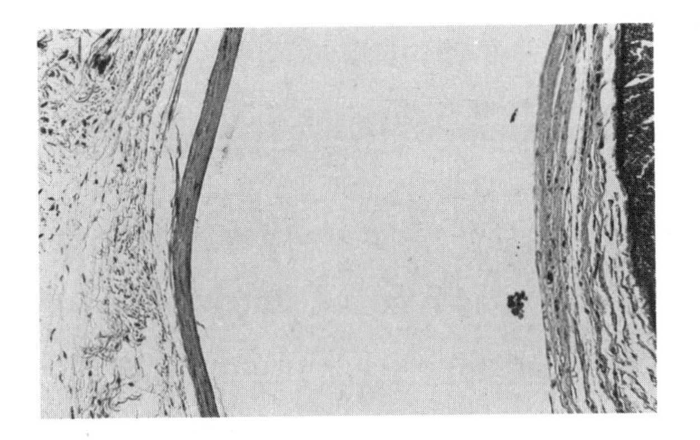

Figure 3(b). Tissue Response to Silastic Tubing at 9 Months X100

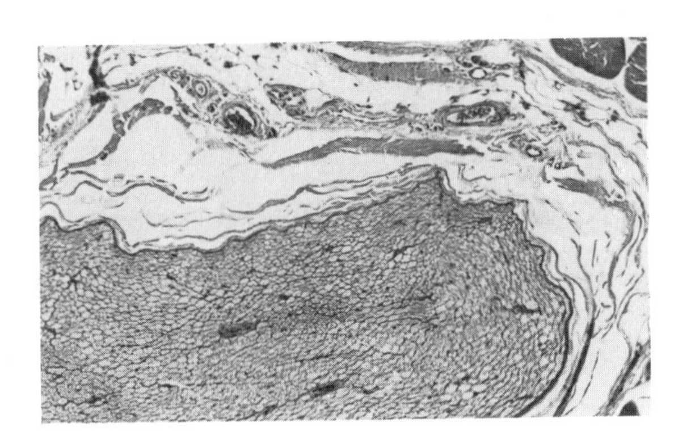

Figure 3(c). Tissue Response to Sham Operation at 9 Months X100

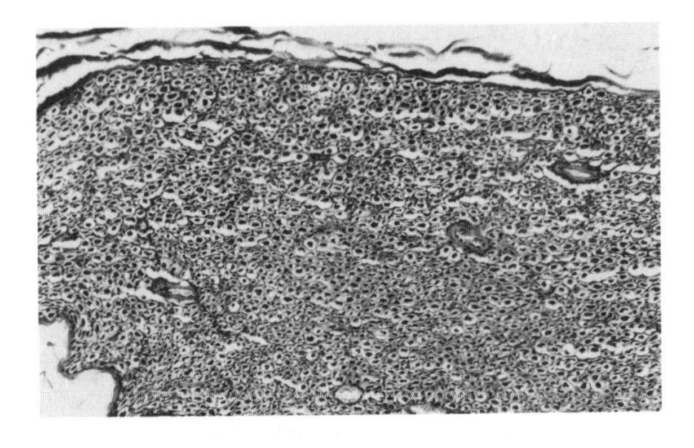

Figure 4. Sciatic Nerve at 1 Month X150

constant with no pathological changes.

## DISCUSSION

These experiments demonstrate that the degradation products, namely glycolic acid and relatively low oligomers of polyglycolic acid, are not cytotoxic to nerves at least when surrounded by an intact perineurium. This is confirmed by primate experiments in which the PGA resorbable device was used to repair a severed median nerve, and it was shown that the repair led to normal conduction velocities across the repair (5). The foreign body giant cell reaction adjacent to the fiber appears to be the principal inflammatory response to the degradation products. However, FBGC's may also be formed as a response to fibrous non-resorbable synthetic materials. The end result is that when the last vestiges of the particles produced during degradation have disappeared the FBGC's also disappear. Whether the stimulus for the FBGC's is phagocytosis of particles or response to a surface remains to be determined. The disappearance of FBGC's coincident with the resorbtion of the last of the material was also noted for the resorbtion of solid implants of polylactides (9).

The amount and distribution of fibroplasia associated with the PGA device is of interest. The absence of a well defined fibrous capsule surrounding the device is related to the effective "texture" produced by the fiber distribution at the surface. This has the advantage that it further reduces the potential trauma to the nerve, since it removes any "contracture" stresses associated with a fibrous capsule (10). The silicone tubing forms results in the formation of a permanent fibrous capsule and the lack of damage to the nerve confirms the data of Mackinnon (7), on which the I.D. of the silicone tubing was based.

The suitability of a resorbable, porous device, for nerve repair has been established. Its efficacy as a device to aid in the repair of severed nerves has also been established by following the innervation of the hand after transection of the median nerve in a one-year study in a primate model. The products of resorbtion of the polyglycolic acid device did not interfere with nerve repair as determined by histology and nerve condution velocity measurements (5).

The authors wish to thank G.S. Nelson for her expertise in preparing the histologic sections and D.M. Grussing for his surgical skill.

REFERENCES

1. Weiss, P. The Technology of Nerve Regeneration: A Review. Sutureless Tubulization and Related Methods of Nerve Repair. J. Neurosurg. 1:400-456, 1944.

2. Gibby W.A., Koerber H.R., Horch K.W. A Quantitative Evaluation of Suture and Tubulization Nerve Repair Techniques. J. Neurosurg. 58:574-579, 1983.

3. Molander H., Engkvist O., Hagglund J., Olsson Y., Torebjork E. Nerve Repair Using a Polyglactin Tube and Nerve Graft.

4. Rosen J.M., Hentz V.R., Kaplan E.N. Fascicular Tubulization: A Cellular Approach to Peripheral Nerve Repair. Ann. Plast. Surg. 11:397-411, 1983.

5. Tountas C.P., Bergman R.A., Barrows T.H., Mendenhall H.V., Gibbons D.F., Lewis T.W., Pyrek J.D>, Stone H.E. Peripheral Nerve Repair: Tubulization vs. Suture (A Twelve Month Primate Study). Presented at the 42nd Annual Meeting of the American Society for Surgery of the Hand, September 9-12, 1987 in San Antonio, Texas.

6. Ducker T.B., Hayes G.T. Experimental Improvements in the Use of Silastic Cuffs for Peripheral Nerve Repair. J. Neurosurg. 28:582-587, 1968.

7. Mackinnon S.E., Dellon A.L., Hudson A.R., Hunter D.A. Chronic Nerve Compression - An Experimental Model in the Rat. Ann. Plast. Surg. 13:112-120, 1984.

8. e.g. Theory and Practice of Histotechnology, Sheehan D.C. and Hrapchak B.B., C.V. Mosby Co., Second Edition 1980.

9. Nakamura T., J. Biomed. Mater. Res., 23: 115-1130, 1989.

10. Baker J.L., Chandler M.L. and LeVier R.R., Plastic and Reconstructive Surg., 68: 905-914, 1981.

QUESTIONS AND ANSWERS

---Upon resuming at 10:00 a.m.

---PRESENTATION BY DR. DON GIBBONS

(3M Center)

DR. LOOMIS: Thank you, Don. And here comes Professor Anderson

DR. ANDERSON: Jim Anderson, Cleveland.

Don, first you showed us a gross photograph which you described at one month and you talked about fibroplasia. But then when you showed us the histology at one month, you said there was no fibrous capsule.

DR. GIBBONS: There is no fibrous capsule surrounding it but the collagen has been deposited. So that all you are talking about, the differentiation that I wanted to make was the organization of the collagen which is formed.

If you have a solid material, one can tend to have a fibrous capsule which has a dense formation whereas in this case, the collagen is less dense, more realer and entwines within that surface and the only difference that that makes would be if contracture was an important factor so that whereas in mammary implants with a solid surface where contracture can cause problems when this form of structure is used, contracture would not be an effect.

DR. ANDERSON: What you are saying then that the fibroplasia which you say grossly at one month was actually scar formation associated with the surgical wound?

DR. GIBBONS: And the presence of the cuff, yes, but not as a thick fibrous capsule.

In other words, it is there but its structure is different. I think that one probably has to conclude that the surface on which the cells are sitting and producing the collagen have an influence on the organization of the collagen. That really is the difference.

DR. ANDERSON: I might remind you that I know that this does not come as a surprise to you, these observations. I recall the experiment that you and I did together some 15 years ago with polyglutamic acid.

DR. GIBBONS: That's right, exactly.

DR. ANDERSON: Thank you.

DR. VERT: Michele Vert.

Don, in the reconstruction of madibular or bone defects in humans using glycolic acid containing lactic co-polymers, okay, our surgeons observed rather strong -- well, let's say, inflammatory response with a lot of liquid around the implants. They were, well, small size implants.

DR. GIBBONS: Yes.

DR. VERT: Okay, and in some cases they failure, not real failure but problems with the worn closure.

DR. GIBBONS: Right.

DR. VERT: But the class last week in Germany told me that they have observed the same things with the bio fix PGA composite, stuff, you know, which comes from Finland and leading to 30 per cent of failures in condyle fracture fixations.

Did you observe something similar with your compound, in some occasions or...

DR. GIBBONS: No. I think the reason is why I went at great lengths to describe to you the morphology and on my overhead suggest that shape and therefore the concentration gradients surrounding the implant, the rate at which it is degrading will all affect the cellular, the biological response.

So I think that in those applications associated with the bone, you have essentially a solid structure and therefore the concentration between that surface and the capillary bed which is the way in which it is being removed from the site gets high enough that you then compromise the cells.

That's why I used the words "cell distress". Sometimes it may not be enough to get a cytotoxic response but it is enough to get edema as the result of changes in the proteoglycans, for example.

So it is very important and I think David Williams mentioned it and we have all mentioned it, that the anatomic site, the shape, all of these are going to lead to differences in response even though the pathways may be all the same, it depends then upon the accumulation of these. So we did not at any time see this but it would not be surprising with a 90 per cent porous material capillaries growing in.

DR. VERT: Thank you, Don.

DR. SHALABY: Shalaby of Johnson & Johnson.

You mentioned after six months or at six months you still see giant cells.

DR. GIBBONS: Yes. Only where you can still visualize a polymer particle.

DR. SHALABY: So there are permanent particles at six months?

DR. GIBBONS: That's right. It was not until the nine months where we could not visualize any polymeric particles whatsoever.

DR. SHALABY: And you see this regardless of whether you sterilize it by gamma or by ethylene oxide?

DR. GIBBONS: That is correct.

DR. SHALABY: How do you explain the longevity, the longer time it took in this case versus what we know of about four months?

DR. GIBBONS: No, I think we do not, I would disagree with you, Shalaby Shalaby, namely that you may lose mechanical integrity in weeks but the polymer itself as particulate lasts for at least six months even in these very rapidly degrading particles and I think that's well-known.

I mean, I don't think that any of the people here who would always like to argue with me, either Jim Anderson or David or anyone else, we have all seen it. I mean you have to differentiate between mechanical loss of strength and the presence of volumes of the material.

DR. SHALABY: So you are saying that beyond four months you still have some PGA at the site?

DR. GIBBONS: Yes, somewhere between, in this particular application, somewhere between six months and nine months the last vestiges of solid material disappeared. And that's quite common so that if you have a material that degrades more slowly, the material will be there for a year, year and a half.

DR. SHALABY: Then if I heard it correctly, if use five matured for sterilization or around five point matured?

DR. GIBBONS: 1.5, excuse me, yes.

DR. RUSE: Dorin Ruse, University of Toronto.

Just a question. I see that you made that cuff to look at the healing of the nerve. It must have been interesting to test actual dissolution of your polymer when the nerve is healing. The situation might be completely different. It doesn't affect your nerve.

DR. GIBBONS: That is not true because we have done the experiments over two years on a severed nerve in the Macque monkey. I was only presenting the biological responses and the safety. In fact, it was equivalent to suture.

DR. LOOMIS: I would like to ask a quick question.

I would like to congratulate you in the way you characterize chemically the material after you have sterilized it. I think everyone should do that.

You found chlorohydrin. Is the chlorohydrin present in the original ethylene oxide sterilant?

DR. GIBBONS: It is my understanding that it is one of the intermediary degradation products. Some people who are experts on ethylene oxide may be able to be more precise but it is my understanding that it is one of the intermediaries.

DR. LOOMIS: It is produced during the sterilization?

DR. GIBBONS: That's right.

DR. LOOMIS: I'm not sure where, I don't see where the chlorine comes from if we are starting with ETO and your sterilizing polyglycolic acid.

Are you saying the PGA has chlorine in it?

UNIDENTIFIED SPEAKER: Some, a trace.

DR. LOOMIS: Well, not if it is made with stannous actuate it doesn't.

DR. HELMUS: If there's any chlorine present from any source whatsoever you'll get ethylene chlorohydrin and it's almost impossible during handling or any other ways to get some of the chlorine.

Michael Helmus, Harbor Medical Devices.

DR. GIBBONS: In fact, it probably is the result of the way it bondeds with the hexachloro isoproponal.

# STUDIES ON THE ENVIRONMENTAL DEGRADATION OF STARCH-BASED PLASTICS

Gene Iannotti, Nancy Fair, Mike Tempesta

Howard Neibling, Fu Hung Hsieh, and Rick Mueller

New Products Group

University of Missouri - Columbia, Missouri

ABSTRACT

Over-zealous promoters of degradable plastics may have lead consumers to believe that these products are suitable for every use and unilaterally will solve this nations's solid waste problems. On the other hand, members of the established plastic manufacturing industry, with a vested interest in maintaining the status quo, have encouraged consumers to believe that degradable plastics will pollute the environment, thwart recycling efforts, create unstable landfills, and generally cause much more harm than good. Most likely, the truth lies somewhere between these two extremes. Our research suggests that disintegration of starch-based plastics will occur in properly managed landfills, soil and anaerobic waste systems. Starch is biodegraded by microorganisms, the plastic material loses strength, and chemical analysis indicates changes in the structure of polyethylene. Addition of both starch and promoters of oxidation results in a faster degradation than with either alone. The rate is dependent upon complex interactions with the environment which need to be studied further; we have seen degradation rates in terms of weeks and those that will probably take years. The changes which are occurring would be expected to lead to eventual disintegration of the plastic. We have not followed the changes long enough to known if there is ultimate conversion to carbon dioxide and water or methane; but after six months; there is a significant amount of starch removed and polyethylene remaining is changing. An average of 25% of the starch is lost in six months; there is a shift in molecular weight of approximately 5% in the polyethylene under anaerobic conditions. Degradation appears to be simply an acceleration of the process that occurs in typical plastics.

## INTRODUCTION

Plastics Industry. Plastic manufacturing involves two distinct processes and two separate industries, the plastic resin-production industry (composed primarily of the large chemical companies which convert industrial organic chemicals into plastic resins) and the plastics-processing industry (composed of smaller companies that extrude, inject, blow-mold and form the plastic resins into end products). The 1982 Census of Manufacturers listed over 10,000 processors in the U.S. with 11,653 plants.

The plastics industry estimates that consumption of plastic resins will grown from 48 billion pounds a year in 1985 to 76 billion pounds by the year 2000. The single largest use of plastics is in packaging. Plastic packaging falls into four main groups: films, bottles, other rigid containers, and coatings and closures.

Together, low density polyethylene, high density polyethylene and polystyrene account for 75% of the plastic resins that are used in packaging (Anon., 1987).

Concern has grown over the extensive use of these disposable materials. Plastics contribute to unsightly litter throughout the environment, urban and rural. A recent study conducted for the EPA by Franklin Associates Ltd. concluded that plastics accounted for 9.6 million tons, or 7.2% by weight, of municipal solid waste in 1984. About 5 million tons of this was packaging. But because plastics are so light-weight, it is estimated that plastics make up from 25 to 32% of the volume of solid waste. To remedy this problem incineration of plastics as a source of energy is being promoted. However, concern has been expressed that incineration of these materials will produce harmful air emissions and ash. Consequently, there is considerable interest in recycling plastics and in the development of biodegradable plastics (Anon., 1988).

Degradable Plastics. Plastics are generally considered one of our greatest pollution problems in the U.S. today, and local legislation has already restricted the sale and disposal of plastic products. In response to these concerns, "biodegradable" plastics are being produced by several companies. However, the extent of deterioration of such plastics in various environments has not been sufficiently studied. Companies making very similar products are making greatly different claims, and obviously exaggerated statements are being made by some. The longer unbiased data remains unavailable, the greater the potential that this emerging technology will not be utilized correctly and that false claims will result in a deterioration of consumer confidence.

Additionally, as degradable plastics have become more widely publicized, questions have been raised by environmental and consumer groups about the effects the degrading products will have on soil and ground water. These concerns must be addressed if degradable plastics are to have a prominent place in the packaging market.

The development of new biodegradable polymeric materials has centered mainly around the following four areas: 1) synthesis of new biodegradable polymers (Reed and Gilding, 1979; Takiwa, et al, 1979); 2) modification of natural polymers (Bradman and Divine, 1981; Dennenberg, et al, 1978; Ferruti, et al, 1972; Kumar, et al, 1981: Yuen, 1974); 3) modification of synthetic polymers (Kopecek and Rejmanova, 1979); and 4) biodegradable polymer composites (Comerfard and Kapur, 1976; Griffin, 1977a; Griffin, 1977b; Otey and Mark, 1976; Young, 1975).

To be a viable competitor in a high-volume market like packaging, the biodegradable material must be cost competitive with existing resins and must also be easily fabricated by injection molding or melt extrusion into films. However, most of these products, especially bulk plastics, will not be developed for years because of failure to be cost competitive with existing resins, problems in fabrication by injection molding or melt extrusion into films. The ones that are being marketed tend to be low volume specialty products. The exception is composites made by direct addition of treated starch to plastics such as polyethylene; this is the product that we are studying.

Polymer composites incorporating corn starch are among a few of the biodegradable materials developed that have the potential to secure a real share of the packaging market. Corn starch is competitive in price with the plastic resins with which it is compounded to make the biodegradable material. Additionally, the composites can be run on the same equipment as the pure resin products. These composites offer the processor of plastic resins the opportunity to expand their product lines with very limited capital investment.

Testing of starch-based degradable plastics has been limited, which poses a problem for the suppliers of the starch masterbatch materials. Knowledge of the performance properties and degradation under a range of typical environmental conditions is holding up emergence of a biodegradable plastics industry and use of a large amount of renewable starch. The small companies must be able to ensure that the end products will perform as stated when they are manufactured into plastic goods. Thus the improvement of functional, environmental or economically advantageous qualities must be defined. The changes in both starch, polyethylene, oils, etc. must be determined. The study of biodegradable plastics must ensure that the environmental impact is reduced, that ecosystems are viable, stable and within normal ranges, and the toxic substances are not produced during degradation.

There are still many questions to be answered about these composite materials, particularly, their mechanisms of degradation, how degradation can be controlled, and what effect the products of degradation have on the environment. A variety of techniques have been reported to study biodegradation (Kumar, et al, 1982). However, a systematic approach of assessing the quality of the environment while following degradation of the polymeric material has not been reported.

METHODS

Plastic Samples. Two major manufacturers of starch-based degradable plastics (Archer Daniels Midland of Decatur, Illinois and St. Lawrence Starch Ltd. of Ontario, Canada) have supplied experimental samples. In initial experiments, low density polyethylene films with varied starch concentrations of 0, 3,6, and 9% were examined. In subsequent experiments, samples with 5.5% starch, with and without pro-oxidant addition were tested. Samples with varying levels of pro-oxidant have also been provided for studies on the thermal oxidation of these films.

Exposure Environments. Four exposure environments were selected to represent the conditions under which plastics would be typically stored or properly disposed. These environments were extreme air (high humidity, high temperature), soil burial, refuse burial, and anaerobic waste treatment digester. The air, soil and refuse samples were placed in an environmental chamber in which temperature and humidity were controlled at 28°C and > 85%, respectively.

The aerobic refuse burial was carried out with relatively heavy organic loads with the composition of municipal solid waste excluding metal and glass. The plastic sheets were buried as prescribed by ASTM methodology. the initial moisture content was adjusted to 60% (based on total weight). The inoculum was from materials saved from previous experiments; the first experiment was started with material from commercial and local composts. In the soil burial, Mexico silt loam, representative of a large number of soils found in the midwest, was used. The initial moisture content of the soil was adjusted to 40% (based on dry weight of soil). The air exposure samples were stored in the dark at 28°C and > 85% relative humidity. since they were in the same environmental chamber as the soil and refuse material, with recirculation of air, there was also a higher microbial activity in the air than one would normally encounter. Thus this air exposure must be viewed as an extreme air exposure and represents the worst case air environment. The anaerobic exposure was carried out at 35°C in the Municipal Waste digester for Columbia, Missouri. The samples were placed in bags of polyester fabric (40 micron pore size) and suspended in the digester.

Thermal studies designed to determine the effect on temperature on the degradation of plastics in a non-biologically active environment were conducted in a convection oven at 50, 70 and 90°C. Continuing studies are being conducted at lower exposure temperatures. In some cases, samples were placed in anaerobic jars under an argon atmosphere to determine the effect of extremely low oxygen levels on the degradation process.

Environmental Monitoring. Characteristics of the aerobic and anaerobic systems were monitored throughout the duration of the experiments. Humidity and temperature in the air were monitored with probes. Moisture content of soil and the moisture, organic matter percent (as volatile solids/total solids) and temperature of the refuse were determined at each sampling time. The anaerobic digester was monitored for temperature, alkalinity, total solids, volatile solids, ash and volatile fatty acids.

Sampling Interval and Sample Storage. Samples of plastic were taken at set intervals, typically at 0, 1, 2, 4, 8, 12 and 24 weeks. The rate of sampling was increased in thermal studies based on the composition of the plastic and the exposure temperature. Samples were stored in the dark under argon to prevent further deterioration while awaiting analysis. These conditions exclude light, excess heat and oxygen.

Measurement of Degradation. Changes in mechanical properties (elongation, breaking strength, energy to break) were assessed according to ASTM Standards. Tests were carried out on a constant-rate-of extension tensile tester in a standard laboratory atmosphere of 23 ± 2°C and 50 ± 5% relative humidity. FT-IR spectroscopy was used to determine starch removal and oxidative changes in the plastic. Scanning electron microscopy (SEM) was used to monitor surface microstructural changes due to starch removal and microbial interactions with the films. Viscometric determination and gel permeation chromatography (molecular weight averages and molecular weight distribution changes), and differential scanning calorimetry (glass transition) were conducted on control and final samples.

## RESULTS AND DISCUSSION

Effect of Starch Addition Alone. Our first studies were conducted with polyethylene films containing starch at 0, 3, 6, or 9% levels, without pro-oxidant addition. These studies were designed to ascertain the effects of granular starch addition alone on the degradability of plastic films. In these films, the most rapid loss of starch occurred in the first four weeks, however, starch removal continued throughout the 24 week exposure period (Figure 1).

The plastic films exposed to the more active soil, refuse and anaerobic environments lost approximately 25 to 30% of their original starch content over the 24 week exposure when averaged across all starch levels. There were significant starch level by environment interactions, however, with the maximum starch removal being near 50%. Approximately 15% of the original starch was also removed under the extreme air exposure.

Scanning electron microscopy confirmed removal of starch. Bacteria could be seen in the cavities left after starch removal. Different types of coatings that included bacteria were associated with the surface of the plastic film following exposure. Physical changes in the polyethylene between starch areas were not readily apparent in these films after 24 weeks of exposure.

Mechanical property changes and changes in the molecular weight of the polyethylene in these samples <u>without</u> pro-oxidant were not significantly different between the 0% control and the starch containing samples in the aerobic environments (air, soil, refuse), but there was a significant change in these properties under anaerobic conditions (Figure 2). There was a corresponding reduction in molecular weight in the anaerobic samples as determined by GPC.

Synergism of Starch and Pro-oxidant. Additional studies have examined the synergistic effect of combining starch and pro-oxidant on degradability of polyethylene films. The four environmental exposures had little overall effect on the mechanical properties of

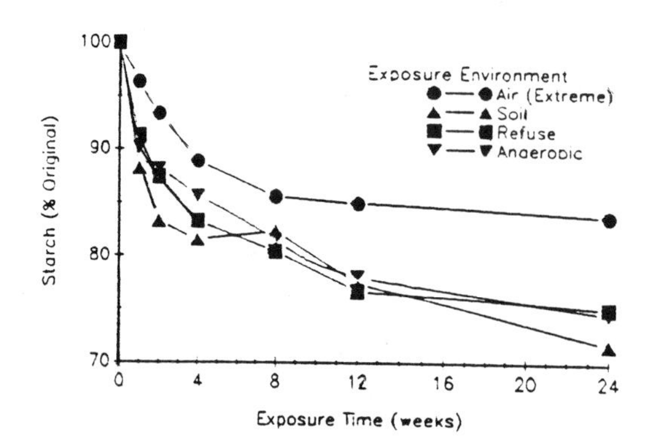


**Figure 1** Rate of starch removal in various exposure environments. Each data point represents the mean of 3, 6, and 9% starch films.

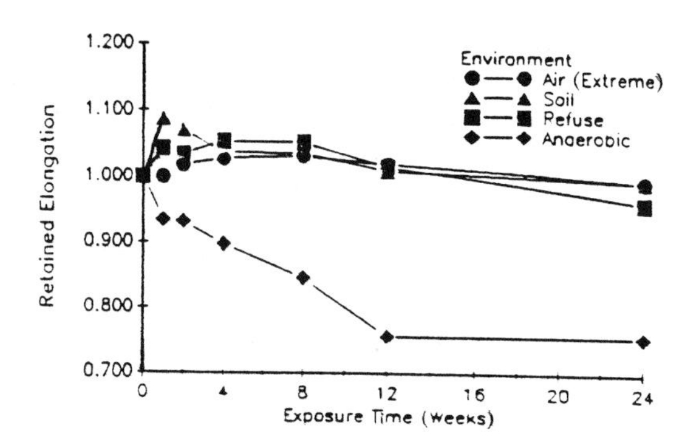


**Figure 2** Loss of elongation in various exposure environments. Each data point represents the mean of 3, 6, and 9 % starch films.

the control film or on the film with only the pro-oxidant added (Figure 3). The addition of starch produced a significant reduction in breaking strength, elongation, and energy to break. The addition of pro-oxidant to the starch containing films caused more embrittlement, resulting in a further loss of elongation and, consequently, energy to break. The synergistic effect was also evident in starch removal, where significantly more starch was removed from samples containing the pro-oxidant than from the samples containing only starch.

Changes in the molecular weight of the polyethylene in the starch/pro-oxidant films after 12 weeks of environmental exposure are given in Table 1. There was a reduction in the molecular weight of the polyethylene in every environment, with the greatest change occurring under anaerobic conditions.

Thermal Studies. Temperature is a critical factor in both the biological activity of an environment and the rate of reactions catalyzed by pro-oxidants. In an effort to understand the temperature dependent degradation of the starch/pro-oxidant films oven studies have also been conducted. The rate of degradation was significantly affected by oven temperature (Figure 4). Higher temperatures produced the most rapid change in mechanical properties. However, exposure at 50°C, which is within the thermophilic range common in aerobic composting processes, resulted in nearly total embrittlement after 28 days.

Priority Pollutants. A survey of commercial plastic films for priority pollutants for the most part did not reveal these compounds. Significant amounts of heavy metals were found in some

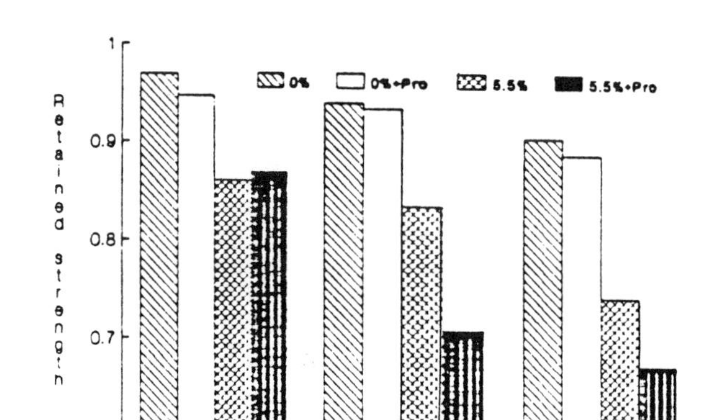


**Figure 3** Overall effect of starch and pro-oxidant addition on mechanical properties after 24 week environmental exposure.

**Gel Permeation Chromatography (GPC) Data on 5.5% Starch/Pro-oxidant Films**

| Exposure Environment | Exposure Time(wks) | $\overline{M}_n$ | $\overline{M}_w$ | $\overline{M}_z$ | D |
|---|---|---|---|---|---|
| Control | 0 | 16,730 | 92,390 | 271,100 | 5.50 |
| Soil | 12 | 15,180 | 89,230 | 267,000 | 5.88 |
| Refuse | 12 | 15,380 | 88,300 | 264,500 | 5.74 |
| Anaerobic | 12 | 3,884 | 86,110 | 276,300 | 22.18 |

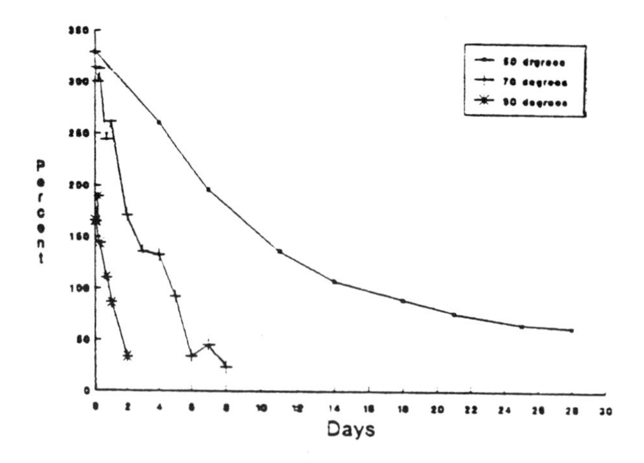


**Figure 4** Effect of exposure temperature on loss of elongation in 5.5% starch/pro-oxidant containing films.

"first-generation" bags. These metals had been added to the films as part of colorant and were associated with the polyethylene. The rate of release during environmental exposure was not evaluated. The assumption is that the metals would be released with disintegration of polyethylene, but not necessarily with biodegradation of starch. Release would be increased due to any factor that increases disintegration of the polyethylene. Since the difference is rate, the most appropriate long term solution is to use more environmentally benign additives in all plastics.

CONCLUSION

The results of our studies to date have demonstrated that significant amounts of starch are removed in a relatively short period of time by microorganisms in different environments. However, some starch always remained and disintegration of the polyethylene matrix may be necessary for total starch removal. Starch removal was significantly greater in all environments when pro-oxidant was present. Mechanical properties also demonstrate a synergistic effect of combined addition of starch and pro-oxidant. The growth of microorganisms on starch could stimulate a longer term biodegradation of polyethylene or change the environment and thus the physical-chemical deterioration. The chemically induced changes in the polyethylene due to pro-oxidant addition could also make the starch more accessible to enzymatic breakdown.

It must be emphasized that the is preliminary information and much more must be known about the changes under different environmental conditions before definitive statements can be made. Mechanisms of degradation, factors that effect rates and intermediary products must be systematically defined. The understanding derived from determining the relationship between composition and functional properties will permit logical approaches to improving the next generation of biodegradable plastics.

REFERENCES

1. Anon. 1987. Heading for 50 Billion Pounds. Modern Plastics January: 66-65.
2. Anon. 1988. A Plastics Packaging Primer. Environmental Action, July/August: 17-20.
3. Bradman, B.W., and M.P. Divine. 1989. Microbial Attack of Nitrocellulose. J.Appl.Polym.Sci. 26:997-1000.
4. Comerfard, T.M. and C. Kapur. 1976. U.S. Patent 3,952,347.
5. Dennenberg, R.J., R.J. Bothast, and T.P. Abbott. 1978. A New Biodegradable Plastic Made from Starch Graft Poly (methyl Acrylate) Copolymer. J.Appl.Polym.Sci. 22:459-465.
6. Ferruti, P., A. Bettelli, and A. Fere. 1972. Linear, High Molecular Weight Poly (2-alkyl-4-vinyl-6-(dialylaminomethyl) phenols) and Poly (2,6-cis(dialkylaminomethyl)-4-vinylphenols). Polymer 13:184-186.
7. Griffin, G.J.L. 1977. U.S. Patent 4,021,388.
8. Griffin, G.J.L. 1977. U.S. Patent 4,016,117.
9. Kopecek, J. and P. Rejmanova. 1979. Practice Copolymers of N-(2-Hydroxypropyl) Methacrylamide with N-Methacrylated Derivatives of L-Leucine and L-Phenylalanine. II. Reaction with the Polymeric Amine and Stability of Crosslinks towards Chymotrypsin.

J.Polym.Sci., Polym. Symp. 66:15-32.

10. Kumar, G.S., V. Kaopagam, U.S. Nandi, and V.N. Vasantarajan. 1981. Biodegradation of Gelatin-g-Poly (ethyl Acrylante) Copolymers. J.Appl.Polym.Sci. 26:3633-3641.
11. Kumar, G.S., V. Kalpazam, and V.S. Nandi. 1982. Biodegradable Polymers: Prospects, Problems and Progress. J. of Macromol.Sci.-Rev. Macromolec Chem. Phys. C22:225-260.
12. Otey, F.H. and A.M. Mark. 1976. U.S. Patent 3,949,145.
13. Reed, A.M., and D.K. Gilding. 1979. Biodegradable Polymers for use in Surgery Polyglycolic/Poly(lactic acid) homo-and copolymers. Polymer 20:1454-1464.
14. Takiwa, Y., T. Suzuki, and T. Ando. 1979. Synthesis of Copolyamide Esters and Some Aspect Involved in Their Hydrolysis by Lipose. J.Appl.Polym.Sci., 24:1701-1711.
15. Young, D.W. 1975. U.S. Patent 3,903,029.
16. Yuen, S. 1974. Pullulan and its Applications. Process Biochem. 9:7-9.

QUESTIONS AND ANSWERS

---PRESENTATION BY DR. GENE IANNOTTI

(University of Missouri)

DR. NARAYAN: Ramani Narayan, Michigan Stare.

I would like to address your statement that I wouldn't like to meet a microorganism that would make such a big hole in the starch. And the mechanism of degradation of starch is pretty well established. It is not the microorganisms that degrade starch. It is the enzymes, the amylase, the pullulanase which breaks the starch and that would create the hole.

I mean, if you look at the systems, the large complex starch molecule is not going to be going into the cell of the microbial system. So you get localized enzyme secretion at the site. So when those enzymes are secreted, they break the starch down and that's why you see the large crater-like hole. And in some cases where the enzymes have problems of diffusion, penetration, you get less break down so the questions or the issue is not that we do not understand how starch breaks down.

I mean, you can get ten volumes on how starch is broken down by enzymes, the various enzymes like 1-4 amylase, pullulanase, how they act, what is the synergy.

The issue here is in a system of starch and plastic, how do these enzymes penetrate into the matrix, how do they diffuse into it, what are the sites of attachment.

So that the statement that we do not know what is happening to the starch is, I feel, incorrect and what we are really looking at is how do enzymes act on these starches.

DR. IANNOTTI: What I meant was the overall interaction between the polyethylene and the starch definitely is not understood. There are cases where we can look and see a collapse of the polyethylene so that in essence what you had is a removal of the starch within that cavity itself.

Now, we can't determine whether microorganisms or enzymes penetrated. We have added enzymes to these types of materials and we were not highly successful. We didn't work at it.

DR. NARAYAN: That's again a question of penetration. I mean, that's what Paul Allenza's talk showed you that if you have the right, you know, it is essentially diffusion into the system but it's the enzymes that break down the starch, not microorganisms. Maybe you get cracking with fungi and algae and all those things but he key mechanism is enzymatic, not microbial, this going into the sample.

DR. IANNOTTI: Well, microorganisms carry with them enzymes. You cannot separate microorganisms and the enzyme itself. With our evidence, we have seen the evidence that there is penetration by the microorganisms of the polyethylene. So it is not simply here's the microorganism out here in the environment and the enzyme is breaking free of the microorganism and diffusing through there.

At times we see evidence of penetrations, very obvious particularly with the filamentous forms so what I'm trying -- the point that I was trying to make is that we could visualize a whole range of different mechanisms I don't think it is one.

DR. HAMMER: Jim Hammer from 3M Company.

Have you taken the polymers and hydrated them and then dehydrated them to see whether you get this hydration and bursting of the polyethylene surface without the enzyme and then the enzyme and the microbe has a hole to operate through. So just take a simple experiment of hydration and dehydration.

DR. IANNOTTI: We did a little bit of that initially. We did it with a light microscope.

DR. HAMMER: So you observed a crack then on the surface?

DR. IANNOTTI: We didn't see too much of a change and that's when we did the enzymes. But a lot of the variation, a lot of the things that I'm now able to stand up here and talk about, know that there is breakdown, those types of statements, came from looking at a range of different things.

I'm not sure that the materials that we were working with was at that point of the game a very select, non-typical type of plastic so why we did it, I don't place much value in our observations at that time.

DR. ADEWOLE: You talk about the significance of the process types. Did you invoke any kind of disparity of experimentation that might enable you to look at interaction effect between product formation process parameters and things of the nature?

DR. IANNOTTI: Yes. My strategy is now going to be a graduate student running our own extruder who's going to control much more than what we were with these initial experiments.

DR. GIBBONS: Don Gibbons, M & M.

One comment you make brings it into the same field that we have with implants and namely your concern with seeing the surface, namely the glycolics that is formed on the surface and also presumably precipitation of other materials from your soil.

We see exactly the same thing on the surface of heart pumps, on intraocular lenses. And we still have the problem. We don't know quite yet how to remove it so we can't really look at the surface without in fact further damaging the surface.

Do you have any suggestions or comments on how you are going to tackle this because we find it very difficult.

DR. IANNOTTI: I wish you would have talked to me about a year and a half ago or two years before we really started looking with the SEM. I feel like I know now less than when I started.

I would like to go back and even begin to look at different washes, different ionic compositions to begin to look at this. We can find starch that seems to be very visibly exposed on the surface. We don't know whether we opened it up in our technique, but that does not have any associated microorganisms to it. So there's just -- I have seen too much now to...

DR. SHALABY: Shalaby Shalaby from J & J.

I have two quick comments and a question. We have been talking quite a bit about starch polyethylene. I haven't heard any definition of what starch this is. Is it corn, is it potatoes, or is it tapioca? I think there is some relevance in how starch behaved.

We also talked quite a bit about starch as if it is soluable but we never said anything about how soluble starch is. We do have a component that's soluble in water and we haven't heard anything at all about if you put polyethylene and starch in water alone, do you get any migration of the amylase out into the medium?

DR. IANNOTTI: They're good points. I can even add some more

questions to that type of thing.

It is corn starch first off, okay, and we thought we had some better idea. Just in the drying of the corn we can see a lot of the corn starch particles. We can see a lot of variation in what's going on, just changes in colour, texture, that. We need to know more about some of those aspects.

DR. SHALABY: So not everything happens inside, it could also migrate outside and you can get some degradation outside the matrix?

DR. IANNOTTI: True:

DR. LOOMIS: Dr. Albertsson won't take no for an answer.

DR. ALBERTSSON: Yes. When we are talking about starch here we have to remember that all of you are talking about different treatment of the starch to get it in a good mix with a polyethylene because some of you use this Silane treatment and I think that in your case you had some other kind of treatment but all of you have to do something because starch is hydrophylic and polyethylene hydrophobic and you need a good contact between them.

DR. MATLOCK: Mark Matlock, ADM.

I can respond a little bit on some studies we have done with the extraction of starch from the polyethylene matrix.

As part of the background looking at the potential of trying to work with the FDA, we have conducted the FDA-type extraction test with the 95 per cent water, five per cent alcohol and vice versa, the elevated temperatures and so forth. And the extractables of all types are over an order of magnitude lower than the requirements for the FDA.

So the extraction, even in those types of environments over the period of time that they specify in those tests, are in order of magnitude below what you might or below the FDA requirements and are typical of what you might find just for the other types of products that are in plastics, plastisizers and so forth.

Also the swelling of the granules has been talked about several times. I don't believe that the starch granules can physically swell enough to burst the polyethylene surface. As you dry them down below one per cent moisture, they have a tendency to become more hollow but they retain the same size pretty much regardless of what the moisture is.

It is simply a hydration thing. It does not affect their size until you actually gelatinize and solubolize it and then you start changing things.

DR. BARENBERG: Barenberg.

That's an in inappropriate answer because people want to know what are the chemistries of the starches, not just an advertisement for ADM and say it is a food content that meets the requirements. In other words, what's the material, what are the extractables, how was it prepared. So let's talk the chemistry and the science, not food testing.

DR. MATLOCK: I'm simply stating that under those conditions, the experiments that we had done by an outside laboratory did not show the extractables. In that particular test, we did not try to analyze how much of it was sugars and so forth, but we are in the process of getting some of those answers. It just takes time and I'm just taking about starch in plastics in general, as starch as a filler.

DR. LOOMIS: Thank you.

DR. IANNOTTI: I just want to re-emphasize that what I emphasized in my talk was some of the problems, some of the variability. I think we have learned a lot over the last year. There's really people that are very knowledgeable putting out some excellent products out there that we know will break down itself and it has been a learning curve but don't take all the I have said and then say: Gene Iannotti cam along and said, well, you can't get degradation.

DR. LOOMIS: Thank you, Gene.

# PRELIMINARY RESULTS OF SCREENING STUDIES ON EFFECTIVENESS OF CERIUM SALTS IN ACCELERATING PHOTODEGRADABILITY OF THERMOPLASTICS

Lucinda K. Ballinger
Rhone-Poulenc Inc. Fine Organic Chemicals Division,
Monmouth Junction, NJ

An important aspect of managing plastics solid waste is the control of urban plastic litter. With continuing growth of plastics in packaging applications and a growing trend towards single-use packaging, the litter problem is beginning to draw the attention of environmentalists and legislators. Once disposed of outdoors, plastic materials slowly undergo light-initiated oxidative breakdown resulting in eventual embrittlement. Water and biotic factors play little, if any, direct role in the breakdown of the material. It is the slowness of the oxidative degradation process and the consequent persistence of plastics under outdoor exposure conditions that brings about the litter problem.

The oxidation of polyolefins under ambient conditions is a free-radical autooxidation reaction (Reich and Stvala 1966, Grassie 1979) which can be initiated by short wavelength ultraviolet light. The regions of the UV spectrum most damaging to polymers are the UV-B (280-315nm) and, to a lesser extent, the UV-A (315-340nm) regions. The primary product of oxygen reacting with a hydrocarbon polymer under such conditions is polymer hydroperoxide. Light induced or thermal decomposition of these hydroperoxides provide additional free radicals to the system which results in autocatalysis. This oxidative process is accompanied by a chain scission reaction (and sometimes a crosslinking) which rapidly destroys the useful mechanical properties of the polymer (Allen 1983), (Davis and Sims 1983).

Transition metal ions are able to photolytically produce radicals initiating the oxidation reactions (Scott 1982). Furthermore, they are also well known catalysts of hydroperoxide decomposition (Reich and Stvala 1969) and are therefore capable of accelerating the oxidative degradation process. The effectiveness of metal carboxylates, metallic halides, and other compounds in bringing about such accelerated degradation is well-known (Mellor et. al 1973, Scott 1972). The effect of transition metal stearates on the degradation rate of polyolefins has been studied extensively (Osawa et. al. 1979, Mellor et. al. 1973, Scott 1972). Iron and cobalt carboxylates are reported to be particularly effective prooxidant additives in polyethylene and iron stearate has been used to obtain rapidly photodegradable agricultural mulch compositions (Kodak, 1969). Cerium is also among the metals particularly effective in promoting rapid photodegradation of polyolefins. The use of cerium compounds as one of several additives for rendering plastics photodegradable has been mentioned in several patents and publications on the subject (Gilead et. al. 1984, Scott 1978, Gratani et. al. 1977, Boberg 1976). To be effective photodegradants, cerium compounds need to be used at specific concentrations in commodity thermoplastics.

Fatty acid salts of cerium are well-suited as additives because of their processability, lack of color, and cost-effectiveness.

For routine use in industry, additives need to be stable enough to be compounded into the polymer during the processing when the resin is in a melt stage. This is not only convenient from a processing point of view, but also allows good dispersion throughout the plastic matrix. In most applications it is also important to obtain either a surface which is either transparent or one which can be readily opacified and colored. Cerium salts being white will not interfere with colorants added to the compound. Cerium salts are also generally non-toxic. While additives for obtaining rapid photodegradability are used at very low concentrations in the resin, it is still advantageous to use an additive of minimal toxicity.

The present experiments were undertaken to determine the effectiveness of cerium carboxylates in several commercially available plastic materials. The data is limited to plastic films as this particular class of products are good candidates for rapid photodegradability. Experimental work cited here was carried out by Rhone-Poulenc, Inc., at Research Triangle Institute, North Carolina under the direction of Dr. Tony L. Andrady.

## Experimental

Commercially available polyethylene, polypropylene and polystyrene resins were used for the study. The origin and the melt flow indices for these resins are listed in Table 1.

TABLE 1: Resins used for the study and their characteristics.

A) <u>Low Density Polyethylene Dow LDPE 503</u>

Melt Index (D1238) = 1.9 g/10 min. Density (D792)=0.923 g/cc.

B) <u>High Density Polyethylene Dow HDPE 50075P</u>

Melt Index (D1238) = 0.65-0.85 g/10 min. Density (D792) = 0.945-0.949 g/cc.

C) <u>Linear Low Density Polyethylene Dowlex 2045A</u>

Melt Index (D1238) = 1.0 g/10 min. Density (D792) = 0.920 g/cc.

D) <u>Polypropylene Himont PD064</u>

Melt Index (D1238) = 2-5 g/10 min. Density (D792) = 0.903 g/cc.

E) <u>Polystyrene Styron 685D</u>

Melt Index (D1238) = 2.4 g/10 min. Density (D792) = 1.04 g/cc.

i) <u>Compounding and Processing</u>

The additives were compounded into the respective base resins in a 40 mm twin screw extruder with a screw L/D of 20:1, to obtain a masterbatch containing 2 phr of the additive. The extrudate was water cooled, pelletized, dried and stored under ambient conditions until use in film blowing process. The master batch was blended in appropriate proportions with virgin resin to obtain mixtures containing 0.1 to 1.0 phr of the additive.

Extrusion blowing of the polyethylene and polypropylene films were carried out in a one inch extruder with a screw L/D ratio of 24:1. The melt was directed to a 2" x 0.40" spiral fed blown film die and blown into a film with a lay-flat width of 6-8 inches. The films were stored in the dark under ambient conditions. Polystyrene compounds were extruded using the same extruder fitted with a ribbon die.

ii) Light Exposure

Outdoor exposure was carried out in Miami, Florida during July to October of 1988 and the samples (approximately 8" x 10", mounted on frames) were exposed at 45° facing south in standard exposure racks. Samples were collected at weekly intervals and stored in the dark under ambient conditions. The total sunlight and the ultraviolet part of the sunlight spectrum reaching the exposure site was monitored on a daily basis. In interpreting the variation of tensile properties with outdoor exposure, it is preferable to consider the changes as a function of the amount of light energy received by the sample. However, if the cumulative solar radiation is a linear function of the exposure time, it is reasonable and convenient to interpret the data in terms of days or weeks of exposure. Figure 1 illustrates that this is indeed the case for the present experiments.

CORRELATION BETWEEN DURATION OF OUTDOOR EXPOSURE AND TOTAL SOLAR RADIATION

FIGURE 1

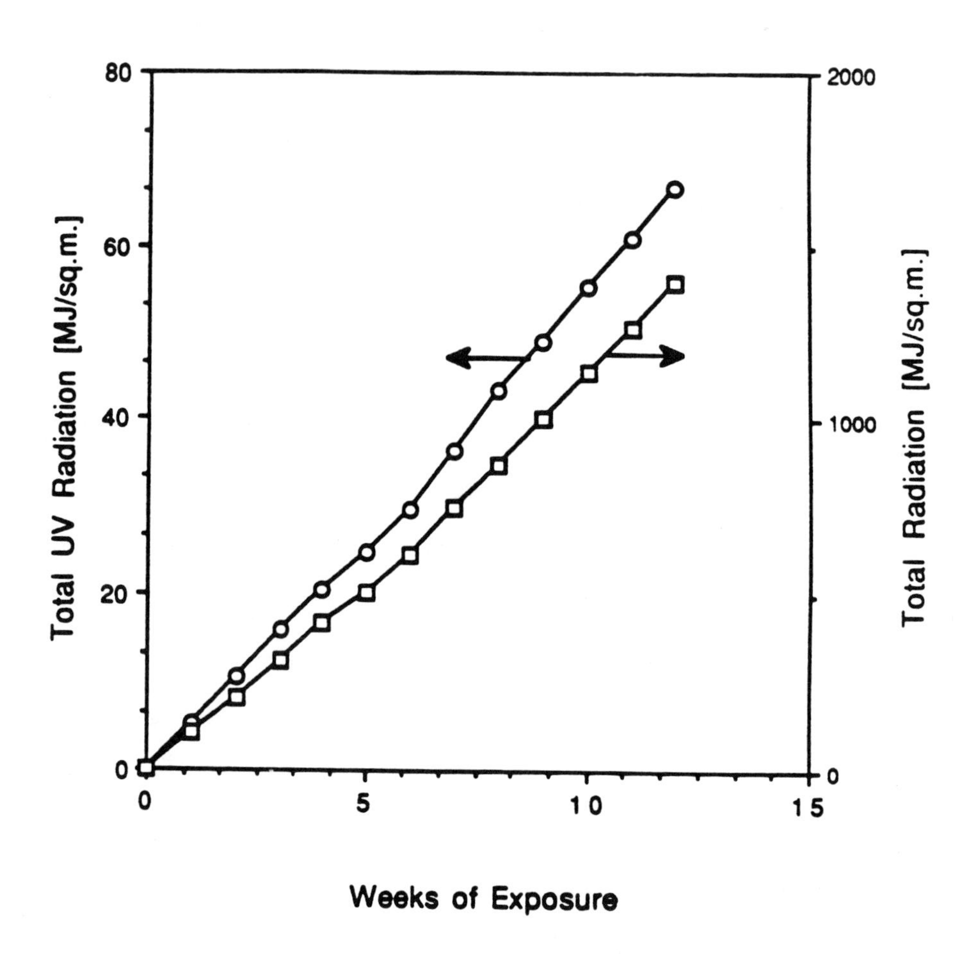

Weatherometer exposures were carried out in an Atlas (Ci) weatherometer equipped with a 6500 watt xenon-arc source and borosilicate inner and outer filters. The use of these filters results in illumination approximating that of outdoor sunlight. Exposure cycle consisted of a 90 min light period folllowed by a 30 min light with water spray, carried out at a black panel temperature of 70°C and a relative humidity of about 60%. Samples were collected at 50 hr or 100 hr intervals depending upon the material.

iii) Tensile Property Determination

Tensile properties were measured on an Instron Model 1122 generally in accordance with ASTM D882. Test strips used were 0.5 inches wide and about 5.0 cm long. The thicknesses ranged from 0.05 to 0.09mm. The rate of crosshead separation was 500mm/min for all samples except the polypropylene samples. A slower rate of strain (i.e. 100mm/min) was more appropriate for the very brittle exposed polypropylene. A crosshead rate of 10mm/min was used for polystyrene. Data points indicated in the figures are average values based on 4-6 individual tests.

iv) Gel Permeation Chromatography

Water System GPC equipment with a model M6000-A pump, a set of five columns (Ultrastyragel with pore sizes ranging from $10^2$ to $10^6$ A.) and a R401 Differential Refractometer Detector. Methylene chloride was the eluent and the flow rate was 1ml/min.

Additional GPC data were obtained at a different laboratory

using a Water System GPC and two columns (Polymer Laboratories, 5 x $10^4$ A mixed bed). THF was the eluent and the flow rate was 1ml/min.

## RESULTS

The primary consideration in assessing the effectiveness of a rapidly degradable material should be its rate of embrittlement to a point where the plastic film fragments due to disturbances caused by wind and rain. The reduction in particle size makes the litter less obvious and also promotes faster oxidative and biological degradation. If this process is rapid enough, the plastic film upon exposure outdoors disintegrates, being separated into fragments to small to be distinguishable from background soil and is dispersed by the wind. Therefore, the time taken by the plastic film to be weathered to a point where sample collection is not possible due to nonavailability of sample in the frame (or extreme fragility of any available sample which prevents it collection) is a useful approximate indication of the effective degradability of the material.

Of the various samples exposed, all except the High Density Polyethylene films reached this stage within the 12 week period of exposure. The control samples containing no cerium carboxylate additive were compounded with no antioxidants or photostabilizers, and did not represent an optimum composition with respect to weatherability. Both polystyrene and polypropylene films which contained no additive also embrittled during the experiment. Polystyrene film was by nature quite brittle and was possibly torn and lost due to wind action.

Samples containing various levels of the cerium compounds disintegrated very much faster than the untreated material (and certainly faster than a stabilized film of the same base resin would under comparable exposure) under present exposure conditions. Figure 2 and 3 show the data graphically. In the case of Low Density Polyethylene, the material was collectable even after 12 weeks of exposure and its lifetime is therefore not known.

A more accurate means of establishing degradation of a plastic film is by measurement of the tensile properties, particularly the elongation at break, of the exposed test piece. Various investigators have this criterion for establishing degradation and embrittlement of films and laminates. However, there is little agreement as to the elongation at break of an "embrittled" material. The figure of less than 2-5 percent seems to be appropriate at least for photodegradable materials. Present experiments yielded the following data on the variation in tensile properties plastic films containing the cerium salt additives.

a) Low Density Polyethylene (LDPE)

Figure 4 shows the percent elongation at break of LDPE samples 0.1-0.3 phr of the cerium (III) stearate (Additive A1, A2, and A3, respectively) with the duration of outdoor exposure. Cerium stearate was effective in accelerating the degradation of polyethylene at all levels of testing. Furthermore, the drop in tensile elongation at break is abrupt rather than gradual, lowering the value to around a hundred percent in four weeks of exposure, a

OUTDOOR SURVIVAL OF LLDPE AND LDPE FILMS CONTAINING CERIUM

FIGURE 2

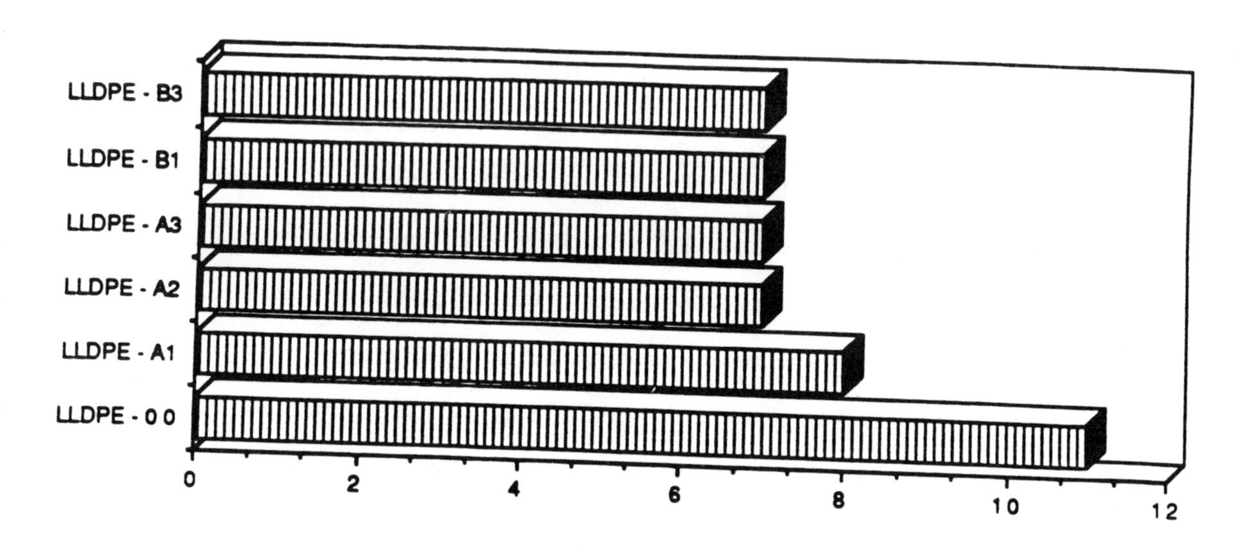


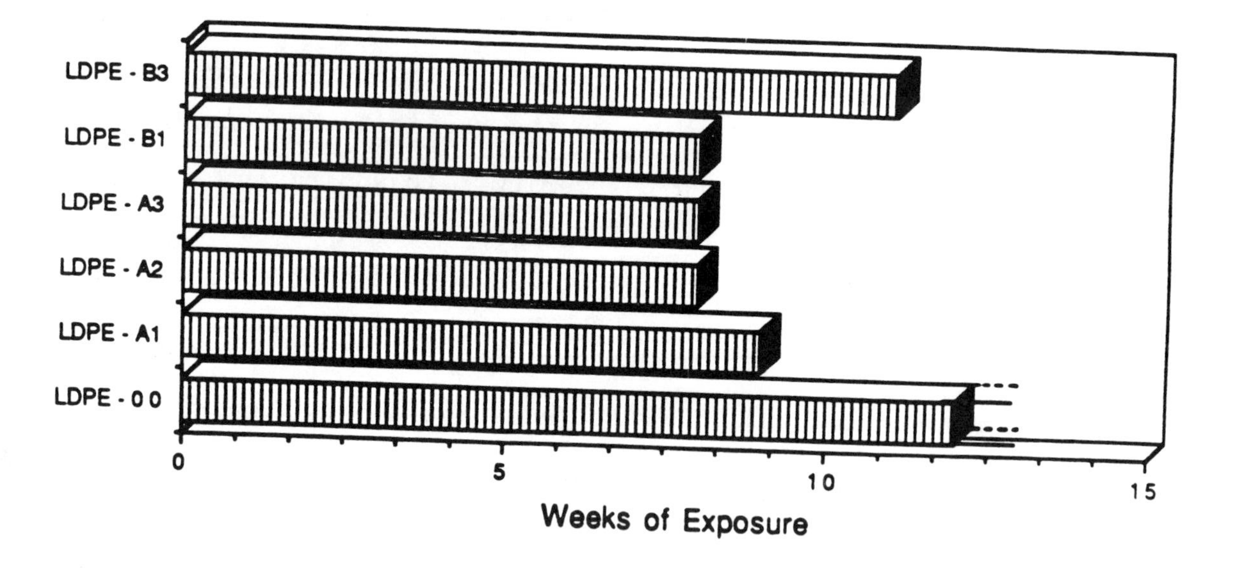

OUTDOOR SURVIVAL OF PP AND PS FILMS CONTAINING CERIUM

FIGURE 3

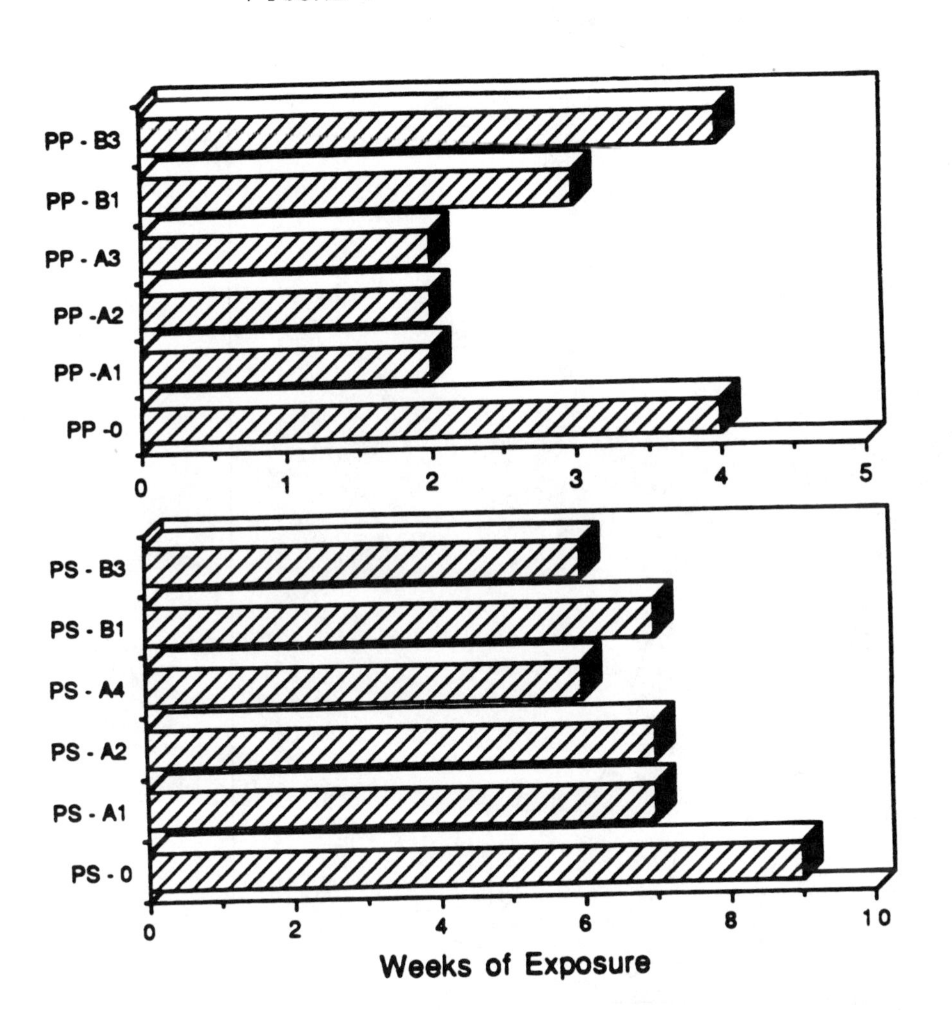

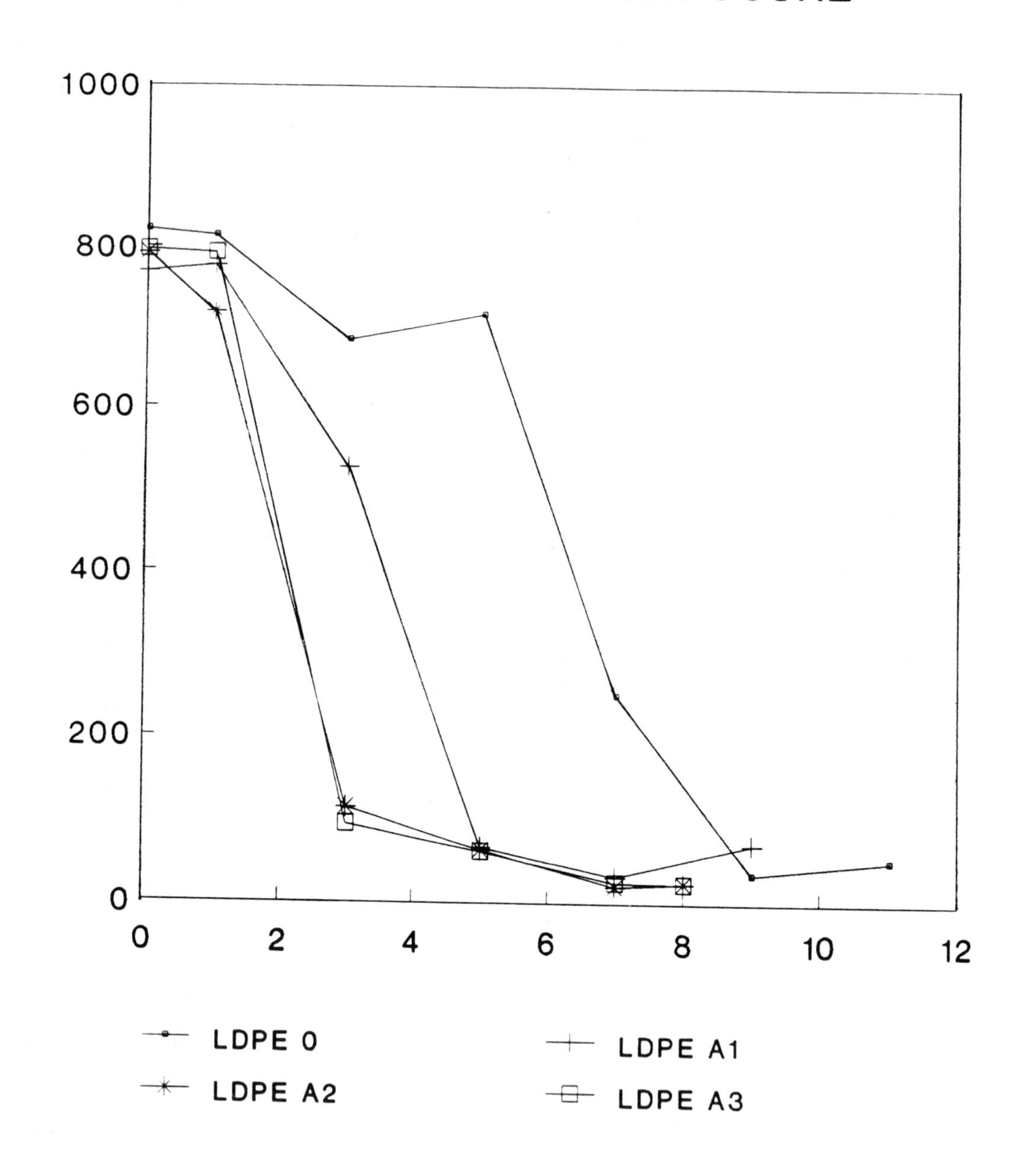
FIGURE 4
PERCENT ELONGATION AT BREAK
LDPE OUTDOOR EXPOSURE
1000
800
600
400
200
0
0
2
4
6
8
10
12
LDPE 0
LDPE A1
LDPE A2
LDPE A3
LKB 11/3/89

reduction which takes at least eight weeks of exposure in the sample without an additive. LDPE containing 0.1-0.3 phr of cerium caprylate (Additive B1, B2, B3, respectively) was also tested and the caprylate was found to be effective in accelerating the degradation. The caprylate was not as effective as the stearate at concentrations of 0.1 phr (Figure 5).

The above observations were confirmed in weatherometer studies of the LDPE materials. However, the lower concentrations of Additive A performed better than the higher concentrations. Data on samples exposed in weatherometer were essentially similar, with 100 hours of exposure amounting to about the same as two weeks of outdoor exposure under present experimental conditions. Film samples comtaining 0.1, 0.2, and 0.3 phr of both cerium stearate and cerium caprylate reached complete embrittlement in about 400 hrs of exposure in the weatherometer.

In both types of exposure, the tensile strength at break also decreased with the duration of exposure. The reduced strength of the material due to chain scission results in facile embrittlement of the plastic on long term exposure.

b) High Density Polyethylene (HDPE)

Both cerium stearate and the caprylate, at concentrations of 0.1 phr, 0.2 phr, and 0.3 phr caused stabilization rather than enhanced photodegradation in the high density polyethylene samples exposed outdoors. This finding was confirmed by the weatherometer test data which also showed slower decrease in the tensile elongation at break for samples containing the additive. As such,

FIGURE 5

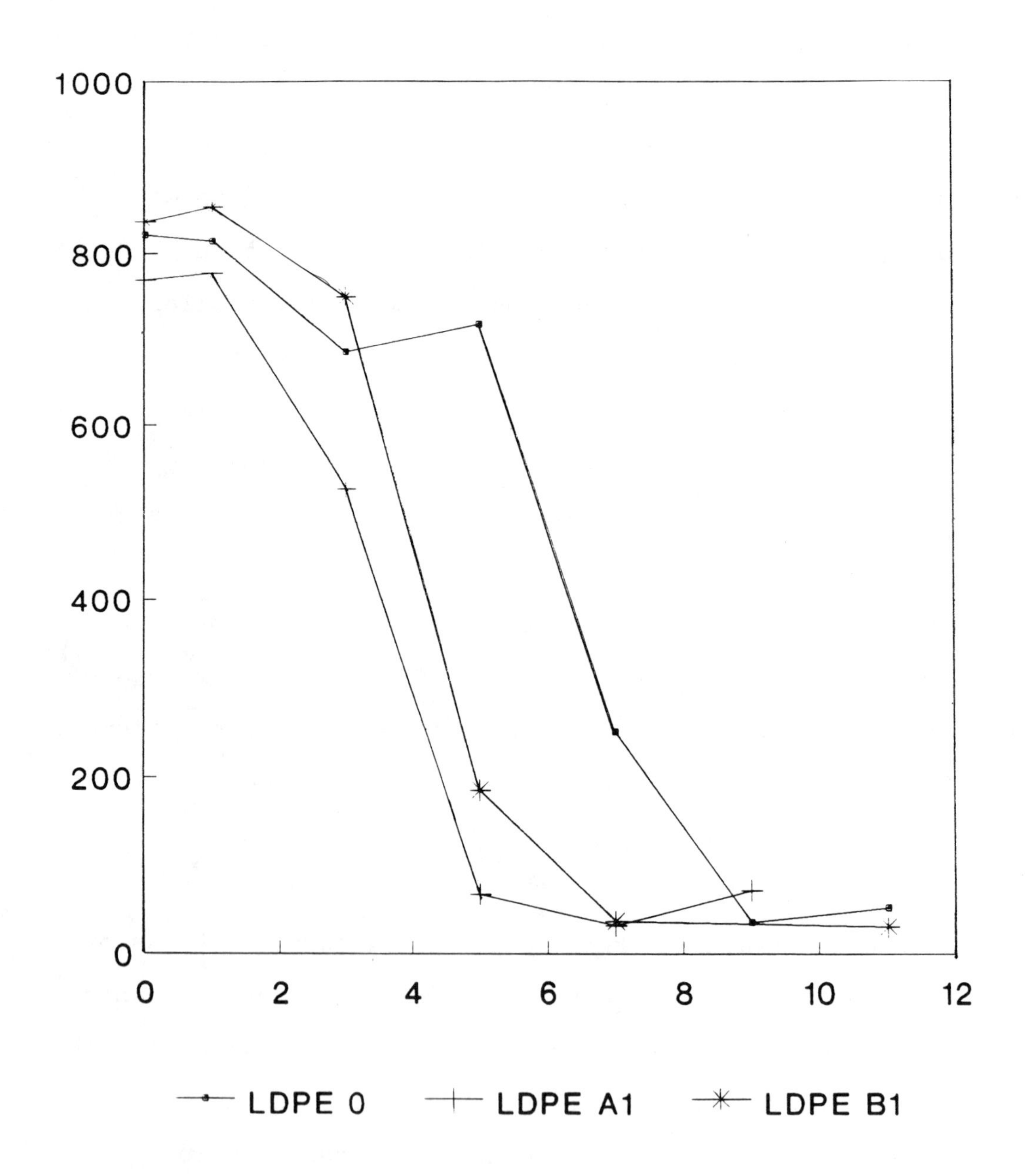


LKB 11/3/89

the test data relating to HDPE will not be discussed here. This observation was consistent with the finding that unlike the other types of films exposed, all HDPE films survived the full 12 weeks of exposure and could be collected for testing at the end of that period outdoors.

However, the possibility of these additives acting as photodegradants at different concentrations in HDPE cannot be ruled out. Prooxidant activity of metal compounds is well known to be concentration-dependent.

c) Linear Low Density iPolyethylene (LLDPE)

Outdoor exposure data for LLDPE films containing cerium stearate is shown in Figure 6. The elongation at break decreased more rapidly in those samples containing both types of cerium carboxylates compared to the control. In the case of the stearate, a very sharp drop in elongation at break to less than 95 percent of the initial value was observed in 3-4 weeks of exposure. The prooxidant activity was not significantly concentration-dependent. No significant dfifference in the effectiveness of stearate versus the caprylate was observed (Figure 7). Data from the weatherometer exposures agreed with the above results based on outdoor exposure. Again, the tensile data for the caprylate did not show any concentration dependence.

The tensile strength of the exposed test pieces decreased during the early weeks of outdoor exposure but tended to increase after about the 5th week suggesting that crosslinking plays a role during extended irradiation of these samples outdoors. On

FIGURE 6

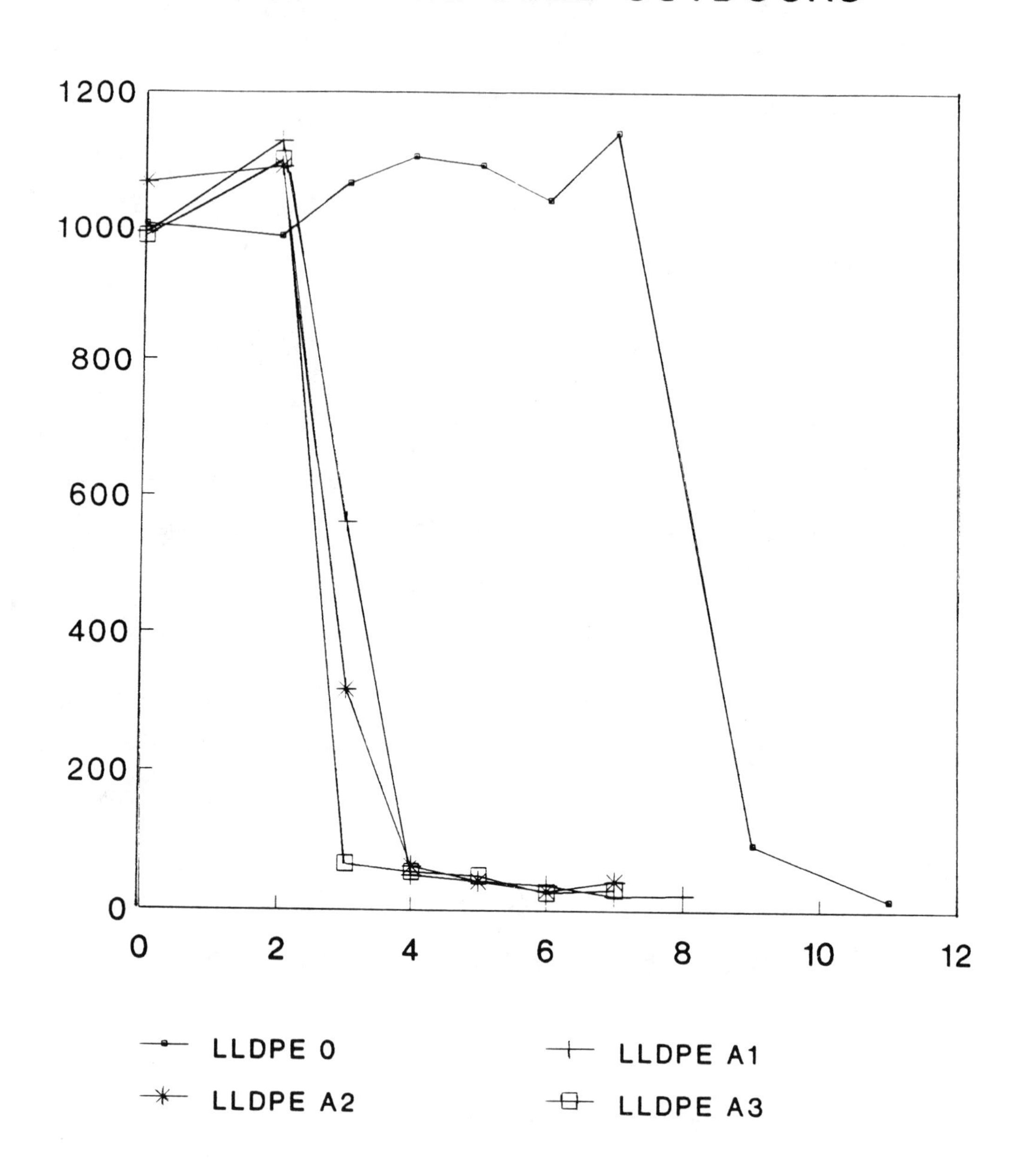


LKB 11/3/89

FIGURE 7

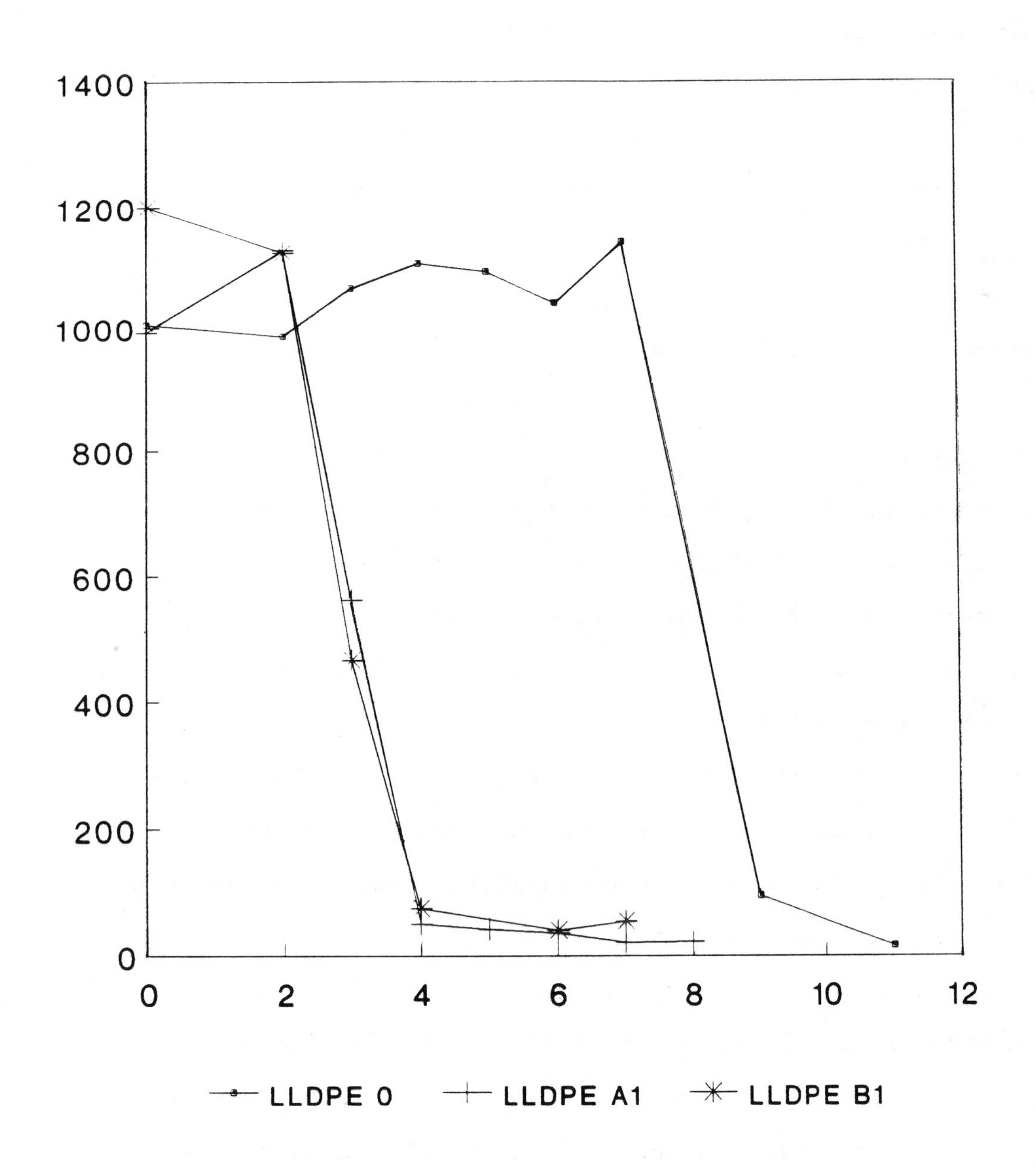


LKB 11/3/89

weatherometer exposure, the tensile strength of the film samples decreased to less than 10 kg/sq. cm. within 300 hrs, a decrease not obtained even with 12 weeks of outdoor exposure. The same was true of ultimate elongation as well, suggesting that the weatherometer conditions represented an acceleration of the degradation process as it takes place outdoors. Film samples containing 0.1, 0.2, and 0.3 phr of both cerium stearate and cerium caprylate reached complete embrittlement within 250-300 hrs of exposure in the weatherometer. Results of tensile tests suggests that 4, 7, and 9 weeks of outdoor exposure corresponded to 150, 225, and 300 hrs of exposure in the weatherometer.

d) Polypropylene

Polypropylene is generally more susceptible to photooxidative degradation than polyethylene. As seen in Figure 3, polypropylene films survived only about 4 weeks outdoors under Miami exposure conditions. Figure 8 shows the change in absolute elongation at break of polypropylene films containing cerium stearate with duration of exposure. The rate of degradation of the sample containing cerium carboxylates was much faster than that of polypropylene film containing no additive. Cerium stearate was found to be more effective than the caprylate at equivalent concentrations tested (Figure 9).

Weatherometer exposure data confirmed these findings and indication the effectiveness of stearate as a prooxidant to be essentially concentration-independent for the three specific concentrations tested. Complete embrittlement (<5% elongation at

FIGURE 8

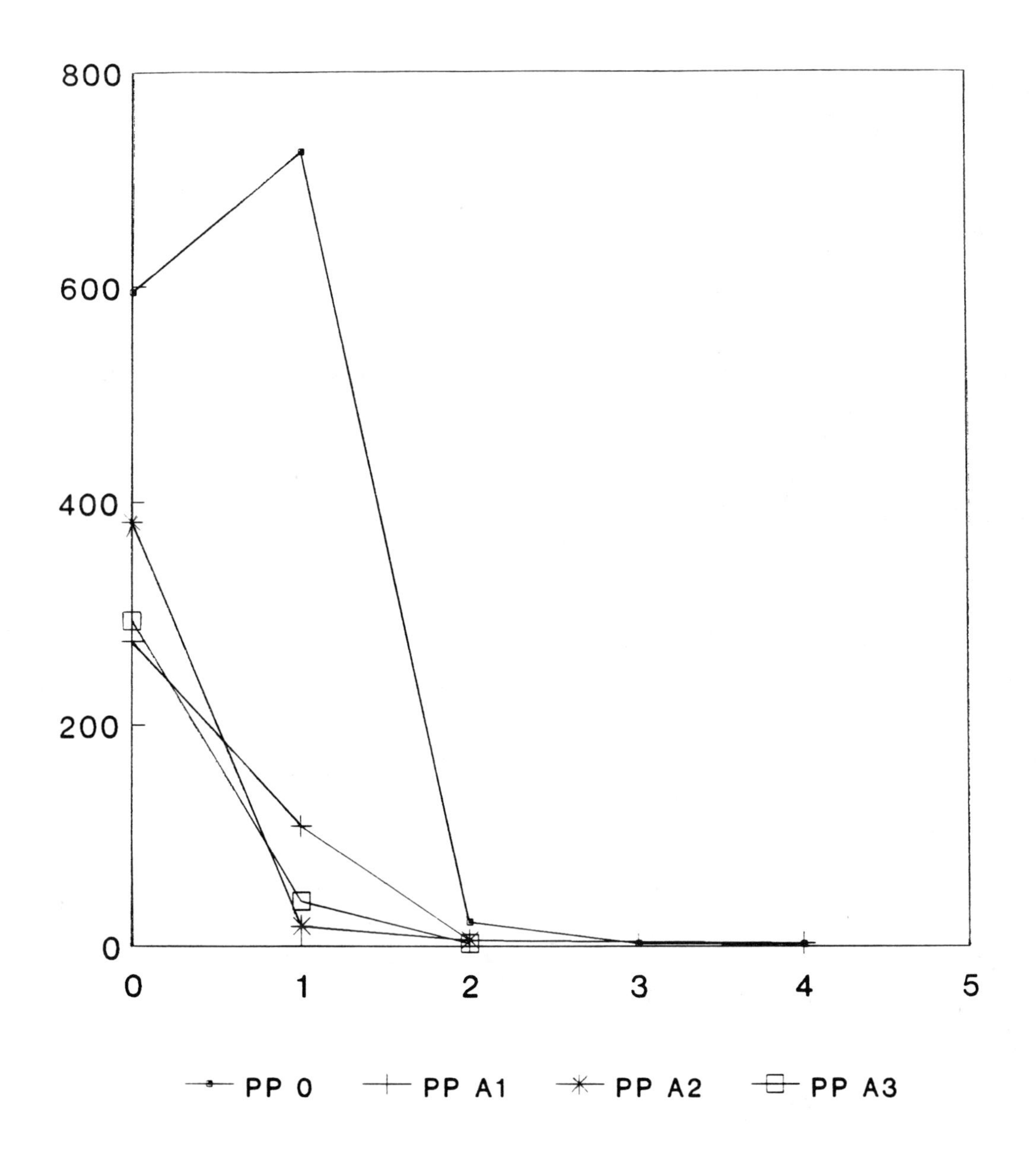


LKB 11/3/89

FIGURE 9

# PERCENT ELONGATION AT BREAK
## PP EXPOSED OUTDOORS

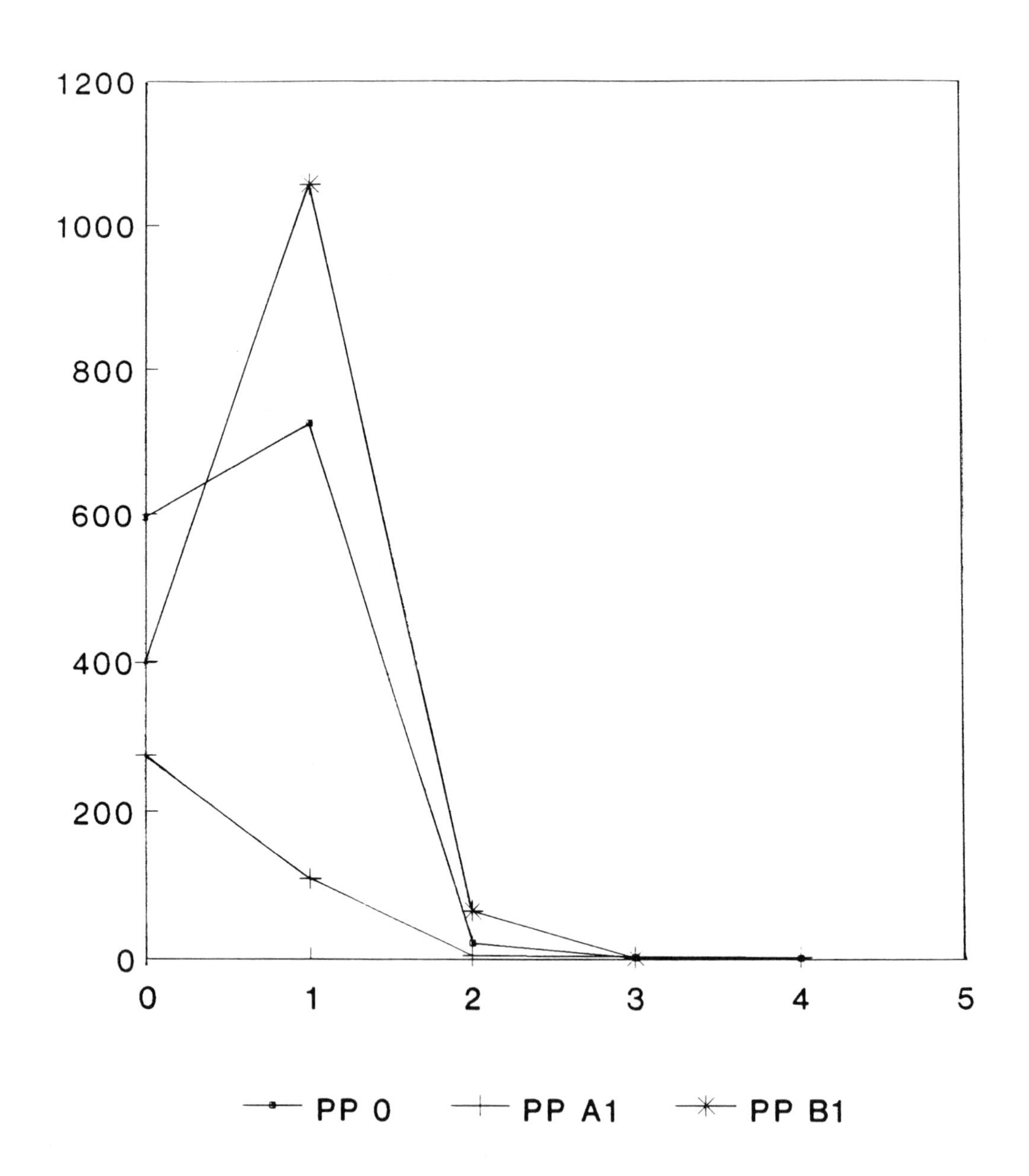


LKB 11/3/89

break) was obtained after 40 hours of exposure of films containing 0.1 phr cerium stearate and after 100 hr for the untreated control. On the basis of changes in the elongation at break, 150 hrs of exposure in the weatherometer had about the same effect as 4 weeks of outdoor exposure, under the present experimental conditions.

e) Polystyrene (PS)

While polystyrene foam is the target of plastic bans and restrictions documented in the U.S. media and legislation, the present study was limited to extruded polystyrene film samples. These crystal PS films were somewhat brittle with very low initial elongation at break values. Thus, measurement of tensile properties and paraticularly, percent elongation at break, was not adequate for monitoring the degradation of these films on outdoor and weatherometer exposure. Therefore, gel permeation chromatography was performed in order to quantitate the loss of physical properties as a decrease in weight average (Mw) and number average (Mn) molecular weight as a function of exposure time. Samples after extended exposure was found to be readily soluble in chloroform or THF and could therefore be analyzed by this technique.

The control polystyrene film survived outdoors in Florida to about nine weeks while films containing 0.1 phr of cerium stearate or cerium caprylate persisted for 7 weeks and 5 weeks respectively. Molecular weights obtained at 5 weeks (cerium caprylate) and 7 weeks (cerium stearate) of exposure are given in

Table II.

TABLE II: Gel Permeation Chromatographic Data on Polystyrene Samples. Chloroform Solvent.

| | Control | Cerium Stearate (0.1 phr) | Cerium Caprylate (0.3 phr) |
|---|---|---|---|
| Mn 0 weeks | 132,900 | 142,959 | 126,456 |
| 5 weeks | 100,000 | - | 81,000 |
| 7 weeks | 73,450 | 54,800 | - |
| | | | |
| Mw 0 weeks | 245,963 | 267,200 | 246,200 |
| 5 weeks | 206,700 | - | 179,800 |
| 7 weeks | 189,600 | 137,700 | - |

The change of number average molecular weight (Mn) from the original, unexposed value (Mo) as a function of exposure time is presented in Figure 10.

The polydispersity, Mw/Mn, increased with exposure for all samples from an initial value of 1.85-1.95, to as high as 2.5, suggesting a broadening of the molecular weight. Average number average molecular weight were also significantly decreased with the extent of exposure, confirming oxidative degradation in these samples.

Alternate GPC data obtained for samples exposure outdoors

CHANGE IN NUMBER AVERAGE MOLECULAR WEIGHT UPON OUTDOOR EXPOSURE

FIGURE 10

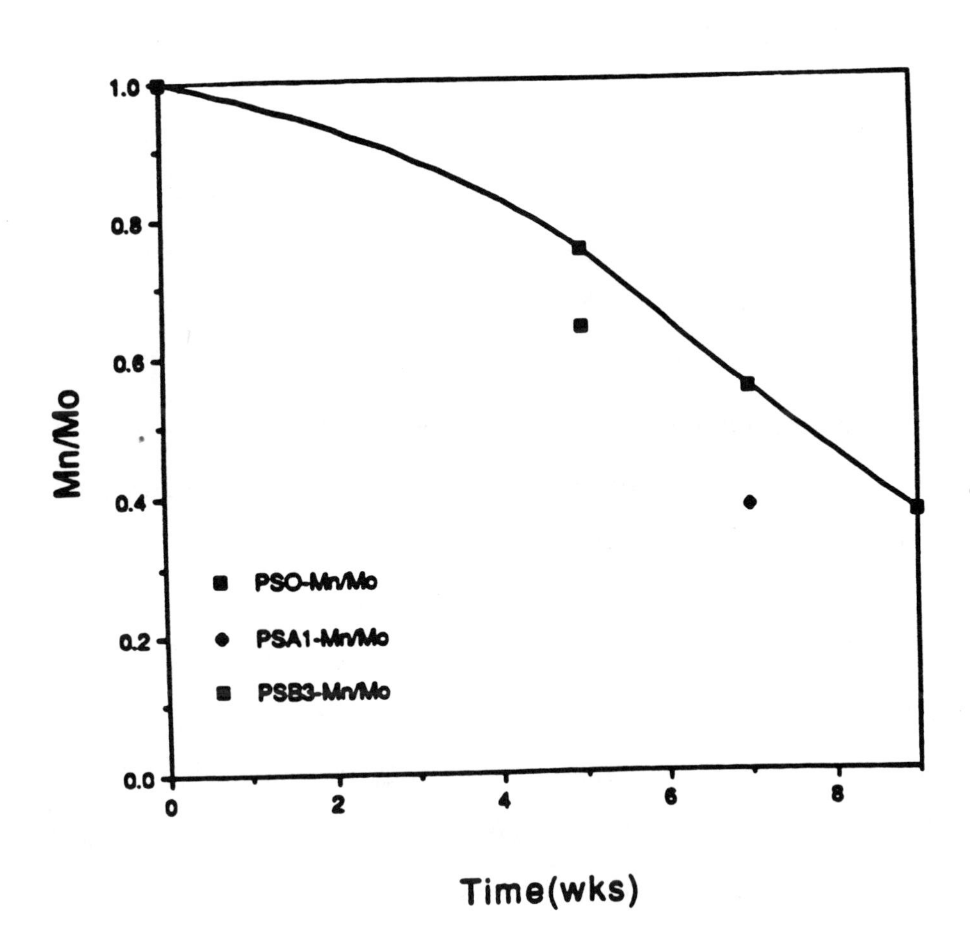

containing cerium stearate (A1, A2, A3, and A4 corresponding to 0.1, 0.3, 0.6, and 1 phr, respectively) shows that films containing cerium stearate had a significantly greater decrease in number average molecular weight compared to the control after seven weeks. This decrease does not appear to be concentration dependent. GPC data for films containing selected concentrations of cerium stearate and caprylate weathered outdoors or exposed in a weatherometer, are presented in figures 11 and 12.

Percentage change in tensile strength as a function of exposure time for films containing cerium stearate or cerium caprylate versus control is graphically depicted in Figures 13 and 14.

## CONCLUSIONS:

The preliminary exposure studies conducted demonstrate the cerium (III) stearate and caprylate to be effective prodegradant additives in polyethylenes, polypropylenes and polystyrenes. A definite acceleration of the loss in elongation at break compared to that in the unstabilized film of the same resin used as a control was observed. The acceleration factor relative to a fully stabilized type of film will be much higher. In the case of polystyrene, the preliminary GPC data indicate a decrease in the molecular weight and an increase in the polydispersity of the polymer. The degree of enhanced degradation could vary according to the source of base resin and could be effected by the presence of other additives in the formulated product. Therefore, additional work is needed to optimize the concentrations of cerium compounds according to the resin type, compounded formulation and end product.

FIGURE 11

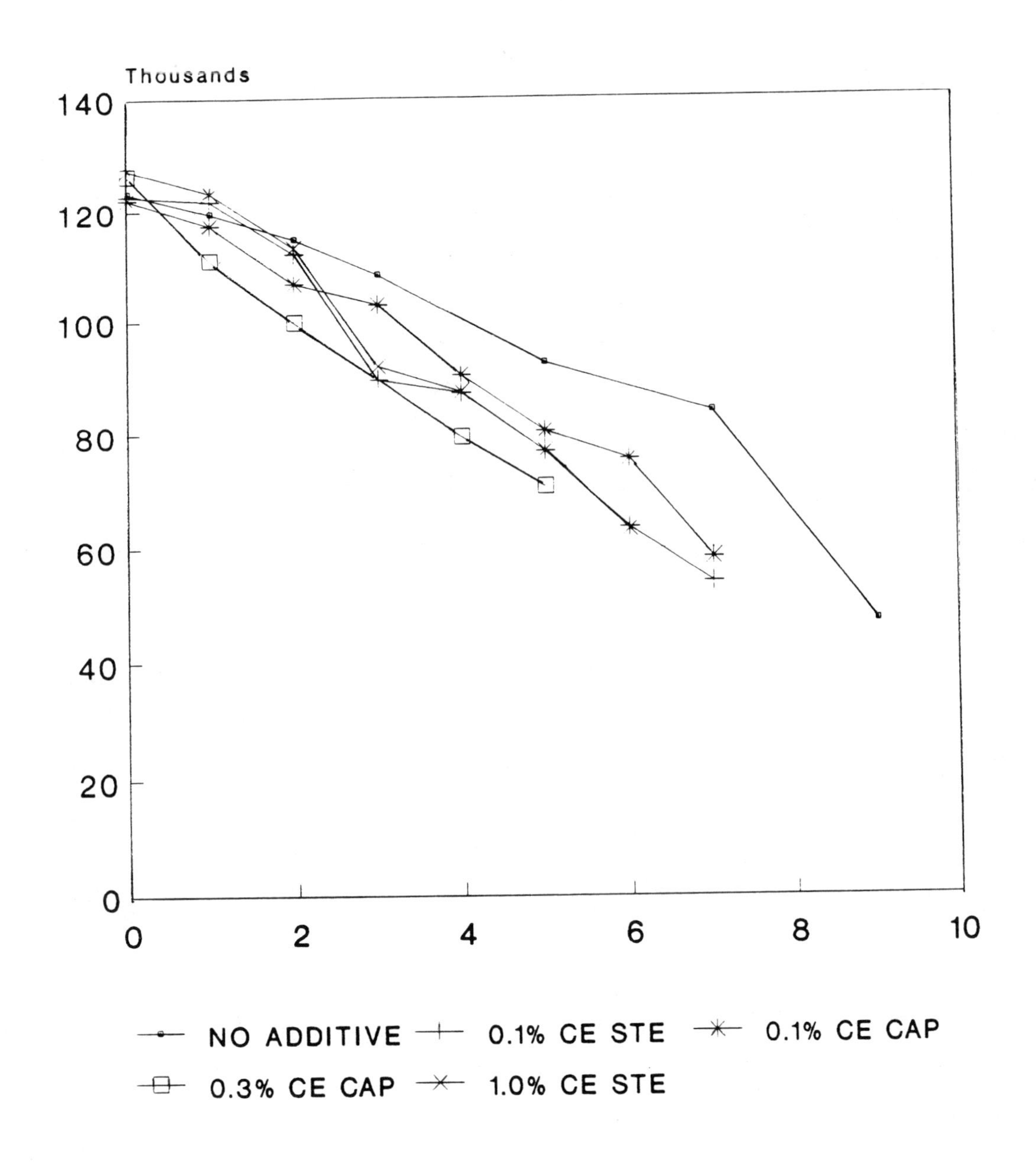


LKB 11/3/89

FIGURE 12

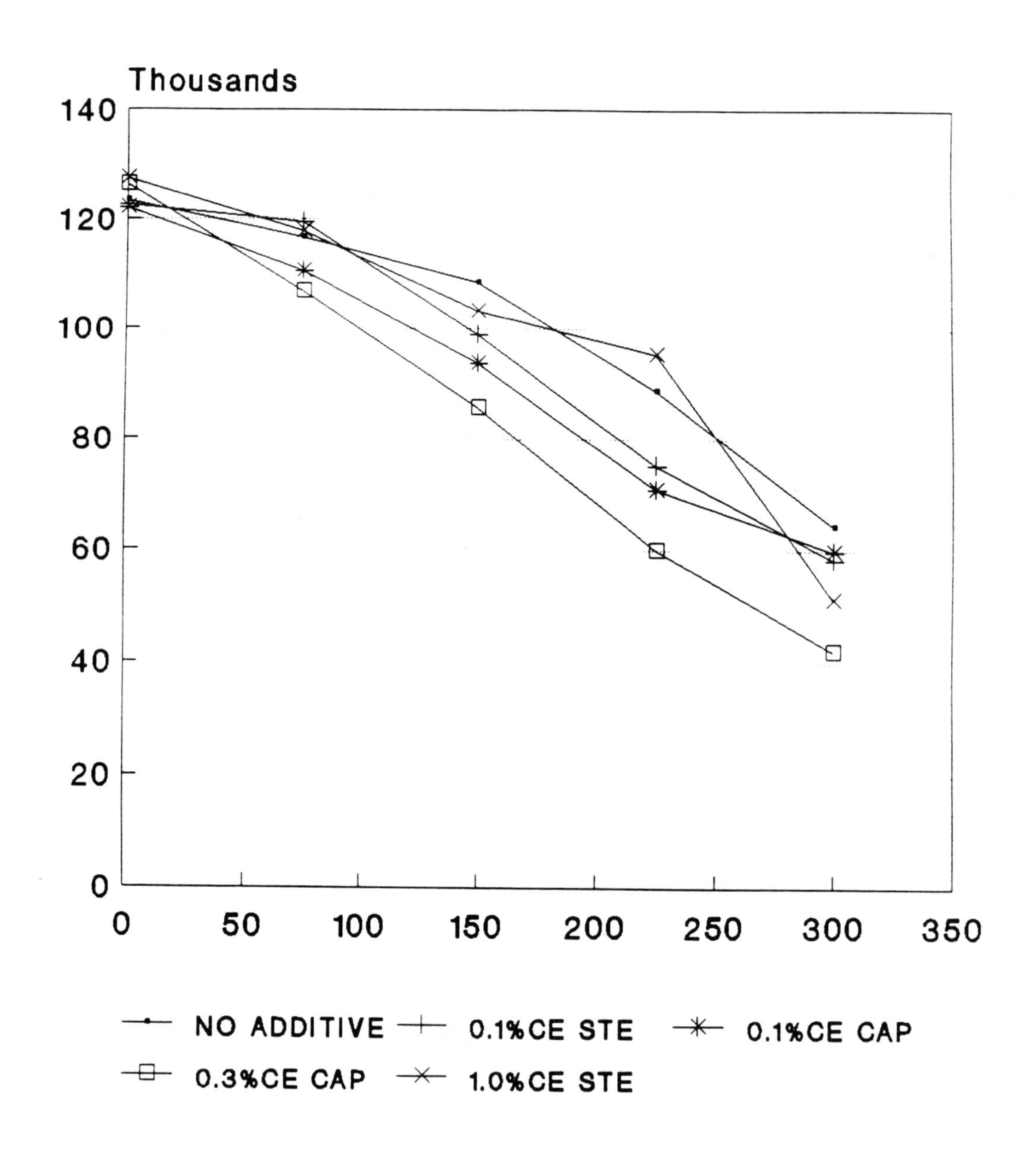


LKB 11/3/89

FIGURE 13

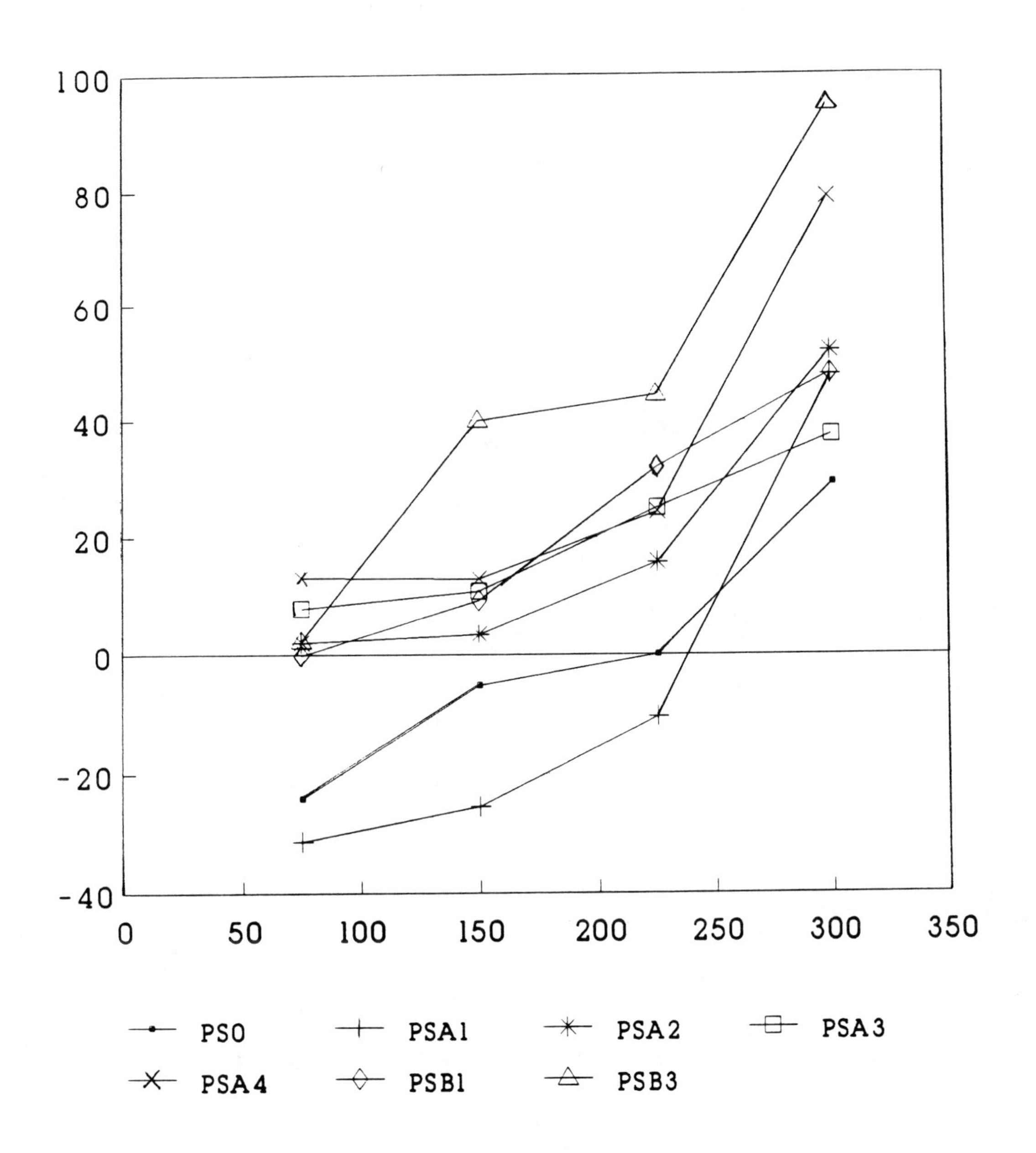


LKB 11/3/89

FIGURE 14

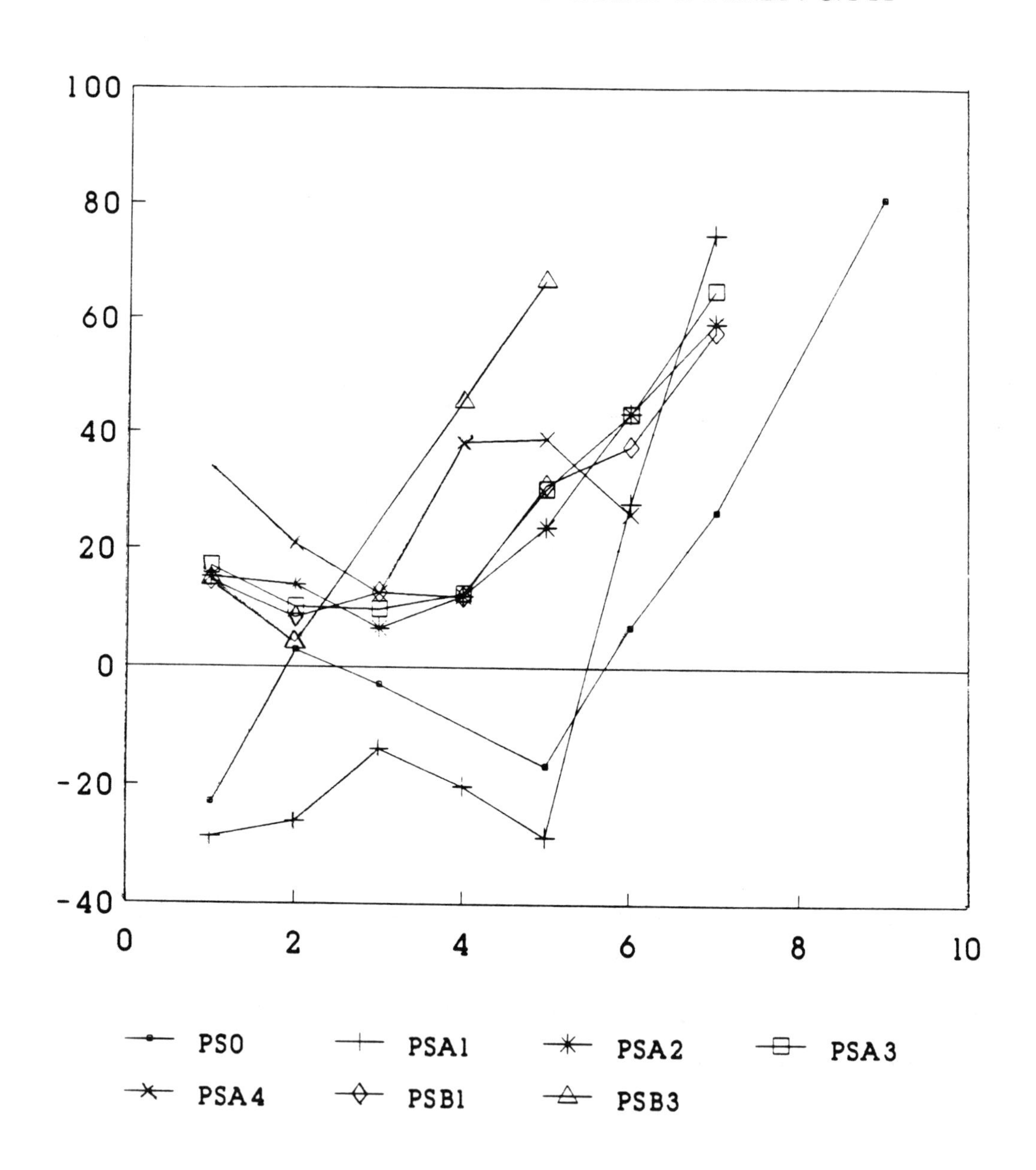


LKB 11/3/89

REFERENCES

Allen, N.S. (Ed.) (1983), "Degradation and Stabilization of Polyolefins", Applied Science Publishers, England.

Boberg (1976). U.S. Patent 3,994,855.

Davis, A. and Sims, D. (1983), "Weathering of Polymers", Applied Science Publishers, England.

Gilead et al. (1984) U.S. Patent 4,461,853.

Grassie, N. (Ed.) (1979) "Developments in Polymer Degradation - 2", Applied Science Publishers.

Gratani et. al. (1977) U.S. Patent, 4,024,764.

Eastman Kodak Company, (1969), U.S. Patent 3,454,510.

Mellor, D.C., Moir, A.B., and Scott, G. (1973), Eur. Polym. J. 9,219.

Osawa, Z. Kurisu, N. Nagashima K. Nakano. (1979), K.J. Appl. Polym. Sci., 23, 3583.

Reich, L. and Stvala, S.S. (1966), Reviews in Macromol. Chem. 1,249.

Reich, L. and Stvala, S.S. (1969), "Autooxidation of Hydrocarbons and Polyolefins", Marcel Dekker.

Scott, G. (1982), Dev. in Polymer Stab, Chaper 4, p72.

Scott, G. (1972), Macromol. Chem., 8,319.

Scott, G. (1978), U.S. Patent 4,121,025.

QUESTIONS AND ANSWERS

---PRESENTATION BY DR. LUCINDA BALLINGER

(Rhone Poulenc)

DR. REDPATH: Yes, Tony Redpath, EcoPlastics.

There's some very interesting data there. You see for the polyolefins there seems to be no concentration effect for the additive. I will say we have seen similar results with our vinyl ketone systems in polypropylene, not on polystyrene. I will be talking about that this afternoon but we have seen similar behavior and I can't explain it. I don't know, can you?

DR. BALLINGER: No, I'm sorry, I can't.

No, No.

DR. REDPATH: I just point that out. It goes contrary to what one might expect in terms of the ability to absorb light which should be going up. One has to look for saturation effects, but I don't think in our case that's not there but there obviously seems to be triggering mechanism and beyond that, additives at least in this case don't seem to have any further effect.

DR. BALLINGER: Yes.

DR. REDPATH: Thank you.

DR. BALLINGER: Yes, Doctor.

DR. SCOTT: Gerald Scott.

I have a question and a comment. The question is what was the colour of the cerium stearates in the polymer, were they intensely coloured or was there not much?

DR. BALLINGER: No, the films were clear, colourless.

DR. SCOTT: Yes, thank you.

Now, the comment really follows up really what Tony Redpath was saying. We found that if you have a fairly insoluble additive, then varying concentration has little effect on photo stability and following up that, did you see any evidence of diffusion of the additives to the surface? Do you see balloon developing, for example, in the one per cent concentration which is a very high concentration by most standards and I wondered if you see it coming out to the surface?

DR. BALLINGER: I don't think I can answer that question accurately, Dr. Scott. Tony? Did you observe, did any of the people that were extruding the film, the cerium stearate containing polystyrene film at one per cent, was there any problem with ballooning to the surface as far as you could see?

DR. ANDRADY: No, not during the...

DR. LOOMIS: Could you use the microphone.

DR. BALLINGER: Thank you, Tony.

DR. ANDRADY: I'm Tony Andrady, Research Triangel Institute.

I oversaw the extrusion and the testing procedures that she presented. No, I did not see any ballooning. We still have some samples stored in our warehouse and still they seem to be clear. I think maybe it's concentration-dependent too and that the concentrations we are working, we do not see discretional nor ballooning to a significant extent. Thank you.

DR. BALLINGER: Thank you.

DR. MITRA: Sam Mitra from 3M. The cerium stearate and caprylate that he used, what's the oxidation state of that as you use it?

DR. BALLINGER: Cerium 3.

DR. MITRA: And what's the absorption profile of that in the Xeon arc, do you know what the --

DR. BALLINGER: We haven't done that yet but we need to do that, yes.

DR. MITRA: Thank you.

DR. BALLINGER: Thank you.

DR. CURTO: It is nice to see someone staying within the bounds of a usable polymer system. I was very happy to see your data in that I have been highly frustrated by working with several systems, both with the ketones and carbonyl groups as an initial activator and saw basically the same thing where we are testing all of these combination systems and they all degrade nicely and your control also degrades nicely and what the question that this brings up that I think this group ought to address is what's timely degradation.

In all of the studies that I have run and they're multiple, my high density polyethylene standard has been driven to what I choose to call the degradtion and again I have a more earthy pragmatic definition, it' when I can take it in my hand and crush it on the desk of a legislator.

Again, the point that I think that this group ought to address is, wouldn't it be wonderful if we could show people that some of these polymers with absolutely nothing in them do in fact degrade in a timely way, if you are willing to say that year or a year and a half is timely. I think that would be something worthwhile for us to look at.

DR. REDPATH: Tony Redpath again.

Just a couple of comments I would like to make, Cindy.

First of all, I would like to say thank you from the organizers. You represent one of the large commercial organizations and thank you for giving a talk with a lot of meat and technical data in it, that's what we are looking for.

And secondly, straight administration, we are about to adjourn for lunch and we come back at one o'clock. Thank you.

DR. LOOMIS: A few things sort of jumped out from some of the talks and I think we are all guilty at various times of presenting scientific data and not characterizing some part of it well and I think it is obvious from this that we have to characterize the polymers to the hilt and that we have to learn to characterize the biological systems as well, particularly those of us who are not biologically trained. And that's basically what I want to say is if our experiments cannot be repeated by someone else in their lab, then I'm not sure they're worth presenting

---Luncheon recess at 11:30 a.m.

# BIODEGRADATION OF POLY (BETA-HYDROXYALKANOATES)

David F. Gilmore[1], R. Clinton Fuller[1], and Robert Lenz[2]

Dept. of Biochemistry[1] and Dept. of Polymer Science and Engineering[2], Univ. of Massachusetts, Amherst, MA 01003

## INTRODUCTION

Poly beta-hydroxybutyrate (PHB) was first described by Lemoigne in 1925 (Lemoigne, 1925) as a lipid inclusion in the bacterium Bacillus megaterium. Subsequent research identified it to be a high molecular weight polymer involved in the storage of energy and reducing power (Macrae and Wilkinson, 1958). It has been realized only within the last 20 years that pure PHB is rarely found in bacteria (Findlay and White, 1983); instead, an entire family of copolyesters, the poly (beta-hydroxyalkanoates) (PHA's) have been found. These copolymers are generally produced by a variety of bacteria (Table 1) under conditions of metabolic stress, where nitrogen, phosphate, oxygen, or other nutrient is limiting in the presence of excess carbon (for review, see Dawes and Senior, 1973 and Merrick, 1978).

It is now appreciated that the PHAs are all aliphatic polyesters with the same basic type of unit. Fig. 1 shows the structure of the principal repeating units found in these polymers. A variety of PHA's can be synthesized by bacteria depending on the species of bacterium and the compounds suppled as carbon sources for growth. PHA's are thermoplastics, and as such can be melted and molded.

Table 1: Limiting Compounds Leading to PHA Formation

| Compound | Organisms |
|---|---|
| Ammonium | Alcaligenes eutrophus<br>Alcaligenes latus<br>Pseudomonas oleovorans<br>Pseudomonas cepacia<br>Rhodospirillum rubrum<br>Rhodobacter sphaeroides<br>Pseudomonas sp K<br>Methylocystis parvus<br>Rhizobium ORS571 |
| Iron | Pseudomonas sp K |
| Magnesium | Pseudomonas sp K<br>Pseudomonas oleovorans<br>Rhizobium ORS571 |
| Manganese | Pseudomonas sp K |
| Oxygen | Azotobacter vinelandii<br>Azotobacter beijerinckii<br>Rhizobium ORS571 |
| Phosphate | Rhodospirillum rubrum<br>Rhodobacter sphaeroides<br>Caulobacter crescentus<br>Pseudomonas oleovorans |
| Potassium | Bacillus thuringiensis |
| Sulfate | Pseudomonas sp K<br>Pseudomonas oleovorans<br>Rhodospirillum rubrum<br>Rhodobacter sphaeroides |

PHA: Poly(beta-hydroxyalkanoate)s

$$\left[ -O - \underset{\substack{| \\ H}}{\overset{\substack{R \\ |}}{C}} - CH_2 - \overset{\substack{O \\ \|}}{C} - O - \right]_n$$

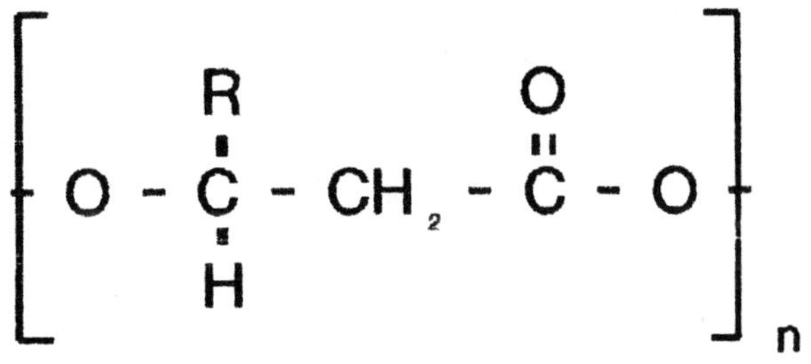

R = n-alkyl pendant group of variable chain length

beta-hydroxybutyrate, R = methyl
beta-hydroxyvalerate, R = ethyl
beta-hydroxycaproate, R = n-propyl
beta-hydroxyheptanoate, R = n-butyl
beta-hydroxyoctanoate, R = n-pentyl
beta-hydroxynonanoate, R = n-hexyl
beta-hydroxydecanoate, R = n-heptyl

Figure 1. Structure of the poly (beta-hydroxyalkanoates). All members of the family contain a backbone consisting of three carbon units connected by ester linkages.

Depending on the nature of the pendant group, the polymer can be a dense, brittle plastic or a rubbery elastomer (Brandl et al., 1988; Brandl et al., 1989), and part of the research in our laboratory involves supplying polymer-producing bacteria with organic compounds not found in nature in an attempt to produce plastics with unusual characteristics or with functional groups for subsequent synthetic modifications.

It is expected that most of these polymers will be biocompatible and biodegradable. As such, there are a number of potential applications for these materials, both in medicine and for industrial applications as indicated in Table 2. As landfills all over the United States reach capacity, the clamor for the need for truly biodegradable plastics has increased, but considerably more research on the production of these polymers is necessary before full scale industrial production can produce consumer disposable items (e.g. plastic bags) at an acceptable cost. ICI Ltd. of England currently markets a small number of plastic bottles made from a copolymer of hydroxybutyrate and hydroxyvalerate (PHB/V).

## BIODEGRADABILITY

As mentioned above, it is expected and often stated without supporting information, that all PHA's are biodegradable. Microbiology has a simple rule, that for every material naturally produced there are one or more microorganisms which will break it down. One of the important aspects of this rule involves stereochemistry. Many natural polymers contain chiral centers,

Table 2: Paractical Applications of PHA

Medical Applications:

- Surgical pins, sutures, and stables
- Wound dressings
- Blood vessel replacements
- Bone replacements and plates
- Biodegradable carriers for long term dosage of drugs and medicines

Industrial Applications:

- Biodegradable carrier for long term dosage of herbicides, fungicides, insecticides, or fertilizers
- Packaging containers, bottles, wrapings, bags, and films
- Disposable items such as diapers, or feminine hygiene products

regions of the macromolecule which can exist in either of two possible configurations which are mirror images of each other. Chemical syntheses generally produce polymers containing an equal mixture of both possible configurations, but in nature, most biological polymers exist in one configuration or the other depending on the type of polymer. Thus enzymes have evolved to catalyze the degradation of polymers containing only one configuration, and polymers containing both configurations (or only the wrong one) are generally not degraded biologically. Because PHA's are biologically produced, their stereochemistry does not pose a problem to their biodegradability, but chemically synthesized PHA's could suffer from this limitation.

Another important consideration in degradability is the nature of the chemical bond between monomer units. Ester bonds seem to be fairly susceptible to cleavage by hydrolysis reactions which are catalyzed by microbial enzymes. Several kinds of synthetic polyesters have been studied (Huang, 1989) and found to be enzymatically or spontaneously degraded.

Invoking our "Rule of Microbiology", it would seem a surety that PHA's will be fully biodegradable. They are produced internally by bacteria as storage products, which means that ultimately the bacteria themselves will degrade them, but individual polymer molecules are most often much too large to be transported directly into a bacterial cell. That is, it is not possible for a bacterium to ingest discarded plastic and degrade it using the same enzyme system that operates on its own storage material. Polymer biodegradation therefore must occur through the synthesis and secretion of exoenzymes, enzymes released into the environment where they can attack a polymer and degrade it into chemical compounds small enough to be transported into a microbial cell and metabolized. Thus, only microbes which have evolved enzymes capable of degrading the polymer produced by other microbes will be able to degrade plastic materials in this manner, and the ability of an organism to store polymer is not necessarily correlated with its ability to degrade polymer that appears in its environment.

Nevertheless, there is considerable hope that many microorganisms will be able to degrade any polymer that is microbially produced in a natural environment. Recent studies in our laboratory and elsewhere (Doi et al., 1987) have shown that bacteria can be made to synthesize unusual polymers, presumably because the PHA synthetase systems of these bacteria are not restricted to the formation of polymers with n-alkyl groups (Fig. 1) and will polymerize atypical monomer units. This provides for the production of plastics with a wide variety of physical properties; however, there is no guarantee that the microbes responsible for environmental degradation of PHA's will have enzymes that recognize these unusual polymers and carry out their degradation. Our laboratory has undertaken a study of this problem.

Yet another problem to be considered is the changes that occur in the physical composition of a plastic during the manufacturing process. In nature, PHA molecules are probably released from dying cells as small particles with a high surface to volume ratio and are surrounded by a moist environment. Such conditions are optimal for enzymatic degradation. During manufacturing, polymeric materials are dried and compressed or aligned (Huang, 1989) into products that are generally at least an order of magnitude thicker. The possible effects of such changes on biodegradation need to be addressed.

## MECHANISM OF BIODEGRADATION

Although there have been some tests of the biodegradability of PHB/V (Williams, 1989), most studies on the biodegradation of PHA's have dealt with PHB specifically. Delafield et al. (1965) reported on the isolation and partial characterization of a number of soil bacteria capable of degrading PHB. Additionally, a large number of pseudomonads from culture collections were tested and also found to degrade PHB. As part of that work they isolated Pseudomonas lemoignei, which very efficiently degrades PHB. The exoenzyme produced by this bacterium was subsequently isolated and characterized (Lusty and Doudoroff, 1966, Nakayama et al., 1985) and was found to be a mixture of two enzymes with similar molecular weights and similar properties. Both enzymes could act on the hydroxyl end of a PHB chain as shown in Figure 2; one enzyme cleaved predominantly dimers from the chain while the other enzyme produced predominantly trimers. When the hydrolysis was allowed to proceed to completion, both enzymes subsequently degraded the trimers, leaving only monomers and dimers as end products. When trimers were added to a preparation of PHB and enzyme, few trimer molecules were hydrolyzed until nearly all the PHB was degraded; it was concluded that the enzyme remained attached to the polymer and hydrolyzed it in a procession fashion.

The extracellular depolymerase responsible for PHB hydrolysis by

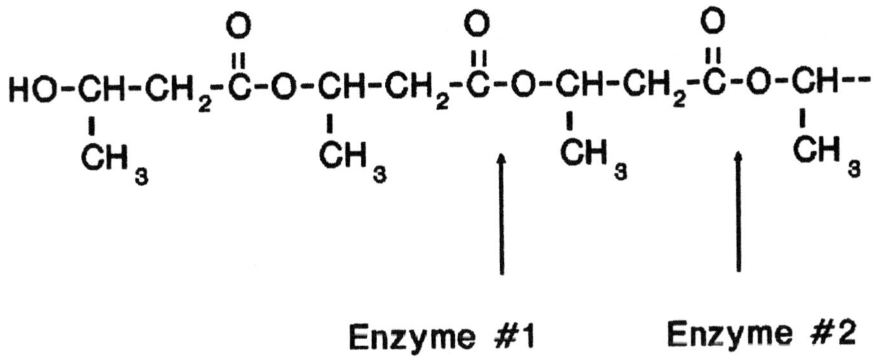

$$HO-\underset{CH_3}{CH}-CH_2-\overset{O}{\overset{\|}{C}}-O-\underset{CH_3}{CH}-CH_2-\overset{O}{\overset{\|}{C}}-O-\underset{CH_3}{CH}-CH_2-\overset{O}{\overset{\|}{C}}-O-\underset{CH_3}{CH}--$$

Enzyme #1 Enzyme #2

Figure 2. Mode of attack by depolymerases from *P. lemoignei*. One enzyme (enzyme #1) cleaves primarily dimers from the hydroxyl end of the polymer chain while the other (enzyme #2) cleaves primarily trimers.

Alcaligenes fecalis has also been studied (Tanio et al., 1982). Depolymerase activity was shown to reside in a single enzyme, similar in many respects to the P. lemoignei enzymes. This depolymerase produced mostly dimers cleaved from the hydroxyl end of the polymer. The substrate specificity of this enzyme was also studied. The three carbon backbone, which is common to all of the poly (beta-hydroxyalkanoates), was found to be required for appreciable enzymatic activity, and other biodegradable polyesters with a different number of carbons in the repeating unit (such as polycaprolactone) were not hydrolyzed by this enzyme.

## ASSAYS OF BIODEGRADATION

The initial studies in our lab utilized the "clear zone method". This method is an adoption of the method used by Delafield et al. (1965) to confirm the isolation of PHB depolymerizing organisms. In our procedure, powdered polymer is incorporated into an agar medium containing a complete basal medium minus a carbon source as illustrated in Figure 3. Wells are then bored into the agar, and these wells are filled with a test inoculum or other source of enzyme. Polymer degrading organisms produce exoenzymes which diffuse through the agar, degrade the polymer to water soluble materials, and produce a zone of clearing as shown in the photographs in Figure 4. This technique has the advantage that any inoculum, from a soil suspension to a pure culture, can be tested for its ability to degrade any polymer that can be ground into a fine powder and incorporated into an agar matrix. Information can also be obtained,

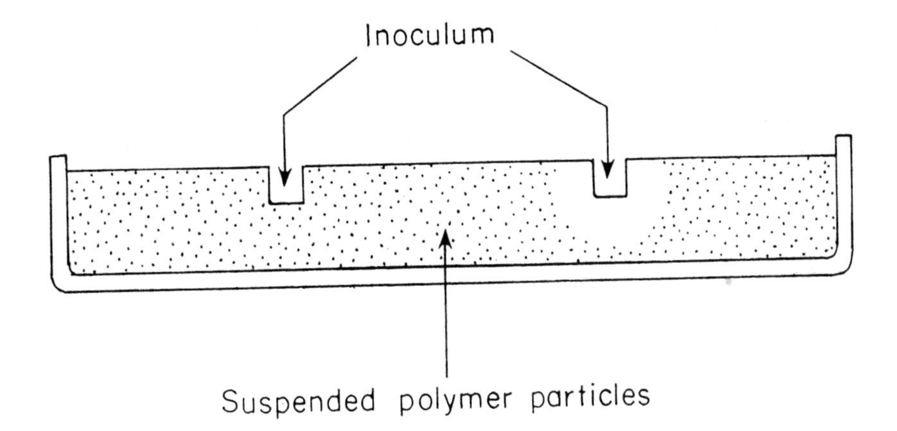


Figure 3. The clear zone method. Powdered polymer is suspended in an agar matrix containing all the necessary nutrients for microbial growth with the polymer as the only carbon source. After wells are bored into the agar, they are filled with an inoculum and the plates are incubated. The well on the left represents a negative or control sample; the well on the right depicts a culture capable of degrading the polymer.

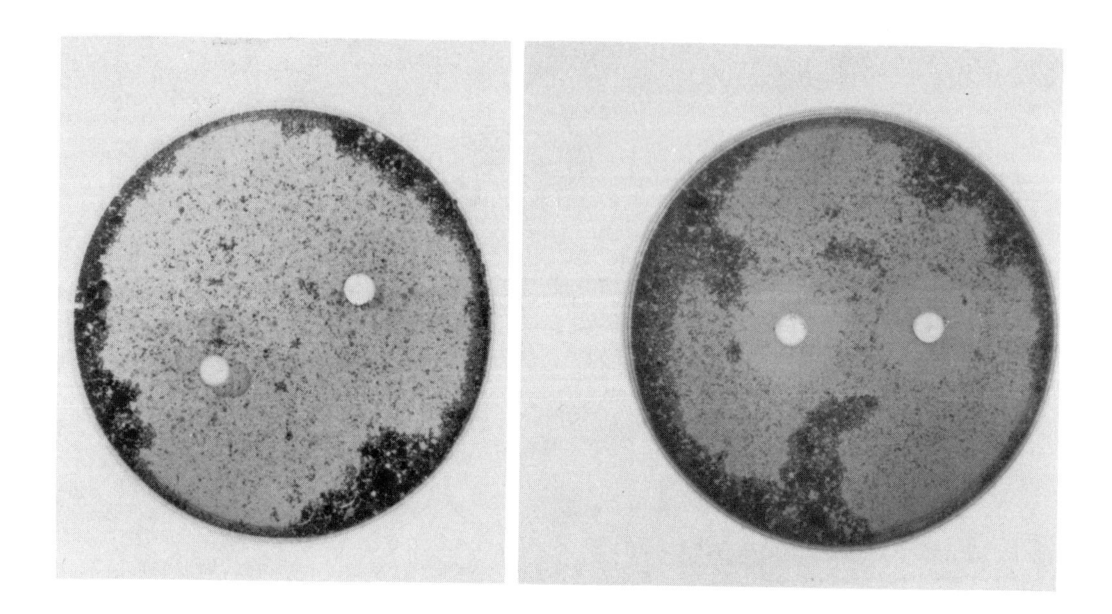

Figure 4. Results obtained using the clear zone method. a) a culture incapable of degrading PHB. b) a culture producing exoenzymes capable of degrading PHB; note the clear zones around the sites of inoculation.

after a lengthy extraction process, on the products of the hydrolysis. However, it is difficult to obtain quantitative data concerning the rate or extent of degradation using this method, nor is it useful for studying the mechanism of degradation.

For these reasons we have developed a modification of the turbidometric assay which has been used by others to follow the purification of PHB depolymerases (Lusty and Doudoroff, 1966; Tanio et al., 1982). In this method, shown in Figure 5, a finely powdered polymer suspension, prepared by sonication, is combined with a source of enzyme in a covet. The polymer particles, which would have a tendency to settle during the course of an assay, are maintained in suspension by the addition of 55 C melted agarose. The resulting 0.25% agarose solidifies within 90 seconds to produce a gel with little background turbidity. The turbidity of the preparation, measured as the optical density (O.D.) at 660 nm, decreases during degradation as the particles become smaller and scatter less light as a result of depolymerization. Figure 6 shows a typical result of this assay. The initial rate of degradation is linear over the first 9 to 15 minutes then slows with time. A value for the total extent of degradation can also be approximated by comparing the O.D. at 3 minutes (the first time point at which turbidity is measured) with the final O.D.

## OUR RESEARCH FOCUS

Our interest is to determine the biodegradability of various PHA's and the mechanism of their biodegradation. These experiments

**Each sample contains:**

**2.0 ml enzyme in buffer**

**0.1 ml PHA suspension**

**0.75 ml 1% agarose (melted)**

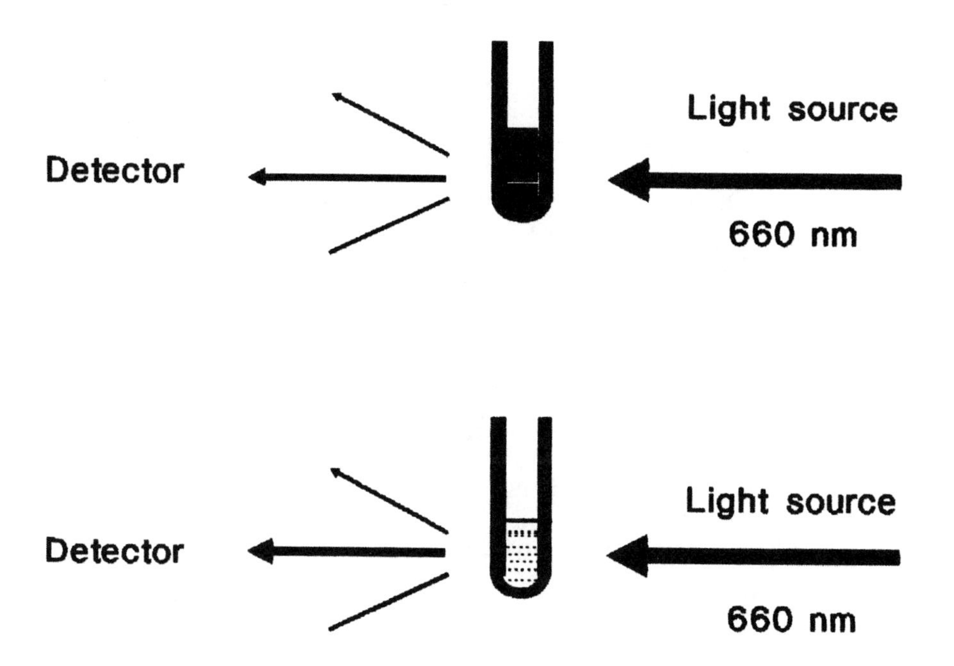


Figure 5. The turbidometric assay. Finely powdered polymer is mixed with a solution of enzyme in buffer (10 mM Tris-HCl, 1 mM $CaCl_2$, pH 7.4) and kept in suspension by addition of agarose to a final concentration of 0.25%. As the polymer is degraded, the observed optical density decreases.

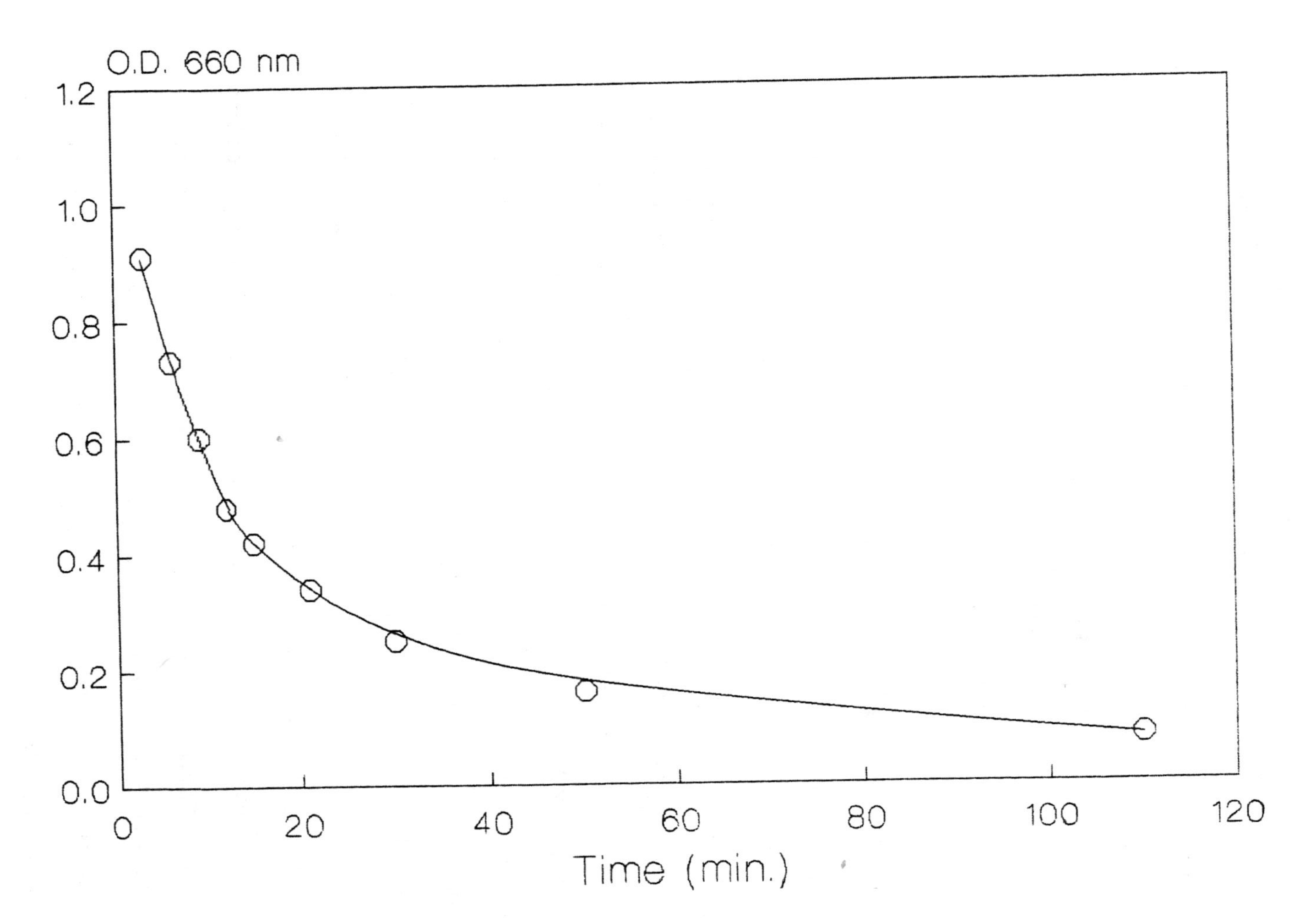


Figure 6. Standard turbidometric assay; degradation of PHB.

have employed the turbidometric assay described above and will be broadened to include other techniques such as determination of particle size changes, molecular weight changes, and changes in crystallinity. Biodegradability of PHA's will be tested using the purified, extracellular depolymerases of P. lemoignei.

Table 3 lists the reasons for using the enzyme. Pseudomonads, which are common in soils, appear to be the predominant PHB depolymerizing bacteria (Delafield et al., 1965), so it seems highly likely that the extracellular PHB depolymerizing enzymes of these pseudomonads are similar in structure and mode of action. For example, the enzyme isolated from a related species, Alacaligenes fecalis, has been shown to have the same mode of action (Tanio et al., 1982). Thus the P. lemoignei enzyme may be one of the primary mechanisms for PHA degradation in soils, and should serve as a realistic model. But purification of the enzyme will be required because, as discussed above, the depolymerization occurs through the action of two separate enzymes with the same mode of attack but different products, and analysis of the degradation mechanisms would be complicated by using a mixture of enzymes.

It is also important to note that PHB is known to adsorb proteins non-specifically (Delafield et al., 1965), and these proteins could interfere with degradation. Figure 7 compares two enzyme preparations during purification of the P. lemoignei depolymerase: the concentrated culture supernate and the purified enzyme. As shown in Figure 7, as the amount of impure enzyme added to the assay is increased, the amount of measured activity falls dramatically. This type of result is typically seen when an inhibitor is present. We

Table 3. Reasons for using the purified P. lemoignei enzyme

1. Many PHB decomposers are pseudomonads.

2. The only other enzyme characterized is from the related Alcaligenes faecalis and has the same mode of attack.

3. The P. lemoignei enzyme(s) is a mixture of two activities.

4. Non-specific adsorption of proteins to PHB is a potential problem.

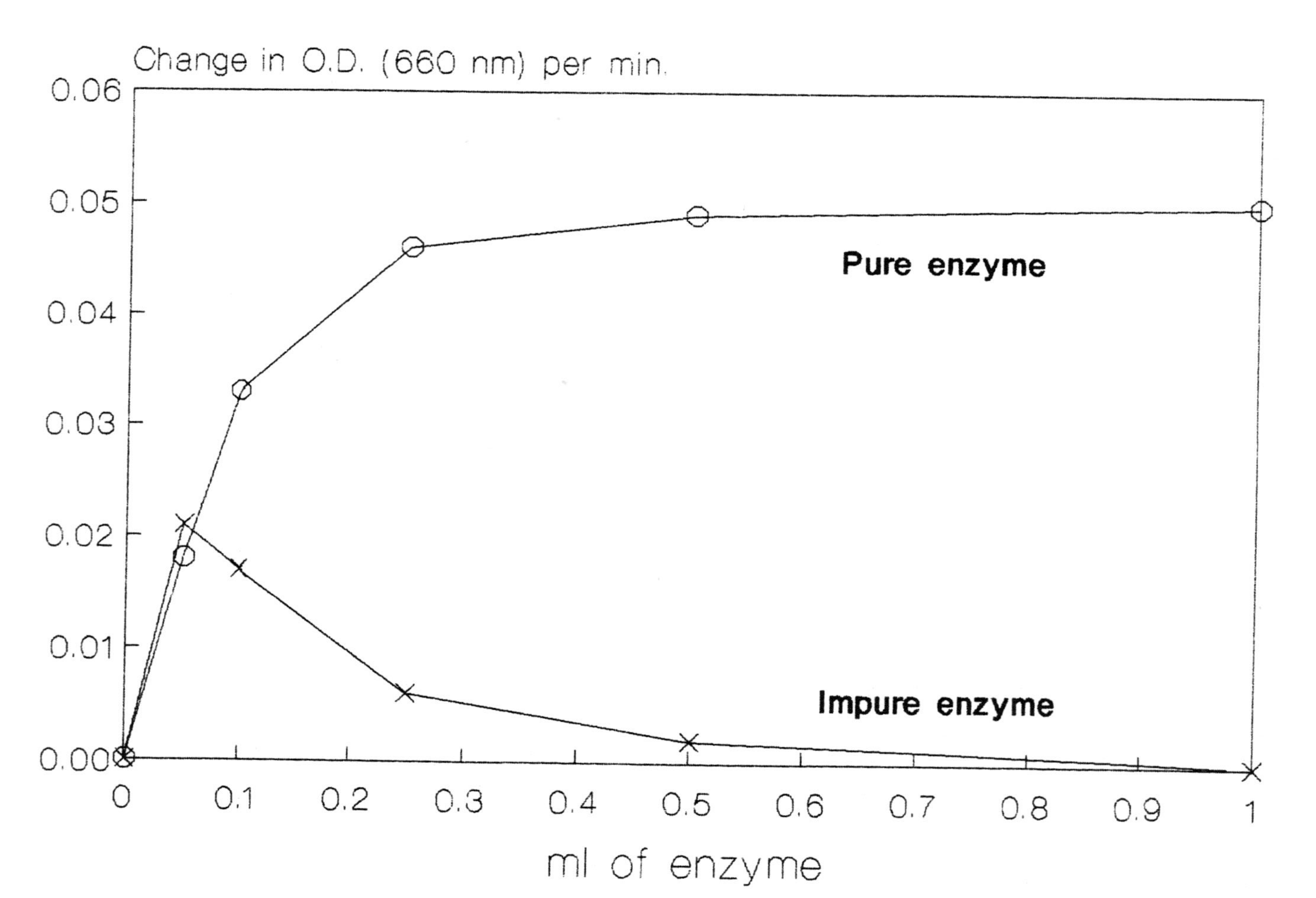


Figure 7. Effect of enzyme purity on activity. A concentrated *P. lemoignei* cell supernate shows decreased activity when increasing amounts of solution are assayed. After purification, the enzyme solution shows typical saturation kinetics.

believe, but have not demonstrated, that this inhibition is a result of the high non-specific protein content of this preparation.

RESULTS

Although stereochemistry is not a problem with naturally occurring PHB, the PHB which is chemically synthesized can have units with either [R] or [S] configurations. For this reason it was of interest to address certain questions concerning the effect of the optical purity of the polymer repeating units on biodegradability. Two polymer samples were powdered and tested for their degradability using the P. lemoignei enzyme in the turbidometric assay. One preparation consisted of stereoblock PHB molecules (synthesized chemically using an aluminum-based coordination catalyst; Gross et al., 1988) which were comprised of both [R] and [S] configuration monomers (50:50) arranged in long blocks along the polymer chains. The other preparation consisted of equal masses of polymers, one having only [R] monomer units (this sample was a PHB of natural origin of 150,000 MW) and the other having only [S] monomers (this sample was chemically synthesized from a lactone using a zinc-based coordination catalyst and was 22,000 MW). These two kinds of polymers were physically blended together. A pure control polymer of natural origin [R] PHB was also tested.

There was a decrease in turbidity of the stereoblock polymer of less than 10% over a period of three days, suggesting little degradation took place. The pure [R]-PHB control was rapidly degraded, the hydrolysis essentially complete by three hours as shown in Figure 8. The physical blend of [R] and [S] polymers showed unusual degradation kinetics. The rate of turbidity decrease was comparatively slow, changing little over the first 3 hours, and there was no further decrease after 15 hours. The final O.D. was about one half the initial O.D. While it may be intuitive that degradation of the [R] form and not the [S] would leave 50% of the total polymer, there is no reason to expect a direct correlation between mass and turbidity, the latter being a function of the surface area of the particles. It remains to be seen, therefore, whether the arrangement of polymer chains in the particles physically prevents access of the enzyme to its substrate or whether the presence of the [S] isomer exerts an inhibitory activity on the enzyme.

We are also studying the biodegradation of PHA's other than PHB. Our first experiments have been on copolymers of hydroxybutyrate and hydroxyvalerate (HV). Copolymers with two different compositions (13.5% and 24% hydroxyvalerate) and molecular weights around 500,000 were obtained form ICI Ltd. as technical grade powders, and suspensions were prepared by sonication. These preparations were assayed fro biodegradability using the P. lemoignei enzyme in the turbidity assay. Pure PHB (150,000 MW) was used as a control.

The turbidity of the 13.5% HV polymer, shown in Figure 9, decreased at about the same rate as the control over the first 10 min., but it quickly reached a minimum value with an O.D. just less than 0.4. The control sample, on the other hand, had nearly disappeared by two hours. The 24% HV polymer was clearly degraded at a slower initial rate, and the turbidity stopped decreasing after about 30 minutes. Clearly, the depolymerizing enzyme was suddenly

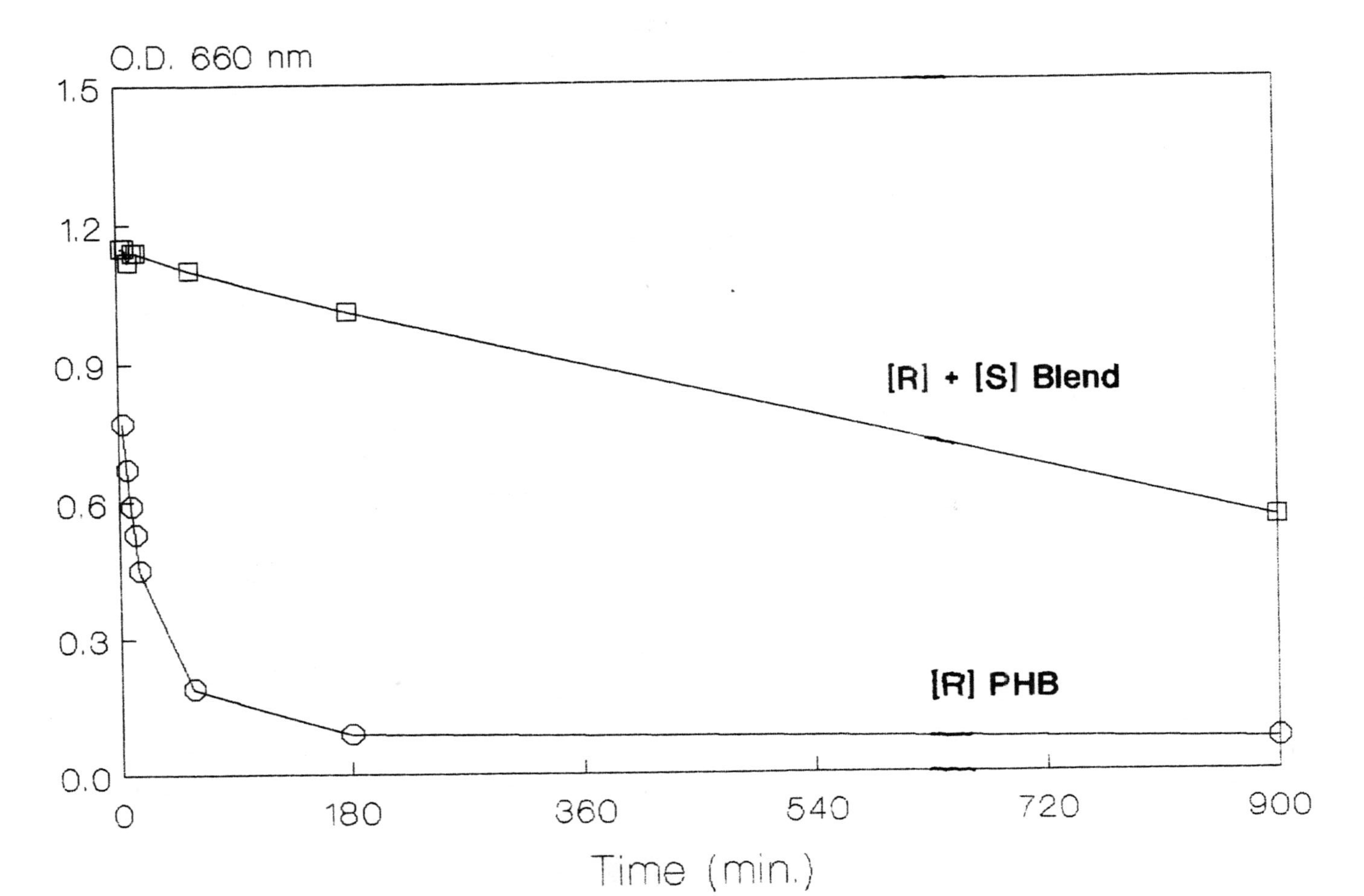


Figure 8. Effect of stereochemistry on enzymatic degradation. Natural PHB is rapidly degraded, while a blend of natural [R] and chemically synthesized [S] PHB is degraded at a much slower rate.

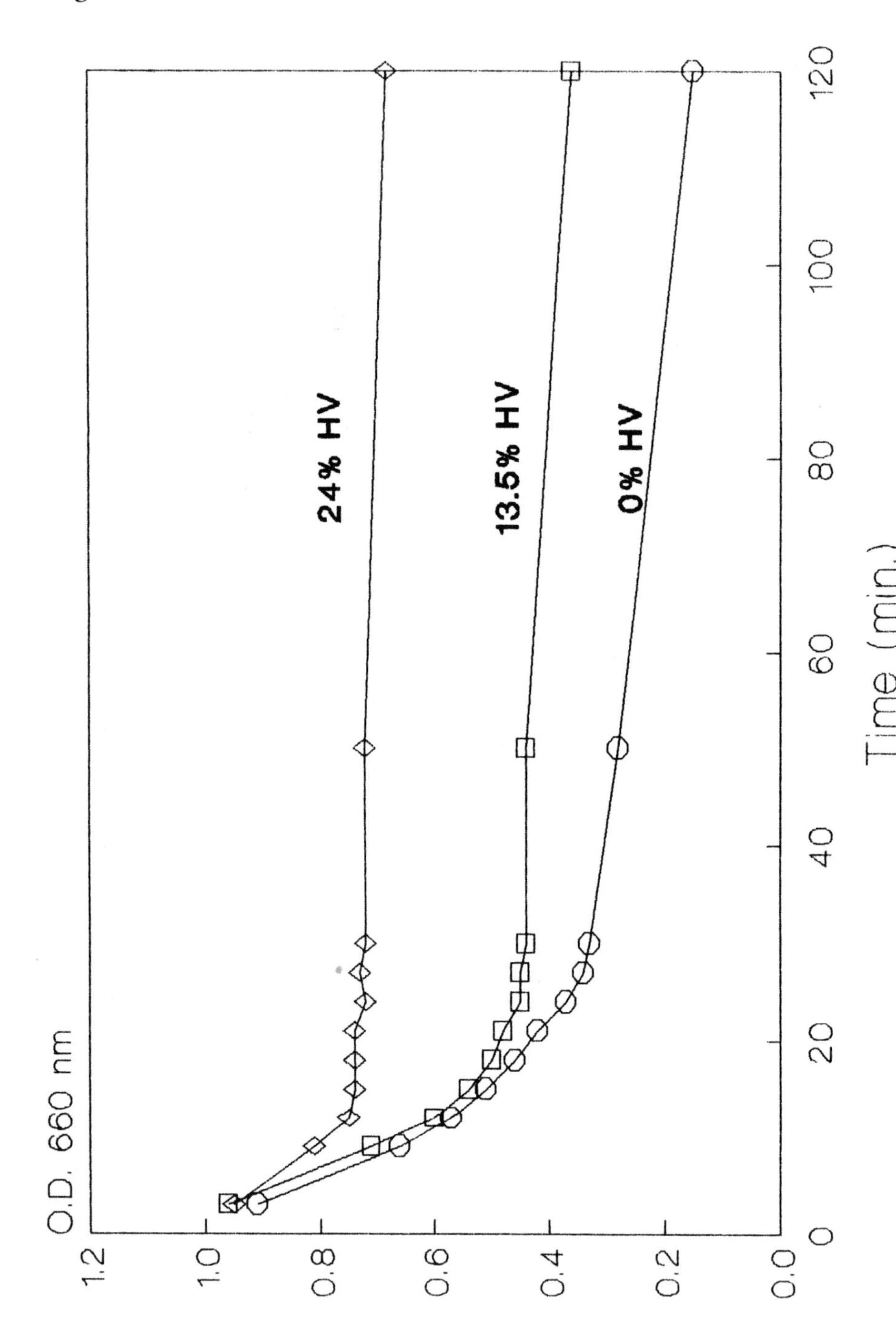


Figure 9. Comparison of the degradation rates of PHB/HV copolymers containing different amounts of hydroxyvalerate.

encountering something in the 24% HV sample that was inimicable to further degradation.

The degradation kinetics of a 7% HV copolymer did not follow this pattern as shown in Figure 10. The initial rate of turbidity decrease was not linear and was very slow; however, the rate increased with time before leveling off again. The final O.D. value was comparable to that of the pur PHB control. Thus, the extent of the degradation was unaffected by the hydroxyvalerate content, and the rate was affected by some property of the preparation other than HV content.

These are preliminary data, and they raise more questions about degradation than they answer. To fully utilize the turbidometric approach, it will be necessary to obtain other kinds of information. The initial particle size distribution and the subsequent changes during degradation will be extremely valuable for interpreting the O.D. changes. Additional information on polymer molecular weights and the degree of crystallinity before and during the degradation process will also complement the turbidometric measurements. These studies are in progress and should greatly aid our understanding of the physical and biochemical mechanism of biodegradation.

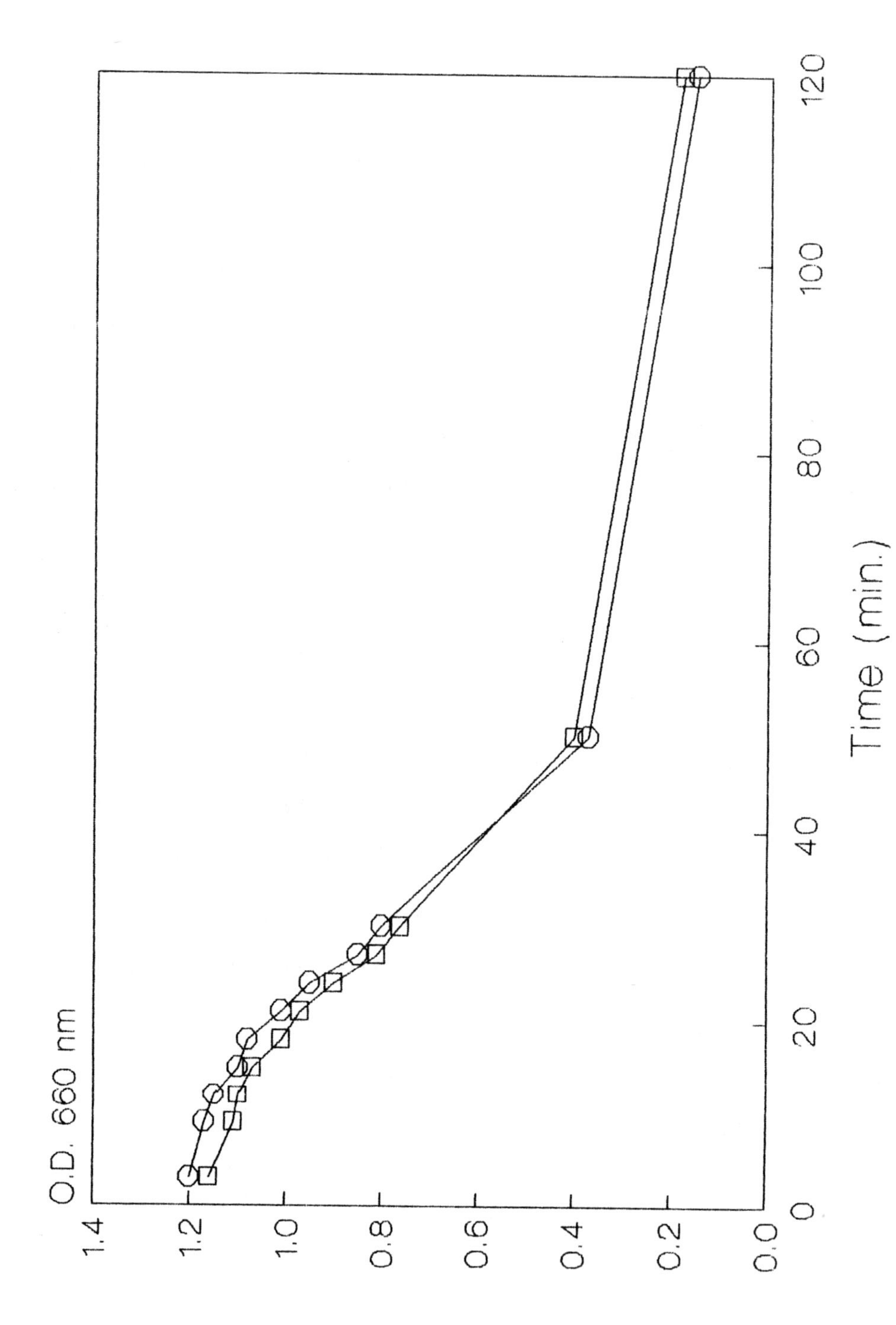


Figure 10. Enzymatic degradation of a 7% HV PHB/HV copolymer. Results from duplicate samples are shown.

References

Brandl, H., R.A. Gross, R.W. Lenz, and R.C. Fuller. 1988. Pseudomonas oleovorans as a source of poly(beta-hydroxyalkanoates) for potential applications as biodegradable polyesters. App. Environ. Microbiol. 54:1977-1982.

Brandl, H., R.A. Gross, E.J. Knee, R.W. Lenz, and R.C. Fuller. 1989. Abiltiy of the phototrophic bacterium Rodospirillum rubrum to produce various poly(beta-hydroxyalkanoates): potential sources for biodegradable polyesters. Int. J. Biol. Macromol. 11:49-55.

Dawes, E.A. and PlJ. Senior. 1973. The role and regulation of energy reserve polymers in micro-organisms. Adv. Microbial Physiol. 10:135-266.

Delafield, F.P., M. Doudoroff, N.J. Palleroni, C.J. Lusty, and R. Contopoulos. 1965. Decomposition of poly-beta-hydroxybutyrate by pseudomonads. J. Bacteriol 90:1455-1466.

Doi, Y., A. Tanaki, M. Kunioka, and K. Soga. 1987. Biosynthesis of terpolyesters of 3-hydroxybutyrate, 3-hydroxyvalerate, and 5-hydroxyvalerate in Alcaligenes eutrophus from 5-chloropentanoio and pentanoic acids. Makromol. chem., Rapid Commun. 88:631-635.

Findlay, R.H. and D.C. White. 1983. Polymeric beta-hydroxyalkanoates from environmental samples and Bacillus megaterium. Appl. Environ. Microbiol. 45:71-78.

Gross, R.A., Y. Zhang, G. Konrad, and R.W. Lenz. 1988. Polymerization of beta-monosubstituted-beta-propiolactone using trialkylaluminum-water catalytic systems and polymer characterization. Macromolecules 21:2657-2668.

Huang, S.J. 1989. Biodegradation. In: Comprehensive polymer science: the synthesis, characterization, reactions and applications of polymers, v. 6, G. Allen and J.C. Bevington (eds.), Pergamon Press, New York, pp. 597-606.

Lemoigne, M. 1925. Etudes sur l'autolyse microbienne. Acidification par formation d'acide beta-oxybutyrique. Ann. Inst. Pasteur 39:144.

Lusty, C.J. and M. Doudoroff. 1966. Poly-beta-hydroxybutyrate depolymerases of Pseudomonas lemoignei. Proc. Natl. Acad. Sci. 56:960-965.

Macrae, R.M. and J.F. Wilkinson. 1958. Poly-beta-hydroxybutyrate metabolism in washed suspensions of Bacillus megaterium. J. Gen. Microbiol. 19:210-222.

Merrick, J.M. 1978. Metabolism of reserve materials. In:

Photosynthetic Bacteria. R.K. Clayton and W.R. Sistrom (eds). Plenum Press, New York, pp: 199-219.

Nakayama, K., T. Saito, T. Fukui, Y. Shirakuru, and D. Tomita. 1985. Purification and properties of extracellular poly(3-hydroxybutyrate) depolymerases from Pseudomonas lemoignci. Biochem. Biophys. Acta 827:63-72.

Tanio, T., T. Fukui, Y. Shirakura, T. Saito, K. Tomita, T. Kaiho, and S. Masamune. 1982. An extracellular poly(3-hydroxybutyrate) depolymerase from Alcaligenes faecalis. Eur. J. Biochem. 124:71-77.

Williams, D.F. 1989. Polymer degradation in biological environments. In: Compreshensive polymer science: the synthesis, characterization, reactions and applications of polymers, v. 6, G. Allen and J.C. Bevington (eds.), Pergamon Press, New York, pp. 607-619.

QUESTIONS AND ANSWERS

DR. HAMMER: Jim Hammer for 3M Company. I am trying to get a sense for how broad this enzyme works.

Have you tried it on poly-e-caprolactone? Will it chew off other esters?

DR. GILMORE: To my knowledge, no one has done substrate specificities on this enzyme. When the alcoligonese enzyme, which operates extremely similarly, was first isolated they did do some substrate specificity studies and found the answer to that is no. It has to have those three carbon atoms in the backbone, and if you add or subtract carbon atoms in the backbone the enzyme doesn't act on it.

So it seems to be a very specific esterase, which may either be a blessing or a curse depending on how it might be used.

DR. HAMMER: So are there esterases out there that don't attack caprolactone?

DR. GILMORE: Probably, but it is not this one.

DR. HAMMER: Yeah.

DR. CHIELLINO: Emo Chiellini of University of Pisa. You showed us a comparison between the degradation of PHB, the way they are, as absolute configuration, and the blend of PHB obtained by the "R" and "S" altimer.

I would like to know, how did you get the "S" altimer of polymers?

DR. GILMORE: Bob Lenz' group in polymer science has been very active in staying ring-opening polymerization fo the sub-units and were able to produce pure "S" polymers and that was the source of the "S" polymer for the blending.

DR. CHIELLINI: So that blend is coming out from one bacterial-type polymer and the other from a synthetic way. So it migh be that -- so the blend has been obtained by blending PHB of our configuration made by a bacteria-type?

DR. GILMORE: Yes.

DR. CHIELLINI: So that's it. I would say from a homogeneity point of view, that blend is not -- you know, that's -- it is -- it is a blend that probably the results that you get is not directly comparable with the data that you get with the polymer because there it is not homogeneous indeed, because --

DR. GILMORE: We don't know the actual morphology of the particle and how these chains may be lining up with each other that they wouldn't otherwise.

THE CHAIRMAN: Are they put in solution?

DR. GILMORE: Yes. They were both dissolved in chloroform and combined and reprocipitated together, so it is not known how the "R" and "ES" chains are interacting in this particle.

DR. SHALABY: Shalaby from J&J. Actually, the blend data makes sense because you do lose about 50% of your turbidity which is due to the "R".

DR. GILMORE: That is actually a bag of worms which is not put.

DR. SHALABY: This is actually a comment and then I have a question. The blend data do make sense because you leave about 50% of your turbidity remain after you disgest the polymer.

DR.GILMORE: Yes. I want to comment on that comment. It is not that obvious because turbidity is a function of light scattering.

DR. SHALABY: That's right.

DR GILMORE: Okay? So if you have a particle that has "R" molecules and "S" molecules all wound through and you even say tunnel out the "R"s and leave the "S", you're still going to have a particle of a similar size, so there is no reason to expect that a 50% decrease in mass will give you a 50% decrease in turbidity because turbidity depends on surface area, not mass.

And also, the molecular weights of these polymers unfortunately were not the same. I think it was a 50/50 weight distribution, but one was a very small molecular weight and one was bigger, and the more -- let's see.

You get a faster degradation per mass with smaller molecular weight because you have more ends to operate on, so it is much more complex than it looks like up on the slide.

DR. SHALABY: Actually, I could probably argue because I see -- this was nice, 1 .2 and 1.6.

DR. GILMORE: I know.

DR. SHALABY: But as you form the particles, you may actually have selectively a form of crystalline "R" that will be a separate particle from the other "S".

DR. GILMORE: Yes. That could be measured by doing crystallinity.

DR. SHALABY: So my question really was relating to the mechanism. You proposed that the enzyme will start from the end into the chain, and if this is the only criterion, your numbers make sense. You have 24% each way, and you get about 24% drop on the way.

On the other hand, anytime the enzyme sees the hydroxy of a valerate, it will stop. And actually, the effect should be much more dramatic than what I've seen here. Can you explain this?

DR. GILMORE: Not yet. Again, that's one of the things that -- being able to sequence one of these polyesters would be fantastic. It would make life much easier.

But I did a little computer simulation trying to figure out how frequently three valerates in a row would occur in any of those polymer preparations. And that's not a lot of information, but intuitively looking at the kind of distributions you generate, it might make sense that if the enzyme tries to cut off a dimer of valerate next to another valerate, that it can't handle that and stop.

But the distribution I got, just plugging these things in on the basis of random distribution of a certain per cent of buterate and valerate, one or two valerates doesn't make sense in terms of mechanism to explain these results; three in a row might.

DR. SHALABY: Would it be possible that the enzyme may be adapting as they go along?

DR. GILMORE: I don't think so.

DR. DOI: Doi from the Tokoyo Institute of Technology. I have a comment on the question.

The rate of biodegradation is strongly dependant on the surface of particle and also crystallinity of PHB. So, I suppose that you use a PHB particle and also copolyester particle?

DR. GILMORE: Mm-hmm.

DR. DOI: So how do you adjust -- how do you control? The

surface area of particle was crystallinity of particle?

DR. GILMORE: Not doing it yet. I don't know yet whether I need to. We are supposed to -- I was supposed to have some data to have here from some people we are collaborating with who are doing particle sizing for us, and I would know whether or not the particle sizes in these different preparations were the same or whether I needed to do something to control them.

Obviously, if they are very spread out and they are a lot different, that's something I'd very much like to do, is use some sort of mesh and control the particle size because this would make the assay much more reproducible, but I'm not sure yet whether I need to.

DR. DOI: If we use the PHB as a standard for a sampling for testing, in that case the most important thing is a control over surface area --

DR. GILMORE: Yes.

DR. DOI: This is my comment.

DR. GILMORE: -- And also crystallinity or all physical factors.

DR. GILMORE: Okay.

DR. DOI: This is my comment.

DR. GILMORE: Yes. I agree. To use that as a standard, we would definitely want a narrow distribution of particle size.

DR. RUSE: Dorion Ruse, University of Toronto. Just a question, I might have missed it.

Did you look at controls, exposing the PHB only to buffer, these actual enzymes?

DR. GILMORE: I'm sorry. I was thinking about that this morning, that I didn't have any of those shown. Yes, there's no change. There are just straight lines across the top of the page, so I didn't bother to put them down. Yes, I ran controls.

DR. FEIJEN: Jan Feijen, University of Twente. Just a remark about your sequence analysis.

Did you ask Bob Lenz whether he did some NMR studies with these copolymers, carbon NMR, to see what is going on in terms of sequences? You can use some spectroscopic methods to do it.

DR. GILMORE: Yes, that's -- okay. I --

DR. FEIJEN: So don't mingle around with what you can expect on a random basis, but try to measure the spectro immediately and see what you get.

DR.GILMORE: Okay. That's something I'll have to look into. I'm new at polymer science still.

DR. ALLENZA: Paul Allenza, Allied-Signal. You mentioned in the mechanism that it cuts off dimers and trimers from the end of the polymer.

But over a time, wouldn't you expect to see monomers and dimers since the trimer could act as a substrate for the dimer-cleaning enzyme?

DR. GILMORE: Apparently, the affinity of the enzyme for the entire polymer molecule is extremely high, and once most of the polymer molecule is gone both of the two ammonia enzymes will go back and trim trimers, and you do get monomers and dimers. But it is a mix.

The enzymes predominantly make dimers and trimers initially until the polymer is essentially gone, but then they look around for whatever is left to chop up farther.

DR. ALLENZA: And have these two enzymes been purified separately?

DR. GILMORE: I try to. One of them didn't come out very will. The one I have been using is the one that chops off mostly dimers, and I haven't finished characterizing the fractions that I have got from purification.

DR. ALLENZA: Thank you.

# DEGRADATION MECHAN1SMS IN POLYETHYLENE=STARCH BLENDS

R. P. Wool, J. S. Peanasky, J. M. Long and S. M. Goheen

Department of Materials Science and Engineering

University of Illinois

1304 W. Green Street, Urbana Ill 61801 USA

Presented at the First International Scientific Consensus
Workshop on Degradable Plastics, Toronto Canada

## Introduction

Polymers can be made which will degrade in many environments by a combination of one or more of the following mechanisms:

1) MICROORGANISM DEGRADATION: The material is consumed by microorganisms such as fungi and bacteria in aerobic and anaerobic environments. Carbon dioxide, methane and other natural products are derived from the degradation process. Materials which completely degrade by this mechanism are usually considered to be biodegradable.

2) MACROORGANISM DEGRADATION: The material is consumed by insects, animals and other living creatures. Degradation occurs by mastication, digestion and post ingestion processes. This mechanism usually provides the fastest degradation rate.

3) PHOTODEGRADATION: Ultraviolet radiation leads to photoinitiated radical reactions which may propagate to cause chain scission and degradation of the material. The material may embrittle and eventually biodegrade depending on the concentration of chemical groups to initiate and sustain chain scission reactions.

4) CHEMICAL DEGRADATION: Chemical reactions result in bond scission, e.g., via oxidation of double bonds, to cause molecular weight degradation of the material. Natural rubber degrades in this manner. Chemical degradation usually occurs independently of microbial activity and is distinguished from enzymatic and hydrolytic degradation.

Plastics can be designed to degrade in a controlled manner by a combination of these mechanisms depending on the application, processing conditions, performance requirements and method of disposal. Different combinations of mechanisms will be required for plastics in landfills, composting of yard waste, agricultural applications, marine and terrestrial environments and in many applications where the degradation of the plastic is an important feature of the materials performance.

In this paper we focus on the Microorganism and Macroorganism degradation mechanisms in polymer/starch blends and comment on the Photo- and Chemical degradation mechanisms. A major objective of this work is to examine the biodegradability of these blends via accessibility and connectivity of starch particles in a polyethylene matrix by the use of a percolation model. Examples and applications of the degradation mechanisms in several terrestrial and aquatic environments will be given. Critical scientific issues for degradable plastics in solid waste management, recycling, composting, litter and marine environments are discussed.

It is generally accepted that PE could be catabolized (degraded) by microbes in a manner similar to that of fatty acids[1,2]. In this process, the polyethylene chains are degraded stepwise at their chain ends. Several investigators used starch as a filler in plastics to speed degradation[3-8]. When the plastic is placed in a biologically active environment, the starch is thought to accelerate the degradation rate. Microbes invade the plastic by consuming starch, creating pores in the process. This provides greater surface contact between microbes and PE, speeding biotic reactions and promoting abiotic reactions in the synthetic plastic. Macroorganisms such as crickets, roaches, slugs and larger animals, can speed degradation through consumption of these blends[8,9]. Photodegradation via Norrish-Smith and other reactions may also occur[10,11].

In this paper, a computer simulation of a percolating microbial invasion system on a cubic lattice is reviewed and compared to acid hydrolysis and other degradation studies with PE/starch blends. The results of this work have application to materials where one component is dispersed in a matrix of a second component, such as recycled plastics containing blends of degradable and nondegradable material, and interactions of photo- and chemically reactive species dispersed in a polymer matrix.

## Percolation Theory

Scalar percolation theory[12,13] concerns the connectivity of one component randomly placed in another. For example, in Figure 1a, starch particles at 58% concentration (1% below the percolation threshold in two dimensions) are placed at random in a 512 x 512 square lattice. With simulated bilateral invasion of microbes from the top and bottom borders, we examine the connectivity or accessed starch. All accessed squares connected to the border are transformed into black ones and the rest of the starch is not shown. The percent connected (16.6%) is defined as the number of black squares divided by the original number of occupied squares. When the concentration is increased by 1% to 59% (the percolation threshold), the black squares span or percolate through the entire lattice as shown in Fig 1(b) and 63.62% of the starch is accessed. The minimum concentration at which such an occurrence could happen on an infinite lattice is known as the percolation threshold. This is analogous to the gel point in polymerization of multifunctional monomers and the point at which one observes electrical conduction through metal particles dispersed in a non conducting matrix. At $p = 60\%$, Figure 1c shows that 76.97% of the starch has been accessed.

The percolation threshold is an example of critical phenomena, and thus many properties of the system follow certain scaling laws of the form

$$H(p) - (p - pc) . \qquad 1)$$

where is some critical exponent as listed in Table 1[7], p is the percent occupied, and $p_C$ is the concentration corresponding to the percolation threshold. Percolation concepts in biodegradable plastics are explored by experiment in the next Section.

## Experimental Section

### Computer Simulations

The three dimensional simulation was run on a simple cubic lattice populated using a random number generator and a known occupation probability p, corresponding to the volume fraction of starch in the PE matrix. Accessibility is defined as the total number of connected sites invading from both the top and bottom surfaces divided by the original number of occupied sites. The simulations were done on a Cray XMP supercomputer with 50 simulations per occupation percent which ranged from 5-60%. Periodic boundary conditions were used such that the configuration simulated the invasion of a thin plate or film from both top and bottom surfaces. Further details of the simulation are contained in reference 7.

## Material Preparation

Low density polyethylene (LDPE) supplied by Quantum Chemical Co. was mixed with corn starch supplied by Cargill Inc. in concentrations ranging from 5 to 60 volume percent. The fractional volume of starch p, in these blends was determined using the formula, $p=1/[1+yqs/xq_p]$ where x and y are the weight fractions of starch and PE, $q_s$ and $q_p$ are the densities of starch and PE, respectively. The density of starch was determined using pyncnometry. The melt blending was done in a Brabender Data Processing Plasti-Corder model PL2000 with a 350 cc capacity mixing head attachment. Samples were mixed at 140°C and 60 RPM for 30 min. The blends were then compression molded into plates of thickness 0.085-0.2 " using 5" X 5" window frame molds. Samples were cut from this plate using an ASTM D-412C (D-638 IV) half size dogbone cutter.

## Acid Hydrolysis

Acid hydrolysis of starch in plastics involves the random cleavage of the glycoside bond through the addition of a water molecule to the oligosaccharide[7]. Glucose, a monosaccharide, eventually evolves during this reaction and dehydrates in the presence of a strong acid to form hydroxymethylfuranose which has an intense amber hue. The acid hydrolysis experiments were carried out by refluxing samples in HCl. When the color of the solution was noticeably amber the reacted solution was replaced with a new solution. Solutions were changed over a period of 24 to 72 hours until no color change was visibly detected. The samples were placed under vacuum in an oven at 60°C for another 72 hours, removed to a desiccator under vacuum for 2 months and a final weight was then obtained.

Soil Environments:

Several soils were obtained from farmland top soil in March before planting. They were sifted with a 1/8 inch screen to remove clumps, plant debris and macroorganisms. Three inches of soil was placed in a plastic box, 15 x 10 x 6.5 inch, which was lined with stainless steel cloth to allow air circulation. The soils were kept moist but not wet with deionized water and stored in a room at ambient temperature (70-78 F) and humidity (37-90%). Thin film (1-3 mil) samples, 2 inches in diameter, were buried in a 4 x 5 array at a depth of 2 inches. Samples were removed once per month for evaluation of mass change and infrared analysis.

Outdoor soil burial experiments were conducted to compare field tests with the indoor laboratory experiments. Samples and soil were placed in perforated plastic cups which were cut to allow access of macroorganisms and moisture when buried 6-8 inches in the soil. The cups also permitted easy retrieval of the degrading samples from the ground. The soil was subject to ambient conditions of moisture and temperature.

Bilateral Invasion P = 58% / 16.60% Accessed

a

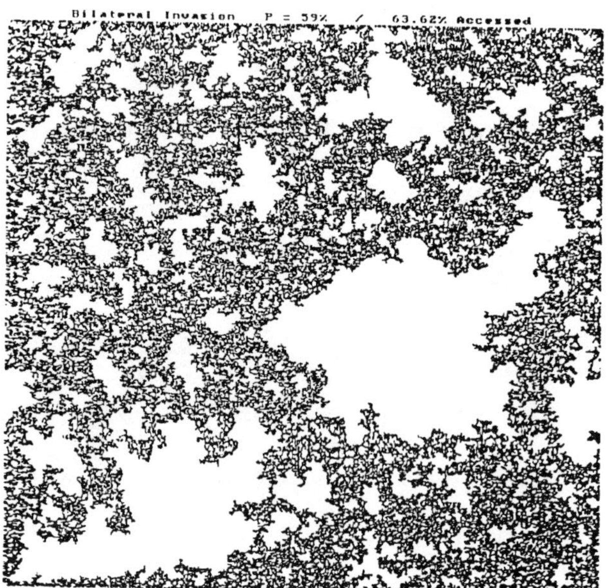
Bilateral Invasion P = 59% / 63.62% Accessed

b

Bilateral Invasion P = 60% / 76.97% Accessed

c

Fig 1

## Macroorganism Environments

Plastic materials were exposed to crickets under several conditions involving 1) isolated crickets, 2) groups of crickets and 3) plastics with feeding stimulants. The mechanism of degradation and the fate of the ingested plastic is discussed.

## Marine Environments:

Polymer samples were placed into and suspended on the outer walls of cages with mesh sizes 1.0 and 1/4 inches. The cages were placed in 15 feet of ocean water off the coast of Massachusetts, USA. The mechanism of degradation is discussed and compared with fresh-water fish tank experiments.

## Results and Discussion

The percolation threshold for the three dimensional cubic lattice simulation was determined from the maximum of the second moment of the cluster distribution as pc = 31.17% volume fraction. The critical exponents obtained from the simulation were found to be in good agreement with the theoretical values for the $128^3$ lattice and are compared in Table I[7]. Results from finite sized scaling are reported elsewhere. The deviations of the critical exponents from other reported values are small[7]. The percolation threshold of 31% by volume corresponds to about 41% by weight in PE/starch blends when corrections are made for densities of 0.928 and 1.44 g/cc in PE and starch, respectively.

The accessibility as a function of occupation percent and lattice size or thickness is shown in Figure 2 For thick samples, ($z = 128$), the transition is quite sharp at 31% occupation. At $p < p_c$, the surface effects are small compared to the bulk and very little degradation occurs. For thin specimens, surface effects dominate and the percolation transition is broadened. The zlattice thickness effect can be roughly interpreted in terms of the starch particle size compared to the film thickness. Obviously if the starch particle diameter is equal to the film thickness, then all the starch is removed by invasion from the surfaces. For example, for a thick PE film ($z = 128$) containing 5% starch (see Fig 3), only about 2% of the starch (0.1% of total material) would be removed. However, if $z = 4$ starch diameters, then over 50% (2.5% of total material) of the starch would be removed. When $p > p_c$, connected pathways through the material are established, large amounts of starch are removed and thickness effects are not as important.

The acid hydrolysis results are compared to the computer simulation results using $64^3$ and $128^3$ lattices in Figure 3. The results are consistent with the percolation model such that a sharp increase in the accessed starch occurs in the vicinity of 31% by volume corresponding to the theoretical percolation threshold. Another important aspect of starch removal in the PE/Starch blends is the amount of new internal surface area A, of PE exposed to microbes, which is given by $A = 6\ u/d$, where u is the total volume fraction of material removed and d is the starch diameter. For example, at $p = 31\%$, $u = 0.19$ and for $d = 10$ micron starch particles, the internal surface area generated is about 1000 $cm^2/cc$ The latter value increases rapidly at $p > p_c$.

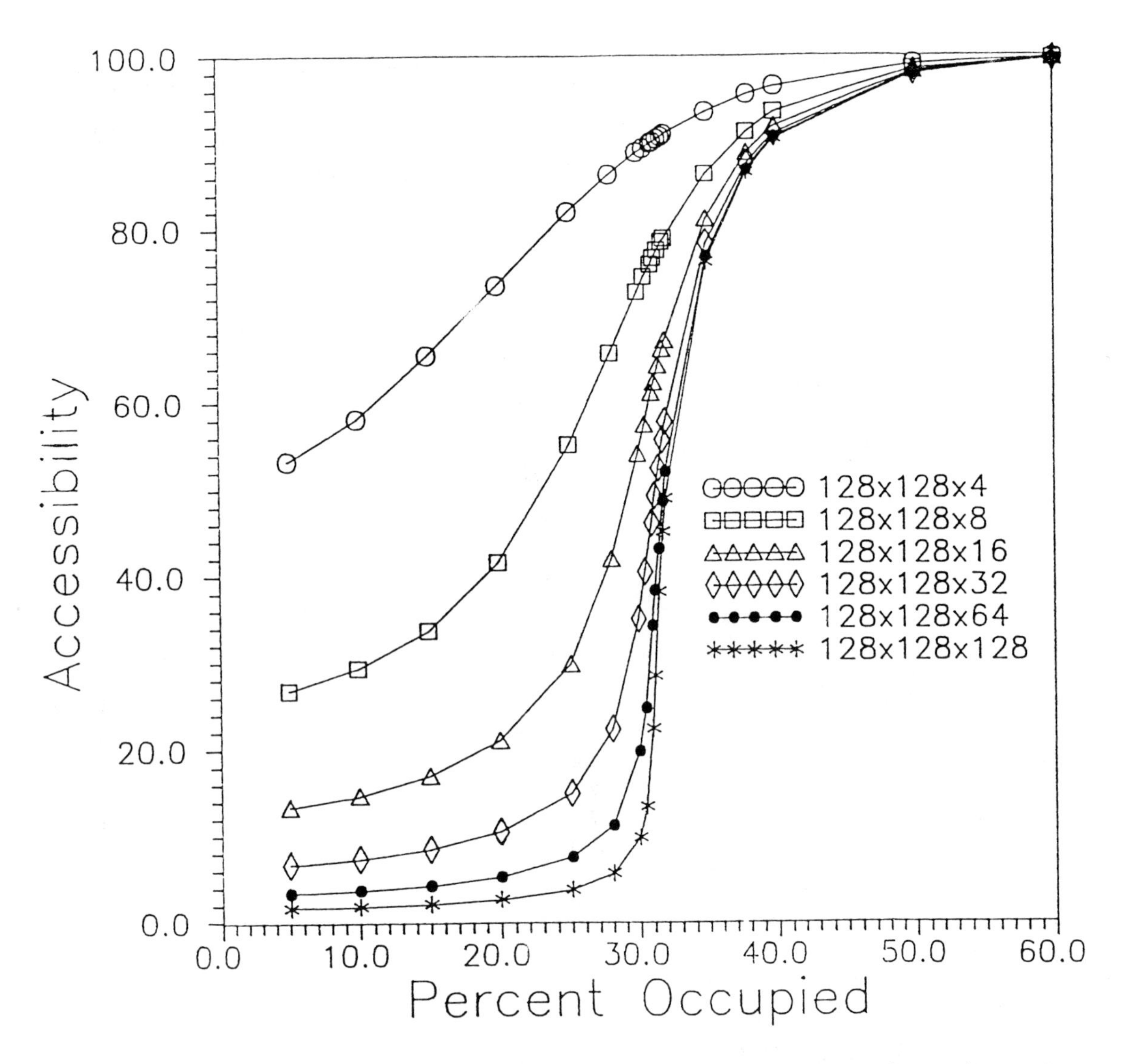


Figure2. .) The accessibility, modeled by bilateral invasion, is plotted against the occupation probability for different lattice sizes. The effect of lattice size is the same as in connectivity.

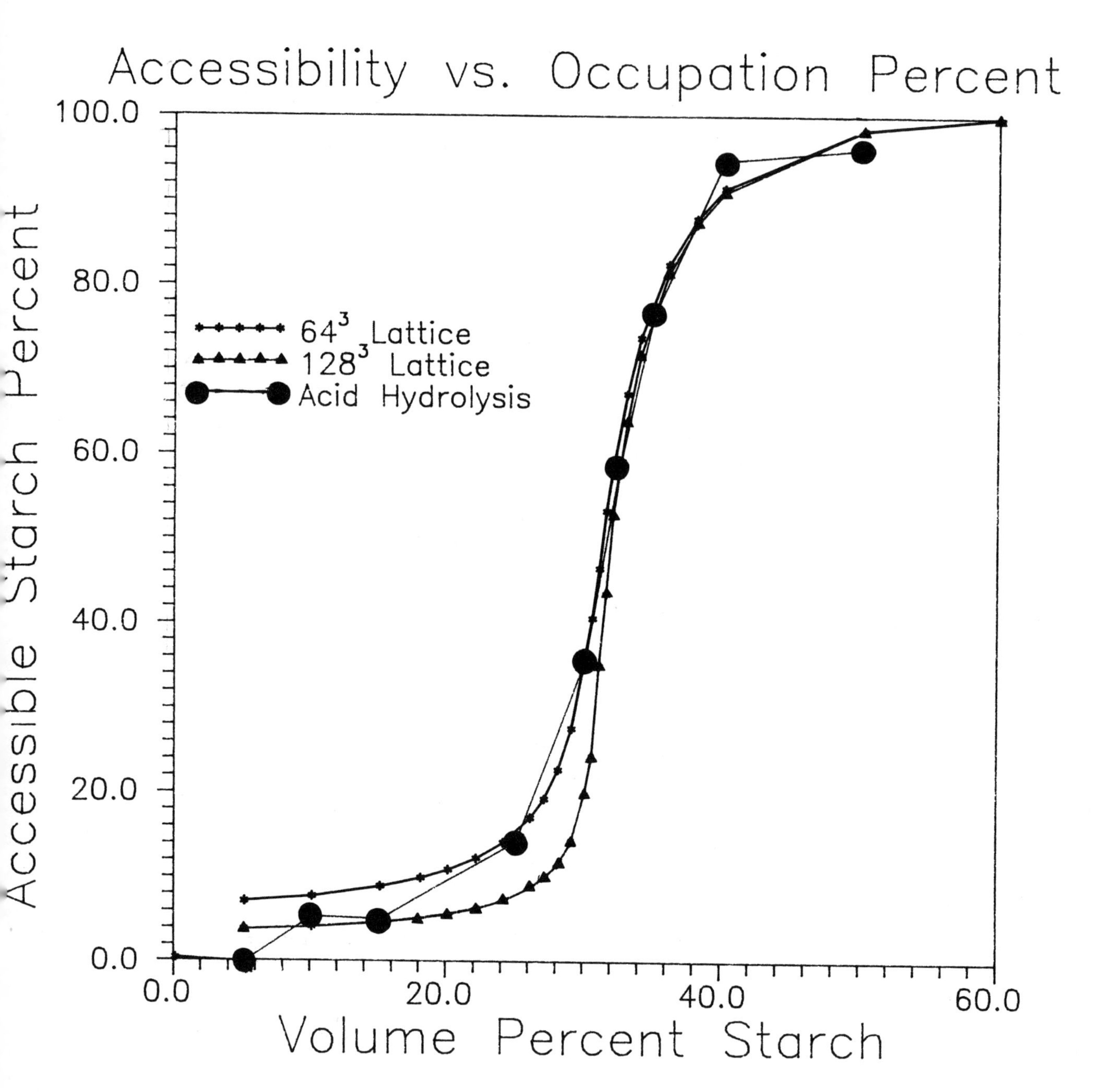
Accessibility vs. Occupation Percent
Accessible Starch Percent
Volume Percent Starch
64³ Lattice
128³ Lattice
Acid Hydrolysis
0.0
20.0
40.0
60.0
80.0
100.0

The advantage gained from gelatinizing the starch to produce sub-micron sized particles is clearly obvious in terms of the enhanced internal surface area. The internal surface features at $p_C$ also have fractal characteristics (see Fig 1b) which are described by the authors in another paper. The new chain ends exposed on the surfaces can increase the rate of biodegradation and other abiotic chemical degradation processes. A critical issue to be resolved is the extent of transference of microbial activity from the percolation network to the PE matrix.

The effect of soil burial on weight loss as a function of starch content is shown in Figure 4 for PE/starch blends containing 0, 29, 52 and 67% starch by mass (0, 20, 40, 55% starch by volume). The plastic films were buried to a depth of 2 inches. The slight loss of mass in the LDPE control is attributed to removal of low molecular weight constituents. After 120 days, the sample with 29% by mass starch shows about a 10% change in mass which is consistent with the surface removal of starch at concentrations below the percolation threshold of 31% by volume. Above pc, both the 52 and 67% by mass starch blends show extensive and continued degradation.

FTIR analysis of the degraded samples during the 6 month burial period showed that the starch (10 micron particles) was preferentially removed leaving the PE network. This removal of starch is expected to weaken the plastic and leave the PE network with a higher internal surface area which promotes further degradation by both biotic and abiotic mechanisms. However, the crucial question of the fate of the PE matrix is still under investigation. The rate of removal of starch is consistent with

CHANGES IN THE MASS OF
BINARY BLENDS IN DRUMMER SOIL

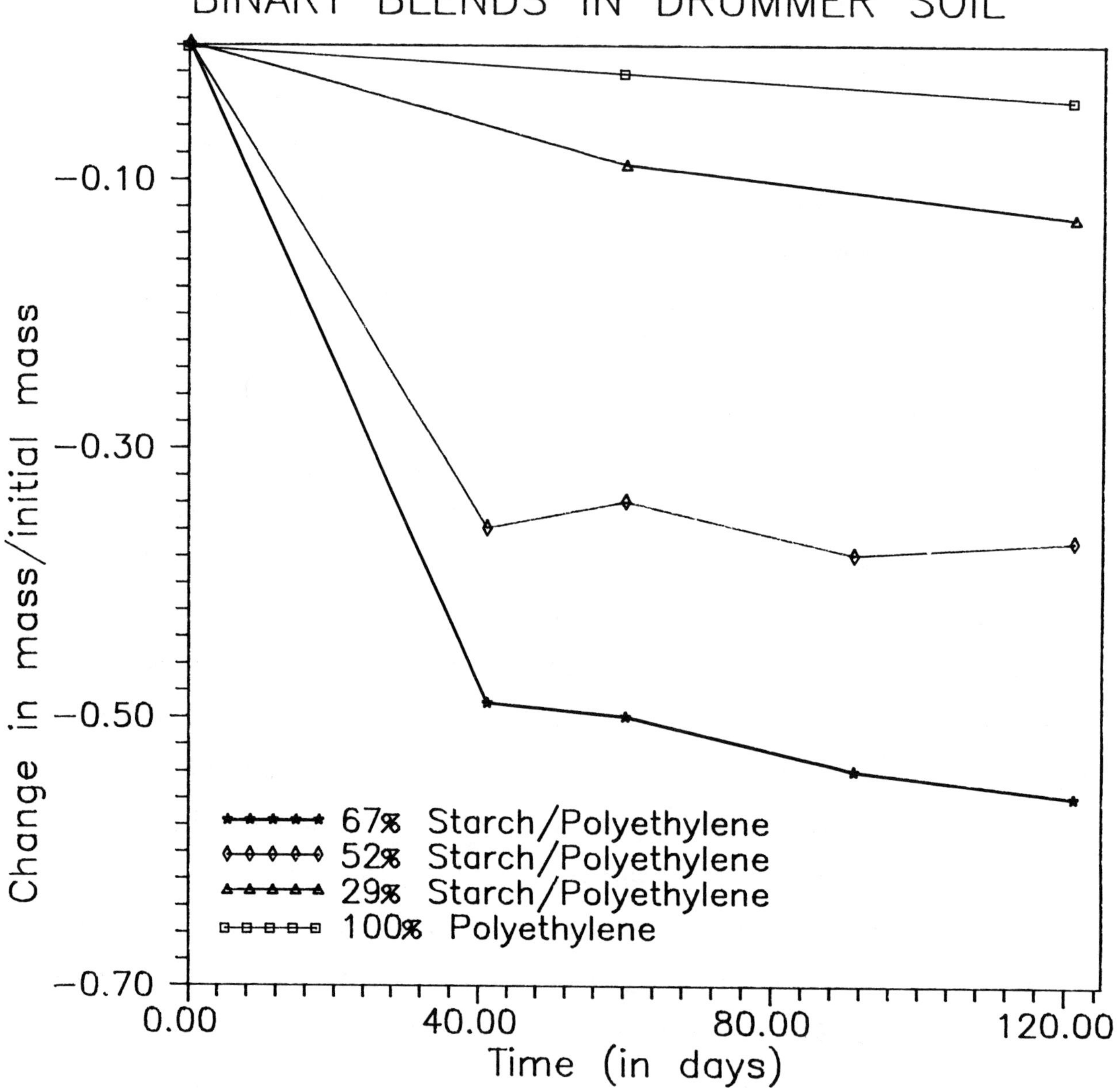
-0.10
-0.30
-0.50
-0.70
0.00
40.00
80.00
120.00
Time (in days)
Change in mass/initial mass
67% Starch/Polyethylene
52% Starch/Polyethylene
29% Starch/Polyethylene
100% Polyethylene

previous studies by Wool and Cole[8] where carbon dioxide evolution from PE/starch blends was quantitatively measured with time. In these and other soil studies, the fastest rate of degradation was found with sandy soils compared to rich loam soils high in organic matter.

Little or no weight change was observed for commercial film samples made with 6% starch and oxidizing agents to promote chemical degradation[3,4], when they were buried in soil for 6 months and placed in leaf composting environments for a year. Similar results were obtained for carbon dioxide evolution experiments from the same samples mixed with soils[8]. The chemical degradation mechanism currently being used in commercial films containing 6% starch is easily perturbed by both the processing conditions and the environment.

Degradation of PE/starch blends by crickets and other macroorganisms was extensively studied[8,9]. Both thick compression molded and thin film samples containing greater than 30% starch were consumed by crickets in times ranging from hours to weeks. Similar results were obtained for both single and group cricket experiments. Samples with 20% starch could be consumed by crickets in a few days with the addition of feeding stimulants such as linseed oil, wheatgerm and fructose. This mechanism of degradation is the fastest observed by these investigators and is a potential contributor to terrestrial pollution problems. Degradation of plastics by insects should also be considered when designing plastics for agricultural applications.

In leaf composting studies at the Urbana Illinois Yard Waste Reclamation site., PE/starch films containing 40% by weight starch were degraded by both insects and microbes. Consumption of plastic by insects in leaf composting is a surprise considering the abundance of natural food for the insects in the leaf pile and could be used to considerable advantage in designing compost bags.

The degradation of plastics by insects occurs via three stages. The material is first degraded by the mastication process of the insect. Size Exclusion Chromatography of sliced HDPE indicated significant molecular weight changes such that $5 \times 10^{15}$ broken bonds per square centimeter, or $10^{16}$ new chain ends/ $cm^2$ were formed by the simulated mastication process. Secondly, the masticated degraded material is ingested by the insect. Amylase and acid digestion are efficient in removing the starch and the remaining PE particles pass through the insect. FTIR analysis of the fecal material shows that most of the starch is removed and the PE is largely unaffected by the digestion process. Finally, the fate of the polymer in the fecal material remains to be determined. The micron sized particles are further subjected to both chemical and microbial degradation and should eventually be metabolized. This important issue is being further studied.

Degradation studies of PE/starch blends and commercially available degradable plastics were conducted in both marine and freshwater environments. Blends containing greater than 20% starch would sink in marine environments. For example, neutral buoyancy of 1 g/cc is achieved with 20% by weight starch with density 1.44 g/cc, blended with LDPE with density 0.928 g/cc. Complete disintegration of a 40% by weight PE/starch blend processed using Otey methods[5,6], was achieved in 12 months in a tropical fish tank. An algae coating developed on the plastic which was found to be proportional to the starch content. In marine environments in Massachusetts, the growth of algae on plastics on the ocean floor at a depth of 15-20 feet was found to be strongly influenced by the exposed surfaces and local environment. The greatest growth occurred under the most sheltered conditions. For example, in a 2 x 1/2 inch thin walled hollow cylinder made with a 40% PE/starch blend, extensive algal growth occurred on the inside with very little growth on the outside of the cylinder after one month. The smaller mesh cage also provided a better algal growth environment than the larger mesh cage.

The growth of algae has important consequences for plastics which are designed to photodegrade since the UV radiation can be screened out by the algae and the lifetime considerably prolonged. In marine environments we recommend the use of plastics which sink and degrade harmlessly on the ocean floor. Current commercially available photodegradable plastics used in six pack loops and films float on the ocean surface and continue to present hazards to mammals and birds before and after they slowly embrittle and fragment. Ideally, we would like to have a combination of both photo- and biodegradable plastic resins to address marine pollution problems. A critical issue to be addressed is whether plastics can be made to degrade harmlessly in deep ocean environments.

Conclusions

Invasion of PE/starch blends was found to be well described by a percolation process such that a threshold level of starch, 31% by volume, was required to promote degradation of the plastic. Below 31%, the starch granules were essentially encapsulated in the PE matrix and only a surface layer was accessed. The surface layer could promote a growth of fungi and bacteria but would not result in the degradation of the bulk of the material. The surface growth, according to ASTM G-21 and G22, test methods for determining resistance of plastics to microbes, could be misconstrued and suggest that the material is biodegradable. However, complete degradation will not occur without the surface growth first occurring. Above 31% starch content by volume (approx 42% by weight) rapid invasion of the material was observed. The extent of degradation of the PE matrix by transference of microbial activity and abiotic reactions remains to be determined as a function of starch particle size and concentration.

The percolation analysis indicates that for recycled plastics, up to 31% by volume of a biodegradable plastic can be safely encapsulated in a non degradable plastic matrix, with consideration for thickness effects. Manufacturers of benches, garden posts, playground material, etc., from recycled plastic might consider the addition of starch at concentrations $p < p_c$ to promote a fungal growth on the surface and provide their material with a more natural patina. However, it is interesting to note that in such recycled products containing up to 30% or more impurities, with slow cooling following extrusion of thick objects (approx. 2-6 inches), the PE crystallization mechanism forces the impurities into the center of the material leaving a fairly pure PE skin on the outside. It could be considered a form of radial "zone refining". This material behavior during processing is also a problem for PE/starch blends and plastics containing chemical additives designed to promote chemical degradation. It is a typical scale-up problem from bench to mass production levels, for degradable plastic technologies.

Consumption of plastics by insects is an important feature of PE/starch blends and results in the fastest rate of degradation. This mechanism could be utilized in plastics which are found in the litter stream and will need to be controlled for plastics used in agricultural mulch film and food packaging applications. Several additives are available which will accelerate or retard the rate of attack by macroorganisms. For certain agricultural applications involving feedstock, the ingestion of plastics by farm animals (ruminants) needs to be studied. Consumption of discarded plastics by macroorganisms is a solution to persistent litter problems but should not be a substitute for litter control.

For plastics in marine environments, we recommend using densities greater than sea water (achieved with approx 20% by weight starch) and designing for both biodegradation on the ocean floor and photodegradation for shore litter and buoyancy factors. The photodegradation mechanism alone applied to plastics which float on the ocean surface is not the most efficient means of addressing marine pollution problems. The extent of the consumption of PE/starch plastics as food by marine macroorganisms and the rate of biodegradation on the ocean floor is currently unknown and needs to be evaluated.

Solid waste management problems can be constructively addressed by several solutions involving degradable plastics. The conversion of the entire plastic waste (7% by weight, about 20% by volume of the landfill) to biodegradable materials is not a very technologically feasible or efficient solution. Recycling, reuse, reduction and energy recovery methods should be attempted in parallel with the following contributions from degradable plastics.

1) The use of degradable composting bags for yard waste facilitates the recovery of approximately 17-20% by weight (national average) of the landfill. This solution can be implemented quickly and is approximately equal in impact to removing all of the plastics currently being placed in landfills.

2) Degradable plastic films can be used as a substitute for soil cover on the daily deposition of garbage in many landfills. Removal of the "land" from landfills is estimated to recover approximately another 15% of the landfill. This solution is readily implementable in many kinds of landfills but requires further research and EPA approval.

3) Degradable plastics used as solutions (sometimes mandated) to other problems involving marine and terrestrial pollution can also be designed to degrade in landfills. For example, six-pack loops are required to be photodegradable to minimize marine pollution problems. However, more than 99% of this product will be deposited in landfills (if not recycled) and it makes sense to design for both anaerobic degradation in landfills as well as photodegradation in marine environments. Other products should be treated similarly.

4) The extent to which discarded plastics should be made biodegradable for landfill recovery depends largely on the management of the landfill. If landfills are going to be made biologically active, then it is reasonable to make biodegradable plastics for disposal in landfills. About 70% of the landfill (50% paper) consists of degradable material and considerable energy recovery can be accomplished through the harvesting of methane as a fuel. (New ASTM methods are being developed to test for the anaerobic degradation of plastics in landfill environments). However, if the landfill is maintained in a biologically inactive state then the conversion to biodegradable plastic trash is inefficient in terms of landfill recovery. For certain plastic products in the health care and personal hygiene area, the use of degradable plastics is considered constructive in many cases. However. the disposal of plastics, metals and glass in landfills must be considered in general a waste of valuable material and efforts should be made to resolve the problem at its source, i.e. not to place the plastic in the landfill.

Constructive solutions to solid waste management problems involving yardwaste composting and landfill cover offer immediate relief to overburdened landfills, which combined have greater impact than removing all the plastic waste from landfills. These solutions should be actively pursued in addition to recycling, reuse, reduction and incineration programs by the Plastic Industry.

## Acknowledgments

The authors thank the Illinois Corn Marketing Board and the IBM Corporation for financial support of this work. Appreciation is expressed to Joshua Folk for his assistance with the cricket experiments and to Professor May Berenbaum, Department of Entomology at the University of Illinois, for her constructive discussions on Macroorganism degradation.

## Bibliography

1.) Albertsson, A. C., Andersson, S. O., and S. Karlsson, Polymer Degradation and Stability, 18, 73-87, (1987)

2.) Albertsson, A. C., and Z. G. B nhidi, J. App. Poly. Sci., 25, 1655-1671, (1980)

3.) Griffith, G. J. L., Proceedings of SPI Symposium on Degradable Plastics, pp. 47-50, June 10, 1987, Washington D. C.

4.) Griffith, G. J. L., ACS Advances in Chemistry Series, 134, (1975)

5.) Otey, F. H., Westhoff, R. P., and W. M. Doane, Ind. Eng. Chem. Prod. Res. Dev., 19, 592-595, (1980)

6.) Otey, F. H., Westhoff, R. P., and C. R. Russell, Ind. Eng. Chem. Prod. Res. Dev., 16, 305-308, (1977)

7.) Peanasky, J. S., Long, J. M. and Wool, R. P., "Percolation Effects in Polyethylene-Starch Blends", J. Polym. Sci., Polym. Phys. Ed., in press (1990).

8.) Wool, R. P. and Cole, M. A., "Microbial Degradation", ASM Engineering Handbook, Vol 2, 783-787 (1988)

9.) Sahlin, K., M.S. Thesis, "Biodegradable Plastics: Macroorganism Degradation", University of Gothenburg, Sweden (1988); work performed at the University of Illinois with R. P. Wool and M. A. Cole.

10.) Guillet, J., Polymers and Ecological Problems. pp. 1-26, Guillet, J., Ed., Plenum Press, New York, (1973)

11.) Scott, G., Polymers and Ecological Problems. pp. 27-44, Guillet, J. Ed., Plenum Press, New York, (1973)

12.) Stauffer, D., Introduction to Percolation Theory, Taylor and Francis, (1985)

13.) Broadbent and Hammersley, Proc. Camb. Phil. Soc., 53, 629, (1957)

Table I: Critical Exponents for Three Dimensions

| Exponent - Related Property | Theoretical[12] | Experimental[7] |
|---|---|---|
| Alpha -Number of Clusters | -0.60 | -0.64 |
| Beta -Strength of Infinite Network | 0.40 | 0.41 |
| Tau -Divergence of Second Moment | 1.8 | 1.9 |
| Sigma -Average Cluster Size | 0.45 | 0.45 |
| t -Cluster Number at $p = p_c$ | 2.2 | 1.1 |

## QUESTIONS AND ANSWERS

THE CHAIRMAN: The stenographer has a little trouble following the discussion, so be sure you enunciate you name clearly, and the question clearly and slowly.

DR. FEIJEN: Jan Feijen, University of Twente. I saw there is a small shift in the curve based on the series of the percolation approach, and the actual research you get is hydrolysis.

Is that caused by the uneven distribution of particles at the surface in the extrusion process?

DR. WOOL: No. this is a finite size effect. The percolation period applies to an infinitely large object. When you're below the percolation threshold in a thin film, then assuming first of all that you have a uniform distribution, you will access the material.

For example, if you take a film which has a starch particle straddling that film which accessed both from the top and the bottom, then that obviously comes out. This is like an artificial, a finite-sized percolation effect. You can get skin effects.

DR. FEIJEN: That's what I meant. Explain the same, actually.

DR. WOOL: The type of skin effect that can occur with these materials is when crystallization occurs from the surface, as it will do usually.

If this occurs - and that depends delicately in your processing history - once the polyethylene starts to crystallize, it will, if you like, exude out the starch particles towards the center. And I can make materials that will have a beautiful inner layer of starch in the interior by controlling the processing history and have a complete skin of polyethylene in the surface, which will not be at all degradable.

DR. FEIJEN: Yes. And, therefore, also your FTIR measurements do not explain always that you have, let's say, starch exposed to the surface because the penetration depth is about 4,000 angstrom. If you want to see what really is at the surface, you have to do XPS measurements.

DR. WOOL: Okay. The FTIR results that I gave you were for transmission. We have also done the ATR, Attenuated Total Reflectance measurements and several other measurements.

I guess the point I want to make here was that the mass reduction correlated exactly, or almost exactly, with the FTIR change in the amount of starch in the material. These changes are very large when we have materials containing 60 per cent starch and that material is being removed.

The mass changes are large. The FTIR signal changes are also large.

DR. FEIJEN: My only remark is that if you want to tackle what is really still covered by even a very thin polyethylene layer, XPS would be much more appropriate than the Attenuated Total Reflection Infrared.

DR. WOOL: I agree. The XPS is good and neutron reflectivity is much better, we use thins ourselves.

DR. ANDRADY: My name is Tony Andrady from Research Triangle Institute.

DR. WOOL: Yes?

DR. ANDRADY: I have a question and something I want to clarify.

My experience with starch containing plastic films, particularly low-density polyethylene films, have been that if it has more than about 15 per cent starch it is very soft and you cannot make anything like a plastic bag out of it.

Did I understand you right when you said that in your technology now, which you did not want to discuss further, you can add a high percentage of starch like 40 to 50 per cent and still maintain good tensile properties?

DR. WOOL: Okay. Most of my results today were on a Model 2 component system. Of course, when you add the percolation threshold level of starch in there, 40 percent by weight, the properties are miserable.

Now, the technologies that have been produced to rescue that are the following: The USDA technology in which you are using a compatibilities and you gelatinize the starch, that gives you a considerable recovery of the elongation to break, but nothing close to the original material -- to the original.

The new approach that we have used at the University of Illinois will give you about 200 per cent elongation at around 70 per cent starch content, and we are currently evaluating it as far as films are concerned. This material costs exactly half the price of polyethylene by the way, for those of you who are interested.

DR. ANDRADY: And now for my question: What you describe as the "macroorganism degradation" is what the environmentalists call "ingestion".

So have you looked at the amount of plastic that enters the animal as opposed to what comes out? Is any plastic retained by the animal over the long term?

DR. WOOL: Okay. We looked at the fetal material that came out, and we looked to see how much of the starch had been removed from the fetal material. We did not do an exact mass balance. I want you to appreciate some of the difficulties of doing these experiments.

DR. ANDRADY: I know. Yes, I appreciate it fully. [Laughter].

DR. WOOL: I'm not saying it can't be done.

DR. ANDRADY: Thank you.

DR. SCOTT: General Scott, Aston University.

DR. WOOL: Yes?

DR. SCOTT: I think you may have misunderstood what happens when plastics photobiodegrade. You said that they lay on the surface of the sea, for example, and birds ingested them.

This is not so because they are very highly modified with carboxylic acid groups, they are now very hydrophilic materials, and they lose tensile strength in that way. So they haven't got the brittle characteristics of plastics found on land.

They furthermore will support macroorganic growth. In other words, they will act as a medium for sea organisms which take them to the bottom of the sea. Now, admittedly, they don't provide the nice, nutritious environment that the starch situation does, but these materials do disappear by a very similar procedure, and I think the presence of starch is probably a bit irrelevant.

DR. WOOL: The technology that you showed us this week seemed to be rather interesting in that regard, and the extent to which one passes over from a photo to a chemical degradation mechanism at that point whereby microbial degradation might be allowed, I'm certainly not going to exclude that.

There are technologies in existence today and there are companies within my own state of Illinois who are not necessarily using the technology that you are prescribing, which appears to be a very good technology.

So, if the photodegradation act is simply to split the bond apart and embrittle the material, then I don't think that is a very useful solution to the marine problem.

If on the other hand, the photolytic event leads to chemical reactions, which further leads to microbial enhancement of the degradation process, then you've probably got a winner.

DR. SCOTT: Okay. Thank you.

DR. ENNIS: Bob Ennis, Plastigone Technologies. We have heard a lot of discussions during the meeting about how increasing the surface area of plastics will enhance its ability to degrade.

I'm just wondering whether you did any control experiments, Richard, to separate the mechanical-surface-increasing effect of the starch - that is, the honeycombing of the plastic with starch particles - to separate the mechanical change in the starch from the actual starch effect itself.

In other words, can you make a honeycombed plastic without starch in it as a control to test whether that plastic will degrade as rapidly as a plastic with starch in it?

We said that the starch is acting as a filler. Is that all that it is doing or does the starch have some particular property, and can you separate that in your experiments?

DR. WOOL: You can separate out most of these effects, and it depends on the way the starch is incorporated into the plastic.

If you have a gelatinized starch, of course the surface roughness effect almost goes to zero.

If you have a simple bimodal system of spherical starch particles in the polyethylene matrix, then the stiffness is affected. The Young's modulus is affected by the presence of the starch; however, the elongation to break is not very much affected by the presence of the starch.

In fact, when you take out the starch by acid digestion, the elongation to break hardly changes at all. The fracture stress is somewhat modified by the type of microfracture event involving the Young's modulus.

So, you can separate out these effects. It is a complex issue, and the only technology...

That's a little bit more interesting because the starch particles can be bonded to the polyethylene matrix even though the ethylene acrylic acid is a very inefficient compatibilizer.

If the ethylene acrylic acid is free-floating in the polyethylene, then it is incompatible almost with the polyethylene, and in fact leads to a deterioration of the properties.

DR. ALBERTSSON: My name is Ann Christine Albertsson, and I would like to add that also ordinary polyethylene will sink in water, and you can make these experiments easily if you put in surfactants and have enough surface area. And if you have it UV-degraded, that means oxidized, which will go quicker.

So changing the surrounding, you will also have this to sink. So I think it is not a special property just for starch and polyethylene. I have mentioned it in a paper that is called "Surfactants".

DR. WOOL: there are nylon materials that will sink and there are many ways of changing the density; you do not just have to add in starch. Most of the ceramic-type fillers, silicates and so on, will give you a very nice density increase, and it can cause most materials to sink.

THE CHAIRMAN: Thank you. The next paper is entitled, "Degradable Plastics in the Marine Environment," by Anthony Andrady from the Research Triangle.

# DESIGN OF POLYESTER-BASED BIODEGRADABLE MATERIALS

T. Tokiwa, T. Ando, T. Suzuki and K. Takeda

Fermentation Research Institute
1-1-3, Higashi, Tsukuba, Ibaraki, Japan 305

## Introduction

It is generally known that among synthetic condensation polymers, aromatic polyesters and polyamides are nonbiodegradable although aliphatic polyesters are susceptible to biological attack [1~4]. We were the first to show that synthetic aliphatic polyesters are hydrolyzed by several commercial lipases and hog liver esterase[5]. But aliphatic polyesters are not widely used owing to their low melting points (Tm). Thus, to find both new biodegradable synthetic polymers and the reasons why aromatic polyesters and polyamides are not biodegradable, we synthesized copolyesters (CPE) and copolyamide-esters (CPAE) and studied their susceptibility to hydrolysis by lipase.

## Materials and Methods

MATERIALS: Polycaprolactone (PCL) and polypropiolactone (PPL) were prepared by the ring opening polymerization of ε-caprolactone[6] and ß-propiolactone, respectively, in benzene in a nitrogen atmosphere at 60°C with a diethylzinc-water catalyst system. Other saturated aliphatic polyesters were synthesized by melt polycondensation[7] and unsaturated polyesters were synthesized by high temperature solution polycondensation[8]. All aromatic polyesters were from Nihon Chromato except poly(ethylene terephthalate) (PET) from Asahikasei, poly(tetramethylene terephthalate) (PET) from Union Carbide, copolyester (PETG) of PET and poly(cyclohexylenedimethyl succinate) from Eastman Chemical Products. CPE and CPAE were synthesized by transesterification reaction[9] and amide-ester interchange reaction[10], respectively. All nylons were from Aldrich Chemical except nylon 6, obtained from Toyokasei and nylon 11 and nylon 12, from Nihon Lilsan.

Ultracentrifugally homogeneous preparation of R. delemar lipase was from Seikagaku Kogyo and a partially purified preparation of R. arrhizus lipase was from Boehringer Mannheim Yamanouchi. One unit of enzyme liberated one mole of fatty acid from olive oil per min at pH 7.0 and 37°C. Assay of enzymatic hydrolysis of synthetic solid polymers. Hydrolysis of solid polymers was measured by the rate of solubilization, and measurement did not necessarily involve complete hydrolysis into constituent parts. The rate was determined from the water-soluble total organic carbon (TOC) concentration at 30°C in reaction mixture using a Beckman TOC analyzer (Model 915-B). In the casu of CPE and CPAE, the reaction mixture contained 100 $\mu$ mol of phosphate buffer (pH 7.0), 0.1 mg of surfactant plysurf A210G, CPE (CPAE) powder or its films ( 20 mg as polyester moiety), and 0.2 mg of R. delemar lipase in a total volume of 1.0 ml. In the substrate and enzyme controls, the enzyme or substrate was omitted from the reaction mixture. Reaction mixtures were incubated on a shaker at

150 rpm at 30°C for 16 hours. After incubation, TOC concentration in the filtrate of the reaction mixture was measured. Formation of Water-soluble TOC was in proportion to substrate amount (up to 50 mg as polyester moiety) in this reaction system. CPE and CPAE were powdered by grinding or cut into film (0.20-0.27 mm thick). The biodegradability of CPE and CPAE was determined assuming water-soluble TOC in the reaction mixture to be formed through polyester blocks alone.

DETERMINATION OF MOLECULAR WEIGHT: Number average molecular weight ($\overline{Mn}$) was measured by the vapor pressure equiplibrium method using Hitachi Molecular Weight apparatus (Model 117).

MEASUREMENT OF MELTING POINT: Melting points of polymers (Tm) were measured with Yanaco Micro Melting Point apparatus (Model MP-S3).

## Results and Disscussion

THE RELATIONSHIP BETWEEN Tm AND BIODEGRADABILITY OF POLYESTER BY LIPASES: This relationship is shown in Fig. 1. For the same series polyesters, biodegradabilities decreased with increase in Tm.

Tm could generally be represented as follows:

$$Tm = \Delta H/\Delta S$$

where $\Delta H$ is the change in enthalpy in melting and $\Delta S$, the

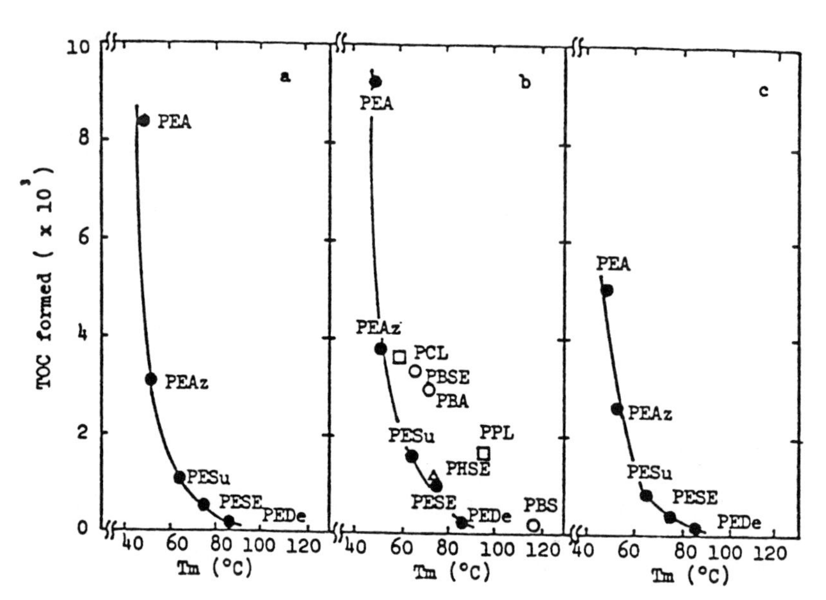


Fig. 1 Relationship between Tm and the Biodegradability of Polyesters by *R. delemar* (a) and *R. arrhizus* (b) lipases, and PEA-Degrading Enzyme from *Penicillium* sp. Strain 14-3[11] (c).
PESu: poly(ethylene suberate); PEAz: poly(ethylene azelate); PESE: poly(ethylene sebacate); PEDe: poly(ethylene decamethylate); PBS: poly-(tetramethylene succinate); PBA: poly(tetramethylene adipate); PBSE: poly(tetramethylene sebacate); PHSE: poly(hexamethylene sebacate).

Change in entropy in melting. Interactions among polymer chains mainly affect ΔH and internal rotation energies corresponding to the rigidity (flexibility) of the polymer molecule remarkably affect the ΔS value. The high Tm of aliphatic polyamide (nylon) is caused by large ΔH due to hydrogen bonds among polymers chains. The high Tm of aromatic polyester is caused by small ΔS with increase in the rigidity of the polymer molecule due to an aromatic ring. Nylon and aromatic polyesters are not biodegradable.

HYDROLYSIS BY LIPASE OF COPOLYESTERS (CPEs) CONTAINING AROMATIC AND ALIPHATIC ESTER BLOCKS.[12]: CPEs were synthesized by transesterification reactions between aromatic polyesters

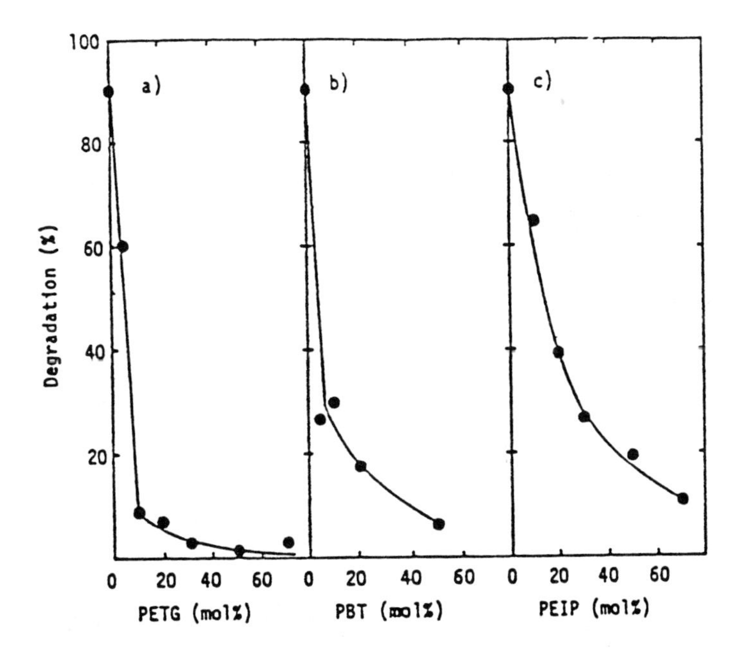


Fig. 2 Effect of molar ratio of PCL and aromatic polyester on the biodegradability of CPE by Rhizopus delemar lipase.

(a),(b),and (c) indicate PCL-PETG, PCL-PBT, and PCL-PEIP systems, respectively. The reaction conditions for each CPE synthesis were (a) 270°C for 2 h, (b) 270°C for 2 h, and (c) 170°C for 2 h.

(PET, PBT, PETG) and aliphatic polyester (PCL). The susceptibility of CPEs to hydrolysis by R. delemar lipase decreased rapidly during the initial stage of transesterification and increase gradually as it proceeded. Susceptibility to hydrolysis decreased with increase in aromatic polyester content (Fig. 2). Susceptibility to hydrolysis by the lipase of CPEs consisting PCL and poly(ethylene isophtalate) (PEIP), the latter used as a low-Tm (103°C) aromatic polyester, exceeded that of other CPEs, as shown in Fig. 2. The rigidity of the aromatic ring in CPE chains was thur concluded to strongly influence their biodegradability by this lipase.

HYDROLYSIS OF COPOLYAMIDE-ESTERS (CPAEs) BY LIPASE.[13]: CPAEs were synthesized by an amide-ester interchange reaction between polyamide and polyester. Polyamide blocks in CPAE were shortened with increasing reaction time and polyester content (Table 1).

Table 1 Polymerization Degree of the Main Component of the Polyamide Blocks of CPAEs 1-6*

| Kinds of standard for determination of polymerization degree | CPAE-1 | CPAE-2 | CPAE-3 | CPAE-4 | CPAE-5 | CPAE-6 |
|---|---|---|---|---|---|---|
| Polystyrene | 94 | 32 | 53 | 16 | – | – |
| Each nylon oligomer | – | 9 ~10 | – | 2 ~6 | 7~8 | 2~4 |

*The polymerization degree of each block was determined from the position of the peak top on each GPC chromatogram. CPAE-1: nylon 6-PCL (50/50 mol%); reaction time for synthesis, 1 h. CPAE-2: nylon 6-PCL (50/50 mol%); reaction time for synthesis, 4 h. CPAE-3: nylon 12-PCL (50/50 mol%); reaction time for synthesis, 1 h. CPAE-4: nylon 12-PCL (50/50 mol%); reaction time for synthesis, 4 h. CPAE-5: nylon 6-PCL (20/80 mol%); reaction time for synthesis, 4 h. CPAE-6: nylon 12-PCL (20/80 mol%); reaction time for synthesis, 4 h.

The susceptibility of CPAEs to hydrolysis by R. delemar lipase decreased with shortening of the polyamide blocks and increasing polyamide content (Fig. 3). Simple blends of nylon and PCL at 270°C for 10 min retained high biodegradability of PCL. It wa concluded that the number and distribution of hydrogen bonds, based on amide bonds, in CPAE chains strongly influences their biodegradabiltiy by this lipase.

The new biodegradable synthetic polymer, CPAE, can be made into any desired shape. A transparent thin film (about 0.02 mm thickness) was made from CPAE. The physical properties are shown in Table 2.

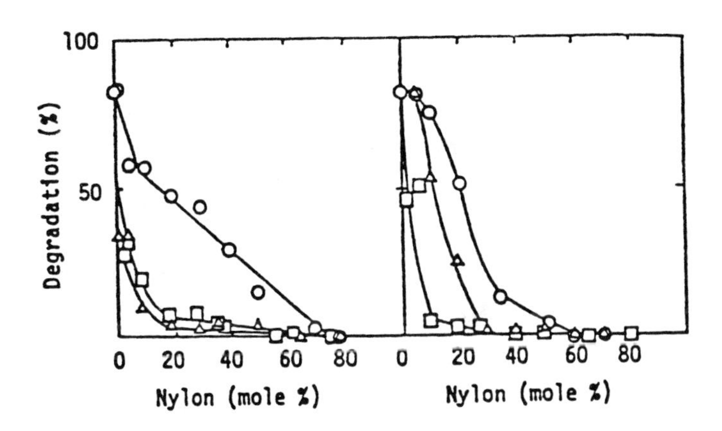


Fig. 3 Effect of the molar ratio of nylon and PCL on biodegradability of CPAE by R. delemar lipase.

The reaction time for each CPAE synthesis was 4 h. The basic structures of nylon were of two types. One was [-NH$(CH_2)_m$CO-]n (left); the other was [-NH$(CH_2)_6$ NHCO$(CH_2)_m$CO-]n (right). Left: -○-, nylon 6; -△-, nylon 11; -□-, nylon 12; right: -○-, nylon 66; -△-, nylon 69; -□-, nylon 612.

It is very important that various types of interactions among macromolecular chains, related to Tm, be taken into consideration when designing biodegradable solid polymers.

Table 2 Tm and Mechanical Properties of CPAEs

| Composition (wt %) PCL | Nylon12 | Tm (°C) | Mechanical properties Ty | Ey | Tb | Eb |
|---|---|---|---|---|---|---|
| 20 | 80 | 175 | 245 | 12 | 453 | 369 |
| 20 | 80 | 170 | 266 | 11 | 518 | 435 |
| 30 | 70 | 170 | 214 | 16 | 312 | 413 |
| 50 | 50 | 167 | 141 | 20 | 135 | 109 |

Ty: Tensile stress at yield point ($Kg/cm^2$)
Ey: Elongation at yield point (%)
Tb: Tensile stress at break ($Kg/cm^2$)
Eb: Elongation at break (%)

REFERENCES

1. R.T. Darby, and A.M. Kaplan, Appl. Microbiol., 16, 900 (1968).
2. J.E. Potts, R.A. Clendinning, W.B. Ackart, and W.D. Niegisch, Am. Chem. Soc., Polymer Preprints, 13, 629 (1972).
3. R.D. Fields, F. Rodrigues, and r.K. Finn, J. Appl. Polym. Sci., 18, 3571 (1974).
4. M.J. Diamond, B. Freedman, and J.A. Garibaldi, Int. Biodetn. Bull., 11, 127 (1975).
5. Y. Tokiwa, and T. Suzuki, Nature, 270, 76 (1977).
6. R.D. Lunberg, J.V. Koleske, and K.B. Wischmann, J. Polym.Sci. A-1,7, 2915 (1969).
7. W.H. Carothers, and J.A. Arvin, J. Am. Chem. Soc., 51, 2560 (1929).
8. H. Batzer, H. Holtschmidt, F. Wiloth, and B. Mohr, Makromol. Chem. 7, 82 (1951).
9. W.H. Charch, and J.C. Shivers, Text. Res. J., 29, 538 (1959).
10. T. Kiyotsukuri, K. Takada, and R. Imamura, Chem. High Polym. Jpn. 27, 410 (1970).
11. Y. Tokiwa, and T. Suzuki, Agric. Biol. Chem., 41, 265 (1977).
12. Y. Tokiwa, and T. Suzuki, J. Appl. Polym. Sci., 26, 441 (1981).
13. Y. Tokiwa, T. Suzuki, and T. Ando, J. Appl. Polym. Sci., 24, 1701 (1979).

BIOGRAPHY

DR. Y. TOKIWA

Dr. Y. Tokiwa is a senior researcher of chemical ecology division at Fermentation Research Institute (FRI), Agency of Industrial Science and Technology, MITI, Tsukuba, Ibaraki, Japan. Since he joined the FRI in 1972, his research work has been concerned with microbial degradation fo synthetic polymers.

QUESTIONS AND ANSWERS

THE CHAIRMAN: This paper is open for discussion, comments.

I was wondering on the correlation between melting point and rate of biodegradation. We have reached a conclusion that it is more the last transition of the polymer in aqueous medium that determines the rate of biodegradation rather than the melting point, because things like poly(glycolate) has a melting point of 240, and that picks up water in the flask transition of the amorphour phase, then drops quite low and the material was biodegradable.

But polyesters, the aromatic polyesters, have quite a high flask transition and they don't pick up any water, so they are not biodegradable.

DR. ANDRADY: So in the case of a water soluble polymer, synthetic polymers, we can consider just on the molecule, chemical molecule, that in the solid state, for example, in soluble polymer, a aliphatic polyester, on factor -- "Tm" is, I think, one factor.

DR. ALBERTSSON: My name is Ann Christine Albertsson, and I'd like to thank you for your lecture.

Since I know that you have been working in this field more than 15 years, I would like to ask you, have you compared different test methods when you have been testing biodegradation? Because here, you just mentioned this very shortly, and I think you should say some details about your test methods and how you get good contact area and how you increase the rate and so forth.

INTERPRETER: Presently, he says, that he is actually comparing the results with the field test using the soil test and the enzyme test, and he hasn't got the results for the soil tests at this point.

DR. ALBERTSSON: Oh, but what I mean is, that you have shown earlier that if you, for instance, test polyethylene, you can show that if you stir it and you mix it better, you get a better contact area and you increase degradation.

And you have also shown that if you increase - you call it mechanical - at the same time as you do this, you also increase the rate.

I hope, if you can't give it now, next time I would like to hear about the test methods because I know that you're an expert in this.

DR. TOKIWA: Thank you.

DR. ANDERSON: Jim Anderson, Case Western Reserve. This transesterification technique is molecular-weight-dependent.

What can you tell us about the starting molecular weights and the final molecular weights after the thermodynamic equilibrium has been achieved.?

INTERPRETER: The starting material, the molecular weight was between 25,000 and 40,000.

DR. ANDERSON: Both polymers?

INTERPRETER: Nine of them.

DR. ANDERSON: Nylon 12?

INTERPRETER: Nylon will be lower. And the final one was just known as to present in poligomers, so it hasn't been traced.

DR. ANDERSON: The residue hasn't been measured.

DR. ANDERSON: Thank you.

DR. VERT: I guess you said that the degradation rate was decreasing with the melting point of the polymers of the polyesters and the polyester series.

What kind of melting point did you take? I mean, what were the molecular weights of these polyesters, and where are you on the plateau of the melting point, or were you, before the plateau?

INTERPRETER: The molecular weight did not affect the rate of hydrolysis by a stigma lipases when molecular -- number of average molecular weight was more than about 4,000.

In contrast, when average molecular weight was less than about 4,000, the rates of the enzyme hydrolysis was faster with the smaller number of age, molecular weight. This corresponded to the fact that the Tm was lower with the smaller molecular weight of polyester.

DR. VERT: Yes. but for some aliphatic polyesters, the molecular weight was low for sure because -- the molecular weights were low for sure because the condensation polyesters don't give you high molecular weight, let's say. So the melting point is difficult to give when the molecular weight is very small.

Which one did you take in this case?

DR. BRUCK: Could you tell us a little bit about the crystallinity of the polyamide and polyester and the polyester amide? The crystallinity, did you measure it? Did you estimate it?

INTERPRETER: Crystallinity itself hasn't been measured, but when it is measured with DSC, the broadening of the peak is observed.

DR. SHALABY: You did the blending of Nylon 6 with poly-e-caprolactone at temperatures well-above the ceiling temperature of poly-e-caprolactone and very close to the ceiling temperature of Nylon 66, which could lead to the formation of monomers of both polymers, cyclic oligomers.

I was wondering, how would this affect the biodegradability?

INTERPRETER: Depending what's done for five minutes.

DR. SHALABY: But they also showed longer than that.

INTERPRETER: Longer?

DR. SHALABY: We had periods that were much higher than five minutes. Five minutes, you can get away with it, but more than five minutes you tend to get much more monomers than that.

INTERPRETER: This study was done mainly to compare between the control of enzyme or without the enzyme, so it hasn't been shown therefore.

DR. HENDERSON: Alex Henderson, Ethicon Canada. What catalyst did you use and did you remove it from the polymer?

DR. TAKIWA: Unhydrolized acid zinc.

DR. HENDERSON: Okay. And did you remove it?

DR. TOKIWA: Just prepared with precipitation.

DR. HENDERSON: Thank you.

# Effects Of Photodegradants On The Environmental Fate Of Linear Low Density Polyethylene

By

F.C. Schwab
Mobil Chemical Company, R&D
P.O. Box 240
Edison, NJ 08818

## Introduction

Most traditional applications of synthetic polymers are based on their inertness towards outside factors like heat, chemicals, radiation and microorganisms. Only recently has this "inertness" been singled out as a major contributor to the serious and urgent problem of solid waste disposal. Worldwide research efforts are underway to make polymers more susceptible to natural degradation modes like sunlight and microorganisms through the use of additives or polymer modification.

Much of what is being used today was actually developed in the early 1970's by such researchers as Guillet [(1)], Scott[(2)], and Griffin[(3)] to name a few. Today, this technology is being exploited commercially by companies like Ampacet, Polysar, Ecolyte and Plastigone leading to a variety of commercial degradation additives principally for polyolefins and polystyrene. It should be noted that polyethylene alone accounts for about 60% of the total 6 million tons of plastics used in packaging[4].

Polyethylene, in a pure state, is insensitive to sunlight in that its absorption maxima occurs below 200nm which is considerably below the lower cutoff of natural sunlight (~290nm). Most polyethylene will photodegrade to some degree due to impurities which act as chromophores, e.g., catalyst residues or oxidation products from thermal processing[5]. In order to assure the controlled, reliable photodegradation of polyethylene, chromophores are deliberately added either into the polymer structure itself (carbonyls) or as a mixture (metal salts).

Photodegradation, in and of itself, will not lead to the ultimate disappearance of the material. Photodegradation usually results in the oxidation and embrittlement of the polymer leading to the creation of material which is highly oxidized and of a low molecular weight resulting in disintegration into small pieces. The ultimate disappearance of the polymer will depend on its assimilation by microorganisms. However, photodegradation and biodegradation are not necessarily synonymous in that one will lead to the other. many researchers have demonstrated that the ultimate degradation (assimilation) of a polymer by microorganisms depends on many factors like morphology, surface area, chemical composition and molecular weight. In this work we have studied the degradation of linear low density polyethylene (LLDPE) using a variety of commercial photodegradants. This includes effects on chemical composition and molecular weight of the matrix polymer and its eventual biological assimilation.

## EXPERIMENTAL

### 1. Materials

#### a. Linear Low Density Polyethylene (LLDPE)

The matrix resin (LLDPE) used in this study was an ethylene/butene copolymer containing about 3.5 mole % butene. The resin has a density of 0.918 gm/cc and a melt index (190$^{o}$C, 2.16 Kg) of 1.2. It is a fully formulated resin containing ~450 ppm of both primary and secondary antioxidants.

#### b. Photodegradants

All photodegradants were added as a masterbatch of the photodegradant in polyethylene. The following masterbatches were used in this study:

Table I

| Masterbatch | Amount Masterbatch Used, % |
|---|---|
| Ampacet Polygrade I | 2 and 5 |
| Plastigone 221 | 2 |
| Ampacet Anatase | 6 |
| Ecolyte II | 10 |

2. Polymer Preparation

All photodegradabel compositions were prepared by melting the LLDPE in a Brabender at 185°C and 50 rpm. The proper additive level was added and the mixing was continued for three additional minutes. The samples were compression molded at 190°C in a Carver press to 10 mil plaques.

3. Degradation

The polymers were photodegraded using QUV fluorescent/condensation tester with UV-A lamps. Samples were exposed to continuous radiation for up to 500 hurs. at 60°C.

4. Analytical Techniques

a. FTIR - Infrared analyses were carried out using a Perkin-Elmer Model 1600 FTIR.

b. NMR - NMR analyses were carried out using a JEOL GX400 spectrometer. Spectra were run at 130°C with tetrachloroethane-$d_2$ as the solvent. Experimental conditions were quantitative.

c. Thermal Analysis - Oxidative stability temperatures were determined in oxygen using an Omnitherm DSC. The sample size was 10mg and was heated at a rate of 10°C/min.

d. GPC - Polymer molecular weights and molecular weight distributions were run in trichlorobenzene on a waters 150C GPC using four Phenogel colums at 140°C.

5. Biological Testing

Biological assimilation testing was conducted using EPA 560-6-82-003 shaker flask test. The sample was the only carbon source in the medium to which was added an inoculum of soil and sewage microorganisms. Testing was done in an aerobic manner and the $CO_2$ evolved was trapped in alkali and titrated to determine conversion to $CO_2$.

RESULTS AND DISCUSSION

Commercial photodegradants can be divided into two basic groups: additives such as transition metal salts and pigments and copolymers which contain a chromophoric moiety such as a carbonyl group. Transition metal salts (e.g., acetoacetonates, stearates and dithiocarbomates) in general work by the following scheme:[5]

$$M^{n}(X)_{n} \xrightarrow{hv} M^{n-1}(X)_{n-1} + X \qquad (1)$$

The radical can then abstract a hydrogen from the polymer backbone. The polymer can be further oxidized leading to a series of typical autooxidation steps as illustrated by the following equations:

Initiation: $X + R - H \dashrightarrow X - H + R\circ$ (2)

Propagation: $R^\circ + O_2 \dashrightarrow ROR^\circ$ (3)

$RO_2\circ + R - H \dashrightarrow - R - OOH + R\circ$ (4)

Termination: $R\circ + R\circ \dashrightarrow$

$RO2^\circ - R\circ \dashrightarrow$ (5)

$RO_2\circ + RO_2\circ \dashrightarrow$

Some pigments such as titanium dioxide or zinc oxide can also function as photoinitiators. Titanium dioxide, a commonly used pigment in polyethylene, exists in two crystalline forms - rutile and anatase. It is well known that these two crystalline forms exhibit markedly different photoactivities toward the polymer matrix[6]. The difference in photoactivity of these two forms has been attributed to the difference in the energies of their photoexcited states[7]. The anatase forms of $TiO_2$ has a photocatalytic effect on the degradation of polyethylene whereas the rutile form has a stabilizing effect on photodegradation. The mechanism for the photooxidation of polymers by anatase $TiO_2$ is quite complex, but it is believed tat the reaction scheme involves the formation of a hydroxyl radical (OH) which is believed to be the reactive species.

This species can react with the polymer matrix and begin the autooxidation process.

The second group of photodegradants are polymers in which the chromophore is incorporated into the polymer chain usually by copolymerization. The most common chromophore used commercially is the carbonyl group. In the Ecolyte technology, a carbonyl containing monomer is chemically incorporated into the polymer constituting the additive. On exposure to UV radiation, the carbonyl containing polymer can undergo a Norrish I or II type reaction[8]. In the case of ethylene-vinyl ketone copolymers only, a type II reaction will lead to chain scission. Type I reaction is important only when the environment permits a diffusional separation of the product radical pairs, viz.

$$-CH_2-CH_2-\underset{\underset{\underset{R}{|}}{C=O}}{CH}-CH_2---- \xrightarrow{I} -CH_2-CH_2-CH-CH_2- + \overset{O}{\overset{\|}{C}}-R \rightarrow CO + R \tag{6}$$

$$\xrightarrow{II} -CH_2-CH_2-C-R+CH_2 = CH-CH_2- \tag{7}$$

Oxidation Studies

The rate of oxidation as well as the time required to first observe oxidation (oxidation induction time) in the polymer was followed via FTIR spectroscopy. This was accomplished by measuring the change in absorption of the carbonyl peak at 1717 $cm^{-1}$. The oxidation induction times for the systems tested are given in Table II.

Table II

OXIDATIVE INDUCTION TIMES

| Photodegradant | Oxidative Induction Time (hrs. in QUV |
|---|---|
| Polygrade I - 5% | 2.0 |
| Polygrade I - 2% | 2.5 |
| Plastigone 221 | 6.0 |
| Anatase | 4.0 |
| Ecolyte II | 6.0 |
| LLDPE Control | 58 |

As can be seen from Table II, the photodegradant additives have a dramatic effect on the time required to begin the oxidation process. It must be remembered that the LLDPE matrix resin contains antioxidants and the observed induction times reflect the time required to overcome these stabilizers. The LLDPE without the photodegradant requires about 5-10 times longer for noticeable oxidation to begin.

Once oxidation has started, the rate at which the carbonyl content increases varies with the type of photodegradant as well as its concentration. Table III gives the rate of oxidation as observed by the change in carbonyl index as a function of time over the first 100 hrs. of QUV exposure after oxidation was first observed.

Table III

RATE OF OXIDATION

| Photodegradant | Rate of Oxidation ($\Delta$ ABS/hr.) |
|---|---|
| Polygrade I - 2% | .0095 |
| Polygrade I - 5% | .0318 |
| Plastigone 221 | .0113 |
| Anatase | .0043 |
| Ecolyte II | .0018 |
| LLDPE Control | .0019 |

As can be seen from the data, once oxidation starts, the rate at which it continues is dependent upon the concentration and the nature of the photodegradant used. It appears that even though the Ecolyte II starts the oxidation fairly quickly, its rate of oxidation is essentially the same as the LLDPE control after its induction period.

Oxidative Stability Temperature

A brief study was undertaken to determine what effect, if any, oxidation has on the subsequent thermal stability of the polymer as measured by the oxidation stability temperature. While this may have little effect on the final environmental fate of the polyethylene, it may have a serious effect on the presence of photodegradable materials that find their way into the recycle or reclaim streams. Table IV shows the effect of previous oxidation on the oxidative thermal stabiltiy temperature of the various photodegrandant systems.

Table IV

EFFECT OF OXIDATION ON OXIDATIVE STABILITY TEMPERATURE

| Photodegradant | Oxidative Stabiltiy Temperature, °C 0 hrs. | 100 hrs. | (Carbonyl Index at 100 hrs.) |
|---|---|---|---|
| Polygrade I - 2% | 186 | 176 | (0.95) |
| Plastigone 221 | 186 | 166 | (1.81) |
| Anatase | 202 | 194 | (0.43) |
| Ecolyte II | 200 | 197 | (0.18) |
| LLDPE Control | 219 | 202 | (0.10) |

As can be seen from Table IV, the temperature at which the oxidation of the polyethylene occurs is dramatically decreased by the presence of oxidation in the materila. In fact, the oxidative stability temperature of the samples with no exposure to the UV source all show a decrease over the LLDPE control. This is probably due to oxidative degradation (photo and/or thermal) having occurred during the preparation of the samples even though no detectable carbonyl is observed in the infrared. The dependence of the polyethylene is easily seen in the samples exposed for 100 hrs. in the QUV (Figure 1).

Effect of Photodegradant on Molecular Weight

The molecular weight of the polyethylenes containing the photodegradants was determined as a function of exposure time via size exclusion dhromatography. The results are given in Table V.

The data in Table V clearly shows that the photooxidative degradation results in molecular degradation via chain scission. This is especially indicative in the narrowing of the molecular weight distribution as evidenced by the Mw/Mn ratio. Comparison of the data in Table V with the oxidation data of Table III shows that the rate at which the molecular degradation occurs roughly parallels the initial rate of oxidation up to the first 100 hours of exposure.

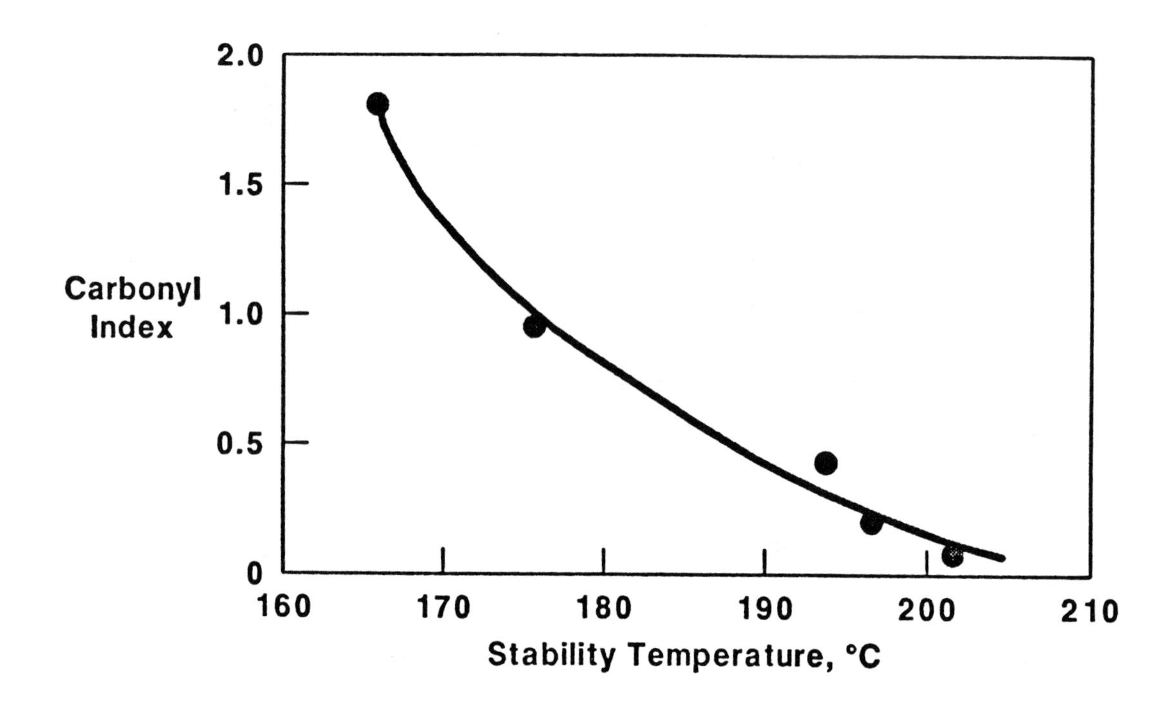


FIGURE 1. DEPENDENCE OF OXIDATIVE STABILITY ON THE PRESENCE OF OXIDATION IN LINEAR LOW DENSITY POLYETHYLENE

TABLE V

MOLECULAR WEIGHT AS A FUNCTION OF QUV EXPOSURE

| Photodegradant | Exposure Time 0 | 100 | 200 | 300 | 400 | 500 |
|---|---|---|---|---|---|---|
| Polygrade I - (2%) | | | | | | |
| Mn | 27,400 | 3,400 | 1,700 | 1,800 | 1,100 | 1,100 |
| Mw | 115,600 | 10,700 | 4,000 | 5,600 | 3,000 | 3,400 |
| Mw/Mn | 4.2 | 3.1 | 2.4 | 3.1 | 2.7 | 3.0 |
| Plastigone 221 | | | | | | |
| Mn | 28,800 | 2,100 | 2,000 | 1,500 | 870 | .950 |
| Mw | 116,700 | 5,600 | 6,300 | 3,200 | 2,000 | 2,500 |
| Mw/Mn | 4.1 | 2.7 | 3.1 | 2.2 | 2.3 | 2.6 |
| Anatase | | | | | | |
| Mn | 29,100 | 8,800 | 5,800 | 1,370 | 3,500 | 2,800 |
| Mw | 121,300 | 46,200 | 28,900 | 4,700 | 18,800 | 10,200 |
| Mw/Mn | 4.2 | 5.3 | 5.0 | 3.4 | 5.3 | 3.6 |
| Ecolyte II | | | | | | |
| Mn | 28,800 | 12,700 | 7,700 | 2,900 | 2,600 | 1,900 |
| Mw | 118,700 | 85,300 | 39,200 | 11,300 | 8,600 | 7,500 |
| Mw/Mn | 4.1 | 6.7 | 5.1 | 3.8 | 3.3 | 3.9 |
| LLDPE | | | | | | |
| Mn | 25,900 | -- | -- | 3,500 | -- | -- |
| Mw | 91,000 | -- | -- | 13,200 | -- | -- |
| Mw/Mn | 3.5 | -- | -- | 3.8 | -- | -- |

comparison of the chromatograms of the various sam;es shows a definite shift toward a bimodal distribution on the part of the Polygrade I and Plastigone 221 samples after about 300 hours of exposure (Figure 2). The Anatase and Ecolyte II degradants generally maintain their original distribution except for a shift to low molecular weights. Since both Polygrade and Plastigone are believed to be metal salt type degradants, the mechanism of degradation is probably reflected in this unusual molecular weight behavior.

As can be observed from Table V, at the end of 400 to 500 hours of exposure, several of the samples are exhibiting average molecular weights that are getting close to the 500 daltons which can be utilized as a food source by microorganisms[9]. The various chromatograms were analyzed as to the percentage of the molecular weight distribution under 500 daltons (Table VI).

As can be seen from Table VI, the Polygrade I and Plastigone 221 have the highest amount of potentially biodegradable low molecular weight materials. Results of the biodegradation of these materials will be discussed in a later section.

## Effect on Mechanical Properties

As can be seen from previous sections, the presence of the photodegradants leads to a rapid decrease in the molecular weight of the polyethylene matrix. This, of course, leads to a subsequent decrease in the mechanical properties of the polymer. Table VII shows the effect of the photodegradants on the elongation of polyethylene matrix which is a measure of the embrittlement of the material.

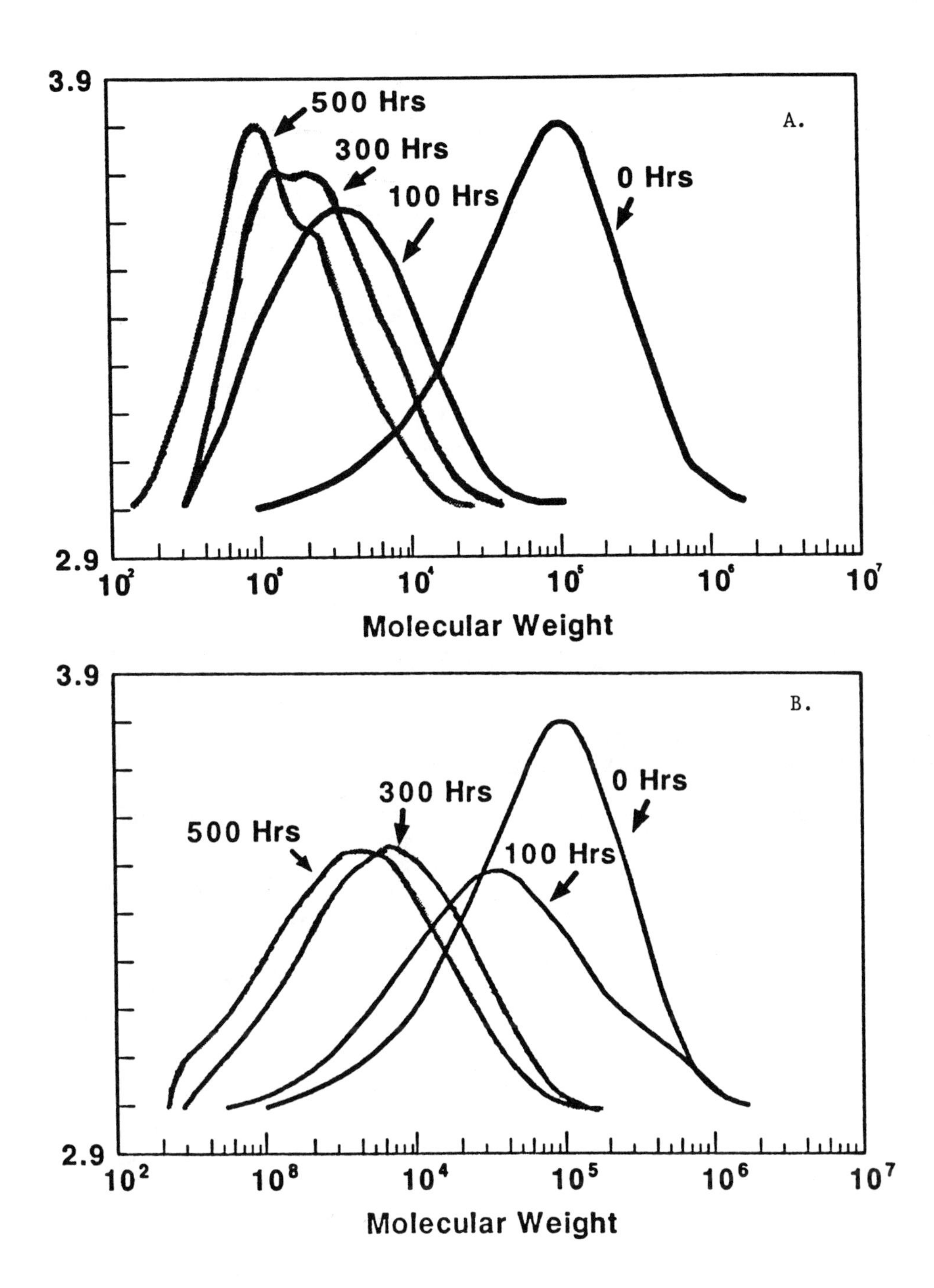


FIGURE 2. GPC CHROMATOGRAMS AS A FUNCTION OF QUV EXPOSURE TIME.
A. POLYGRADE I AND PLASTIGONE 221
B. ANATASE AND ECOLYTE II

TABLE VI

DEGRADED MATERIAL LESS THAN 500 DALTONS (500 HRS. EXPOSURE)

| Exposure Time, Hrs. | % Less Than 500 Daltons | | | |
|---|---|---|---|---|
| | Polygrade I | Plastigone 221 | Anatase | Ecolyte II |
| 0 | 0 | 0 | 0 | 0 |
| 100 | 0.8 | 2.6 | 0.1 | 0 |
| 200 | 3.1 | 2.7 | 0.6 | 0.1 |
| 300 | 3.3 | 3.4 | -- | 2.0 |
| 400 | 10.9 | 15.1 | 1.7 | 2.0 |
| 500 | 11.1 | 13.3 | 1.6 | 5.5 |

TABLE VII

EFFECT OF PHOTODEGRADANTS ON ELONGATION (QUV)

| Exposure Time, Hrs. | % Elongation Retained | | | | |
|---|---|---|---|---|---|
| | LLDPE | Polygrade I | Plastigone 221 | Anatase | Ecolyte II |
| 0 | 100 | 100 | 100 | 100 | 100 |
| 50 | 100 | 5 | 6 | 100 | 78 |
| 100 | 100 | 0 | 0 | 23 | 17 |
| 150 | 81 | 0 | 0 | 23 | 17 |
| 200 | 76 | 0 | 0 | 19 | 8 |

(10 Mil Plaques)

TABLE VIII

OUTDOOR WEATHERING - NEW JERSEY

| Exposure Time, Days | % Elongation Retained LLDPE | % Elongation Retained Polygrade I | LLDPE Mn | LLDPE Mw | Polygrade I (2%) Mn | Polygrade I (2%) Mw |
|---|---|---|---|---|---|---|
| 0 | 100 | 100 | 25,900 | 91,000 | 27,400 | 115,600 |
| 14 | 100 | 32 | 32,000 | 111,000 | 8,400 | 26,000 |
| 28 | 100 | 3 | -- | -- | -- | -- |
| 42 | 100 | 0 | -- | -- | -- | -- |
| 56 | 12 | 0 | -- | -- | -- | -- |
| 70 | 9 | 0 | -- | -- | -- | -- |
| 84 | 8 | 0 | -- | -- | -- | -- |
| 98 | 6 | 0 | 6,800 | 29,400 | 3,200 | 11,800 |

(10 Mil Plaques)

TABLE IX

OUTDOOR WEATHERING - ARIZONA

| Exposure Time, Days | % Elongation Retained LLDPE | % Elongation Retained Polygrade I (5%) | % Elongation Retained Ecolyte II (5%) |
|---|---|---|---|
| 0 | 100 | 100 | 100 |
| 15 | 95 | 79 | 99 |
| 30 | 94 | 79 | 97 |
| 45 | 93 | 6 | 90 |
| 60 | 101 | 2 | 96 |
| 90 | 64 | Gone | 97 |
| 120 | 19 | -- | 10 |

(1 Mil Clear Film)

As can be seen in table VII, all the systems tested were potent degradants when tested using the QUV. A limited amount of outdoor testing was carried out on a few of the photodegradant systems. Tables VIII and IX give the results of outdoor exposure in both New Jersey and Arizona.

It is interesting to note that within the limited amount of data obtained on outdoor weathering, the test results tend to track the highly accelerated QUV tests. It can also be observed from Table VIII that the Polygrade I sample has been severely degraded at the end of 14 days. This result also correlates well with the accelerated QUV data obtained on plaques for Polygrade I (Table X).

## Effect of Photodegradants on Chemical Composition

The composition of the samples exposed for 500 hrs. in the QUV was determined using $^{13}C$ NMR. The compositional features are given in Table XI as a concentration per 1000 carbon atoms.

Table XI shows several interesting features of the results of the photooxidation on the LLDPE. First, there is a dramatic decrease in the amount of ethyl branches for the Polygrade and Plastigone samples as compared to the Anatase and Ecolyte II. Disappearance of the ethyl branches indicates that attack is occurring at the tertiary carbon which forms the branch via:

$$-CH_2-\underset{\displaystyle C_2H_5}{\underset{|}{CH}}-CH_2-CH_2 \quad \xrightarrow{X\,\circ} \quad -CH_2-\underset{\displaystyle C_2H_5}{\underset{|}{C}}-CH_2-CH_2- + XH$$

TABLE X

Polygrade I (5%) IN LLDPE

| QUV Exposure Time, Hrs. | Carbonyl Index | Mn | Mw | % Elongation Retained |
|---|---|---|---|---|
| 0 | 0 | 25,900 | 91,000 | 100 |
| 50 | 1.07 | 7,300 | 22,200 | 7 |
| 100 | 1.48 | 3,600 | 14,400 | 0 |
| 150 | 1.27 | 3,100 | 11,400 | 0 |
| 200 | 1.50 | 2,900 | 9,300 | 0 |

(10 Mil Plaque)

TABLE XI

COMPOSITION OF DEGRADED LLDPE (500 HRS.)

| | Concentration per 1000 Carbons | | | |
|---|---|---|---|---|
| Photodegradant | Ethyl Branches | End Methyl | End Vinyl | Carbonyls |
| LLDPE | 17 | 0 | 0 | 0 |
| Polygrade I | 4.9 | 3.0 | 3.5 | 2.6 |
| Plastigone 221 | 2.4 | 3.5 | 7.1 | 7.9 |
| Anatase | 15.3 | 1.0 | 3.5 | 4.6 |
| Ecolyte II | 13.6 | 3.3 | 4.7 | <0.5 |

Oxidation can then occur at this site and lead to the oxidation products via Equations 1-5. If chain rupture also occurred at this site, this would explain the increase in methyl end groups via:

$$-CH_2-\underset{\displaystyle CH_2-CH_3}{\overset{\displaystyle O}{\overset{|}{\underset{|}{C}}}}-CH_2-CH_2- \quad ---\rightarrow \quad -CH_2-\overset{\displaystyle O}{\overset{\|}{C}}-CH_2-CH_3 + CH_2-CH_2-$$

By contrast, the Anatase and Ecolyte II, samples show very little decrease in the ethyl branch content over the unexposed LLDPE. This implies that degradation, as evidenced by molecular weight decrease,is not occurring at the tertiary carbon of the branch. This is somewhat surprising for the Anatase sample in that if degradation is occurring by a free-radical mechanism, one would expect attack to occur at the tertiary carbon. By contrast, if the Ecolyte II degradation is occurring solely by a Norrish Type Ii mechanism (Equation 7), the ethyl branches might remain intact while the vinyl end groups would increase (Equation 7).

Secondly, the Polygrade I, Plastigone 221 and Anatase are more highly oxidized than the Ecolyte II. In fact, carbonyl type carbons are not detected in the $^{13}C$ spectrum for the Ecolyte II. This might not be totally unexpected if degradation of the Ecolyte II sample is occurring solely by a Norrish Type II process (Equation 7).

Since the degradation process is a very complex one, it does little good to speculate too far. However, the evidence from the $^{13}C$ NMR work tends to show that a different mechanism may be operating for the Polygrade and Plastigone materials. This tends to support the observed differences in the molecular weight changes discussed earlier.

Biological Degradation

As can be seen from the preceding discussion, the photodegraded LLDPE has been drastically changed both chemically and molecular weight-wise due to the photodegradants. The final experiment to determine the ultimate fate of the material is its assimilation by microorganisms. Only one system has been under test to date - Polygrade I (2% in LLDPE). The sample was exposed 500 hrs. in the QUV. Its analysis is given tin Table XII.

The shaker flask test measures the amount of carbon dioxide expired by the microorganisms and is a direct measure of the amount of carbon materila assimilated. Figure 3 and Table XIII show the results of the test to date as well as comparisons with pure cellulose (control) and a starch filled polyethylene (6%).

Table XIII shows that, as expected, the cellulose control is essentially totally biodegraded in about 1 month. The biological activity on the polyethylene film containing 6% starch stops after about 1 month at the level of $CO_2$ evolution which one would expect for the starch alone. It appears from the test the polyethylene matrix is essentially untouched over the total 100 days of the test.

The highly degraded polyethylene sample, on the other hand, shows strong activity at the end of 78 days. Figure 3 shows that the biological activity was somewhat slow to start, but by the end of about 20 days, there was a decided increase in activity and the sample was indeed being assimilated by the microorganisms. At the end of 78 days, the carbon dioxide level of 15.4%, which was calculated from the material which is less than 500 daltons, had been exceeded.

TABLE XII

ANALYSIS OF SAMPLE IN BIOLOGICAL TEST

Polygrade I (2% IN LLDPE)

| | |
|---|---|
| Mn | 810 |
| Mw | 1760 |
| % <500 Daltons | 18 |
| Carbonyl/1000 Carbons | 2.5 |
| Carbonyl Index | 2.05 |

TABLE XIII

SHAKER FLASK RESULTS

| | Carbon Available, % | | % Carbon to $CO_2$ | | | |
|---|---|---|---|---|---|---|
| | Total | Degradable | 4 Days | 34 Days | 50 Days | 78 Days |
| Polygrade I (500 hrs.) | 85.7 | 15.4* | 0.7 | 8.5 | 9.7 | 15.7 |
| Cellulose | 44 | 100 | 54.9 | 98 | 98 | 98 |
| PE/Starch (6%) | 83.4 | 3.1 | 3.3 | 3.8 | 3.8 | 3.8 |

*Material less than 500 Daltons

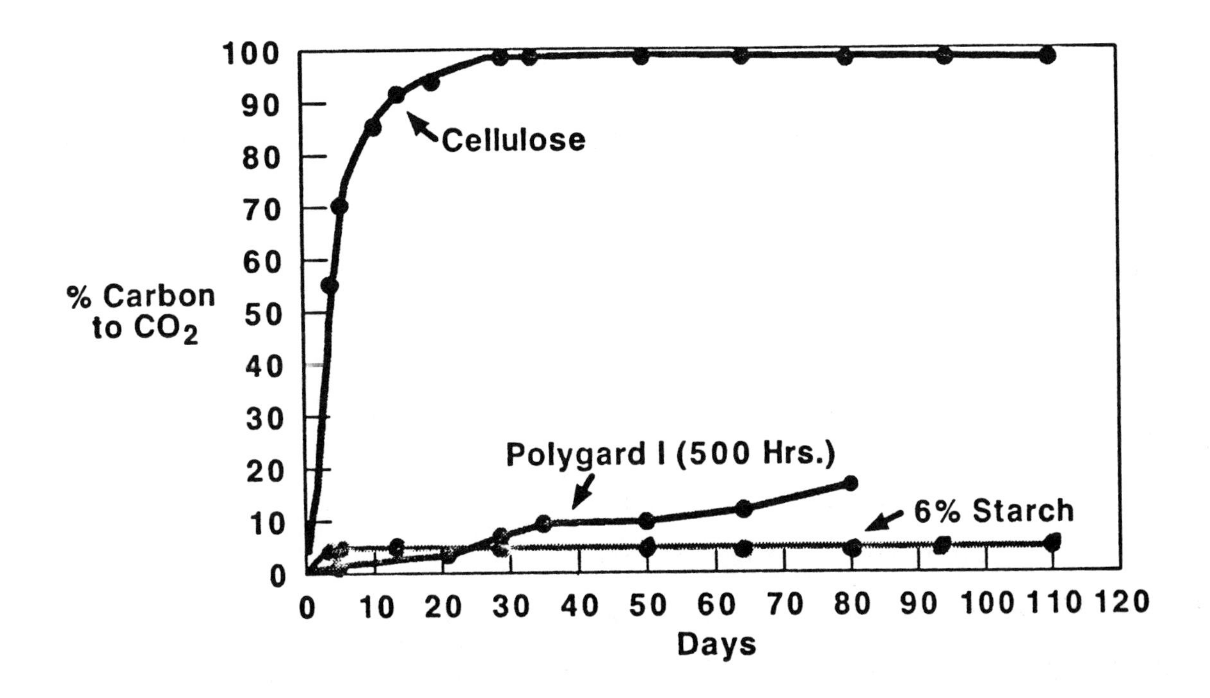


FIGURE 3. CUMULATIVE BIOLOGICALLY DERIVED CARBON DIOXIDE AS A FUNCTION OF TIME

## CONCLUSIONS

In summary, it appears that:

1. The presence of the photodegradant greatly reduces the time required to begin oxidation.
2. The rate of oxidation depends on the concentration and type of photodegradant.
3. The presence of oxidation in the polyethylene tends to reduce the temperature at which oxidation occurs.
4. The molecular weight and molecular weight distribution of the degraded polyethylene depends on the photodegradant used.
5. The accelerated effects of the photooxidative process on mechanical properties tend to track outdoor results.
6. the chemical composition of the degraded polyethylene appears to depend on the photodegradant used.
7. Biological assimilation can be achieved, at least in part, if the polyethylene can be sufficiently altered by the photodegradation process.

BIBLIOGRAPHY

1. Hartly, G.H. and Guillet, J.E. Macromolecules 1 (2), 165 (1968).
2. Scott, G., Mellar, D.C. and Moir, A.B., Eur. Polym. J., 9, 219 (1973).
3. Griffen, G.J.L., British Pat. Appln., No. 55, 195/73 (1973).
4. Johnson, R., Plastic Film and Sheeting, 4, 155 (1988).
5. Omichi, H. in Degradation and Stabilization of Polyolefins (ed. by N.S. Allen), Applied Science Publishers, Ltd., London, P.187 (1983).
6. Allen, N.S. and McKellar, J.F., Photochemistry of Man-made Polymers, Applied Science Publishers, Ltd., London, p.199 (1979).
7. Allen, N.S., McKellar, J.F., J. Polym. Sci. (B), 12, 723 (1974).
8. Li, S.K. and Guillet, J.E., J. Polym. Sci. Chem. Ed., 18, 2221 (1980).
9. Potts, J.E., Clendinning, W.B., Ackart, W.B. and Niegisch, W.D., Polym. Preprints, 13 (2), 629 (1972).

QUESTIONS AND ANSWERS

DR. ALBERTSSON: Ann Christine Albertsson. First, I would like to thank you for your citation. Then I would like to mention that you take up here Potts, his publication about this dotriacontone and hextriacontone component.

And I discussed this with Potts in 1972 since I had made similar, and I got the same results as he did, I got more channels in the nuclear proton, but I got no weight loss of the hextriacontone.

But I didn't believe that there should be a limit between the C32 and the C36. So I tried to contact the companies that made them and tried to find out if there were any toxic things and so forth. Of course, they did't answer me.

But instead, as you know, in the literature, you can find the report about C44 from Heinz and Alexander. I don't remember the year and date. I've shown that is biodegradable.

As I mentioned before, I have also had a lot of discussions with Suzuki and his group, which started in 1975, and he has published 1980, and suggested that biodegradation was dependent on the availability of the substrate and not affected by the molecular weight.

He examined the microbial growth of polyethylene oligomers and especially the effect of dispersion on the microbial rate, and hfe had mixed it with a softener and treatments and so forth, and he showed a degradation of a molecular rate of 600 to 800. And so I don't believe in the limit.

DR. SCHWAB: I don't either, but it gets tossed around. Let's face it. I mean, you hear this all the time with the magic 500. You see it in journals. You see it everywhere. So I was curious to see where that is.

I had no intention -- or I couldn't possibly believe that things would stop at 500 molecular. That would have been a real find, the 500 molecular.

DR. ALBERTSSON: Well, if anyone can tell me anything about how they make this paraffin, C32 and C36, I will be very happy, and also know if there is some special additives or something. I have seen this very big difference between no weight loss and small chance. And if I can, I will make a picture and publish -- not in Swedish; this time in English. [Laughter]

DR. SCHWAB: Thank you.

DR. CARLSSON: Carlsson, Research Council of Canada. I am pleased to see somebody actually using a polymer where we know exactly what is in it; what the catalyst residues are.

There was one point I missed. Did you say what your oxidant level was in your formulations?

DR. SCHWAB: Yes, it was between 400 and 500. It was a mix. I think it was like 250 parts of the primary oxidant which is a phenol, and then the remainder which is about 150 parts of a secondary oxidant which was probably a phosphate.

DR. CARLSSON: That was constant overformation?

DR. SCHWAB: Yes. It was the same polymer.

DR. CARLSSON: Okay. I think you have got a problem in just using the carbonyl index, especially with the Ecolyte material, where you are probably losing as much carbonyl as you're forming; just the way that stuff has to act. So you really need a much more definitive method to --

DR. SCHWAB: Okay. That hasn't exactly been what I have seen.

DR. CARLSSON: Tony Redpath might go after you on that.

DR. SCHWAB: Okay.

DR. CARLSSON: The change in end group concentration is really interesting. We have looked at a whole range of polyethylenes and found no effect of branching to about the level you're working at in your rate of oxidation.

How did you measure your end groups, is the critical question?

DR. SCHWAB: We did it by Carbon 13 and by actual count. I didn't do the work. It was done by a colleague of mine, so you're...

DR. CARLSSON: Okay. This is solution NMR. Did you throw away a lot of absorbable gel? is the critical question.

DR. SCHWAB: No. It isn't really solution anymore. It is almost like a gel form. What we do is put enough solvent in there to soften the polymer to get it out of the -- you know, to overcome the Tm so that we get a nice result.

DR. CARLSSON: I'm just worried that you've got cross-linking and gel formation. they are not showing in the NMR.

DR. SCHWAB: Could be.

DR. CARLSSON: One really little question. Nobody recently converted all that are crushlined to photodegradables. I just wondered why. What is the logic behind that?

DR. SCHWAB: We've got to be, you know, market-driven. I guess that is the easiest way to say it. You're not going to beat them, so you better join them.

DR. CARLSSON: Okay.

DR. SCHWAB: It is a lot easier to do that than to convince somebody that the Arizona tests are good, anyway.

DR. NARAYAN: Ramani Narayan, Michigan State University.

In your biodegradation shake flask experiments, was cellulose the sole carbon source? Were the cells still viable after the experiment? And if they were viable, then part of that carbon and cellulose had to end up in the biodegradable cells. And I see you have converted all of the cellulose into CO(2) --

DR. SCHWAB: Basically, that's correct. We got 98 per cent. I don't know whether, you know, that's and error or not. The answer is yes, cellulose was the only carbon source as was --

DR. NARAYAN: How did the cells survive then? In part of that carbon, it needs first viability of cells, too, right? Isn't that correct?

DR. SCHWAB: That's correct. I will have to --

DR. NARAYAN: If all of the cellulose is converted to CO(2), the cells should have died then.

DR. SCHWAB: I can't --

DR. NARAYAN: Well, but you see what I'm getting at?

DR. SCHWAB: Yes, I know what you're getting at. That when it ran out of food, something should have happened; they should have died.

DR. NARAYAN: The viability part of that carbon is to go to cell mass or biomass things. I'm just confused at the CO(2) numbers that your're getting.

DR. SCHWAB: Again, I'll beg off that because I didn't run that test, but I'll definitely go ask them about that one.

DR. GILMORE: I'm David Gilmore, University of Massachusetts.

Looking at this Pott's data thing. We are working in an enormous vacuum here. The only way that these compounds are getting into the cell is by specific transport systems that the microbes have.

I known very little or nothing about how these particular transport systems work, what their substrate specificity is, and so forth. Until we know something about that, it is very hard to look at these 500 molecular weight numbers and get any sense of what they mean.

Another thing with your data is that you've got a certain number of carbonyl groups and these low molecular pieces that have been transported. You're no longer looking at just the alkanes that Potts looked at, you're looking at modified carbon chains, and they are going to have a different specificity of transport into the bacteria. So there is a lot we don't know to even begin thinking about what data means.

DR. SCHWAB: Oh, Yes, I agree.

DR. LOOMIS: gary Loomis, Dupont. You were comparing all of these various photo options, so that they are adding on a weight per cent basis, right?

DR. SCHWAB: Low. Yes, you're right.

DR. LOOMIS: And also, I think -- I'm not sure why. There are very few people that seem to be looking at anaerobic digestion, since it thereby seems to be aerobic.

Have you considered doing it with anaerobic and monitoring methane rather than CO(2) or --

DR. SCHWAB: Yes. Yes, we have, but I have not done it.

DR. GUILLET: Guillet, the University of Toronto.

First, I'd like to congratulate you on doing a very nice piece of work. Very few people characterize polymers that they use as well as you have and given as much structural data about the chemistry.

I'd like to make two comments.

The first one is that it has never been a problem to make unstable polyolefins. In fact, for the first century of polyolefin manufacture, the real problem is to get one stable enough to use in any practical application. Any kind of metal salt practically that you can think of will make polyolefins degrade photochemically and thermally.

DR. SCHWAB: You're absolutely right.

DR. GUILLET: And so I hesitate to use the words "degradable plastic" about the carbonyl compounds that we develop because we really were trying to develop something with a controlled rate of degradation.

The problem with most metal salts is that they, in fact, are both thermal and photochemical oxidants, and many people have reported the fact that you have -- you can't store the master batches very long if you use metal salts because they won't be any good in a month or two, or they will cause problems during excursion.

So the real problem chemically in terms of this type of product is to be able to control the thing very carefully so it does exactly what you want and not something else.

I would also like to comment on the magic 500 number, so -- the Potts 500, and it is based strictly on the fact that when you increase the carbon, number of hydrocarbons, at some point you start off with liquids, and at magic number of 500 you suddenly have a crystalline solid.

It is quite obvious that bacteria had a much greater difficulty in attacking a crystalline solid than they do a liquid.

On the other hand, when you thermally or photochemically degrade linear low polyethylene, you don't have linear hydrocarbons; you have branch hydrocarbons, and of course they are liquids at much higher molecular weights.

And so the magic number has nothing to do with the capability of these molecules to be degraded, but only with the question of what state they are in at the time they are attacked by bacteria.

I really believe, as I think you do, that you could probably degrade any hydrocarbon as long as the bacteria can get to it. It has to chomp along on the ends of the chain.

Now, if you have a high molecular weight, you have got very few ends, and therefore, there are very few places to attack. Also, if you have oxidized the polymer, it becomes more hydrophilic and --

DR. SCHWAB: Exactly.

DR. GUILLET: But I think we ought to get rid of this straddling of the 500 limitation. It was based on a lousy test and an improper experiment, in my point of view. Thank you.

DR. SCHWAB: Thank you. I put that in as tongue in cheek. I hope I -- I have realized, and I figured that is, you know, that is -- it is a good way to bring it out and maybe, you know, we will get rid of the Potts 500.

THE CHAIRMAN: Potts 500 uses the ASTM --

DR. SCHWAB: Yes. Again --

THE CHAIRMAN: -- their own test, which is considerably different than yours.

DR. SCHWAB: I should really probably go back and see if I can get bacterial growth in it.

DR. SCOTT: Scott, Aston. I am not going to disagree with what Jim just said, and he's absolutely right, that if you put transidual metal lines into polymers they will degrade it under any conditions.

what I do want to impress, though, is that the Plastigone system does not use a metal salt. It is often called a metal salt, but it is not a metal salt; it is metal complex. And there is a world of difference.

When you look at the prechemistry to the very sharp oxidation process that you were talking about there, because these things are anti-oxidants, not pro-oxidants, and there is a very big difference there.

You can't really put transidual metal lines as such into polymers and get way with it for any length of time because it causes degradation during the processing operation.

DR. SCHWAB: Thank you.

DR. SHALABY: I do agree actually with the importance of having crystalline material versus liquid and how this relates to this magic 500.

And what I'm suggesting, that you actually do some thermal analysis work with your material to make your results even more heavy, to see how do they perform in terms of, do they crystallize?

# THE EVALUATION OF DEGRADATION RATES IN PHOTODEGRADABLE ECOLYTE POLYSTYRENE

P.Quan, M.Lemke, A.Sinclair, I.Treurnicht, A.Redpath

ECOPLASTICS LIMITED
518 Gordon Baker Road
Willowdale, Ontario M2H 3B4

The photochemistry and photophysics of polymeric systems are subjects that have attracted much study over the past three decades(1). In particular, the need to stabilize polymers against the effects of exposure, to outdoor sunlight, has generated a considerable body of published information on the mechanisms by which both photodegradation and stabilization take place(2). The effects of commercial stabilizers have been well quantified(3), however, the more recent interest in enhanced degradation of polymers has not yet produced a comparable body of data on the field performances of either prodegradant additives or polymeric systems. It is the purpose of this paper to discuss the evaluation of photodegradation rates in polystyrene samples which have been blended with a commercial prodegradant copolymer, Ecolyte PS2005.

The evaluation of polymer degradation rates is complicated by a number of factors. These include the variety of polymer properties which can be measured, the range of radiation exposure conditions, and by the absence of an accepted definition of degradation. There are, broadly speaking, two types of measurements which may be performed: a) measurements of molecular properties (e.g. chemical changes such as oxidation or molecular weight changes) as a function of exposure, either outdoors or to an artificial UV light source; b) measurements of macroscopic physical properties (e.g. tensile strength or brittleness) as a function of exposure.

The basic photochemistry of vinyl ketone/styrene copolymers, of which Ecolyte PS2005 is a commercial example, have been well studied. The general mechanism of their photodegradation[4], as well as quantum yields, for some systems,[5], have been reported. However, no data pertaining to changes in physical or molecular properties as a function of exposure has been published. Based on this earlier laboratory work, a study was initiated to follow oxidation rates, via infrared spectroscopy, and molecular weight changes (via solution viscometry) of samples which were exposed in either in a laboratory accelerated UV weatherometer or outdoors. In addition, the mechanical properties of the outdoor aged samples were monitored both qualitatively (i.e. appearance) and quantitatively, via a tumbling block friability test. For this study, foamed polystyrene containing various levels of Ecolyte PS2005 prodegradant was investigated. As well, the question of degradability of recycled polystyrene, containing different levels of recycled Ecolyte PS2005 copolymer, was also addressed.

## EXPERIMENTAL

Ecolyte PS2005 vinyl keton/styrene copolymer was obtained from Enviromer Enterprises and used as received. It has a melt index of

3, and a nominal viscosity molecular weight of 350,000. It was blended with Dow 685D crystal polystyrene at let down levels of 0, 5 and 10%. Foam sheets, 2 mm thick, containing low levels (<1%) of a brown pigment were produced on a commercial foam sheet line and fabricated into hamburger boxes. The processing conditions and the presence of a small amount of pigment ensured that these samples were a fair representation of the type of commercial product for which test data is most relevant. Samples were tested either as is, in half hamburger box units, or as strips, approximately 2.5 cm x 3 cm mounted into 35 mm photographic slide holders. Since the IR spectra from foamed PS sheets proved unsatisfactory, 15 mil thick oriented crystal PS sheets, of the same base resin with 0, 5 and 10% Ecolyte loading, were also prepared.

The slide mounted samples were exposed, to UV-B radiation, in an American Ultraviolet NS-1200 Weatherometer equipped with a 1200 watt medium pressure mercury lamp with a Corex-D jacket. Samples were mounted 16 inches from the lamp and the aging was conducted at a constant temperature of 50°C. Samples were also exposed outdoors in accordance with ASTM designation D1435-85, "Standard Practice for Outdoor Weathering of Plastics"[6]. Slide mounted samples were placed in south facing racks inclined at 45°C. (Since Toronto is at latitude 44° this roughly corresponds to an at-latitude exposure). In addition, half hamburger box samples were mounted in nylon mesh net bags (1 cm mesh size) on racks similar to those described above. Both indoor and outdoor samples were removed for analysis at various intervals, typically 400-600 hours for indoor samples and 20-25 days for outdoor samples.

Slide mounted oriented PS samples, aged both indoors and outdoors, were analyzed first by Fourier Transform Infrared Spectroscopy (FTIR) using a Nicolet 5DXC spectrophotometer These samples, as well as the foamed PS samples, were then analyzed via solution viscometry. Viscosity measurements were obtained by dissolving a 0.075 g piece of sample in 25 ml of toluene and then running solution viscosities on a custom built capillary viscometer[7]. Intrinsic viscosities were calculated via the Solomon-Ciuta equation. Viscosity average molecular weights were calculated from the intrinsic viscosities using the Mark-Houwink equation:

$$1) \quad [n] = K\,M^a$$

where K and a values, for polystyrene in toluene, were found to be $1.762 \times 10^{-4}$ dl/g and 01673 respectively[9]. Comparing the initial molecular weight, $M_o$, with that of a degraded sample, $M_t$, permits the calculation of the number of chain breaks per molecule as follows:

$$2) \quad \text{Chain Breaks} = \frac{Mo}{Mt} - 1$$

Half hamburger box samples, aged outdoors, were analyzed qualitatively by a subjective description of the sample appearance. In addition, a limited number of these samples were subjected to physical testing. A variation of the ASTM C421-83 Tumbling Friability test was used[9]. The test samples were cut into twelve 2.5 cm x 2.5 cm pieces and placed in a box containing 24 (1.9 cm x 1.9 cm) oak cubes. The box was tumbled for 600 revolutions and the weight loss of the abraded samples was then calculated.

For the recycling study, blends were prepared using conventional crystal polystyrene extruded sheet and degradable polystyrene materials (for these experiments, the degradable materials chosen were foam and crystal polystyrene containing 10% Ecolyte PS2005). Samples containing 0, 5, 10, 20 and 100% of the degradable material were prepared using a two roll mill. (These samples would correspond to Ecolyte PS2005 loading levels of 0, .5, 1, 2 and 10%). The mixtures were manually mixed on the mill for ca. 5 minutes at a temperature of 180-190°C. They were then pressed on a Carver Press at 200°C to form clear sheets 10 and 20 mil thick. These sheets were cut into 1 cm x 3 cm samples and mounted for exposure and evaluation as described previously.

The matrix of samples tested and condition sued is outlined in Table 1 below:

TABLE 1

Test Method

| SAMPLE | Infra-red | Molecular weight | Physical Appearance | Tumble Test |
|---|---|---|---|---|
| Oriented PS | UV-B aging | UV-B aging | - | - |
| | Outdoor aging | Outdoor aging | - | - |
| Foamed sheet | - | UV-B aging | - | - |
| | - | Outdoor aging | Outdoor aging | Outdoor aging |

RESULTS AND DISCUSSION

A generalized reaction scheme for the photodegradation of (vinyl ketone/styrene copolymer)/(styrene polymer) mixtures can be presented as follows(10):

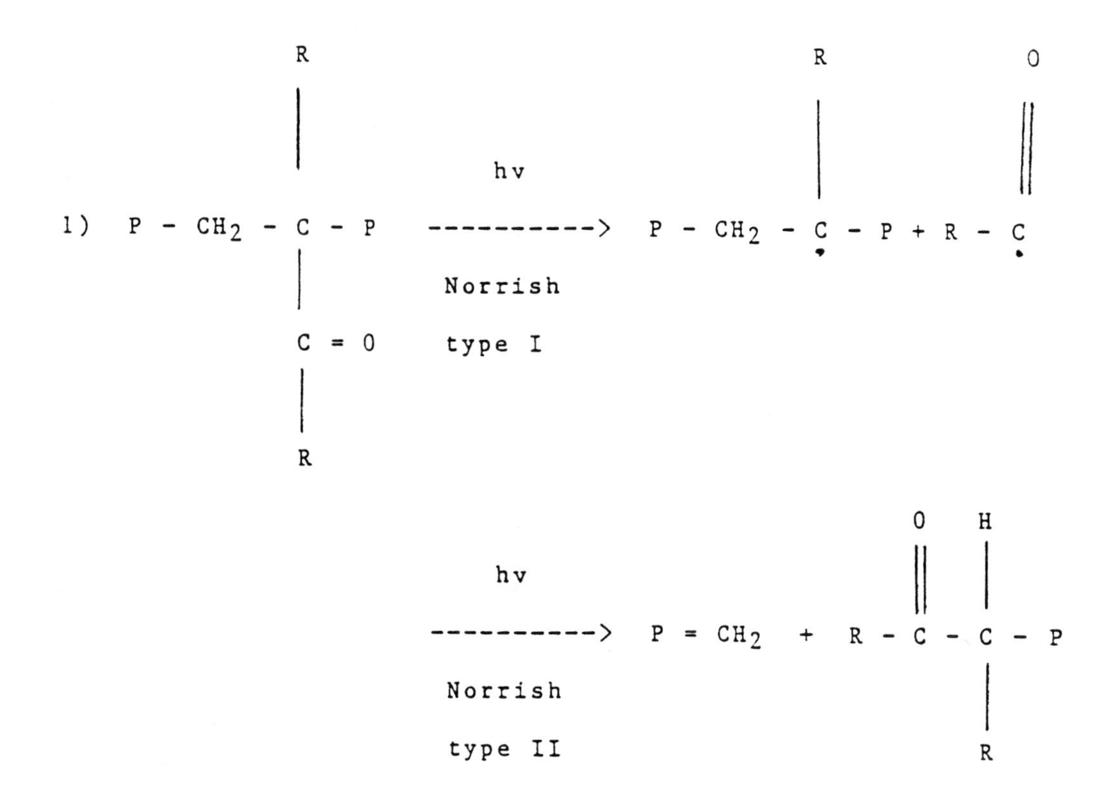


If P represents a polymer chain and R can be any radical, then:

2) $R - \dot{C} = O$ + PH -----> $R' - CH = O$ + $P^{\bullet}$

3) $P^{\bullet}$ + $O_2$ -----> $POO^{\bullet}$

4) $POO^{\bullet}$ + PH -----> POOH + $P^{\bullet}$

5) POOH --hv--> $PO^{\bullet}$ + $^{\bullet}OH$

6) POOH + $R^{\bullet}$ -----> $POO^{\bullet}$ + RH

7) $\cdot OH + PH \dashrightarrow P^{\cdot} + H_2O$

8) $PH^{\cdot} + PH \dashrightarrow POH + P^{\cdot}$

disproportionation

9) $PO^{\cdot}$ -----------------> $P^{\cdot}$+ aldehyde or olefin + Ketone

10) $P^{\cdot} + P \dashrightarrow P - P$

11) $POO^{\cdot} + P^{\cdot} \dashrightarrow POOP$

12) $POO^{\cdot} + POO^{\cdot} \dashrightarrow P - (C = O) - P + PO + O2$

This reaction scheme illustrates the basis for the two traditional indices of degradation that were used here: the carbonyl index, measured by infrared spectroscopy, indicating the presence of oxidation species (ketones, aldehydes, acids, esters, etc., see reaction 9 and 12) and chain scission, measured by molecular weight changes (see reactions 1b + 9).

Infrared Measurements

We chose to monitor oxidation species via the carbonyl infrared absorbency in the 1700-1740 $cm^{-1}$ range. The infrared absorption spectra for the various blends of crystal polystyrene with Ecolyte PS are shown in Figure 1 and Figure 2 for indoor and outdoor aging

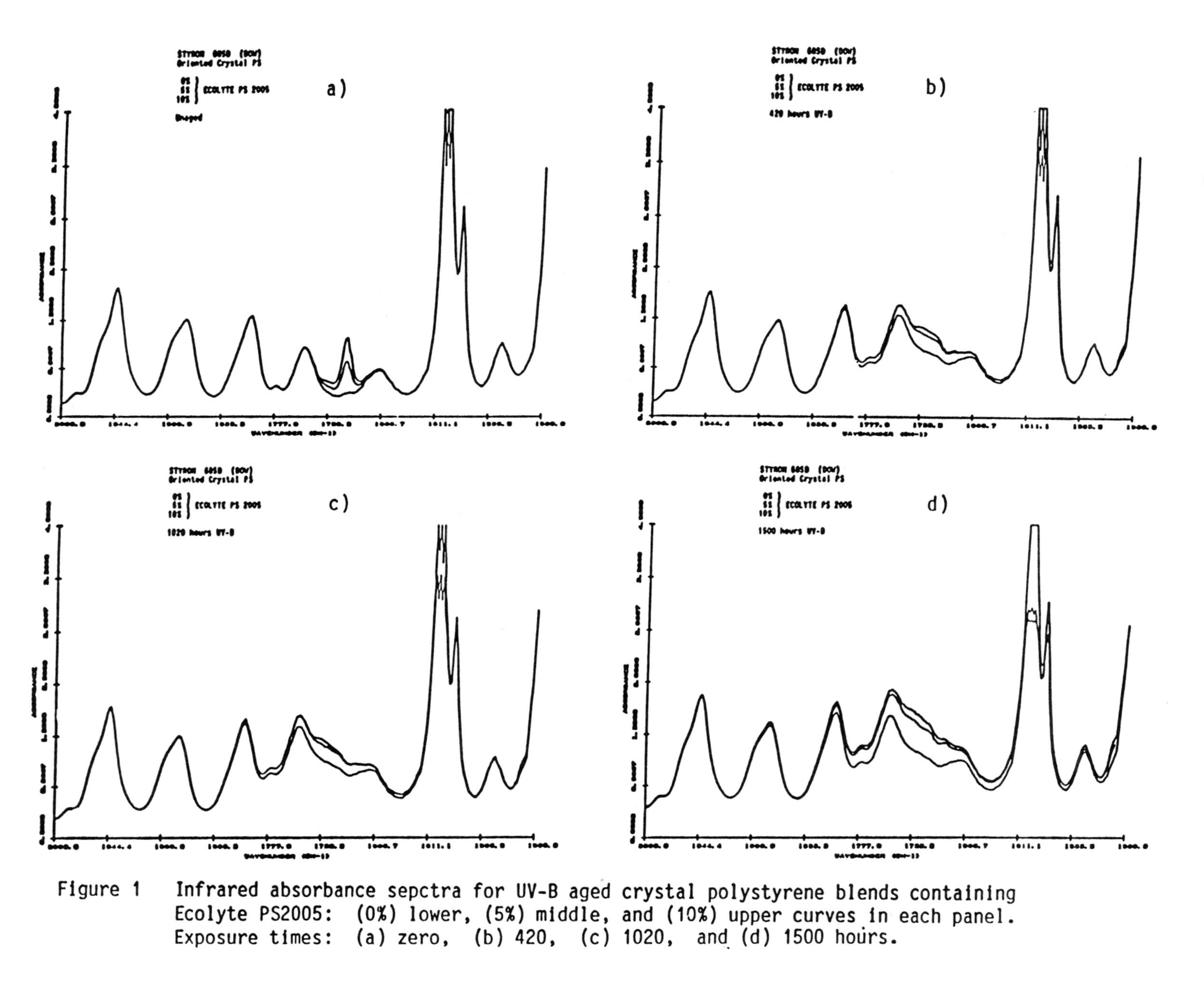


Figure 1 Infrared absorbance sepctra for UV-B aged crystal polystyrene blends containing Ecolyte PS2005: (0%) lower, (5%) middle, and (10%) upper curves in each panel. Exposure times: (a) zero, (b) 420, (c) 1020, and (d) 1500 hours.

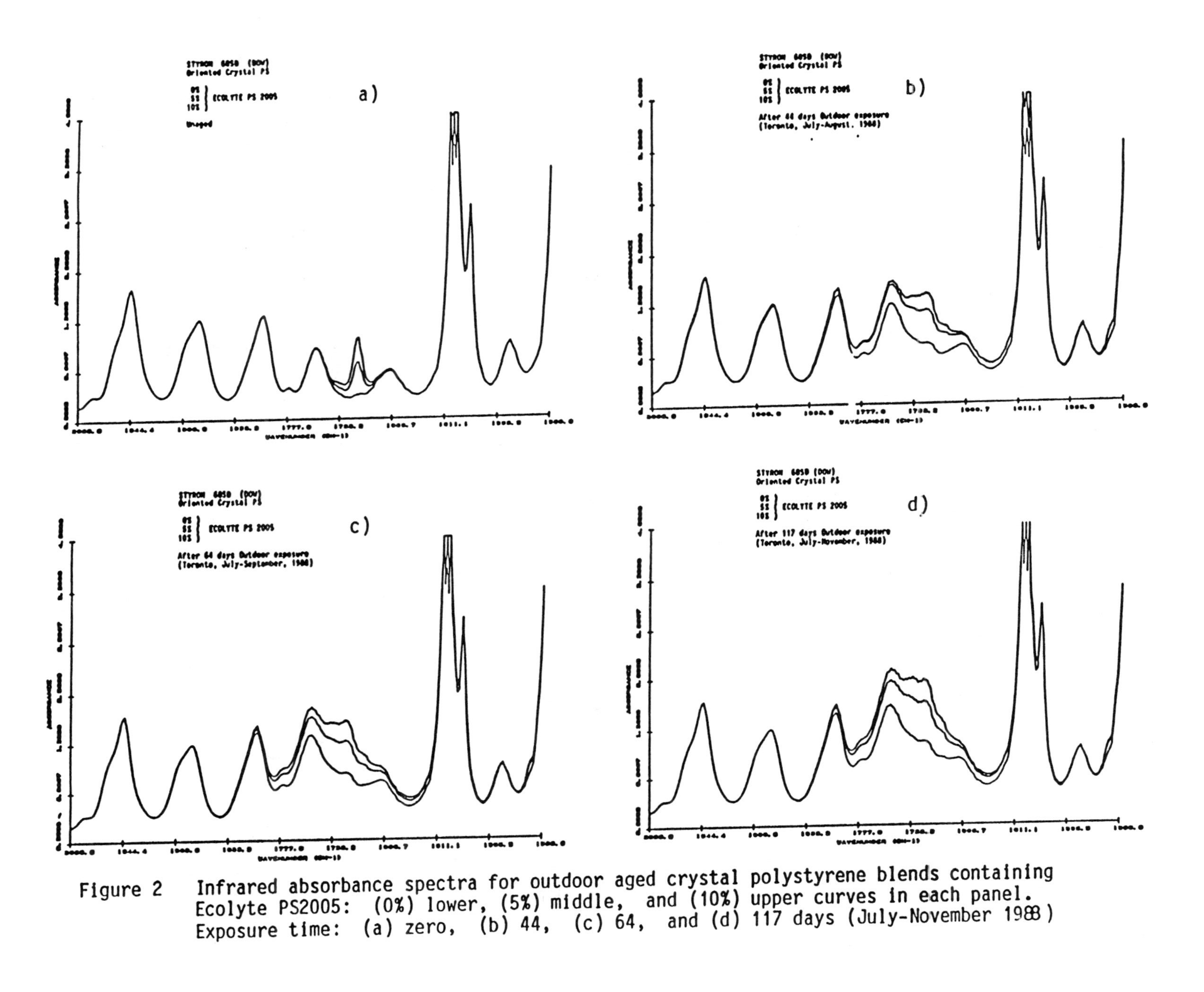


Figure 2  Infrared absorbance spectra for outdoor aged crystal polystyrene blends containing Ecolyte PS2005: (0%) lower, (5%) middle, and (10%) upper curves in each panel. Exposure time: (a) zero, (b) 44, (c) 64, and (d) 117 days (July-November 1988)

respectively. The time zero spectra provide a calibration check for Ecolyte PS loading levels and clearly show the initial carbonyl absorbency at 1702 $cm^{-1}$. As expected, in both indoor and outdoor aging the entire envelope of absorbency, due to oxidized species, grows with increasing exposure time. An unexpected observation is the apparent lack of a linear correlation between the intensity of the overall carbonyl absorbency with exposure time. Increasing levels of prodegradant do not result in corresponding increases in carbonyl index. This is most pronounced for the indoor aged samples. A quantitative carbonyl analysis would involve either integrating the area under the full carbonyl region of the division of the absorbency region into areas representing the different carbonyl containing species present. We have limited our comments to the qualitative observations made above. A more comprehensive analysis is being deferred to a later publication.

## Molecular Weight/Chain Break Measurements

The results of the solution viscosity derived molecular weight measurements are presented in Tables 2 - 5. These tables include data for both indoor and outdoor aged samples of oriented crystal and foamed polystyrene samples. The same data, presented as a plot of chain breaks vs. exposure time is presented in Figures 3-6.

It is evident from Figures 3 and 4 showing the degradation profiles of the oriented crystal polystyrene samples that there are significant differences in the rates of chain scission for the different Ecolyte PS loading levels - differences that were not

Table 2

Molecular weight/chain breaks vs exposure time for 15 mil oriented crystal polystyrene - UV-B aging.

| Exposure Time (hours) | 0% Ecolyte PS Mol.Wt. | 0% Ecolyte PS Chain Breaks | 5% Ecolyte PS Mol.Wt. | 5% Ecolyte PS Chain Breaks | 10% Ecolyte PS Mol.Wt. | 10% Ecolyte PS Chain Breaks |
|---|---|---|---|---|---|---|
| 0 | 300770 | 0 | 300770 | 0 | 300770 | 0 |
| 160 | 277010 | .09 | 292590 | .48 | 172410 | .74 |
| 420 | 267040 | .13 | 195350 | .54 | 166370 | .81 |
| 600 | 201080 | .30 | 172190 | .75 | 165740 | .81 |
| 840 | 185330 | .62 | 185990 | .62 | 139160 | 1.16 |
| 1020 | 236120 | .27 | 182040 | .65 | 135220 | 1.22 |
| 1300 | 229640 | .31 | 139900 | 1.15 | 128260 | 1.35 |
| 1500 | 169550 | .77 | 132540 | 1.27 | 109080 | 1.76 |
| 1720 | 163630 | .84 | 146180 | 1.06 | 131210 | 1.29 |

Table 3

Molecular weight/chain breaks vs exposure time for 15 mil crystal polystyrene - outdoor aging.

| Exposure Time (days) | 0% Ecolyte PS Mol.Wt. | 0% Ecolyte PS Chain Breaks | 5% Ecolyte PS Mol.Wt. | 5% Ecolyte PS Chain Breaks | 10% Ecolyte PS Mol.Wt. | 10% Ecolyte PS Chain Breaks |
|---|---|---|---|---|---|---|
| 0 | 300770 | 0 | 300770 | 0 | 300770 | 0 |
| 23 | 251760 | .19 | 156310 | .92 | 126760 | 1.37 |
| 44 | 209450 | .44 | 120600 | 1.49 | 98720 | 2.05 |
| 64 | 187550 | .60 | 109700 | 1.74 | 89080 | 2.38 |
| 79 | 168130 | .79 | 101160 | 1.97 | 81210 | 2.70 |
| 96 | 153200 | .96 | 93250 | 2.23 | 78030 | 2.85 |
| 117 | 136690 | .92 | 93490 | 2.22 | 76950 | 2.91 |

Table 4

Molecular weight/chain breaks vs. exposure time for 2 mm foamed polystyrene - UV-B aging.

| Exposure Time (hours) | 0% Ecolyte PS Mol.Wt. | 0% Ecolyte PS Chain Breaks | 5% Ecolyte PS Mol.Wt. | 5% Ecolyte PS Chain Breaks | 10% Ecolyte PS Mol.Wt. | 10% Ecolyte PS Chain Breaks |
|---|---|---|---|---|---|---|
| 0 | 265990 | 0 | 265990 | 0 | 265990 | 0 |
| 400 | 153320 | .73 | 106680 | 1.49 | 55560 | 3.79 |
| 920 | 113320 | 1.35 | 78570 | 2.39 | 43530 | 5.11 |
| 1560 | 44130 | 5.03 | 32940 | 7.07 | 19620 | 12.56 |
| 2040 | 26900 | 8.89 | 24280 | 9.96 | 17670 | 14.05 |

Table 5

Molecular weight/chain breaks vs. exposure time for 2 mm foamed polystyrene - outdoor aging.

| Exposure Time (days) | 0% Ecolyte PS Mol.Wt. | 0% Ecolyte PS Chain Breaks | 5% Ecolyte PS Mol.Wt. | 5% Ecolyte PS Chain Breaks | 10% Ecolyte PS Mol.Wt. | 10% Ecolyte PS Chain Breaks |
|---|---|---|---|---|---|---|
| 0 | 265990 | 0 | 265990 | 0 | 265990 | 0 |
| 25 | 209930 | .27 | 149410 | .78 | 99670 | 1.67 |
| 46 | 162100 | .64 | 120570 | 1.21 | 81590 | 2.26 |
| 60 | 153160 | .74 | 110520 | 1.41 | 73840 | 2.60 |
| 84 | 137980 | .93 | 104380 | 1.55 | 72550 | 2.67 |
| 97 | 133680 | .99 | 85290 | 2.12 | 56490 | 3.71 |
| 118 | 117900 | 1.26 | 81837 | 2.25 | 60210 | 3.42 |
| 136 | 98280 | 1.71 | 73580 | 2.61 | - | - |
| 154 | 98520 | 1.70 | - | - | - | - |

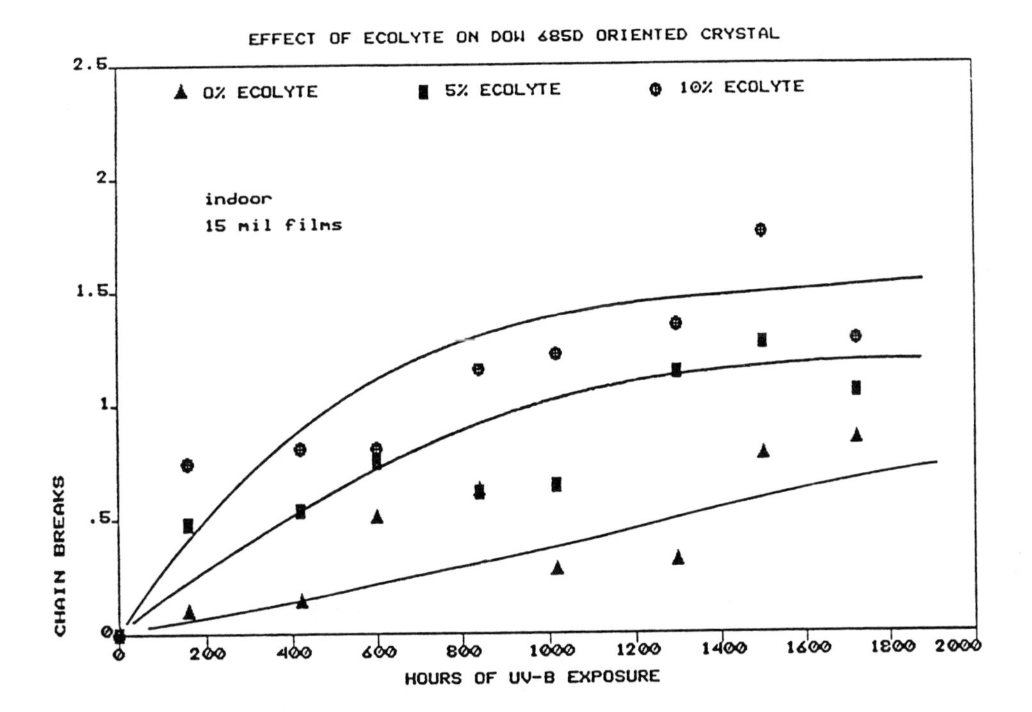


Figure 3    Chain breaks vs. hours of UV-B exposure: crystal polystyrene, 15 mil thick, with 0, 5 and 10% added Ecolyte PS2005

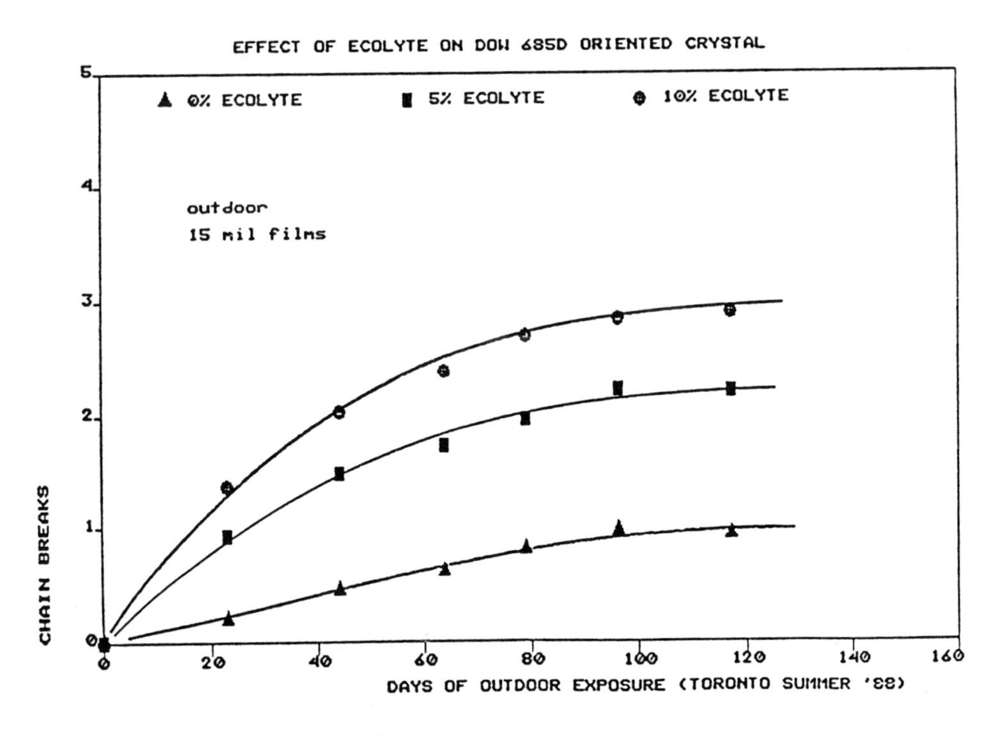


Figure 4    Chain breaks vs. days of outdoor exposure: crystal polystyrene, 15 mil thick, with 0, 5 and 10% added Ecolyte PS2005

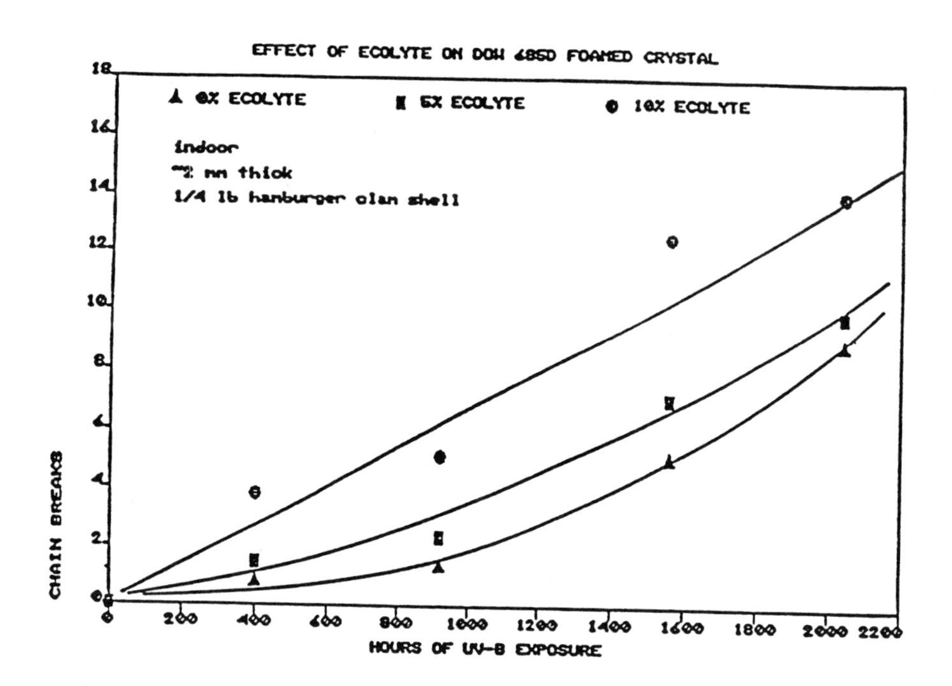


Figure 5 Chain breaks vs. hours of UV-B exposure: foamed polystyrene, 2 mm thick, with 0, 5 and 10% added Ecolyte PS2005.

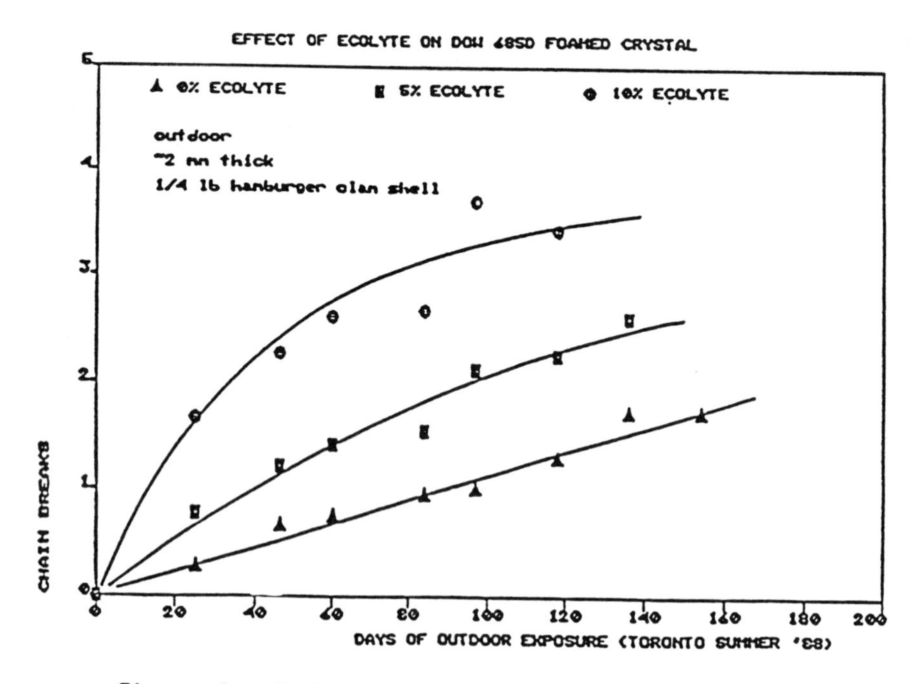


Figure 6 Chain breaks vs. days of outdoor exposure: foamed polystyrene, 2 mm thick, with 0, 5 and 10% added Ecolyte PS2005.

observed in the FTIR curves as discussed earlier. It would appear that either the chain scission process is not paralleled by oxidation or that the measurement of oxidation is in some manner inaccurate. We conjecture here that it is the latter that is true. As equations 1 and 2 in the reaction scheme indicate, low molecular weight carbonyl compounds may be produced in the degradation process. These materials will be sufficiently volatile to escape the samples and thus be lost for purposes of measurement. This explanation is consistent with the observation that the apparent understatement of degradation is greatest for the indoor aging which takes place at a higher sample temperature (50°C) than the outdoor tests (nominally 15-30°C).

Figures 3, 4, 5 and 6 illustrate another potential artifact in the measurement of degradation rates. For the crystal polystyrene samples, the outdoor exposure over 120 days appears to have led to significantly more chain breaks than for the indoor exposure of ca. 1500 hours. The opposite appears true for the foamed samples. The outdoor aged foamed samples show a levelling off in the chain break curve that is not evident in the indoor samples. The indoor aged samples showed evidence of a colored powdery material on the exposed surfaces, presumably highly degraded polystyrene. There was considerably less evidence of this on the outdoor aged foamed samples. This may indicate that the lower apparent rate of chain scission is in fact caused by the ablation by wind and rain of degraded surface polymer in the outdoor samples, leaving behind higher molecular weight residual material (a process far less likely in the solid crystal PS samples.)

Physical Appearance Observations

As discussed earlier, half hamburger box samples were observed during their outdoor exposure in an attempt to describe the physical changes that occur. Some weight loss data was recorded. the observations are reported in Table 6 below:

Table 6

| Days | 0% Ecolyte PS | 5% Ecolyte PS | 10% Ecolyte PS |
|---|---|---|---|
| 35 | Net cutting into bottom edge | Net cutting into all edges | Net cutting into all edges, holes |
| 60 | Surface slightly thinning | Surface thinning | Holes, very thin around sides |
| 90 | Surface thinning | Thinning holes | Broken up, 50% mass gone |
| 120 | Thinning holes around corner, 10-20% mass gone | Thinning, holes, 30% mass gone | 75% mass gone |

Tumbling Block Friability Measurements

Half hamburger box samples which had been aged outdoors for 136 days were tested in a tumbling block chamber. Only samples containing 0% and 10% Ecolyte PS were tested. After tumbling, the weight of the sample was compared to the initial weight of the non-aged sample to calculate a % mass lost due to degradation/mechanical action. Table 7 presents the results

Table 7

Weight loss of degraded foam sheets:

| % Ecolyte | % Mass Loss |
|---|---|
| 0 | 35 |
| 10 | 85 |

While these results are preliminary, they do suggest that this technique agrees with both the qualitative observations (described in Table 6) and the molecular weight data. That the weight loss does not increase substantially, with respect to the weight loss associated with outdoor aging, following tumbling of the outdoor aged samples suggests that most of the highly degraded material had already been removed by wind and rain as proposed earlier to explain the molecular weight data. It is interesting to note from Table 6 that significant weight loss is occurring after approximately 60-90 days for the 10% Ecolyte PS sample and after 120 days for the 5% sample. This would correspond to two to three chain breaks per molecule (see Figure 6) or a molecular weight of about 100,000.

Recycling Measurements

The recycled samples comprising 0, 5, 10, 20 and 100% of the degradable polystyrene feedstock (defined as the 10% Ecolyte PS2005 material) were exposed to UV-B radiation as 10 mil thick clear sheets. The resultant chain break data is presented in Figure 7. (For comparison, the data for 15 mil thick oriented polystyrene sheets was shown in Figure 4.) The 100% recycle sample degrades similarly to the equivalent 10% Ecolyte PS2005 sample. A direct comparison cannot be made as the 15 mil 10% Ecolyte samples degrades slower than the 10 mil 100% recycle sample due to thickness effects. It is equally clear that up to 20% of the degradable polystyrene feedstock has only a small effect in increasing the degradation rate relative to the control. Other samples containing small amounts of pigment (2% $TiO_2$ or 0.4% carbon black) in addition to 10% of the degradable polystyrene feedstock were also tested - results are shown in Figure 8. The chain break scale has been expanded to emphasize the differences. The overall conclusion is that the pigmented samples are considerably more stable than even conventional PS. The final set of samples evaluated the effect of added UV stabilizer. Mixtures of 50% of the degradable polystyrene feedstock with and without 0.3% Tinuvin P (Ciba-Geigy) stabilizer were compared to regular PS controls. The results are plotted in Figure 9 and show both the enhanced degradation of the Ecolyte PS rich sample (equivalent to a normal 5% Ecolyte PS let down) and the enhanced

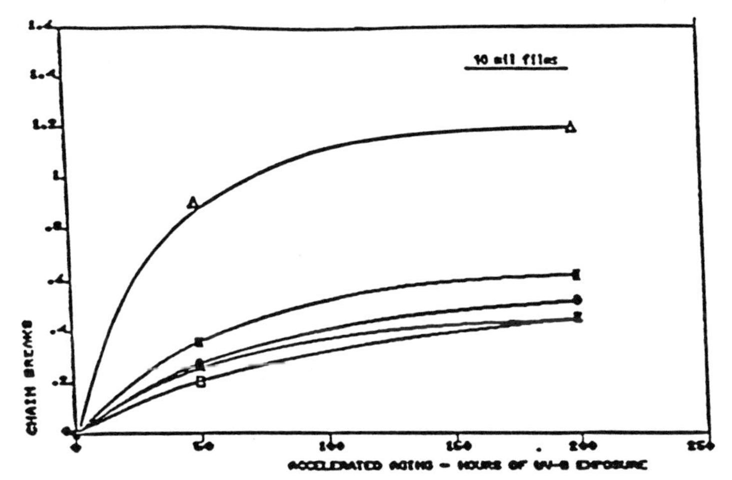

Degradable PS added

- 100%
- 20%
- 10%
- 5%
- 0% (Conventional PS)

Figure 7 Chain breaks vs. hours of UV-B aging: recycled PS with various levels of degradable PS (10% PS2005) added

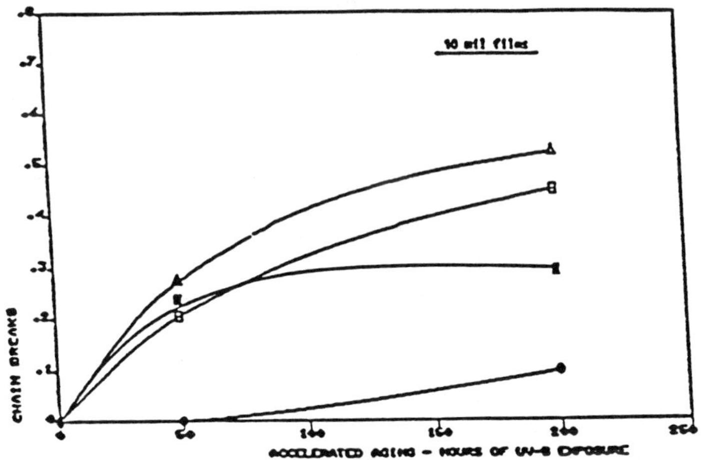

10% Degradable PS
Conventional PS

10% Degradable PS
+2% $TiO_2$

10% Degradable PS
+ 4% Carbon Black

Figure 8 Chain breaks vs. hours of UV-B aging: recycled PS with 10% degradable PS added, with and without pigments.

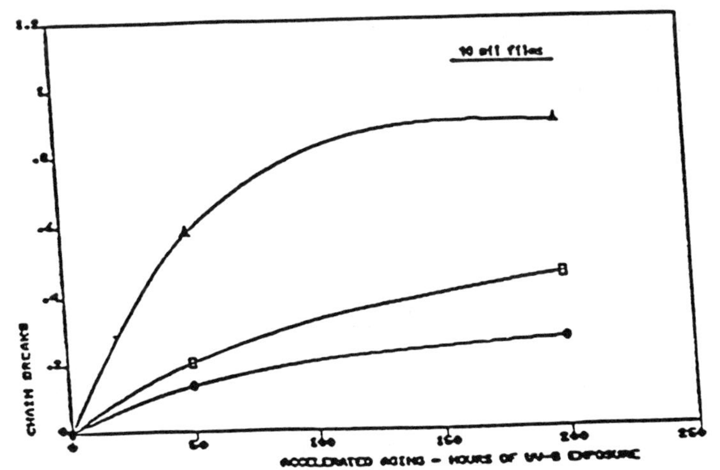

50% Degradable PS

Conventional PS

50% Degradable PS
+ 0.3% Tinuvin P

Figure 9 Chain breaks vs. hours of UV-B aging: recycled PS with 50% degradable PS added, with and without UV stabilizer.

stability of the stabilized Ecolyte PS sample.

## Conclusion

Based on our experiments to date, it appears that molecular weight measurements, obtained via solution viscometry, offer the simplest accurate measurement of degradation rates in polystyrene systems. FTIR data, at least in the carbonyl region, may not accurately track the degradation process. Other wavelength regions (due to peroxy groups, terminal vinyl groups, etc.) need to be explored to test their potential in this regard. For crystal polystyrene, UV-B indoor aging appear to offer an acceptable correlation with outdoor exposures. However, in view of the experience of other researchers as well as our own with polyolefin degradation testing, it seems likely that UV-A aging will become the method of choice. Some of the experiments described within are currently being repeated under UV-A aging conditions. In addition, the effect of lower exposure temperatures is being examined.

the question of evaluating the outdoor degradation of foamed PS, is still problematical. Changes to the samples, such as would occur under field conditions, reduce the correlation between indoor and outdoor testing. It would appear that there is still no acceptable substitute for real time outdoor testing.

REFERENCES

1. See, for example, J.E. Guillet, "Polymer Photophysics and Photochemistry", Cambridge University Press, Cambridge, (1985)
2. See, for example, "Polymer Stabilization and Degradation", P.P. Klemchuk, Ed., ACS Symposium Series 280, American Chemical Society, New York (1984)
3. Carlsson, D.J., Garton, A., Wiles, D.M.," Developments in Polymer Stabilization", vol. 1, G. Scott, Ed., Applied Science, Banking (1980)
4. M. Heskins, T.B. McAneney, and J.E. Guillet in Ultraviolet Light Induced Reactions in Polymer, S.S. Labana, ed., American Chemical Society, New York, (1976)
5. E. Dan and J.E. Guillet, Macromolecules, 6, 230, (1973)
6. "Standard Practice for Outdoor Weathering of Plastics", ASTM, D1435-85
7. T. Kilp, B. Houvenaghel-Defoort, W. Panning and J.E. Guillet, Rev. Sci. Instrum., 47, 1496 (1976)
8. "Standard Practice for Tumbling Friability of Polystyrene Foams", ASTM
10. J.E. Guillet in "Stabilization and Degradation of Polymers, D.L. Alhara and W.L. Hawkins, Ed., American Chemical Society, New York (1977)

QUESTIONS AND ANSWERS

DR. OWENS: Owens, Illinois. We have been admonished to pay attention to testing that we think might be liable in the development of our consensus, and so I'm here on my horse to actually come to the defense of infrared spectroscopy.

It is such a simple, quick test, and I can understand some of the confusion that was generated both in the last talk and in this talk.

I think the problem comes from two sources: No.1, you didn't start looking at the polymer fast enough. Both with ECL and with the Ecolyte what happens is you start off with a very sharp peak, like what you showed there, and what happens initially is that sharp peak decreases.

DR. REDPATH: Yes.

DR. OWENS: And it virtually disappears.

DR. REDPATH: Yes.

DR. OWENS: And then what appears now after that in the next sequence is a broadened peak, two peaks really - the middle peak is now gone - and then that thing increases monotonically.

But the other problem is that the reason we switched from the UVB to the UVAs is because when you use the UVBs, you develop a very clear-cut doubling, and that doesn't appear either with our work from Whitman, Arizona Outdoors, Ann Arbor, Michigan Outdoors, or with the QV -- the QV -- with the "A" lamps.

DR. REDPATH: Yes.

DR. OWENS: The "A" lamps duplicate what we see outdoors and the shape is different. Again, all that I would is ask is, let's not dump the spectroscopy yet because, like I said, it is very quick and it offers a lot of other fringe benefits, too.

With the low levels that you're working at, it helps you to tell whether or not you have got a good mixture.

DR. REDPATH: Oh, and you found this to be --

DR. OWENS: We found this to be absolutely key.

DR. REDPATH: We use that -- the spectra that I showed you there times zero - zero, five and ten - that's our routine measurement for analyzing the amount, the let-down material, because we find that most bag manufactures don't put in the amount that you tell them to.

So from a purely analytical viewpoint, a very powerful tool.

I would agree, there's a tremendous amount of information available from IR. If anything, I think looking at that, I guess I shy away and think there is almost too much information to be useful as a quick screen to tell me what's going on.

Maybe I'm being too lazy. Maybe we need to sit down with enough information about mechanisms and start dividing that up into an aldehyde region, a ketone region. We can pull out a lot more detailed information.

I certainly don't dismiss IR. All I wanted to do was point out that there are potential artifacts. If I simply looked at that and said, gosh, it isn't growing anymore, I would miss the fact that I'm probably losing volatile compounds; not that the molecular weight is perfect because, as I said, for outdoor, I may be missing -- it is not a volatile compound but it is fragments that fall off.

I guess I don't advertise any one of these as perfect, but both of them as being useful.

DR. OWENS: I will make one more point. The two months that you exposed your -- the people might say, well, gee, you should have seen that problem with Arizona. It should have gone slower, it was outdoors, you should have been able to see this developing.

You just happened to pick the two months in Arizona that were all-time record highs ever --

DR. REDPATH: Mm-hmm.

DR. OWENS: --for temperature. [Laughter]

And we had samples in Arizona which -- we measured the temperature of the sample, not the amine. Our samples actually reached 142 degrees; they were hot to the touch.

I know that sounds incredible, but there were also Corvette bumpers there that I backed into with shorts on. I burnt my leg. [Laughter]

DR. MITRA: Sam Mitra, from 3M. Tony, you didn't mention the compatibility with Ecolyte polystyrene with the Dow resin.

Is there any morphological etymologation to see where the concentrate limits are, that you can mix them?

DR. REDPATH: No, we haven't done any of that data. As I said, the Ecolyte material is a crystal polystyrene itself but melted into X3; nominal molecular weight, 300,000. It is very, very similar. It is made by a very similar process to the Dow, but we have not done any testing on morphology.

DR. CURTO: Nick Curto, Owens-Brockway. Tony, I'm wondering how you did your molecular weight analysis in terms of when you have a solid polystyrene sample and you mechanically work it into a powder so you can resolve it in toluene, that act itself generates quite a number of broken bones. This is something we have quantified to our own level of satisfaction.

How did you prepare your solvents?

DR. REDPATH: Okay. Our samples were prepared, I guess, in the age-old way of polymer ID measurement. We put them in a -- you know, in a volumetric flask. You know, you put about .3 grams in, made it up to the volume, let it sit for a few hours, shook until it was dissolved.

I should -- we did -- so there may have been some bond breaking. Again, at the molecular weights we are looking at here, I don't think we are seeing much change.

I would like to point out, we've since learned from some comments from colleagues at Dow by the ASTM Committee that we use toluene as a solvent. It is an excellent solvent for polystyrene, but clearly what we are dealing with here is an oxidized polystyrene, and it appears that THF is probably a better solvent. And again, we are looking at changing to that. That's a change that we will make.

Some of the curving-over in those plots may be attributable to the change in polymer with time, essentially causing a change in the nonvocane and alpha values that one should use. It is early days in the measurement of this. Those are first order and second order corrections that we will get into.

DR. SCHWAB: Fred Schwab, Mobile. Just a comment.

When you are dealing with foams, you have basically picked a UV. We have done a lot of testing in the foam area on these similar systems, and what you find is you are correct; it is an inflation process, it is a surface oxidation. Then if you've got wind and rain and other phenomenon around, the surface blades and then you get -- you actually watch the samples get thinner as you degrade.

We found a big difference when we tested in a quiescent system like a weatherometer and out in the field, because the weatherometer data never showed what you saw in the field. The reason was unless you had water sprays and other things going on in that weatherometer, you got quite different data.

The other thing is, if you would section your samples and then run your molecular weight say as a surface versus an interior, you will find that surface is a very, very low molecular weight.

DR. REDPATH: We did that. We put a sample up , held it with a thumbtack so it couldn't move around, aged it for 60 days, and took a section off. We basically took a camel hair brush and swept off the surface material, and then we also sectioned out as thin a section as we could of the back.

I wouldn't agree that it was completely UV, okay? What we saw on that was again, nominal 260,000 molecular weight start.

The top was 10,000/20,000, a meaningless number. It depended on how hard you swept. The back was down around 150,000.

So there had been some light penetrating through to the back. But, yes, it is a process that occurs from the top down.

DR. SCHWAB: there is a big difference in how you test, you know, weatherometer or otherwise. It makes a difference.

DR. REDPATH: Yes.

DR. SCHWAB: You've got to have some mechanism to get the surface cleaned off.

DR. REDPATH: Agreed.

THE CHAIRMAN: Thank you.

# THE OUTLOOK FOR ENVIRONMENTALLY DEGRADABLE PLASTICS[1]

Michael N. Helmus, Ph.D.

Harbor Medical Devices, Inc., Boston Massachusetts

Social awareness of litter and the growing problem of waste disposal has been driving legislation mandating the use of environmentally degradable plastics, particularly for packaging materials and garbage bags. This legislation is, in turn, driving the development of new resins and additives for use in packages that mitigate deleterious effects on the environment. Materials for degradable applications must meet all the design, processing, and safety criteria of traditional products. In addition, they must not be accidentally mixed with recyclable plastic materials. Key technology approaches to plastics degradation include photochemical action, microbiologic attack, physical stress, and chemical attack. The four applications that appear to hold the greatest potential for degradable plastics are trash bags, personal care products, agriculture mulches, and encapsulated agricultural chemicals. Six-pack container rings and some trash bags are currently available in degradable formulations. Other highly visible plastic waste includes food containers, but food applications will require longer development times due to FDA requirements for food contact applications.

## INTRODUCTION

Much effort in the past has been to stabilize plastics and make them more resistant to ultraviolet light and other environmental conditions that tend to make them degrade. The army had a major program to determine which plasticizers would not promote fungus growth in tropical environments (16). Ironically, the success of these efforts has made the disposal of plastic waste a major problem:

- o Plastics that do not degrade in the environment may result in litter, barriers to the degradation of other materials in landfills (e.g., garbage contained in plastic bags), and potential health threats to wildlife.
- o Recycling and incineration are applicable only to materials that can be - and are likely to be - collected. In addition, the separation of plastics from other waste products for recycling or incineration has been both costly and time-consuming; also, when improperly incinerated, some plastics can emit toxic fumes.

Plastic materials that degrade into environmentally innocuous end products has been proposed to alleviate the plastic waste disposal problem. This approach faces a number of technical and commercial hurdles. It can, in some cases, even thwart another environmental goal - that is, recycling. If mistakenly mixed in with recyclable material, degradable material could give the entire batch degradable qualities. Recent studies indicate that many landfills are aneorobic in nature and many organic substances may not degrade (17). Core samples have demonstrated undegraded foodstuffs after 15 years. This has implications for the use of microbiologically degradable plastics that may be buried in landfills. Legislative pressure is building to

Table 1
Examples of Environmentally Degradable Polymers

| Material | Company/Institution | Degradation Mechanism |
|---|---|---|
| **Microbiologically/Hydrolytically Degradable Resins** | | |
| Polyester-Based Resins | | |
| Polyhydroxybutyrate (PHB) and copolymers with hydroxyvalerate (PHBV) | Marlborough Biopolymers (ICI) | Hydrolysis, microbiologic, and surface |
| Polycaprolactone | Union Carbide | Hydrolysis and microbiologic |
| Polyactide copolymers | DuPont | Hydrolysis |
| Cellulosics | | |
| Polyethylene/starch mixture | U.S. Dept. of Agriculture<br>St. Lawrence Starch<br>AgriTech<br>Epron Industrial Products (U.K.) | Microbiologic |
| Cellulose acetates (low acetylation levels) | | Hydrolysis and microbiologic |
| Biologics | | |
| Gelatin graft polymers | Indian Inst. of Science (India)<br>Purdue University | Microbiologic |

**Table 1 continued**
**Examples of Environmentally Degradable Polymers**

| | | |
|---|---|---|
| **Photodegradable Resins** | | |
| Carbonyl-containing polyethylene | Dow Chemical<br>Du Pont<br>Union Carbide<br>Illinois Tool Works<br>(using Union Carbide resin) | Photodegradation, physical |
| Ketone carbonyl copolymers | EcoPlastics, Ltd. (Canada)<br>Princeton Polymer Laboratory | Photodegradation, physical |
| Photosensitizer Additives | | |
| Photoactivating material and a prooxidant for polyolefins and polystyrene | Princeton Polymer Laboratory | Photodegradation, physical |
| Polyethylene -- containing an anti-oxident and a photoactivator | Ideamasters, Inc./Plastopil Corp. (Isreal) | Photodegradation, physical |
| Polyolefins -- poly-grade (containing photosensitizers) | Ampacet | Photodegradation, physical |
| **Autooxidants** | | |
| Polyethylene, starch, unsaturated oil | Colorall<br>St. Lawrence Starch<br>Epron Industrial Products | Autooxidation |
| **Solubilizable Resins** | | |
| Ethyl acrylate/maleic anhydride copolymers, as well as polyacrylic acid and its copolymers with polyalkylacrylates and dimethyl-amino methylmethacrylate | Belland (Switzerland) | Solubilization |

Source: Arthur D. Little, Inc.

favor the use of degradables in certain high-visibility applications - and degradability could provide a functional benefit in some cases, such as land litter, marine litter and agricultural applications.

The main driving forces behind the increased interest in developing degradable plastics are (2, 14, 17):

- o The growing use of plastics in packaging;
- o The increasing concern over where to dispose of more solid wate;
- o Fewer landfills that are nearing capacity;
- o Increasing amounts of land and marine litter;
- o Concern over harm to the marine environment; and
- o Legislation limiting the use of nondegradable plastics.

Awareness has been growing among the general population concerning the use of degradable plastic materials and can be assessed by media stories, letters to the editors and advertising by consumer product companies. This general interest in degradable plastics did not exist two years ago. The evening news portion of Cable Network News of October 16, 1989 highlighted a company that manufactures biodegradable baby diapers and garbage bags. THe passions and controversy surrounding the environmental issues of using degradable plastics can be seen in a letter written by Everrett B. Carson of the Natural Resources Council of Maine in the Boston Globe of September 30, 1989. He concludes that "a clear cut ban is the only effective way to ensure that plastic connectors do not remain a threat to the marine wildlife or continue to contribute to significant litter problem...Sixty-day degradable yokes lull consumers into thinking that it is not so bad to toss them in the ditch..." Maine has banned six-pack rings effective July 1, 1991. An article in the Boston Globe of October 15, 1989 noted that during a recent cleanup of 200 miles of Massachusetts coastline 30 tons of trash was collected, 68% of plastic origin including trash bags, tampon applicators, and fast food containers. Newer six-pack rings, which are now required by Massachusetts law to disintegrate within 60 days, were found to be "cracking and falling apart, as intended" while old rings "turned up intact". Until recently only specialty companies marketed degradable garbage bags. The growing interest in this area can be assessed by the use of degradable plastics by large manufacturers. Glad garbage bags highlight the use of photodegradable plastic on the package. Bold print proclaims "Safe for the environment...And now Glad garbage bags are more photodegradable than ever, thanks to a new additive that promotes degradation without sacrificing strength...".

Plastics account for approximately 15-17% of the $19 billion food packaging market, and this share will likely increase to 50% of a $44 billion market by the year 2000 (11). As a result, the need to deal with disposal problems is becoming more pressing. Between 1979 and 1986 the number of landfills in the United States for disposal of solid waste decreased from 18,500 in 1979 to 6000 in 1988 (2, 17). Inability to meet government standards has been a primary reason for many landfill closings, and public opposition has made it difficult to open new ones. Although plastics make up only about 7% of all solid waste by weight, they account for 30% by volume.

The pervasiveness of plastic products makes them especially

visible when they show up as litter. Roadside and public-area litter is a national problem. Recreational use of lakes and oceans, the commercial fishing industry, and sewage treatment processes that release partially treated sewage into the ocean have all contributed to marine litter. A recent study of marine pollution commissioned by the Environmental Protection Agency analyzed 124 tons of waste recovered from a a 121-mile stretch of texas coastline. Of the 170,000 items recovered, 56% were made of plastic, including plastic bottles, bags, lids and caps, and six-pack retainer rings (11, 14).

Compounding the aesthetic problem, many marine birds and animals are being hurt by the increasing amount of plastic litter in water. They often mistake it for food or get caught in discarded plastic fishing nets or six-pack retainer rings.

State and federal environmental legislation has been initiated to address the problems created by disposing of plastic packaging in the environment. Several measures have been introduced to force manufacturers of plastic packaging to offer degradable versions of their products. Twenty four states already have legislation in place requiring that six-pack beverage rings be degradable[1] while Federal legislation was enacted in 1988 (17).

Much of the proposed legislation would require degradable products for those items which cannot be recycled. Current legislation in Berkeley, California will ban all nonbiodegradable fast food packaging in 1990. Suffolk County, New York has banned all plastic food packaging (17). Also, grass roots efforts are being undertaken by groups like the National Federation of Women's Clubs, which is encouraging grocery stores to offer a choice between plastic and paper grocery bags in an effort to decrease the problem of plastic bags in landfills.

Legislation limiting the use of disposable diapers is being addressed in Iowa, Oregon, and Washington. Nebraska has passed a law banning retail sales of disposable diapers effective October 1, 1993. The majority of the programs to help solve the disposal of 16 billion diapers a year are being formulated on recycling. However, there are programs dedicated to developing biodegradable shells for diapers, while Dafoe and Dafoe of Brantford, Ontario has announced the development of a biodegradable diaper (1).

Moving bagk to paper-based products instead of plastics would not be as beneficial to the environment as it might seem, however. Nearly 10% of the plastics used in packaging is used as coatings on other materials, including papers. Both sides of a typical paper milk container, for example, are coated with polyethylene to pretect the paper and the milk. Thus, most of the paper used in packaging degrade slowly or not at all.

## APPLICATIONS

The use of environmentally degradable plastics will increase as legislation is implemented mandating their use in specific applications. Successful materials for degradable applications must meet all the design, processing, and safety criteria of the traditional product; in addition, in most cases they cannot be mingled with recyclable plastics. However, new technlogies that will address this problem are under development. For example, some plastics that degrade via chemical action, e.g. Belland's solulizable acrylic polymers, while still retaining their intrinsic polymer structure are being investigated as tie layers between plastics that are otherwise inseparable. These degrading tie layers are captured in the chemical reagent and recycled separately (11).

Because manufacturers and packagers cannot control where their products will end up, degradables will first be most successful in highly specific applications, including some where degradability offers a functional benefit. Five potential applications are: 1) trash bags, 2) six-pack retainer rings, 3) personal care products, 4) agricultural mulches, horticultural wrappings and ground covers, and 5) sustained-delivery coatings for agricultural chemicals.

Trash bags typically end up in landfills, where they are buried. It would be advantageous to have these bags open so that degradable contents can be exposed and start to degrade. The darkness and moist soil form a perfect environment for the enzymes necessary for microbiological degradation. Trash bags fabricated using starch fillers are ideal for this application. However, if the landfill becomes anaerobic, these conditions may not be met. Aerated applications such as compost heaps may be more amenable to microbiologic degradation. Photodegradable trash bags are available, but will not be a useful if they are buried in a landfill. However, technology which results in continuing degradation after being initiated by exposrue to sunlight might be suitable. Chemically degradable bags (those that dissolve or otherwise degrade in the presence of an alkali or other chemical reagent) also offer some possible utility.

Degradable trash bags already on the market (such as those

manufactured by the Webster Industries Division of Chelsea Industries), although made of costlier resin, are priced 10% below most nondegradable bags. Nonetheless, speciality brands that highlight the degradable nature have not been selling as well as the national brands (11). Glad garbage bags are now made with a photodegradable formulation and growing public awareness may now be influencing the purchase of degradable formulations.

Photodegradable six-pack retainer rings are also on the market. These rings start to degrade within a week of full exposure to sunlight. Made by Illinois Tool Works, they function well because many six-pack retainer rings are discarded as litter and hence are exposed to sunlight. The legislation regarding these retainer rings states that they must be nonrecognizable as litter after exposure to sunlight for 180 days.

Personal care products made with biodegradable plastics that can be flushed down the toilet are being considered for development, as many of these products are being released into the environment in sewage effluent. Fully degradable products will be difficult to develop because they must be stable in use - often in a wet environment - and yet must degrade rapidly in sewage. Products that do not properly degrade could result in the fouling of septic systems and municipal sewer intakes.

Several degradable plastic materials for mulches used to protect seeds and plantlets from weeds, pests, and inclement weather are now available or under development. Photodegradable materials are available from Ampacet and Ideamasters, and microbiologically degradable materials based on blends of plastic resins with starch are being commercialized by AgriTech. Belland (Switzerland) has developed horticultural film for wrapping root balls of nursery plants. The film dissolves when watered with an aminical fertilizer. But so far these degradable products have captured less than 10% of the $79 million/year market for plastic agricultural mulches (11). The difference in cost between degradable mulches and nondegradable plastics - about $35/acre more for the degradables - is significantly less than the cost of removing the nondegradable mulches. Agriculture is a very cost-sensitive industry, so pricing and overall savings will be critical.

Another promising agricultural application for degradable plastics is as encapsulants for pesticides and fertilizers that can be spread on crops as microspheres. The plastic beads ensure even distribution and long-term release of the active agent. Few systems are commercially available in biodegradable formulations, although several experimental systems exist. High cost is a deterrent to using these systems. An inexpensive system that could be spread by hand without harming the handler would likely find wide acceptance in Third World countries where mechanized application systems are not available.

Some fertilizer formulations (e.g. urea formaldehyde foam) are microbiologically degradable systems based on urea. A wide range of materials are being investigated as degradable encapsulants, including lignin, starch, xanthan gum, and polycaprolactone. A variety of additives such as water-soluble resins can be used to modify release rates.

There is still no FDA approval for the use of any degradable plastics in food packaging applications. Although several companies have applied for this approval, food applications will be limited until the basic formulations are documented with the FDA as being safe for this use. Polyhydroxybutyrate has applications both in food packaging, medical implants and controlled drug release implants. The manufacturer, Marlbourough Biopolymers will be opening a master file at the FDA.

## TECHNOLOGY

While the general technology to develop environmentally degradable materials is currently available, the widespread use of this technology is limited by cost and by the unknown environment that the product will eventually be disposed. Degradation mechanisms will vary depending on that environment - e.g., in the water, on land as litter, or in a garbage dump.

Degradable plastics are made from polymers designed to break down in the environment through a number of mechanisms. In general, materials degrade by physical stress, chemical attack, photochemical action, or biological attack (consumption by enzymes produced by microbes). The materials need to degrade or at lease break up into small particles that can be dispersed harmlessly into the environment (13, 14). Physical breakage and dispersion occurs at the point the material becomes brittle during the degradation process, or becomes soluble in water or other mediums.

Different mechanisms are needed for specific end uses and polymers. For instance, biodegradation is most appropriate in landfills, while photodegradation is suitable for exposed litter.

Chemical degradation is most suitable for applications requiring a trigger mechanism, that is, the materials must be insoluble for their period of use, and then soluble once their functional use has ended.

The polymers that are used most in plastic packaging - low-density and high-density polyethylene, polypropylene, polystyrene, polyethylene terephthalate, and polyvinyl chloride - can be made to degrade photochemically, but they are very resistant to biodegradation.

APPROACHES

Five approaches to environmental degradation exist (Table 1): biodegradation autooxidatin, photodegradation, hydrolysis, and solubilization.

**Biodegradable materials** degrade through the action of enzymes produced by bacterial, fungus, yeast, and molds. Enzymatic degradation involves the catalyzed hydrolysis of the backbone or side chains of the polymer (4,6,7,10,13,14). The degradation can result from a multistep reaction involving both enzyme catalysis and non-enzymatic reactions such as oxidation.

The addition of naturally biodegradable materials such as starch to nondegradable plastic materials can result in biodegradable formulations. The ability to formulate and produce plastics with starch requires processing additives or special resins to make the blends compatible. AgriTech uses a copolymer of ethylene and acrylic acid rather than polyethylene to achieve blends that can be processed by standard thermal methods with starch levels of up to 50%. St. Lawrence Starch, using technology developed by Colorall, uses a silane coupling agent on the starch to make compatible blends (14, 15).

The rate of microbiological degradation is frequently a function of aerobic (oxygen present) or anaerobic conditions. These two conditions generate different types of bacteria and enzymes, which are the agents responsible for degrading the plastic material. Some materials are designed to degrade well in anaerobic conditions that would typically be found in a septic tank (12). But if these materials end up in aerobic sewage, such as is found in most city sewage treatment plants, they will degrade more slowly. Polyhydroxybutyrate (PHB) is an example of a polyester that degrades rapidly under anaerobic conditions and more slowly under aerobic conditions. First isolated at the Pasteur Institute in 1925, the chemical structure of this material is unique because it is produced as an energy reserve by the bacterium Alcaligenes eutrophus (12). Imperial Chemical Industries (ICI) has developed and is currently test marketing PHB and its copolymer with hydroxyvalerate (PHBV). The bacterium creates hydroxybutyrate from glucose, the source of which can be corn or wheat starch, fructose, molasses, or simple alcohols. As the bacterium needs energy, it releases enzymes that break down the polymer so that it can be consumed. Similar enzymes are present in soil, so this material is suitable for packaging that will end up in a landfill after it is consumed. Another important aspect of this process is that glucose source materials are readily available in many Third World countries, so it would offer an inexpensive alternative to petrochemically derived plastics. PHB shows negligible degradation in water and is therefore suitable for use in beverage containers.

The PHB and PHBV polymers currently cost $15/lb with current production at 50 tons a year in 1990. PHB has processing qualities similar to polypropylene, and much of the same procesing equipment can be used. Blow molded bottles made of PHBV will be introduced in Europe this year.

Polycaprolactone, a synthetic polymer from Union Carbide, has properties similar to those of PHB, but its degradation rate is expected to be slower because of its lower concentration of polyester groups. This material costs approximately $3/lb. To be useful in cost sensitive applications, such as agriculture, the raw material cost will need to be in the order of $1/lb. It has been investigated for use in biodegradable seedling containers and herbicide sustained-release delivery systems.

**Autooxidation** of unsaturated fat is another degradation mechanism that can be used to induce degradation of polyolefins. The pioneering work of Gerald Griffin has demonstrated that the addition of unsaturated fat such as soybean oil in plastics generates peroxides by autooxidation (6,7). The level of activation is small in soil, but dramatically higher in compost heaps because they generate temperatures as high as 70°C. The autooxidation results from the reaction of metal salts in the soil with the unsaturated fat. The peroxides that form can degrade polyolefins, most probably through the small amount of carbonyls groups present in the polymer backbone. The St. Lawrence Starch system mentioned before contains unsaturated fat to enhance degradation in addition to the microbiologic degradation of the starch.

**Photodegradable materials** degrade from the energy of light, particularly ultraviolet light from the sun (wave-lengths of 290-320 nanometers). The molecular weight of photodegradable materials is gradually reduced by the sun's rays until the product becomes brittle. It is then broken up by physical forces such as wind.

Photodegradable resins can be created by copolymerizing groups sensitive to UV degradation, such as carbonyls, or by adding resin additives containing photodegradable groups. Many polyolefins contain small amounts of UV-sensitivie groups, particularly carbonyls that result from thermal processing of the resin. Degradation can be further enhanced by the use of photosensitivie additives. These materials can degrade substantially within one month (5,8,9,14).

Union Carbide's photodegradable polyethylene, used in the six-pack beverage retainer rings manufactured by Illinois Tool Works, is an example of a polymer that is already in use and is likely to find additional applications. This polymer includes a small amount of copolymerized carbon monoxide to provide degradable carbonyl groups in the chain. Both Dow and DuPont offer similar resins.

**Hydrolytically degradable materials** degrade when water cleaves the backbone of the polymer, resulting in a decrease of molecular weight. This process can result in a loss of physical properties and eventual solubilization of the backbone. In addition, hydrolytic degradation can involve the cleavage of the side chain to water-soluble groups such as hydroxyl or carboxylic groups. Groups susceptible to hydrolytic attack, such as ester groups, must be present (10).

Polymers can degrade from their surface or in the bulk.

Surface-erodable polymers have the advantage of maintaining bulk properties as degradation proceeds from the surface of the material. These surface-erodable polymers, such as polyanhydrides and polyorthoesters, are hydrophobic. They are also expensive, so they are being investigated mainly for medical applications.

**Soluble materials** degrade by a dissolving action. Dissolution of polymers occurs because of the water-soluble nature of the backbone or the formation of water-soluble side chains (10). These materials are not degraded in a chemical sense. Additives can be combined with the polymer to cause pH changes in the presence of water that enhance solubilization. Belland, for example, as produced more than 1,200 such polymers, and is now commercializing several acrylic/acrylate polymers that can be dissolved at a specific pH level. In addition, polymers with large number of acid or alkali side groups can be produced to be selectively soluble in solutions of varying pH (11).

## TECHNICAL HURDLES

Many technical problems with degradable plastics must be resolved before they are implemented extensively. One of the primary problems is the cost of manufacturing these materials. Several factors add to the cost: the need to keep degradable plastic products separate from nondegradable plastics so as degradable plastic will not contaminate a batch of recyclable material; the potential need to distinguish pohtodegradable products for markets at different latitides because of different amounts and intensities of sunlight; the need for product manufacturers to handle degradable resins separately;

different types of degradable resins will be required depending on the ultimate fate of the product in the environment; and the cost of degradble resins or additives since these are produced in smaller quantitites and may use more-expensive ingredients than nondegradable formulations.

In addition, degradable polymers are still relatively new, so they are not yet well characterized. For example, the control of degradation timing in critical to many potential applications. And the environmental toxicity of intermediate and end products from degrading plastics is not yet well known. Other long-term factors must also be studied; for instance, large amunts of degradable products in a landfill could result in unsafe ground structures that, in time, could collapse.

To be successful, the materials for degradable applications must meet the design criteria of the product, be processable by standard equipment, be cost effective, not contaminate the contents when used as packaging, and be stable on the shelf. The ultimate disposal site of the plastic product will determine the type of degradable material to be used. Certainly a product that exists as litter exposed to the sun would utilize a photodegradable plastic compared to a product that will be buried in a landfill and might utilize a microbiologic degradation or an autooxidation mechanism.

## CONCLUSIONS

Plastic packaging manufacturers that have cost-competitive degradable materials ready as legislation is passed will have a definite competitive advantage in the market. In addition, these companies will be able to gain favorable publicity if they are able to introduce degradable products before the majority of their competitors. However, materials for degradable applications must meet all the design, processing, and safety criteria of traditional products. Selection of the plastic resin based on the degradation mechanism, i.e. photochemical action, microbiologic attack, physical stress, or chemical attack will be based on the ultimate fate of the product (e.g. marine, landfill, or land litter) and time period required for significant disintegration to occur. The four applications will appear to hold the greatest potential for degradable plastics are trash bags, personal care products, agriculture mulches, and encapsulated agricultural chemicals.

Ultimately the use of degradable plastics will address one aspect of the waste disposal problem and the integration of recycling, incineration and environmental degradation will be necessary.

The author gratefully acknowledges the assistance of Stephen Gondert of Arthur D. Little, Decision Resources during the preparation of the original Decision Resource article.

BIBLIOGRAPHY

1. ______, "Disposables Vendors Prepare for Worst", Hospital Purchasing News, Sept. 1989, Vol. 13, No. 9, pp. 1.

2. ______, a) "Plastic Packaging and Degradability", Fact Sheet; b) "Plastic Packaging Recycling", Fact Sheet; c) "Plastic Packaging and the Environment", Fact Sheet; Council on Plastics and Packaging in the Environment, February, 1989.

3. Carlson, W.A., "Degradable Concentrates for Polyolefins", in Degradable Plastics, Proceedings of the SPI Symposium on Degradable Plastics, June 10, 1987, pp. 26-30.

4. Gilding, D.K., "Biodegradable Polymers", in Biocompatibility of Clinical Implant Materials, Vol. II, CRC, pp. 209-228 (1981).

5. Gilead, D., "A New Time-controlled, Photodegradable Plastic", in Degradable Plastics, Proceedings of the SPI Symposium on Degradable Plastics, June 10, 1987, pp. 37-38.

6. Griffin, G.J.L., "Biodegradable Fillers in Thermoplastics", ACS Advanced in Chemistry Series, No. 134, 1975, pp. 159-170.

7. Griffin, G.J.L., "Degradable Plastic Films", in Degradable

Plastics, Proceedings of the SPI Symposium on Degradable Plastics, June 10, 1987, pp. 47-50.

8. Guillet, J., "Vinyl Ketone Photodegradable Plastics", in Degradable Plastics, Proceedings of the SPI Symposium on Degradable Plastics, June 10, 1987, pp. 33-36.

9. Harlan, G.M., "Degradable Ethylene-Carbon Monoxide Copolymer", in Degradable Plastics, Proceedings of the SPI Symposium on Degradable Plastics, June 10, 1987, pp. 14-18.

10. Heller, J., "Biodegradable Polymers in Controlled Drug Delivery", CRC Critical Reviews in Therapeutic Drug Carrier Systems, Vol. I., pp. 39-90 (1984).

11. Helmus, M.N., "The Outlook for Degradable Plastics", Spectrum, Environmental Assurance Issues and Opportunities, Arthur D. Little Decision Resources, February 1988, 3-1.

12. Holmes, P.A., "Application of PHB - A Microbially Produced Biodegradable Thermoplastic", Phys. Technol. Vol. 16, 1985, pp. 32-36.

13. Huang, S.J., "Biodegradable Polymers", in Encyclopedia of Polymer Science and Engineering, Vol. 2, 2nd. edition, 1985, pp. 220-243.

14. Johnson, R., "An SPO Overview of Degradable Plastics", in Degradable Plastics, Proceedings of the SPI Symposium on Degradable Plastics, June 10, 1987, pp. 6-13.

15. Otely, F., "A Starch-based Degradable Plastic Film", in Degradable Plastics, Proceedings of the SPI Symposium on Degradable Plastics, June 10, 1987, pp. 39-40.

16. Rosato, D.V. and R.T. Schwartz, Environmental Effects on Polymer Materials, Vol. 1, Interscience, 1968, pp. 990-1005.

17. Thayer, A.M., "Solid Waste Concerns Spur Plastic Recycling Efforts", Chemical and Engineering News, January 30, 1989, pp. 7-15.

---

1 This article is a revision of an article written for Arthur D. Little, Decision Resources, with permission - M.N. Helmus, "The Outlook for Degradable Plastics", Spectrum Environmental Assurance Issues and Opportunities, Arthur D. Little Decision Resources, February 1988, 3-1.

# NON-ENZYMATIC POLYMER SURFACE EROSION

J. Heller
Controlled Release and Biomedical Polymers Program, SRI International,
Menlo Park , CA 94025

Achieving polymer surface erosion is one of the holy grails of research in controlled drug release because in such systems, drug release in the absence of diffusion is completely controlled by rate of polymer hydrolysis. There are further advantages to surface erosion because in such systems rate of drug release is also controlled by drug loading so that variations in drug loading will corespondingly vary rate of drug release. Additionally, lifetime of the device is controlled by its physical dimensions. However, because rate of drug release also depends on the total surface area of the device, release rate will decline as erosion diminishes the total surface area (1).

Clearly, not all polymers will undergo non-enzymatic surface hydrolysis and, in fact, most known polymers will undergo bulk hydrolysis where the reaction accurs more or less uniformly throughout the bulk of the material. There are, however, a number of polymer systems that under the right conditions can undergo surface hydrolysis and it is the purpose of this manuscript to briefly review these systems.

## PARTIALLY ESTERIFIED COPOLYMERS OF METHYL VINYL ETHER AND MALEIC ANHYDRIDE

Such materials can be can be readilly prepared by the reaction between alcohols and the commercally available copolymer of methyl vinyl ether and maleic anhydride (2). They find applications as enteric coatings (3-5) and can be generally represented as polyacids which in the unionized state are water insoluble but in the ionized state are water soluble. They represent an interesting case where polymer erosion does not occur by chain cleavage and as shown, solubilization only involves ionization of a carboxylic acid group.

$$\left[ -CH_2-\underset{\displaystyle OCH_3}{\overset{|}{CH}}-\underset{COOR}{\underset{|}{CH}}-\underset{COOH}{\underset{|}{CH}}- \right]_n \longrightarrow \left[ -CH_2-\underset{\displaystyle OCH_3}{\overset{|}{CH}}-\underset{COOR}{\underset{|}{CH}}-\underset{COO^- \; H^+}{\underset{|}{CH}}- \right]_n$$

insoluble → soluble

An interesting property of such polymers, shown in Figure 1, is the dependence of pH of solubilization on the size of the alkyl group in the ester portion of the polymer (2). This behavior can be readilly understood by considering the number of carboxyl groups that must ionize to solubilize the polymer. With small ester groups, the polymer is relatively hydrophilic and a low degree of ionization is sufficient to solubilize the polymer. However, as the size of the alkyl group increases, so does the hydrophobicity and progressively more ionization is necessary to solubilize the polymer resulting in an increasingly higher dissolution pH.

Although these polymers were originally designed to dissolve abruptly with an increase in pH, in a constant pH environment they undergo controlled dissolution and have been shown to be useful materials for the controlled release of therapeutic agents dispersed within them (2). Figure 2 shows polymer dissolution rate and rate of hydrocortisone release from thin disks prepared from an *n* -butyl half ester polymer containing dispersed hydrocortisone. Each pair of points represents a separate disk in which the amount of hydrocortisone released was determined by UV measurments and the amount of polymer dissolution was measured gravimetrically.

The excellent linearity of both polymer weight loss and rate of hydrocortisone release over the lifetime of the device provides strong evidence for surface erosion and negligible diffusional release. This latter results was independently verified by placing the disks in a buffer having a pH low enough so that no polymer dissolution took place and determining amount of hydrocortisone relesed. None was found over several days. This system represents the first demonstration of drug release by surface erosion.

An interesting property of such polymers, shown in Figure 3, is the very high dependence of drug release rate on the pH of an external solution (2). This useful property has been utilized in the development of an enzyme-mediated self-regulated drug delivery system (6).

Even though such materials are useful drug delivery systems, they are only generally applicable in topical applications where elimination of a water soluble nondegradable polymer is not a problem. However, they are not generally applicable as

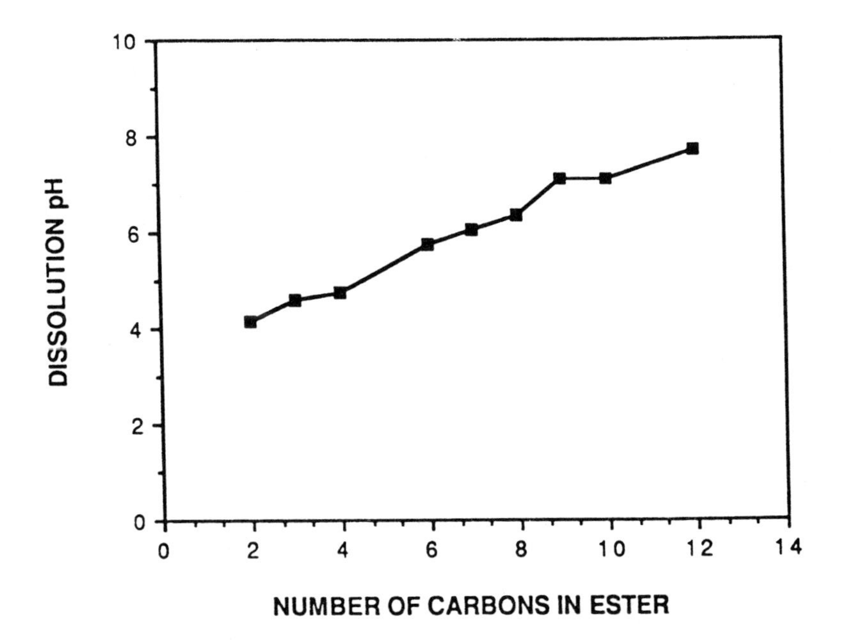


Fig. 1 Relationship between pH of dissolution and size of alkyl group in ester portion in half ester of methyl vinyl ether and maleic anhydride.
[Reprinted from reference 2]

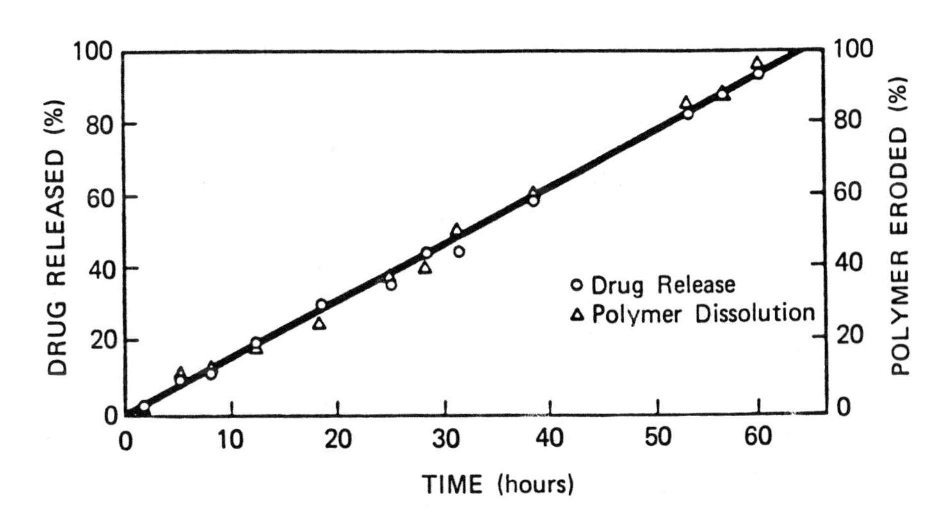


Fig. 2 Rate of polymer erosion and rate of release of hydrocortisone from the *n*-butyl half ester of methyl vinyl ether maleic anhydride copolymer containing 10 wt% drug. Thin disks, pH 7.4 at 37°C. [Reprinted from reference 2]

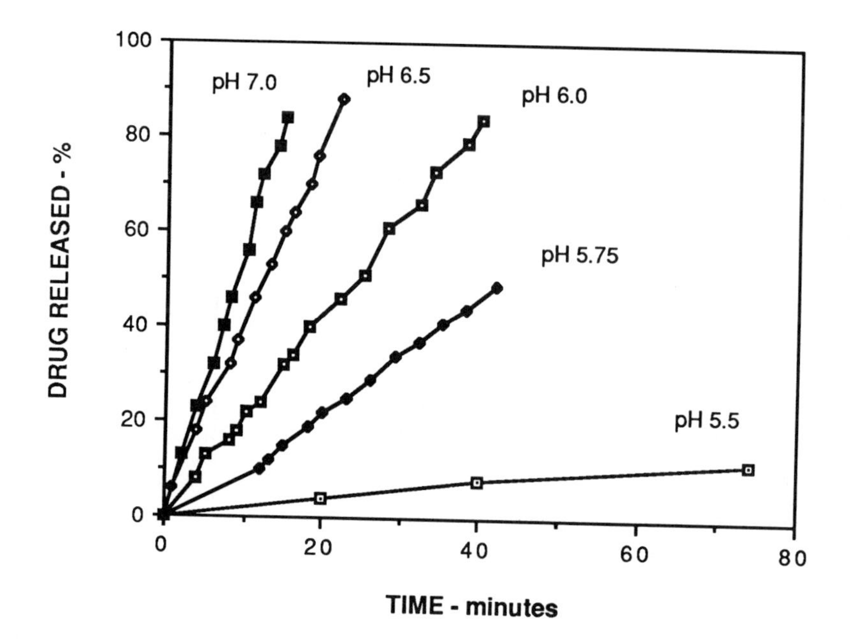


Fig. 3 Effect of buffer pH on rate of release of hydrocortisone from the *n*-butyl half ester of methyl vinyl ether maleic anhydride copolymer containing 10 wt% drug. Thin disks, 37°C. [Reprinted from reference 2]

implants because the all-hydrocarbon nondegradable backbone can represent a serious toxicological problem.

Clearly, polymers generally useful as implants must have degradable backbones and two such polymers, specifically designed as surface erodible materials are poly (ortho esters) and polyanhydrides.

## POLY (ORTHO ESTERS)

Poly (ortho ester) were first prepared at the Alza Corporation and the synthesis described principally in the patent literature (7-10). Although a few manuscripts describe the use of this polymer in controlled drug release applications (11-12), the polymer is treated as a proprietary material and very few details have been provided.

The polymer is prepared by a transesterification reaction between dimethoxytetrahydrofuran and a diol or mixture of diols. The synhesis is shown below.

$CH_3O$ $OCH_3$ (2,2-dimethoxytetrahydrofuran) + HO–R–OH ⟶ [–O–C(ring O)–O–R–]$_n$ + $CH_3OH$

Because poly (ortho esters) are acid sensitive and because one of the degradation products is an acid, the degradation reaction is autocatalytic and in order to prevent catastrophic desintegration, a base must be used to neutralize the acidic degradation product.

[–O–C(ring O)–O–R–]$_n$ $\xrightarrow{H_2O}$ γ-butyrolactone + HO–R–OH

γ-butyrolactone ⟶ $HOCH_2CH_2CH_2COOH$

The polymer has been used in the development of a bioerodible implant for the delivery of levonorgestrel under sponsorship of the World Health Organization and the National Institutes of Health. In this particular implant $CaCO_3$ was used as the base. This polymer has been claimed to undergo surface hydrolysis and concomittant release of incorporated therapeutic agents but very little evidence to support this claim has been presented. Although the exact reasons were never elucidated, this work was discontinued when human trials produced local tissue irritation (13,14).

A second poly (ortho ester) polymer system has been described by SRI International (15). This polymer system has a different structure from that described by Alza and was prepared by the addition of diols to diketene acetals, as shown below:

```
          OCH2       CH2O
         /    \     /    \
  RCH=C         C          C=CHR        +    HO-R'-OH
         \    /     \    /
          OCH2       CH2O

                      |
                      v

 [   RCH2       OCH2       CH2O       CH2R      ]
 [       \     /    \     /    \     /          ]
 [         C          C          C              ]
 [       /     \    /     \    /     \          ]
--    O          OCH2       CH2O       O-R'    --
 [                                              ]n
```

This synthesis is mechanistically similar to the synthesis of polyurethanes and as with that polymer system, dense crosslinked materials can be produced.

```
         OCH2       CH2O
        /    \     /    \
CH3CH=C        C          C=CHCH3      +     HO-R-OH
        \    /     \    /
         OCH2       CH2O

                        |
                        v

         OCH2      CH2O      C2H5  C2H5       OCH2      CH2O
        /    \    /    \    /          \     /    \    /    \
CH3CH=C        C         C                C         C          C=CHCH3
        \    /    \    /    \          /     \    /    \    /
         OCH2      CH2O      O ── R ── O      OCH2      CH2O

                        |
                        |  R'(OH)3
                        v

                CROSSLINKED POLYMER
```

The polymer solubilizes by hydrolysis of the ortho ester bonds to yield a diol, or mixture of diols and pentaerythritol dipropionate which later cleaves to pentaerythritol and propionic acid (16). Because cleavage of a simple ester is slow, initial degradation products are neutral and no autocatalysis is observed.

$$\left[ -O-C(CH_2CH_3)(OCH_2)_2C(CH_2O)_2C(CH_2CH_3)-O-R- \right]_n$$

$\downarrow H_2O$

$CH_3CH_2C(=O)OCH_2$, $CH_2OC(=O)CH_2CH_3$, $HOCH_2$, $CH_2OH$ on C $+$ $HO-R-OH$

$\downarrow$

$CH_3CH_2COOH$ $+$ $C(CH_2OH)_2(HOCH_2)_2$ $+$ $CH_3CH_2COOH$

Because ortho ester linkages are acid-sensitive, hydrolysis rates can be controlled by the incorporation of acidic excipients into the polymer matrix (17). A particularly useful class of excipient are aliphatic dicarboxylic acids or their precursors, acid anhydrides.

Polymer erosion mediated by an acidic excipient can be schematically represented as shown in Figure 4. When a poly (ortho ester) device is placed in an aqueous environment, water penetrates the polymer and dissolves the acidic excipient in the surface layers of the device. Then, as a consequence of the lowered pH in the surface layers, polymer hydrolysis accelerates and an eroding front is created which begins to move through the device.

A detailed understanding of the process requires a consideratin of two phenomena; the movement of a hydration front, $V_1$, and the movement of a hydrolysis front, $V_2$, accompanied by polymer dissolution. Clearly, a hydrolysis process confined predominantly to the outer surfaces of the device can only be sustained if the rates of movement of the hydration front and the hydrolysis front are about the same.

Because even the most hydrophobic polymers are penetrated by water in a matter of days or at best a few weeks, it is clear that this process can only operate with devices that

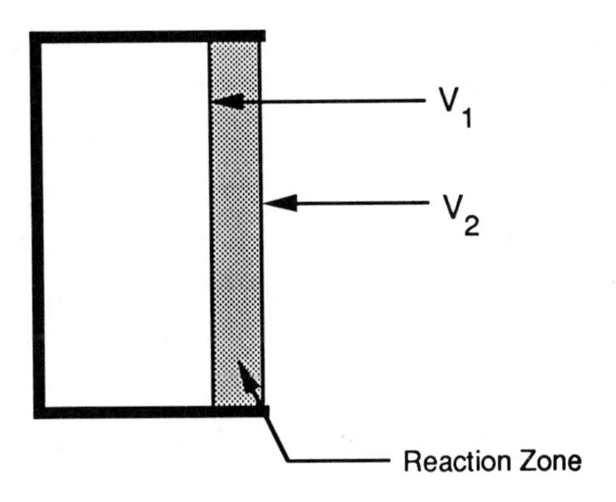


Fig. 4 Schematic representation of water intrusion and erosion for one side of a bioerodible device

have a lifetimes of less than about one month. However, within this limitation, excellent erosion and concomittant drug release behavior can be achieved.

Figure 5 shows the effect of incorporated maleic anhydride on polymer hydrolysis as measured by the rate of release of an incorporated marker drug (18). The data show that release of the drug accelerates as the amount of the incorporated acidic excipient incorporated into the marix is increased and that it does so in a linear manner. Thus, amount of released drug can be readilly controlled by simple variations in the amount of incorporated acidic excipient.

Figure 6 shows rate of release of a marker drug, weight loss of the device and release of the acidic excipient maleic acid which forms by the hydrolysis of maleic anhydride. Because release of the drug, release of the excipient and weight loss of the device occurs concomittantly, erosion of the device occurs by surface erosion. Because thin disks were used in this study, the surface area remains relatively constant so that release rate is not appreciably affected by changes in the total area of the device. Figure 7 shows a relationship between drug loading and rate of drug release. As expected for a surface eroding system, rate of drug release is linearly related to the concentration of the drug in the matrix.

Because erosion occurs by the movement of an eroding front through the solid device, surface erosion can be further verified by noting that lifetime of the device is directly proportional to device thickness as shown in Figure 8. Additionally, data shown in Figure 9, verify that, as expected, rate of drug release is linearly related to the total surface area of the device (18).

The use of aliphatic dicarboxilic acids has been been utilized in the development of bioerodile devices that can release the anticancer agent 5-fluorouracyl (5-FU) (19) and in devices that can release the narcotic antagonist naltrexone (20). Release of 5-FU from poly (ortho ester) thin disks is shown in Figure 10. The intended application for such a device is following glaucoma surgery where an implant close to the incision would provide a continuous release of 5-FU and thus prevent scar formation by inhibiting fibroblast proliferation. As shown in Figure 10, reasonably constant release kinetics can be achieved and rate of release of 5-FU can be readilly controlled by small variations in concentration of the incorporated sebasic acid.

Figure 11 shows release of naltrexone pamoate from a poly (ortho ester). Because naltrexone is a base, its incorporation into the polymer would stabilize the polymer and thus inhibit hydrolysis. However, conversion of naltrexone to its pamoate eliminates this problem and allows preparation of therapeutically useful devices. Because use of naltrexone in opiate addiction treatment requires delivery rates of 3 to 5 mg/day, devices

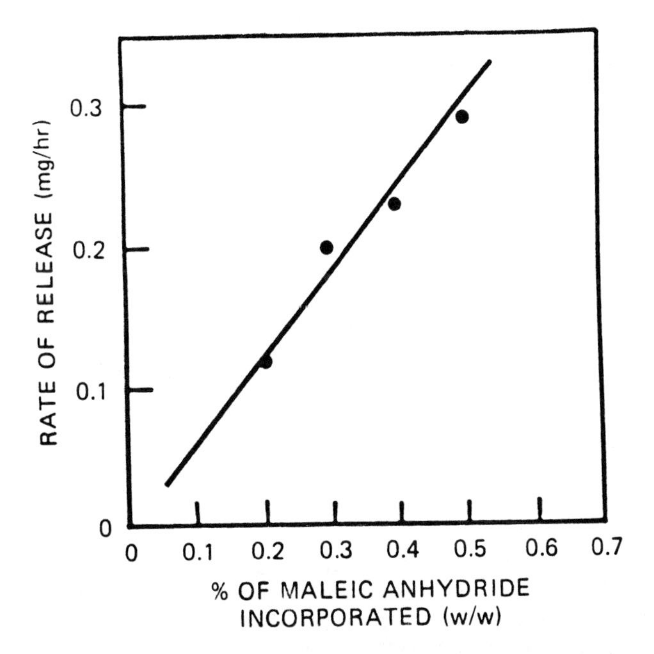


Fig. 5 Effect of amount of anhydride on methylene blue release rate from a polymer prepared from 3,9-bis(ethylidene 2,4,8,10-tetraoxaspiro [5,5] undecane) and a 35/65 mole ratio of *trans* -cyclohexane dimethanol and 1,6-hexanediol at pH 7.4 and 37°C [Reprinted from reference18].

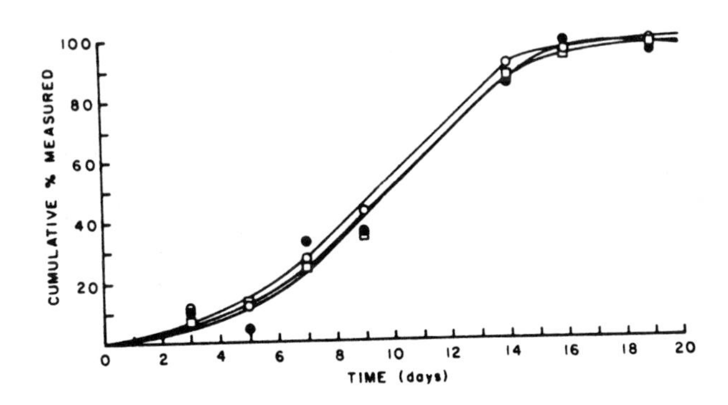


Fig.6 Cumulative release of methylene blue (○), 1,4 -$^{14}$C succinic acid (□) and polymer weight loss (●) from polymer disks prepared from 3,9-bis(ethylidene 2,4,8,10-tetraoxaspiro [5,5] undecane) and a 50/50 mole ratio of *trans* -cyclohexane dimethanol and 1,6-hexanediol at pH 7.4 and 37 °C. Polymer contains 0.1 wt% 1,4 -$^{14}$C - succinic anhydride and 0.3 wt% methylene blue. [Reprinted from reference18].

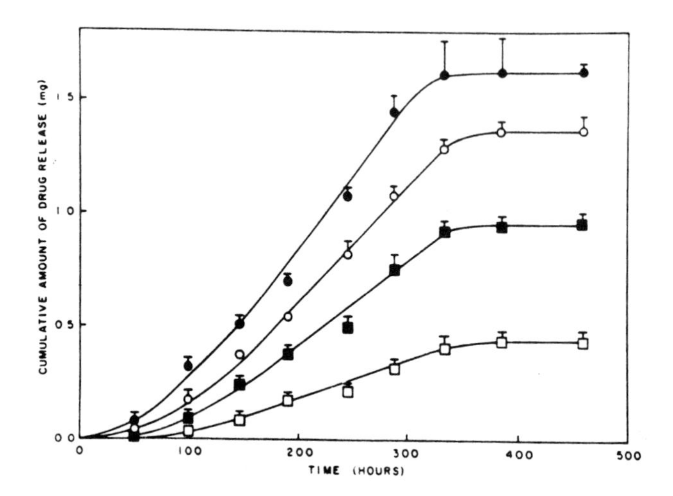


Fig.7 The effect of drug loading on cumulative drug release from polymer disks prepared from 3,9-bis(ethylidene 2,4,8,10-tetraoxaspiro [5,5] undecane) and a 50/50 mole ratio of *trans* -cyclohexane dimethanol and 1,6-hexanediol at pH 7.4 and 37 °C. Drug loading 8 wt% (●), 6 wt% (○), 4 wt% (■) and 2 wt% (□). [Reprinted from reference18].

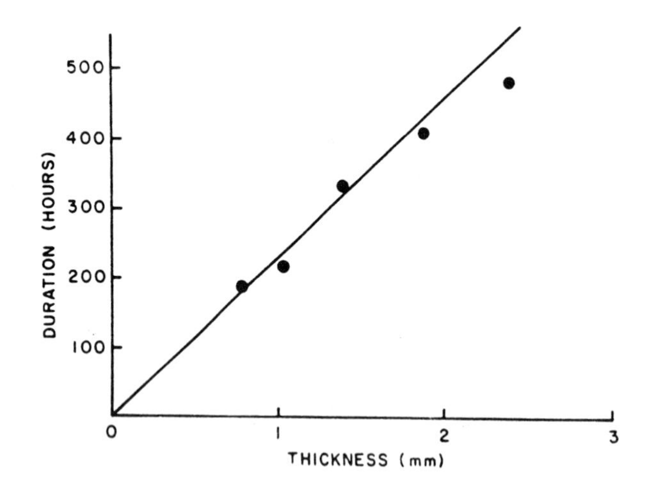


Fig.8 The effect of thickness on duration of drug release from polymer disks prepared from 3,9-bis(ethylidene 2,4,8,10-tetraoxaspiro [5,5] undecane) and a 50/50 mole ratio of *trans* -cyclohexane dimethanol and 1,6-hexanediol at pH 7.4 and 37 °C. Polymer contains 4 wt% drug and 0.2 wt% poly (sebasic anhydride). [Reprinted from reference18].

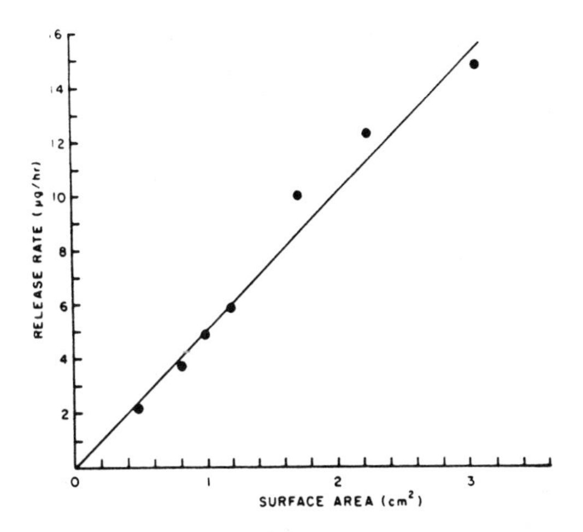


Fig. 9 The effect of surface area on rate of drug release from polymer disks prepared from 3,9-bis(ethylidene 2,4,8,10-tetraoxaspiro [5,5] undecane) and a 50/50 mole ratio of *trans* -cyclohexane dimethanol and 1,6-hexanediol at pH 7.4 and 37 °C. Polymer contains 4 wt% drug and 0.2 wt% poly (sebasic anhydride). [Reprinted from reference18].

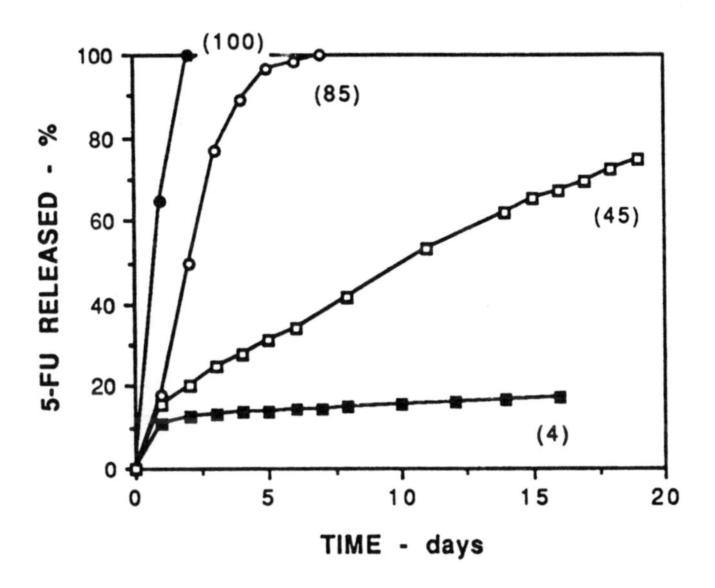


Fig.10 Cumulative release of 5-FU from polymer disks prepared from 3,9-bis(ethylidene 2,4,8,10-tetraoxaspiro [5,5] undecane) and a 35/65 mole ratio of *trans* -cyclohexane dimethanol and 1,6-hexanediol containing varying amounts of itaconic acid (IA). Disks 0.25 x 6.4 mm in a pH 7.4 phosphate buffer at 37°C. Numbers in parenthesis indicate percent weigth loss. Device contains 10 wt.% 5-FU; (●) 0.31% IA, (○) 0.08% IA, (□) 0.02% IA and (■) 0% IA.

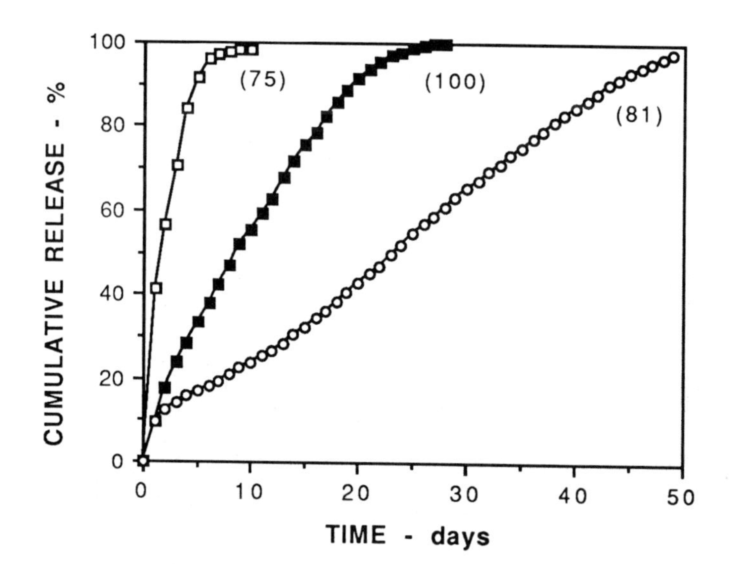


Fig.11 Cumulative release of naltrexone pamoate from a polymer prepared from 3,9-bis (ethylidene 2,4,8,10-tetraoxaspiro [5,5] undecane) and 1,6-hexanediol. Slabs 25 x 4 x 1.25 mm in a pH 7.4 phosphate buffer at 37°C. Number in parenthesis indicate percent weight loss. Devices contain 50 wt% drug and varying amounts of suberic acid (SA). (□) 3 wt% SA (■) 1 wt% SA (○) 0 wt% SA

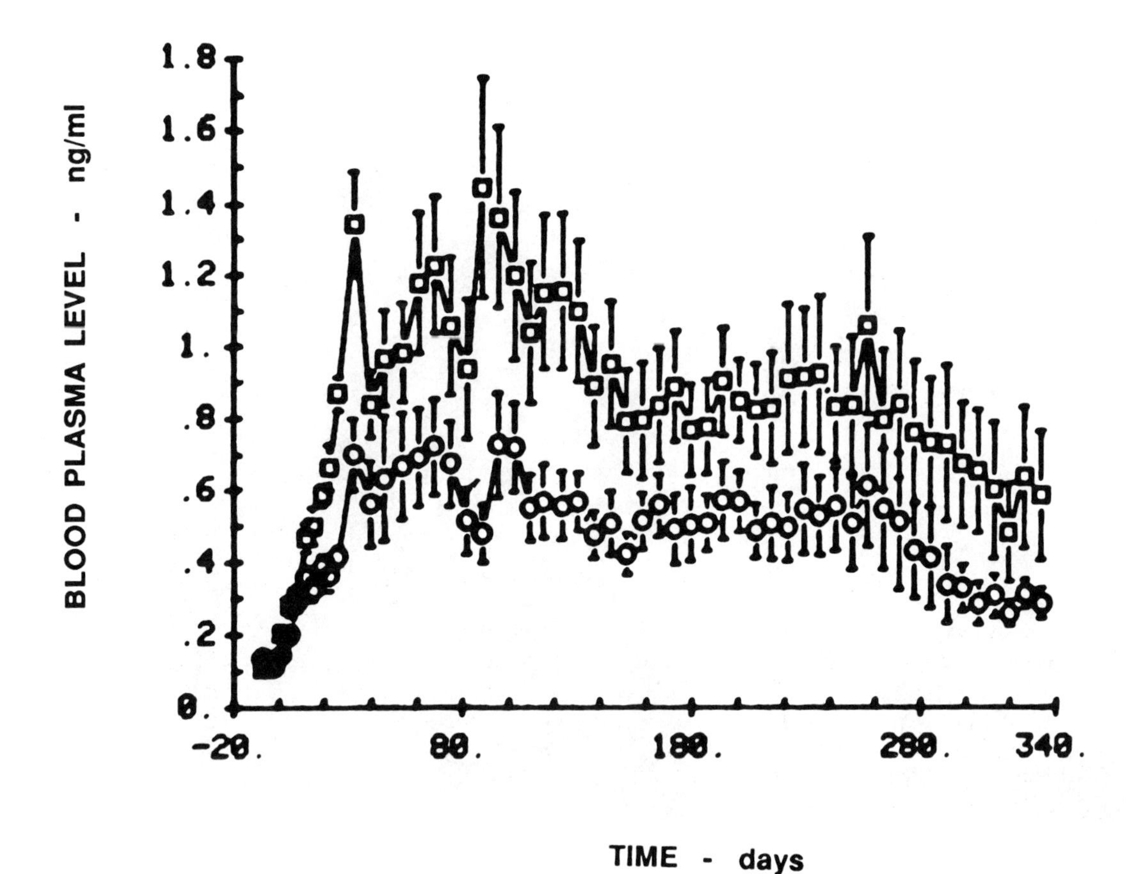


Fig.12 Daily rabbit blood plasma levels of levonorgestrel from a crosslinked polymer prepared from a 3,9-bis(ethylidene 2,4,8,10-tetraoxaspiro [5,5] undecane)/ 3-methyl-1,5-pentanediol prepolymer crosslinked with 1,2,6-hexane triol. Prepolymer containing 30 wt% levonorgestrel and 7.1 mole% $Mg(OH)_2$. Devices implanted subcutaneusly in rabbits.
(○) 1 device/rabbit, (□) 2 devices/rabbit

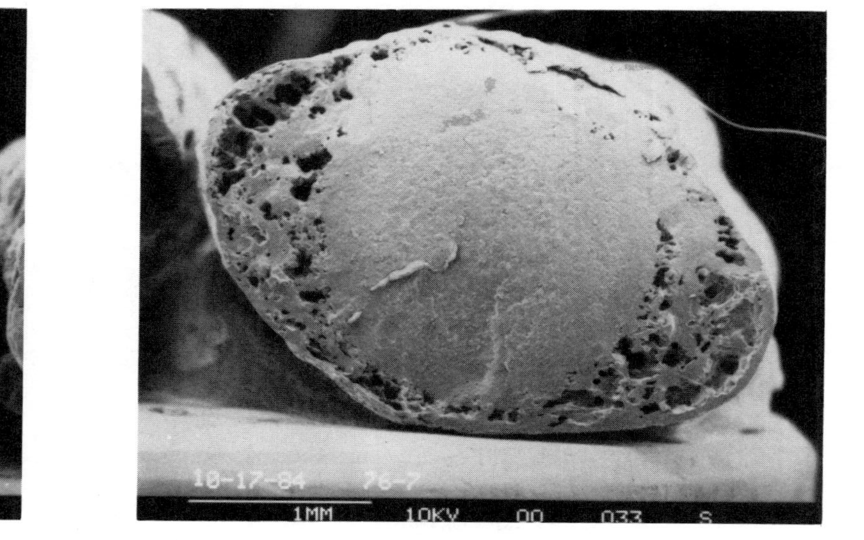


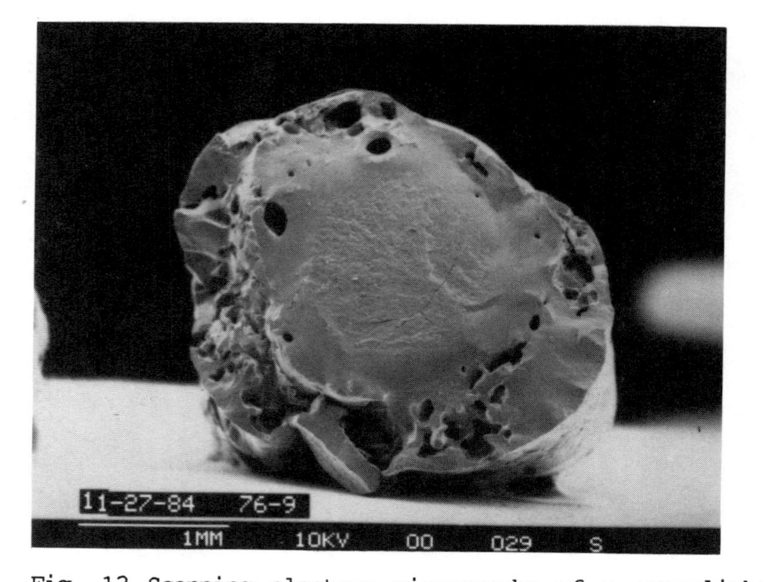


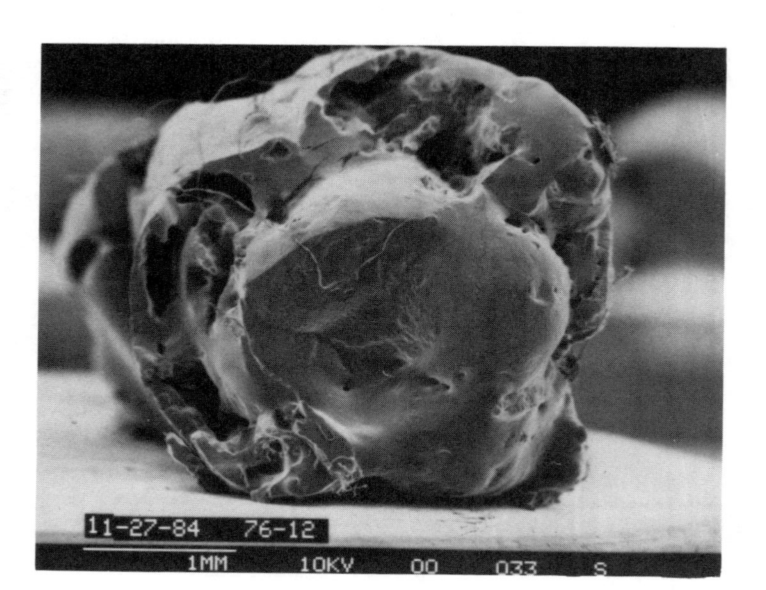


Fig. 13 Scanning electron micrographs of a crosslinked polymer prepared from a 3,9-bis(ethylidene 2,4,8,10-tetraoxaspiro [5,5] undecane)/3-methyl-1,5-pentanediol prepolymer crosslinked with 1,2,6-hexane triol. Prepolymer contains 1 mole % copolymerized 9,10-dihydroxystearic acid. Polymer rods, 2.4 x 20 mm, containing 30 wt % levonorgestrel and 7.1 mole % $Mg(OH)_2$. Devices implanted subcutaneously in rabbits (a) after 6 weeks, 30x; (b) after 9 weeks, 30x; (c) after 12 weeks, 25x; (d) after 16 weeks, 25x. (Reprinted from reference 24.)

with loadings as high as 50 wt% had to be used. The data show that despite the very high drug loading, reasonably linear release kinetics could be achieved and rate of release could be readilly controlled by varying the amount of incorporated sebasic acid.

Use of poly (ortho esters) that undergo surface erosion is not limited to short term delivery devices and work to develop a bioerodible implant that can release levonorgestrel for about one year by an erosion controlled process has been in progress for many years (21,22). In this work the interior of the matrix is stabilized with a slightly basic, low water solubility salt such as $Mg(OH)_2$ which prevents bulk polymer hydrolysis even though water does intrude into the polymer. In such a stabilized system, polymer hydrolysis can only occur in the outer layers of the device where the basic salt has diffused out of the device and where it has been neutralized by the external buffer.

In these studies, a prepolymer was first prepared from the diketene acetal 3,9-bis (ethylidene 2,4,8,10-tetraoxaspiro [5,5] undecane) and 3-methyl-1,5-pentanediol and then crosslinked with 1,2,6-hexanetriol. Prior to crosslinking, 30 wt% levonorgestrel and 7 wt% $Mg(OH)_2$ was incorporated into the mix which was then fabricated into 20 x 2 4 mm rods and these implanted into rabbits. Control of erosion rate was achieved by adding about one mole% 9,10-dihydroxystearic acid to 3-methyl-1,5-pentanediol during prepolymer preparation (23). As shown in Figure 12 constant levonorgestrel blood plasma levels were achieved for many months. Devices were also explanted at various time intervals and examined by scanning electron microscopy. As shown in Figure 13, erosion does occur predominantly in the outer layers of the device and throughout the experiment, an apparently intact core is maintained (24).

## POLYANHYDRIDES

Polyanhydrides were extensively investigated as potential textile fibes but were abandoned due to their hydrolytic instability (25). They are synthesized by first preparing the mixed anhydride from dicarboxylic acids and acetic anhydride and then heating the mixed anhydride to split off acetic anhydride as shown:

$$HOOC{-}R{-}COOH + (CH_3CO)_2O \longrightarrow H_3C{-}\overset{O}{\overset{\|}{C}}{-}O{-}\overset{O}{\overset{\|}{C}}{-}R{-}\overset{O}{\overset{\|}{C}}{-}O{-}\overset{O}{\overset{\|}{C}}{-}CH_3 + CH_3COOH$$

Heat
Vacuum

$$(CH_3CO)_2O + \left[ R{-}\overset{O}{\overset{\|}{C}}{-}O{-}\overset{O}{\overset{\|}{C}} \right]_n$$

Polyanhydrides have been extensively investigated as bioerodible matrices for the controlled delivery of therapeutic agents. Because hydrolysis of aliphatic polyanhydrides occurs at very high rates while hydrolysis of aromatic anhydrides is very slow, good control over erosion rate can be achieved by using copolymers of aromatic and aliphatic anhydrides (26). This is illustrated in Figure 14 which shows weight loss from copolymers prepared from the highly hydrophobic poly[bis(p-carboxyphenoxy)alkanes] and the more hydrophilic sebasic acid (27). Clearly, the combination of aromatic and aliphatic diacids provides considerable control over erosion rates which in this particular case can be varied over three orders of magnitude.

The release of p-nitroaniline from devices prepared from poly[bis(carboxyphenoxy)propane is shown in Figure 15 (27). After an induction period of about two months where no weight loss or release of p-nitroaniline takes place, release with concomitant erosion takes place and continues by good zero order kinetics. Even though the process was only followed to 20% completion, the concomitant release and erosion provides good evidence for surface erosion. As shown in Figure 16, much faster release with concomittant polymer erosion can be achieved when a copolymer containing sebasic anhydride is used.

Data shown in Figures 14-16 indicate that the polymer undergoes a surface erosion process and that by simple variation in polymer composition, changes in erosion rate and concomitant release of incorporated drugs can be varied by many orders of magnitude. However, because the anhydride linkage is very reactive, care must be taken to prevent a chemical reaction between the therapeutic agent and the polymer during device fabrication. Polyanhydrides are currently in human clinical trials for treatment of brain cancer.

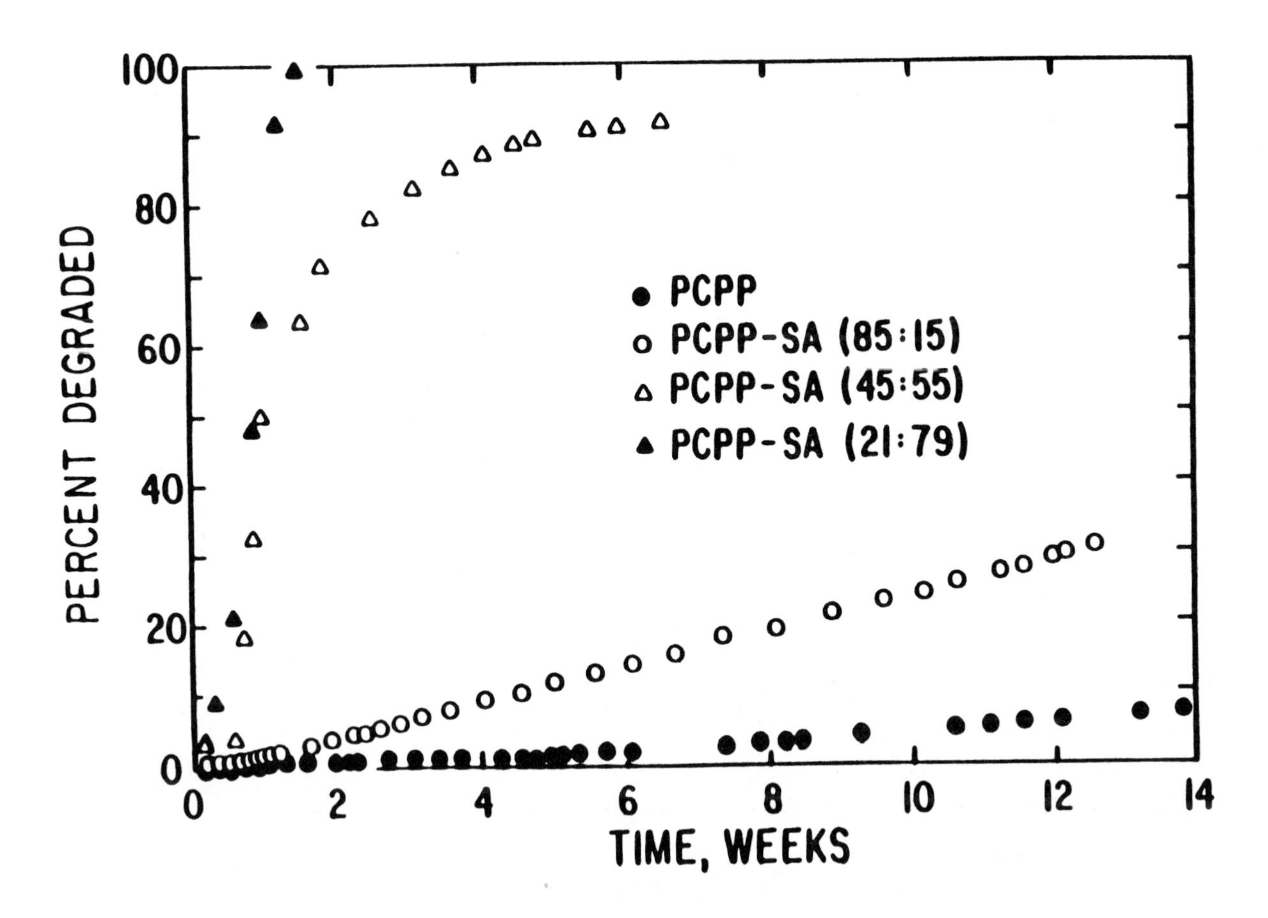


Fig 14 Degradation profiles of compression molded poly[bis(*p*-carboxyphenoxy) propane anhydride] and its copolymers with sebasic acid in 0.1 M pH 7.4 phosphate buffer at 37°C. [Reprinted from reference 27]

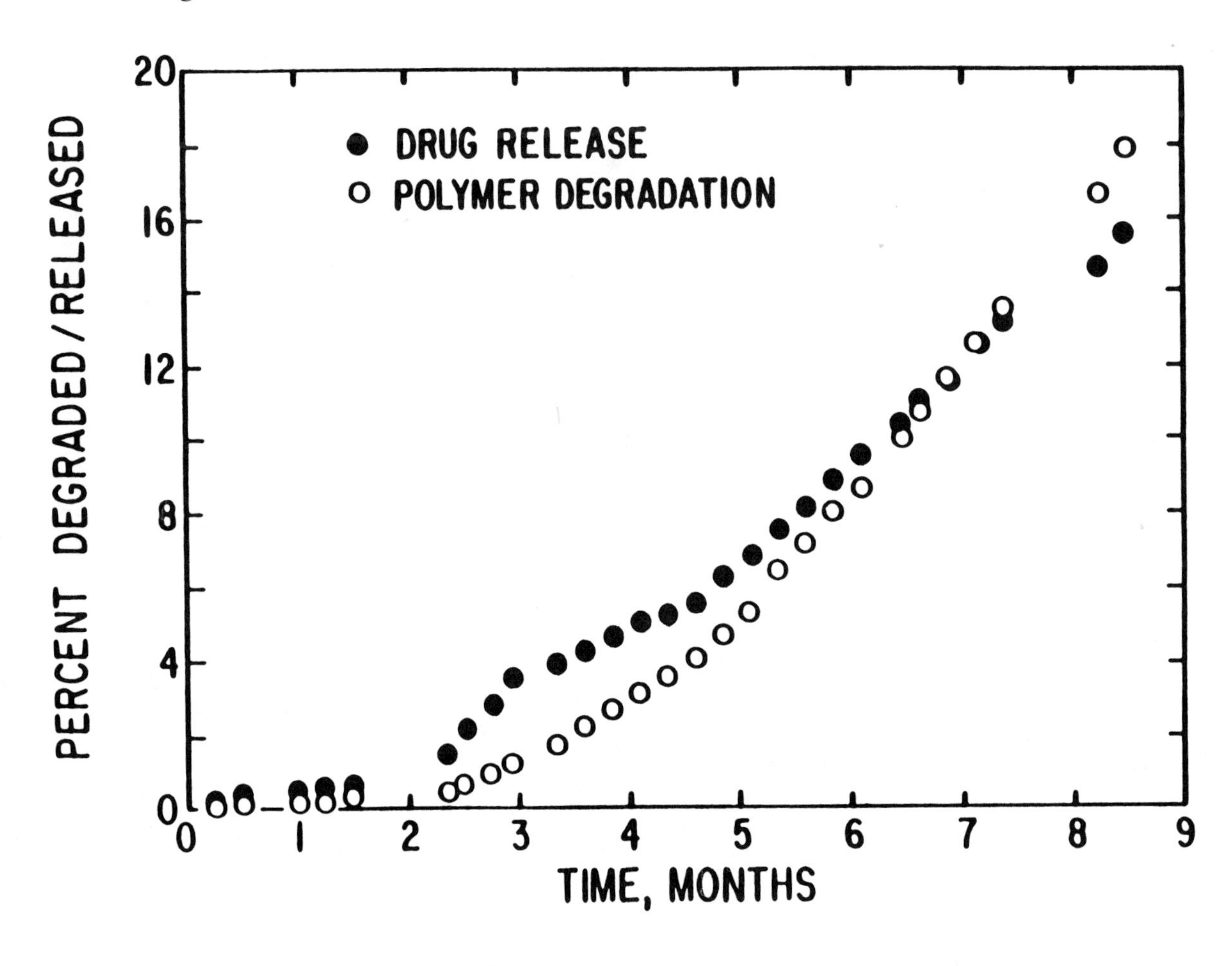


Fig. 15 Release of p-nitroaniline from injection molded poly[bis(*p*-carboxyphenoxy) propane anhydride] in 0.1 M pH 7.4 phosphate buffer at 37°C. Drug loading 10 wt%. [Reprinted from reference 27]

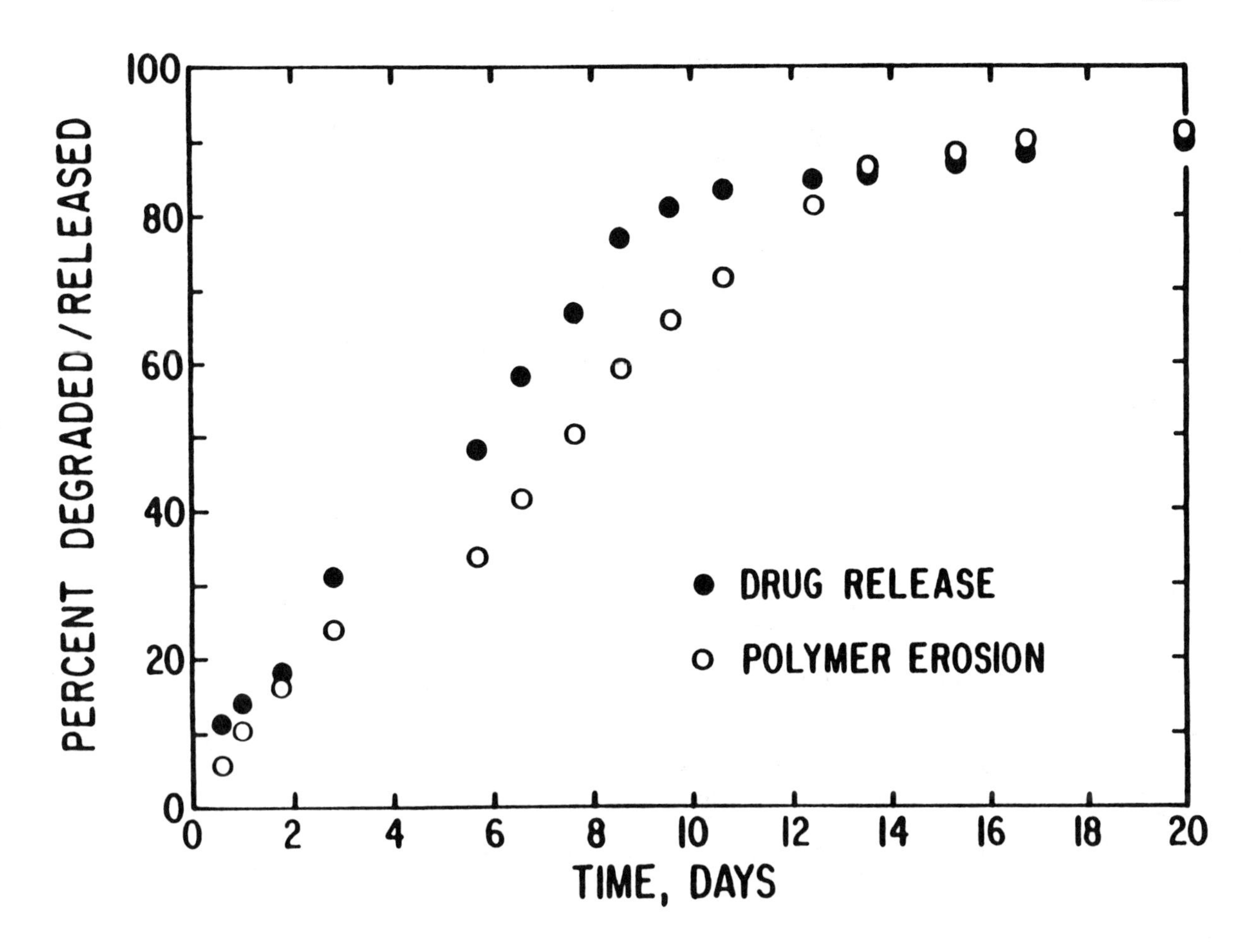


Fig. 16 Release of p-nitroaniline from injection molded 21:79 copolymer of poly[bis(*p*-carboxyphenoxy) propane anhydride] and sebasic anhydride in 0.1 M pH 7.4 phosphate buffer at 37°C. Drug loading 10 wt%. [Reprinted from reference 27]

## REFERENCES


1. Heller, J, Biodegradable polymers in controlled drug delivery, *CRC Crit. Rev. in Therap. Drug Carrier Syst.*, 1, 39, 1984

2. Heller, J, Baker, R. W., Gale, R. M. and Rodin, J. O., Controlled drug release by polymer dissolution I. Partial esters of maleic anhydride copolymers. Properties and theory, *J. Appl. Polymer Sci.*, 22, 1991, 1978

3. Lappas, M. C. and McKeehan, W., Synthetic polymers as potential enteric and sustained release coatings, *J. Pharm. Sci.*, 51, 808, 1962

4. Lappas, M. C. and McKeehan, W., Polymeric pharmaceutical coating materials. I. Preparation and properties, *J. Pharm. Sci.*, 54, 176, 1965

5. Lappas, M. C. and McKeehan, W., Polymeric pharmaceutical coating materials. II. *In vivo* evaluation as enteric coatings, *J. Pharm. Sci.*, 56, 1257, 1967

6. Heller, J, and Trescony, P. V., Controlled drug release by polymer dissolution II. An enzyme mediated delivery system, *J. Pharm. Sci.*, 68, 919, 1979

7. Choi, N. S. and Heller, J., Polycarbonates, *US Patent* 4,079,038, March 14, 1978

8. Choi, N. S. and Heller, J., Drug delivery devices manufactured from polyorthoesters and polyorthocarbonates, *US Patent* 4 ,093,709, June 6, 1978

9. Choi, N. S. and Heller, J., Structured orthoesters and orthocarbonate drug delivery devices, *US Patent* 4,131,648, December 26, 1978

10. Choi, N. S. and Heller, J., Erodible agent releasing device comprising poly (ortho esters) and poly (ortho carbonates), *US Patent* 4,138,344, February 6, 1979

11. Capozza, R. C., Sendelbeck, L. and Balkenhol, W. J., Preparation and evaluation of a bioerodible naltrexone delivery system, (Ed. R. J. Kostelnick), *Polymeric Delivery Systems,* Gordon and Breach Publishers, New York, N. Y., 1978, pp. 59-73

12. Pharriss, B. B., Sendelbeck, V. A.. and Schmitt, E., E., Steroid systems for contraception, *J. Reprod. Med.* 17, 91, 1976

13. World Health Organization *10th Annual Report,* 62, 1981

14. World Health Organization *11th Annual Report,* 61, 1982

15. Heller, J., Penhale, D. W. H. and Helwing, R. F., Preparation of poly (ortho esters) by the reaction of ketene acetals and polyols, *J. Polymer Sci., Polymer Letters Ed.*, 18, 82, 1980

16. Heller, J., Penhale, D. W. H., Fritzinger, B. K., Rose, J. E. and Helwing, R. F., Controlled release of contraceptive steroids from biodegradable poly (ortho esters), *Contracept. Deliv. Syst.*, 4, 43, 1983

17. Heller, J., Control of polymer surface erosion by the use of excipients, (Eds. E. Chielini, P. Giusti, C. Migliaresi and L. Nicolais), *Polymers in Medicine II*, , Plenum Press, New York, NY., 1986, pp. 357-368

18. Sparer, R. V., Chung, S., Ringeisen, C. D. and Himmelstein, K. J., Controlled release from erodible poly (ortho ester) drug delivery systems, *J. Controlled Release,* 1, 23, 1984

19. Maa, Y. F. and Heller, J., Controlled release of 5-fluorouracyl from linear poly (ortho esters), *J. Controlled Release*, in press

20. Maa, Y. F. and Heller, J., Controlled release of naltrexone pamoate from linear poly (ortho esters), *J. Controlled Release*, in press

21. Heller, J., Fritzinger, B. K., Ng, S. Y. and Penhale, D. W. H., *In vitro* and *in vivo* release of levonorgestrel from poly (ortho esters), I. Linear polymers, *J. Controlled Release,* 1, 225, 1985

22. Heller, J., Fritzinger, B. K., Ng, S. Y. and Penhale, D. W. H., *In vitro* and *in vivo* release of levonorgestrel from poly (ortho esters), II. Crosslinked polymers, *J. Controlled Release,* 1, 233, 1985

23. Heller, J., Penhale, D. W. H., Fritzinger, B. K. and Ng, S. Y., The effect of copolymerized 9,10-dihydroxystearic acid on erosion rates of poly (ortho esters), *J. Controlled Release,* 5, 173, 1987

24. Heller, J., Controlled drug release from poly (ortho esters) - a surface eroding polymer, *J. Controlled Release*, 2 167, 1985

25. Hill, J. and Carothers, W. H., Studies of polymerization and ring formation XIV. A linear superpolyanhydride and cyclic dimeric anhydride from sebasic acid, *J. Am. Chem. Soc.,* 54, 1569, 1932

26. Conix, A., Aromatic polyanhydrides, a new class of high melting fiber-forming polymers, *J. Polymer Sci.*, 29, 343, 1958

27. Leong, K. W., Brott, B. C. and Langer, R., Bioerodible polyanhydrides as drug carrier matrices. I: Characterization, degradation and release characteristics. *J. Biomed. Mater. Res.,* 19, 941, 1985

# CONSENSUS SUMMARY SESSION

Introduction Overview

Definitions/Terminology - Dave Willaims
Testing, Goals and Issues - Jan Feijen

Biomaterials

Shalaby Shalaby and Michele Vert

Environmental - Biodegradation

Ramani Narayan and Richard Wool

Environmental - Photodegradation

Tony Redpath and Sam Mitra

Issues and Summary

Ramani Narayan

Plans for the Next Workshop

Michele Vert

Closing Remarks

DR. BARENBERG: I would like to welcome you to the last session of the workshop.

This morning David Williams from the University of Liverpool and Jan Feijen from Twente University, Netherlands, will be chairing the session. The consensus reports, will be presented, followed by a question and answer period. It is imperative during the question and answer period that each person identify themselves for the stenographers.

After the consensus presentations, Ramani Narayan will summarize the issues and the needs. Michel Vert, who will be chairing the next conference, will give an overview of that conference. Then we have an added feature, Dr. Bill Bailey from the University of Maryland will be giving us a postscript with some slides and analytical techniques, and then we will conclude the conference. At this point, I would like to introduce David Williams, University of Liverpool

DR. WILLIAMS: Thanks, Sumner.

As you have heard, I, with my colleaque, Jan Feijen, from the University of Twente, are the co-chairman of this session this morning. I have no idea why the organizers chose us to have this particular task this morning. I have no idea where we are going to go. I would like to think that in my life I have some idea of where my direction is and I have some control over what is going to happen, but this morning I have absolutely no idea where it will lead to, and at 12:00 o'clock we may be anywhere.

It's rather like the story of the aircraft of the future. I don't know if you've heard this story, but the aircraft of the future will have, in its cockpit, just one pilot and one dog. The pilot is there to feed the dog and the dog is there to bite the pilot in case he touches anything. I rather feel like that pilot this morning. I think somebody out there is telling us what to do, but I have no idea what it is and where we are going to go.

We have chose to divide the first 10 minutes of this session, myself and Jan, to just give a few words of introduction before the various chairman of the three individual sessions give their reports.

I'm going to say a few words about definitions and the nomenclature, and Jan will then talk about testing and future goals, future issues, and we understand that the various groups will be addressing these type of issues.

As far as definitions under the nomenclature are concerned, let me show you an overhead. Unfortunately, apparently in Toronto on a Saturday morning there is only one microphone, so I will have to come back here.

Just a few general principles -- and these are my own thoughts, but I hope we can look at these basic principles as we go through.

First of all, I think we are all agreed that there is no need to try to re-define words and terms which have good and standard definitions. One of our problems is we are using words which have been around for a long time and there isn't any need to try to re-define many of these. We can just look in the dictionary where there are very good definitions. I think we have a problem if we try to produce new and different definitions for words which have been and are still are in common and effective use.

Secondly, I really do believe it's important that we have some hierarchy, some structure to our definitions and the nomenclature. We can perhaps choose one word, one term, which is covering a variety of different circumstances and have a clear and simple definition of that term. And then there are going to be subsets of that particular characteristic, that property, whatever, and we can have a number of definitions underneath that heading and so on. We can have a very good hierarchy where each term can be defined in very simple and clear terms and where there is a clear pathway between these.

You might, for example, look upon the hierarchy of classification of animals going from organisms to animals to mammals to primates to humans. If each case, we could have a very clear and simple definition. We just merely have to relate to the previous one if we want to get the detail. And individual terms and definitions don't have to be very complex with that hierarchy. And I believe, as I'll show in the next overhead, that we can do that here when discussing material properties and, specifically, degradation.

And thirdly, let me say that I think it's important that we try not to mix in this hierarchy the nouns with the adjectives and make sure we don't try to unequivocally relate one odjective to one noun. I mentioned this yesterday when talking about biocompatibility and biocompatible, biodegradation and biodegradable.

Because if we have these simple, clear definitions, then it is difficult to say unequivocally that a material is going to be degradable of biodegradable or whatever, unless we define the conditions under which we are operating. I think we have a problem -- and I know the words are already in common usage -- but we do have a problem when taking about biodegradable materials. I think we can much easier -- it is much easier for us to define biodegradation than to define a biodegradable material.

As far as this consensus is concerned here, we are trying to focus on degradation. And in my opinion, degradation is one of the terms we can introduce into such a hierarchy of definitions. This is my own view here, not necessarily that of any particular group, but I hope when the individual groups this morning talk about their definitions, we can bear in mind that we could have some hierarchy like this. We are talking about materials. Actually we are not talking about chemicals or substances, we are talking about materials here. Materials are substances useful for making objects and, therefore, we are -- in terms of classifying the behavior of materials -- we are talking about their properties. And so right at the top we can talk about material behavior or properties, and we can then have a classification, or subsets of those properties depending upon the conditions under which we are operating. That might to, under mechanical energy, thermal energy, electrical energy and so on.

And included in there is going to be an area which we might define as a chemical environment, some term like that. I'm not going to propose any particular term here, but that's, I think, what we are talking about, that area of material properties, material behavior, which refers to the behavior of a material in whatever environment that is.

And when within that, we are going to have a number of different characteristics and properties, and the one we are concerned about here is that of degradation. There may be other ways in which material is going to behave in the enviornment. Some people I know get a little bit upset when we are looking at the difference between degradation and dissolution, for example. And it may be that we have

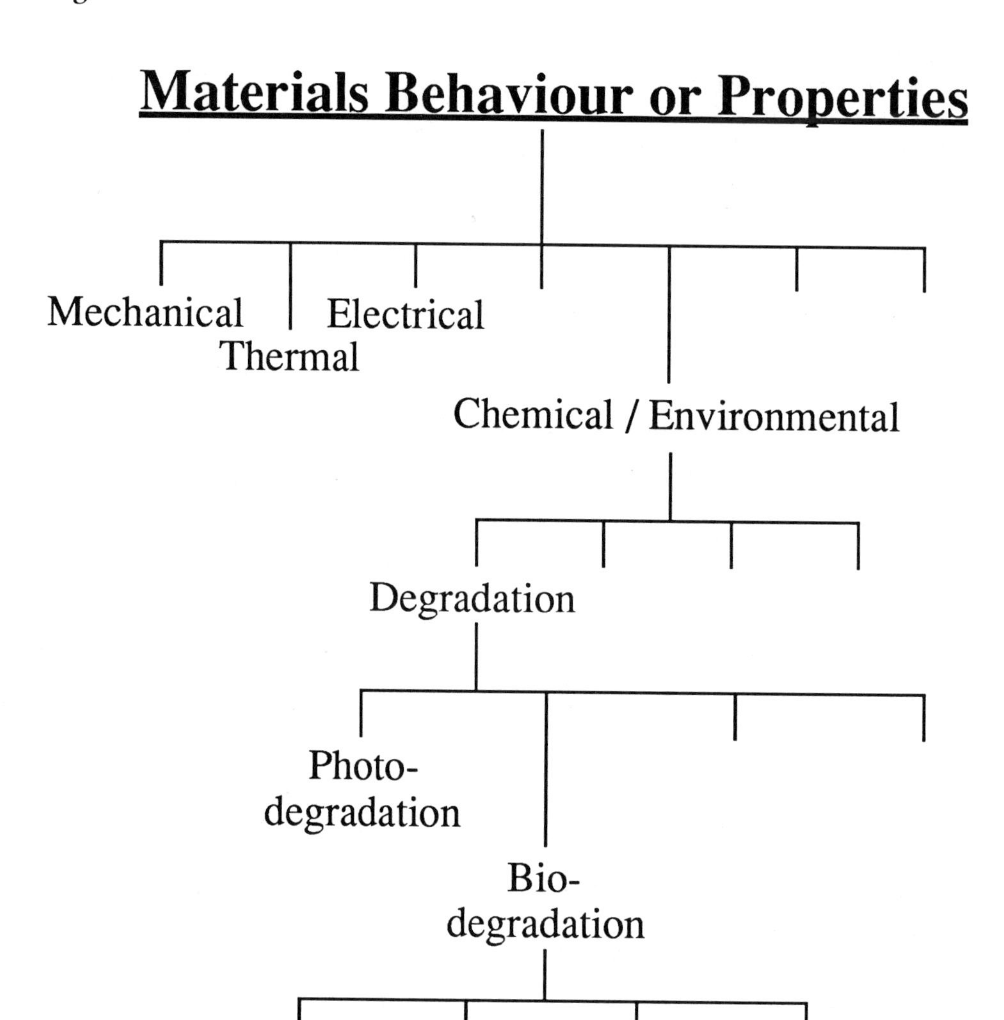
Materials Behaviour or Properties
Mechanical
Thermal
Electrical
Chemical / Environmental
Degradation
Photo-
degradation
Bio-
degradation

a separate line there which refers to dissolution. We could have another separate line there which refers actually to the polymerization process rather than depolymerization, and so on.

We are here, I believe, talking about degradation, and it would be important to define exactly what degradation is. But having done that, we can look at the different mechanisms or different conditions under which degradation can take place. And as far as I'm concerned here, as far as this consensus is concerned, we have considered, or are considering, two specific areas; that relating to photodegradation and that referring to biodegradation.

And I know in the group that I was in last night, we came up with a definition of biodegradation which I think could fit into this hierarchy and, hopefully, there will be a similar definition of photodegradation.

The one difficulty we have there, is where environmental degradation fits in there because personally -- it's my own personal view -- it doesn't fit in on that same line as photodegradation and biodegradation because the environment actually encompasses or includes most of these, and I think that is one area which we may have some heated disucssion about.

And then having defined each of these areas -- perhaps the term biodegradation -- we have to look at subsets of that, defining the conditions, and so on. And there could be a whole series and the behavior of the degradation products, and there are going to be a whole series of definitions underneath that.

If we are going to achieve any consensus I would hope -- it's my personal view -- that we can look at these terms in that type -- not necessarily this one -- but that type of hierarchy.

Those are my personal views, so I'll hand over now to Jan Feijen who will give his personal views on some of the issues especially related to testing.

DR. FEIJEN: Thank you, David.

It's a most difficult task to say something about testing when you are coming from a particular area. As you know, we are involved in biomaterials mainly, and now the testing in terms of biomaterials is not necessarily the same as the testing we have to do for other materials which are used as biodegradable products in the environment or when we study photodegradation. I caught some impression about the way people are doing their research in different areas now and will try to combine a couple of things myself.

First of all, what I would like to emphasize before we start any testing, and which is also part of the testing, is to characterize your materials, not only in the bulk but also the surface. And Michel Vert and Shalaby will go into detail with respect to the biomaterials point of view, but that does not necessarily mean it is not important for the other materials. I think it's very important to know exactly what you are working with.

What is the composition? I heard some presentations where people have studied exactly what are the trace elements, what is present. All these things can have an enormous influence on the degradation properties. And that does not only count for the bulk but especially for the surface.

I have not heard state-of-the-art reports on surface studies and I would like to emphasize that we have to study the surface properties in detail. If you are looking into the interaction of microorganisms with your material, then basically you are looking at surface phenomenon in the first place, and I have not seen that in detail at this conference. So we should move into that area.

Then, carryout standardized in vitro testing. This is important for all the subjects we are dealing with. Let's try to use standardized tests so that we are able to compare definite materials in the same type of tests. I don't have the time to go in detail but this will need some standardized in vitro tests to correlate with the in vivo test research and see what other mechanisms are occurring in the in vivo environment as compared to your in vitro conditions. If we are able to simulate enviornments then we could reduce the number of animal experiments to a certain extent, especially for the biomaterials area.

For the biomaterials area, we have to look into the in vivo testing and, as I have said, we have to correlate -- try to correlate in vitro testing with in vivo testing, and look at the degradation mechanisms and the histology. What is the interaction of an implant with the animal itself?

Toxicity of degradation products is very important. I have not heard too many things about the toxicity of degradation products. It was mentioned that we get degradation products and we have to study them, but I have not heard many things about if we take, for instance, biodegradable packaging materials, what is finally coming out? What is the toxicity of the products coming out?

I'm not a specialist in that field, but I would also be concerned about catalyst residues. If you are combining all the products which are not biodegradable and can put them together, maybe it is possible to collect these catalyst residues. But if you are emphasizing biodegradable materials then they are dispersed and we lose catalyst residues, and some of the catalyst residues are somewhat toxic. So we have to look more into the toxicity of these degradation products.

Finally, there are device-related studies. We have to see what is going on when a particular material is used to make a device. You can change your material properties in the processing of the material.

You cannot necessarily extrapolate the properties of the initial materials you are using to the properties of the materials in the device. And, again, you have to look for the in vitro and in vivo correlations.

Then, we have disucssed reference materials to a certain extent. I believe there will be reports from the different groups about suggestions to use reference materials. It would be good to compare well-defined reference materials for the different types of tests people are setting up and see how results can be correlated.

Finally, there are a several items I would like to emphasize.

We need more fundamental studies dealing with degradation mechanisms. To optimize materials and to find new systems we really need to go in detail into the chemistry and we need more fundamental studies.

As I mentioned before, surface characterization is enormously important. I have only heard one report on microbial interactions with surfaces, and nobody could tell me in detail what is going on. Does the adhesion of bacteria onto the surfaces play an important role? Are these microorganisms activated? To what extent? What is the relation between the surface characteristics and their enzyme production? Are these enzymes which are released free or are they still connected to the surface structure? We have to look into detail as to what is going on there to optimize our materials and see what is really going on.

Then, we have the definition of the requirements. I have heard many reports on the combination of polyethylene with starch. I have not heard too much about its uses, what are the exact requirements. What has to be the lifetime in terms of mechanical properties, in terms of performance? It's very good, to specify that and then see what you can do in terms of development to reach your goals.

Again, the toxicity of the degradation products I think is a very important item which we will face also in dealing with governmental regulations.

These are just a couple of subjects I have picked out of the last two days' discussions, and by no means does it mean that there are not a lot of very important items.

We will now have the reports of the chairmen of the different groups, and then we will provide some time for discussion after each report and try to emphasize important issues. Let's try to emphasize the main items. Let's not go into detail -- very detailed discussions but let's try to emphasize what is really needed and the main items.

Thank you very much.

# Definitions / Nomenclature

1. No need to redefine words with long established use.

2. Need hierarchy of nomenclature: simple/clear
   : obvious line

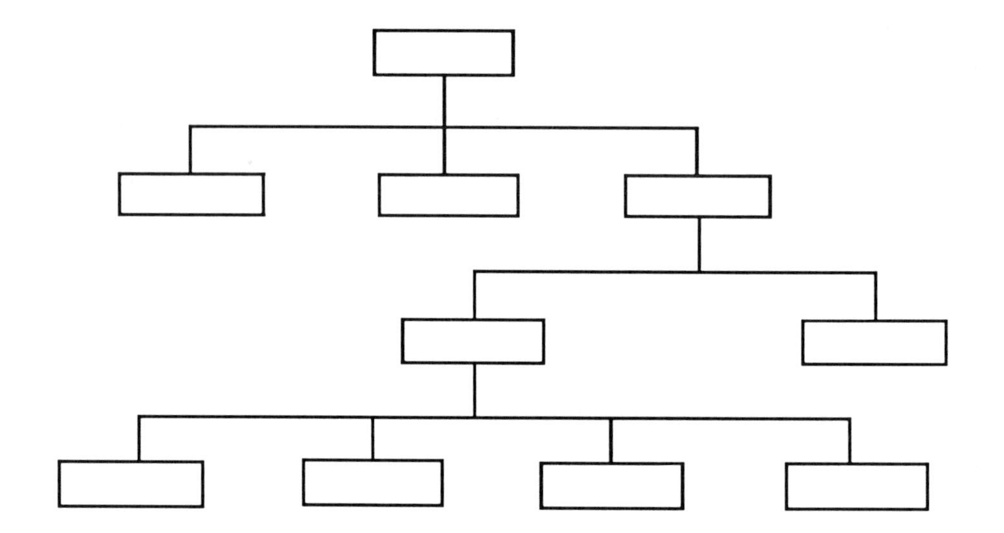

3. Must not mix nouns and adjectives, nor unequivocally relate one to the other.

# Testings

## Identification:

**Molecules**
- Chemical nature; composition; distribution
- MW + polydispersity
  - absolute values
  - SEC relative values
- Presence of desired and non-desired chemicals

**Raw material / Processed specimens**
- Morphology (surface, bulk, crystallinity, microstructures, residual stresses)
- Same as above at the different stages of processing

**Devices**
- Processing
- Sterilization
- Implantation
  - site
  - stress
  - shape

Interdependence

Be sure that techniques are going to reflect what they are supposed to

## Property Changes with Time: Cognitive approach

- Any kind of property → any kind of conditions
- Any kind of property → *in vitro*

## Property Changes with Time: Application driven

- Physiological conditions:
  - temperature
  - ionic strength
  - ptt, etc...
- Sterilization
- Fate *in vivo*

DR. WILLIAMS: Could I ask Michele Vert and Shalaby Shalaby to present the reports of the breakout session dealing with biodegradation specificially related to biomaterials?

PRESENTATION BY MICHELE VERT AND SHALABY SHALABY:

DR. SHALABY: that was a difficult task. It's easier to ask questions that to reach consensus.

We agreed on two definitions and we disagreed, or we could not agree on a few. These are the ones we agreed upon.

The definition of biodegradation from the biomedical perspective would be loss of a property -- loss of a property caused by a biological agent.

The second definition we agreed on is: a biomaterial is a non-viable material used in a medical device intended to interact with a living system.

We were unsuccessful in agreeing on definitions of degradation, bioabsorption and bioresorption.

With this, I'll give the floor to my colleague, Dr. Vert, to continue on the next topic.

DR. VERT: As far as the testing methods are concerned, I've tried to sum up what we said yesterday evening and this morning, and I will give you a list and a few comments and advice.

The first test we have to consider is the identification of the material. This is very important. It's too easy to work with something which is not well defined, and we would like to focus your attention on the identification of the material and, in particular, of polymers.

So there are several levels which have to be considered. The

# Definitions

### Biodegradation

Loss of property caused by a biological agent.

### Biomaterial

A non-viable material used in a medical device intended to interact with a living system.

# Unresolved Definitions

Degradation

Bioabsorption

Bioresorption

first one is the molecular level. And certainly the first parameter is the chemical structure, chemical nature I would say. And we have to be very careful, especially with polymers, because as you know, polymers are long molecules, they are built up with repeating units, and usually we just give the formula of the unit or units. But there can be defects along the chain and these defects are usually very difficult to characterize, especially if there is only a small amount of these defects. But nobody knows what can be the effect of these defects on the biological response.

Chemical nature, composition. The gross composition is important, but what has to be defined in the case of a degradable system is the distribution of the building blocks or units. For the same gross composition you can have dramatic differences in behavior according to the distribution of the repeating untis or blocks.

Second point, molecular weight. The molecular weight is important. Nature usually makes biopolymers of well-defined molecular weight, but polymer chemists cannot do that so far. So the main factor is certainly the polydispersity. You have to give the molecular weight plus the polydispersity -- and polydispersities define the property actually of the molecular weight -- average molecular weight over the number of average molecular weight. It can depend on the way you measure it. Absolute values can be obtained by using special techniques like osmometry or light scattering. But the most common system is GPC or size exclusion chromatography, and in these cases you usually get relative values. We have to be very careful about the way we get these values because -- if you are going to talk about the number average molecular weight, that will depend very much on the baseline you are going to take in the low molecular weight region, and it will be the reverse for the weight average molecular weight.

Another point which has to be specified is the presence of desired and non-desired chemicals. Desired chemicals can be additives, or for drug delivery, drugs, or substances like that.

Non-desired chemicals can be residual monomers, residual initiators or catalysts. It can be a solvent, it can be any kind of degradation product which is formed during the processing which is not removed -- not separated from the polymer. It can be compounds or ions which are present at a trace level.

Let's go now to the assembly of molecules, and it can involve what we can call the raw material or some kind of processed specimen or semi-products.

And in that case, one point which has to be defined is the morphology, and morphology includes several factors. Surface morphology, and it is, as Jan Feijen said, it's very important to characterize the surface but it's also very difficult. Shall we clean the surface before using the material or should we take it as it is normally implanted?

The bulk morphology, and in this case crystallinity is certainly one parameter to be considered, but it's very difficult to control the crystallinity of a material. As I tried to show you the first day, for different crystallinities you can get different properties, dramatic differences in properties.

Microstructures. Microstructures can be amorphous or crystalline, but it can also be a phase separated system in block copolymers or in blends.

Residue traces. This is connected to the type of processing. You can get a peeling effect or some residual stresses. This has to be characterized if we want really to know what we are going to talk about.

For the processed specimens, we have to follow exactly the same characterization as previously described, including the molecular characterization, because processing can be a degradation step and alter the characteristics of the initial system.

And now you have devices and you have to consider the processing step, the sterilization step, when one deals with biomaterials and implantation. Implantation includes the site of implantation, the possibility for stresses and the shape of the system.

In the discussion we were not able to give better definition to some kind of standardization, but what is very important is certainly the interdependence of all these factors. Anytime you change something you have to go back to the beginning to be sure that you are really looking at what you want to look. And to do so, you have to use techniques. I won't give you the different techniques because it's a matter of decision by the researchers or by the analyst. But we have to be sure that the techniques which are going to be used will reflect what they are supposed to reflect. It's not always the case. You can think that you are looking at something and actually you get some artifact. People have to be very careful in this field because we are dealing with a degradable material.

So now we have property changes with time. You can look at them as a cognitive approach, and that is to get information about the behavior of the material. And in that case, we think that any type of property can be considered. But there are two different approaches.

Any kind of conditions. You can put your sample at pH2 or pH13 or at 150 degrees, if you want, just to determine how the material is going to behave. But if you want to use some kind of correlation with in vivo experimentation, then you have to do the investigation in vitro by taking into account the physiological conditions. And now you can study the property change with time, but considering that this research, or this characterization is application driven.

The physiological condition, and the sterilization are very important factors. As far as physiological conditions are concerned, you have to specify the temperature, ionic strength, pH, and I believe it's a difficult thing to do because there is a lot of components in the blood or in body fluids or coming in or coming out of the cells, and it's very difficult to know which one or ones we have to take into account.

Let's assume, what can be the action of lipids on contact with a plastic material? They can be absorbed or not depending on the properties of that materials. Also, we have to have an idea of the fate of the material in vivo. Again, standardization is requested.

The final goal of the characterization, as far as of application is concerned, is certainly in vivo/in vitro correlations. We should start by looking at the behavior in vivo in order to set up experimentation for in vitro investigations and be sure that we are not too far from the reality. This is mostly for insight into mechanisms, and at this level we certainly need basic concepts, and as we are going to deal with not well-defined materials, including the molecules, statistics is probably a theoretical domain, which we will have to develop and use often.

There is another approach which is important. This is accelerated aging. Here again, we have to be very careful because if you are going to accelerate the degradation of a material, put it at a hundred degrees, for example, and if you want to compare it at 37 degrees, then be sure that you take into account parameters like the glass transition and properties like that, which, it is well known, change many physical characteristics of the system, such as diffusion coefficients and so on.

DR. SHALABY: We were given three directives when we started our consensus discussion. One is to agree on certain definitions or develop consensus definitions; second, testing; and third, which is quite important, is actually to develop some thoughts, ideas and consensus on what to do next. These are really topics to address in the next 18 months so when we meet again we will have something to discuss in a much more elaborate or a much more organized fashion.

The next slide will show you some of the key issues and needs -- research needs that we should be dealing again either next week or thereafter.

We feel as a group that we should be addressing or starting or doing basic studies of biomaterials degradation to allow for the design of materials with controlled degradation profiles.

It is most desirable to be able to sit down and draw certain features that you would like to achieve chemically or physically and do it and then test it in the animal to see if it works or not. So that we would be able control the destiny of our devices. This is quite important. We do this in a sporadic fashion in the industry, but there must be some basic research in academia to address this issue.

Second, it's felt that the fate of degradation products and the effect on the biological environment is quite important. We don't follow sufficiently what happens to the products, what do they do in the biologic environment. We worry about what happens to the implant or the device, but we should worry more about what happens to the tissues into which we are introducing the device.

This is one of the topics that we discussed quite a bit, and I have put them in order of priority, which was one of the directives were received from our chairman that we should list our priorities.

The third item we discussed was to develop in vitro test methods for effective modelling of in vivo situation. This is quite important. We rely a great deal on animals, and the animal rights group are quite vocal nowadays on the misuse of animals.

In terms of cost, it's very important to find in vitro methods that you could mimic or model to give you some good idea about what will be the performance of your device in animals. It is very important for screening and, likewise, it's very important in actually selecting your animal model too. So this is an area that we should be addressing both in academia and industry.

The fourth area is the definition of generic steps and processes and needs from material concept to clinical trials with list of criteria for experimental design plus rationale and justification. Let me explain this to you.

Not everybody knows how to go from the point of conceptualizing of an idea, putting an idea on paper and developing a material to the stage whereby you can actually put the material into device and put it through trials. I think it will be worthwhile for a group like ours to put together some generic steps, how are these pursued and how will they do it and also provide justification for what you would do and why should you do it.

Also, it deals with the design of the experiment. And, again, not everybody is an expert in every area, and having some generic or systematic type of listing of protocols which would be helpful.

The next one is identification of potential standard experimental procedures for evaluation of biomaterials. This is a very, very important area because we do have, in the States, the National Bureau of Standards. You need have to have the ability to reproduce experiments from one lab to another. And one of the most difficult things in our business, that is your business, is to reproduce experimental procedures or do the same testing or get the same results on the same device in different labs.

We talked last night about standard experimental procedures, and a collegue of ours said a standard becomes a standard when someone reproduces another person's experiments. That is true, but we need to identify these so people can use them and be able to reproduce this type of experiment.

The next one is basic studies to allow the development of protocols to evaluate degradation profiles of materials which have prolonged in vivo residence. We have experience so far with materials that reside in the living tissues perhaps for three months, two months, six months, and maybe as long as a year, but what if we find some need for material which will survive or reside longer than what we are experiencing at this stage, and we should think ahead. We should be able to develop protocols for materials which are expected to reside longer than the norm today and how would this affect the environment and how would the external environment, be it mechanical or biological, affect these devices.

Last, but not least, is the actual identification of standard reference materials. There was a proposal this morning on picking up one or two or three standards or controls, and we listed that we should be able to recommend to you as researchers to use good judgement in selecting the proper control, and whatever experiment you do you must have a control for it.

In addition to this, we decided it might be a good idea, not just for our group but for this meeting in general, to establish three committees to prepare consensus material for discussion in 1991. That is, before we come to the meeting we have material prepared by a group; that they take our directions from today, put it in some form, develop it over 18 months and present it to the consensus group so we don't waste time to innovate on-site. We will have something prepared before we come. You could not do it here, but this is why we are here the first time, so the second time we will be smarter and more effective.

These committees should address nomenclature, they should address reference materials and the standard test methods, and, third, they should address research needs and long-term planning.

With this I would like to conclude by thanking you for listening,

## Testing

1. Characterization of materials Bulk / Surface
2. Standardize *in vitro* Biodegradation tests
3. Initial *in vivo* testing: degradation and/or histology
4. Toxicity of degradation products
5. Device related studies (functional), *in vitro*/*in vivo*
6. Reference materials

## Research Needs

1. Basic studies of biomaterials degradation to allow the design of materials with controlled degradation profiles.
2. Fate of degradation by-products and effect on biologic environment.
3. Studies to develop *in vitro* test methods for effective modeling of *in vivo* situations.

- Definition of generic steps/processes and needs from material concept to clinical trials with list of criteria for experimental design Plus Rationale and Justification

- Identification of potential standard experimental procedures for evaluation of biomaterials.

- Basic studies to allow development of protocols to evaluate degradation profiles of materials with prolonged *in vivo* residence

- Identification of "Standard Reference Materials"

and I would be glad to answer any questions.

DR. WILLIAMS: Thanks Shalaby and Michele. I'm absolutely amazed that you have been able to distil such a clear list from that shambles that was our meeting last night. That was really very, very good.

Do we have any questions, please, to either Michele or Shalaby?

DR. SHALABY: Can I ask a question?

DR. WILLIAMS: Not of me you can't, no.

We have one question from one member of our group, yes.

SPEAKER FROM THE FLOOR: Since there are no questions to whatever is presented now, is there any need to go farther with the other presentations? Maybe we might be able to reach a consensus now. There were no questions, no objections to whatever we proposed. Is there any need to be present further.

DR. WILLIAMS: That was a good question. The answer is, we need to carry on.

Okay. Let's move on then to the second group titled "Environmental," and I presume Ramani is going to -- Ramani and Richard -- Richard himself is to present.

PRESENTATION BY RAMANI NARAYAN and RICHARD WOOL:

DR. WOOL: I would like to summarize for you our discussions on the environmental consensus breakout session. The session was chaired by Ramani Narayan and myself, Richard Wool.

We had close to 50 people in the room. Those of who are here right now, would you stand up so the crowd can see who was in that session? Thank you.

The agenda that we arrived at this meeting is the following: We gave a brief overview of the ASTM deliberations; we then had a very interesting discussion on the definitions of degradable plastics, and in fact came up with a consensus definition from the group which we will present here; we had a discussion, a brief one, on test environments; we discussed the different kinds of test methods, and concluded with a discussion of the research issues.

The first point on the agenda is the overview of ASTM activities. Several thousand man and woman hours of debate are recently behind us in ASTM D20, and this is a section that is headed by Ramani Narayan and is entitled "Environmentally Degradable Plastics." The Environmentally Degradable Plastics ASTM section is further subdivided into four other sections; one on photodegradation, one on biodegradation, one on environmental fate and one on chemically degradable plastics.

I am the chairman of the biodegradable plastic ASTM section, and up here you see a list of some of our current activities. I just want to briefly review those, and I reviewed then last night in terms of giving our group some idea as to where ASTM was heading. This was for information purposes only and was not intended to bias their own opinions in this matter in any way.

The biodegradable plastic section has the following activities. We have several task chairmen who are working on the following tasks: Graham Swift is working on the evaluation of mechanical property test methods and two biometer test methods, one involving carbon dioxide evolution, the other involving methane gas evolution.

It turns out that the photodegradable plastics group of ASTM has already modified two existing ASTM standards techniques; one, D 882 dealing with films; and the other one, D 883, I believe, dealing with thick plastic sheeting. The basic philosophy being that the mechanical property test method is going to be one of the most generally used methods for evaluating degradation of plastics.

Now, the biometer test method is both an environment and a quantitative test method, and this is currently being evaluated. The results of this evaluation will be presented at the Orlando ASTM meeting this coming week.

We are looking at several different environments. Hal Heck is examining composting environment, principally looking at leaves and possibly other composting methods. This will involve a technique with an apparatus approximating a 50-gallon drum. Hal Heck, in collaboration with Eric Reiner, is looking at simulated landfill environment. This will be about a five-gallon apparatus with a specific recipe that needs to be agreed upon. And soil environments, ASTM tests for soils -- several exist. I'm looking for one task chairman to be appointed in Orlando this coming Wednesday to specifically study this and update it for plastics.

We have made considerable advances in determining test methods for aquatic environments, and this is headed by Frank Ruiz. He is going to present three first documents to the general body this coming Wednesday, and they will be on ocean environments and fresh water environments for plastics. And, finally, I am examining the development of microorganism test environments.

So the general philosophy is that we should have a minimum number of test methods which can be applied to a potentially infinite number of environments. We have broken the number of environments down to the most important number initially.

Moving along, Ramani Narayan gave a status report on the definitions that are currently being debated by ASTM. There are definitions -- general definitions for degradable plastics, there are definitions for photodegradable plastics, there are definitions for biodegradable plastics.

A very lively discussion developed when these definitions were presented last night, and we decided to see if we could arrrive at a consensus on a definition for a degradable plastic. So we presented the two definitions put forward by ASTM for degradable plastics, and there are two.

The first one is of a general nature. It is a plastic material that disintegrates under environmental conditions in a reasonable and demonstrable period of time. That is definition A.

Next slide gives definition B. It is a bit more specific. It reads,

> "Plastic materials that undergo bond scission in the backbone of a polymer through chemical, biological and/or physical forces in the environment at a rate which is reasonably accelerated to a control and which leads to fragmentation or disintegration of the plastic."

A motion was put on the floor that we should vote on this. There was considerable discussion, and the results were as follows: 15 people voted for the general definition A, 22 people voted for the more specialized deifinition B. One would conclude from this that there was absolutely no consensus.

We then proceeded to see if we could thake these two definitions and arrive at a consensus definition. So there was considerable discussion and the final result is given in the next slide. This will be the first ISCW consensus definition for a degradable polymeric material within this consensus workshop.

It is a polymeric material -- oh, let me back off a minute.

The group objected to the use of the term "degradable plastics." They wanted that changed to "degradable polymeric materials," which would involve a wider class of materials involving both rubber and other materials that do not necessarily fall under the name "plastic."

And so the definition then reads, "Polymeric materials that undergo bond scission under environmental conditions at an accelerated rate as compared to a control."

At this point, the meeting was almost over and when the next topic on the agenda on test environments came up, that was a rather brief discussion. But a few important points are as follows:

One is, we recognize that there is a potentially infinite number of environments, and it was the consensus of the group that these test environments should be designed in such a way that they are reproducible and they are suitable environments.

A few points to expand on the word "suitable." Meaning these environments should be appropriate to the test condition, one should not be looking for an in vivo or in vitro test environment as a way of evaluating solid waste management problems.

There was considerable consensus on that, that the test environment should, at least at this point, be reproducible.

Let's discuss the test methods. We agreed that the number of

# Suggested Definitions

### Degradable Plastics

Plastic materials that disintegrate under environmental conditions in a resonable and demonstrable period of time.

### Biodegradable Plastic

Plastic materials that disintegrate under environmental conditions in a reasonable and demonstrable period of time, where the primary mechanism is through the action of microorganisms such as bacteria, yeast, fungi, and algae.

### Photodegradable Plastic

Plastic materials that disintegrate under environmental conditions in a reasonable and demonstrable period of time, where the primary mechanism is through the action of sunlight.

### Degradable Plastics

Plastic materials that undergo bond scission in the backbone of a polymer through chemical, biological and/or physical forces in the environment at a rate which is reasonably accelerated, as compared to a control, and which leads to fragmentation or disintegration of the plastics.

test methods should be a definite number, and the most important ones are as follows: The mechanical test methods. We didn't say which mechanical test methods. There was some discussion of that, but I believe that is fairly straightforward.

The aspiration biometer test methods involving CO2 determination method and methane gas evolution. Molecular weight analysis was considered quite important. There are many tests methods of evaluating molecular weight change, dilute solution viscosity, GPC -- otherwise known as size exclusion chromatography -- light scattering, et cetera.

There was the need to have accelerated test methods which would be reliable and which could be related also to real-time test methods, "real-time test methods" meaning experiments would be conducted in environments and data taken as the test proceeded.

There was a need to have a quantifiable end-point to the experinemt. When do you stop? When is the system considered degradable?

And lastly, we needed to have an evaluation of degradation products. The evaluation of degradation products is an extension over Item 2 in which it could be realized that many other side products, other than CO2 and CH4, could be given off during even a biodegradation experiment.

We concluded with a discussion on the research and issues pertaining to research.

Point No.1 is on reference standards. It is currently felt that there is a great need to develop both positive and negative reference controls -- reference standards and controls.

# Suggested Definitions

### Biodegradable Plastics

Those degradable plastics where the primary mechanism of degradation is through the action of microorganisms such as bacteria, fungi, algae, yeasts.

### Photodegradable Plastics

Those degradable plastics where the primary mechanism of degradation is through the action of sunlight.

# Test Environments (No $\rightarrow \infty$)

Should be reproducible and suitable

Test Methods (No $< \infty$)

1. Mechanical
2. Aspiration, $CO_2$, $CH_4$
3. Molecular Weight
4. Accelerated Tests
5. Real Time Tests
6. Quantified End-points
7. Evaluation of Degradation Products

One suggestion was made that the former National Bureau of Standards should perhaps take responsibility for this. They already have a good head start in terms of having many in-house polymers that are considered as standards, and I know they are looking at the possibility of developing reference standards for different degradable plastics in different environments. This could lead to quite a matrix of reference standards.

Item No.2 is laboratory versus field correlation. When we do tests in the lab, how well do they correlate with experiments on the outside? And the underlined word here is the correlation of the two.

Standardization of lab procedures. It's very important, whether through ASTM or other techniques or in other groups.

Point No.4, degradation mechanism elucidation. This is perhaps the most scientifically exciting for this group because considerable discussion was held during most of the talks as to what the specifics of the degradation mechanism were, and this is something that is of particular interest for further research.

Item No.5 is the inter-relationship between research and policy as determined by either legislative bodies, governmental groups, different agencies. How should research direct policy and how should policy direct research? Obviously there needs to be a very tight inter-relationship in the future between these two areas.

Someone pointed out that we perhaps have 12 different legislative actions pending per day at this point, and how and to what extent should that legislative action be directed by research findings to help in the development of proper policy and how much can policy tell the researchers to give them better direction and better data.

Point No.6 is the characterization of working materials. Most people were struck by a brilliant presentation of Dr. Schwab from Mobil yesterday on the excellent characterization of the starting material. So the characterization of your starting material would be extremely beneficial.

It was also pointed out from a practical standpoint most people are working with black box plastics that had been given to them by people who perhaps don't even understand the basic composition themselves, but it would be very good for standardization procedures to have very well characterized working materials. And that also goes back to perhaps the reference materials that could be developed by the National Bureau of Standards.

Lastly, then, we needed a characterization of the test environments themselves. How well were the test environments understood, what was the level of biological activity, et cetera?

To conclude on the composition of the group. Forty-five people gave their names, there was one person who said she was not there. There were 28 people from industry, four people from government bodies, and 13 from university.

The proceedings amount to 78 pages. Farr & Associates did an incredible job of getting those proceedings out, and we have several volumes of the unedited, unexperigated discussion over there on the table.

If you have any questions of Ramani or myself or any of the group members -- this was a very good working group -- we would be very happy to address them. Thank you.

DR. VERT: Michele Vert, University of Rouen.

Maybe I missed the point, but is there any concern about the possible effect of degradation products on the trees, plants, grass and so on, and where is the connection with -- the possible connection with the biologist dealing with this -- don't you think it should be included in this kind of research?

DR. WOOL: Yes. There is a concern of this in the group. As a matter of fact, the environmental fate issue was one of the items on our working agenda, and due to the pressure of time, that issue was not given the time it deserves. I will add, however, there is an environmental fate group working within ASTM to further carry this debate on.

DR. GUILLET: Yes, I have a little concern about the consensus definition --

DR WILLIAMS: Your name, please?

DR. GUILLET: Guillet, University of Toronto.

The consensus defintion simply says bond scission. Does that mean breaking of an OH bond, an hydrolysis of an ester on the side chain of a polymer, or does that mean the breaking of a backbone bond which reduces the molecular weight?

It seems to me that only one of your test methods actually addresses the question of bond scission and that one refers to the determination of molecular weight. So one might assume that your definition meant breaking the bond in the backbone. Why did you remove the backbone bond concept int he proposed ASTM definition?

DR. WOOL: Thank you for that question.

This was a consensus definition which is akin to developing science by democratic action.

On the issue of whether we should have kept the main chain scission or side group chain scission, this issue was discussed at considerable length and it was agreed in the end that to avoid further confusion and to not directly face the issue of water soluble polymers at that point, that we should just go with the idea of chain scission.

DR. HAMMER: Can I respond to that a little bit? Jim Hammar from the 3M Company.

If you have a graft polymer, how long does the graft have to be until you consider that the graft itself is the backbone? Are we going with just bond cleavage as a general statement?

DR. GUILLET: Obviously any chemical reaction involves a bond scission. I can hardly think of any one that doesn't, so if you don't specify what kind of bond scission --

DR. HAMMAR: Actually, I would like to see covalent in there.

DR. GUILLET: Yes, but ionization of an acid is breaking of a hydrogen from an oxygen, and that is not a covalent bond.

DR. HAMMAR: If you hydrolyze an ester group, even if you removed the alcohol portion in an ester, I think that is degradation. If you clip off part of the graft end of a polymer, that is degradation. If you merely dissolve it, that is not degradation. That is why I say covalent bond cleavage.

DR. FEIJEN: May I add one thing to it? I think it matters what kind of bonds are cleaved, it doesn't matter at all as long as you are cleaving bonds.

DR. WOOL: Can you speak into the microphone?

DR. FEIJEN: Yes. This is Feijen, University of Twente.

In my opinion, I think the definition is not too bad because it doesn't matter what kind of bonds you are cleaving. The process even may lead -- to a re-combination. If you make radicals -- introduce radicals into the structure, you may have re-combination, and we have sometimes seen in terms of viscosity that you may have an initial increase after the first bond scission, but still, I can live with this definition. It's not too bad.

DR. WOOL: Gentlemen, if I could just add the following: The type of bonds that can be broken could involve secondary bonds, primary bonds, covalent bonds, ionic bonds. Perphaps we don't need to define it. But it is the bond breakage in conjunction with the accelerated rate as compared to a control which is where the meat of this definition lies. It's not a redundancy to say just bond scission on its own. By bond scission we didn't want to prescribe any one particular type of bond.

DR. GUILLET: It's seems to me that one of the chairmen of this session said that one shouldn't change the meaning of words that are already well-defined. It seems to me the word "degradation" is rather well-defined, it terms of biochemistry at least. Why do you change it when you relate it to plastics?

DR. WILLIAMS: Jim?

DR. ANDERSON: I would prefer to defer my question. It doesn't specifically pertain to the discussion that is currently ongoing regarding the definition.

DR WOOL: Could we conclude our discussion on the definition? Jan, did you have another comment on that?

DR. FEIJEN: This one comment is the control. What do we think about control? What is the meaning of the control here? Can you tell me?

DR. WOOL: You can have positive and negative controls. The intent in this case of positive control is a material that would degrade, is known to be a degradable material. A negative control would be one which does not degrade.

DR. FEIJEN: But I'm thinking about the accelerated rate as compared to -- we can take stainless steel, for instance, then we always have the control. What is the meaning of control? So accelerated rate, as far as I'm concerned, it's just the polymeric materials that undergo bond scission under environmental conditions. That would be -- that is degradable material.

If you further specify it for a specific application and say, okay, we are talking about an environmental degradable material, then you could maybe further specify and say then we are talking about materials which are designed to degrade in a time interval which is optimal for that application, or something like that, optimal in relation to that application.

So I would suggest when we are going to combine definition for biomaterials area and also for this area in terms of degradable polymer, we don't need, in the first definition, acceleration. And maybe for further specification, it would be nice to say we are talking about environmental degradable polymers and specify that you are looking at polymers which degrade in such a fashion that is optimal to the application and further degradation process. That is just a suggestion.

DR. WOOL: It was necessary for us to introduce some type of a control, and these controls would come from our pool of standards with which we would all agree on. Obviously the definition is not completely all encompassing in that regard. It requires further qualification.

DR. ANDERSON: This definition requires that you define control. Now, I appreciate the fact you just did that. But the point is it's not in the definition and it has to be in the definition.

DR. WOOL: That we define control?

DR.ANDERSON: Absolutely.

DR. WOOL: If we were to expand on that last phase by saying "as compared to a standardized control," would that satisfy you?

DR. ANDERSON: No. I think you should eliminate the entire last two phrases, and, I would couple this with the previous definition and say it's a polymeric material that undergoes a loss of property under environmental conditions. That does not impose a time scale on it, it does not impose a mechanistic or kinetic rate which you are attempting to put in there and it does not requre definition of a control.

DR. WOOL: But it introduces the vagary of the loss of property.

DR. ANDERSON: Of a loss of property? As a chemist you certainly appreciate what I'm saying. By loss of property we don't specify that it be a chemical property as you've defined by bond scission. We appreciate the fact that it could be physical, it could be morphological, it could be mechanical. It's any way that you, as a scientist, want to measure property.

DR. WOOL: Ramani wants to address this point.

DR. NARAYAN: I would like to address this issue about what you said about loss of property and some of the reasoning why this accelerated rate came in and why people said can we live with this.

The rationale -- and before I say what the rationale is, let me say this is not my rationale. Being a chemist, I agree with what you say. But the rationale is that the reason we are gathered and the reason why you are hearing about environmentally degradable polymers, even the very existence of this area and coming, is to address the issue that the persistence, that is the slow rate of degradation.

And so when you are talking design of materials -- the word "design" never came up in any of the discussions -- but you are generally talking about design of materials that are supposed to undergo very accelerated degradation as compared to what exists today in terms of materials like the polyethylene, which is what is considered degradable, but over the period of time. So that is why the kinetic factor, the rate factor came into being.

To address the question you have got to recognize that the research community and then the companies and these, are going to be designing materials to satisfy a demand, and that demand is to design materials, whatever that material may be, which would, quote, unquote, "degrade, disintegrate," whatever you want to use the word, to something which is environmentally -- under environmental conditions, and, again, into materials which are environmentally compatible. That is why the issue of toxicity came up, that you have to do that. That is No.1.

No. 2 is that the definition is generic, and we are hoping that the test method and the test environments will control or will dictate more strongly and more forcefully what the definition would come up with.

DR. ANDERSON: That is exactly what was in my definition. Test method determines how you measure loss of property.

DR. WILLIAMS: Let me speak from the chair here. I think we have two issues here. First of all, whether we are talking about bond scission or we are talking about change of property. I doubt if today we are going to get any consensus on that. I think provided the conference report recognizes there are these two possibilities, both of which are logical, we have to say, depending which way you look at it, I think that is all you can do.

Of more concern is this latter aspect of being compared to a control, because without the definition of that control, the reader of that definition can define any material as either degradable or non-degradable depending on what he chooses as the control. And there will be enormous problems if that is left in. But I don't see there is any problem with our consensus meeting saying that at this stage we can't agree whether we are talking specifically about bond scission or specifically about loss of properties. Both are acceptable, I would have thought.

Jim, a last word and I think we better move on.

DR. ANDERSON: I wanted to go to my next issue.

DR. WILLIAMS: Oh, okay.

Did you want to respond to that?

DR. WOOL: I think that is an appropriate spot to leave the discussion on the definitions. We could carry on for several more thousand hours, I'm pretty sure, on this point.

DR. WILLIAMS: Okay, Jim.

DR. ANDERSON: Don't leave, you'll want to answer this.

I take serious exception to the fact that your committee, your subcommittee, did not address ecological or environmental considerations. I appreciate the fact that both of you are involved in ASTM, and I am too, in various committees. But this is a stand alone group. We are not a subcommittee of ASTM. We don't ponder to what ASTM wants in spite of what you two gentlemen who are involved in ASTM may think. And I would not want us to walk away from this meeting and have some newspaper say these gentlemen did not put down ecological and environmental considerations as being an important research issue, specifically. I think that you should reconsider your list, and I think we have to highlight this. This is really what we are here for.

DR. NARAYAN: I take very strong exception to the fact that the ASTM issue is brought up. If you read -- and this is the suggested definition -- the ASTM -- wait, let me finish, Jim, you had your say.

In fact, when Richard Wool and I put up the definitions, it was open to the house, the consensus definition which was proposed by a member from Dow Chemical -- I don't know where he is -- he is not in even in the ASTM committee, he was not even part of that. And there were 50 people, and you were not one among them, but they were 50 people in that group which you seem to think that agreed or put a consensus saying that is what we would live by. And if I might even drag in Michele or Jan's comment that he could live by the definiton.

So I think you are totally wrong to ascribe to the fact that, you know, this was an ASTM type of thing.

DR WILLIAMS: Wait a moment.

DR. ANDERSON: Richard had the last word on the definition and it was left open. I moved to a new issue.

DR. WILLIAMS: One comment from the chair, one comment here, then back to you.

DR. WOOL: I'm not a member of ASTM. We are using ASTM as a platform to bring all the interested groups in the country, and even abroad, to give us some debate and some consensus on the degradable plastic issue.

I fully appreciate your comment on the environmental fate issue. We didn't get to discuss it in depth, but it was alluded to in Point No.7, which we would have an evaluation of the degradation products.

DR. ANDERSON: Would you please put up the research issues which you had there previously?

My discussion here is not on definition. You had the last word and it was left open, and I agreed with that. I thought I made it perfectly clear that I was moving to a new issue and one in which I didn't see ecology or environmental considerations highlighted with more impact. Now, maybe the audience feels that has been covered appropriately.

DR. WILLIAMS: I think, Jim, that is a point that has been well made. I understand that a gentleman from -- or representing or related to ASTM is going to make some comment about this after the next session report. Is that correct? could we come back to it then? Thank you.

Jan?

DR. FEIJEN: I would like to make one short remark. Jan Feijen. You mentioned that to characterize the material in a very good way is beneficial. I like to change that word, an essential. All experiments which are carried out with materials which have not been very well characterized either in the bulk or also at the surface, I think we can as well throw them away.

DR. WOOL: We cannot what?

DR. FEIJEN: We can throw the research away, because it does not give us any firm insight what is really going on and how to go further and how to modify, optimize your materials for specific applications.

So I just want to stress that point. I'm very happy you mentioned, and we have this one -- couple of examples of very well characterized materials. I was very happy to see that yesterday. So I want to at least stress that again.

DR. WOOL: Thank you.

This is extremely important for research, but there is also the practical issue of working with plastics that many plastics are complex in the way they are formulated and the average industry or user of plastics is often times not in a position to get that kind of a detailed molecular characterization as we saw yesterday, and we'll have to almost take a black box approach to degradation testing.

DR. GRUBY: Yes. I'm Walter Gruby with the Environmental Protection Agency, Office of Research and Development.

I appreciated Dr. Anderson's concern for environmental and ecologic aspect of the topics. I perceive that I'm alone pretty much here this week in representing our types of agencies. I would certainly like to see some of my brothers and sisters that I have not met yet this week.

I think I'm quite comfortable in going back to my agency and reporting on the breath of concern both technical -- from the standpoint of engineering -- biochemistry, biomedical application. I have seen, I think, a lot of references to the ecologic and environmetnal application, the research needs. We have discussed in several of the sessions and subcommittees the question of toxicity of degradation by-products. I agree the definitional aspect -- I think we have such a variety of people here we could talk about that for many thousands of hours. I'm comfortable personally with the general approaches and then segregating it down to individual needs.

We have seen some reports on the different environments of degradation; that I am confortable they are being addressed, but certainly not to the extent that our agency is interested, and indeed we are hoping to pursue some of the by-product fate and toxicity aspects in a number of different environments with a number of different products and their degradation residues, resource dependent, of course.

So I think that Dr. Anderson's concerns are a lonely voice in the wilderness, and were I a chairman I would ask for a show of hands who also thinks that area needs to be moved up the priority list. I think the fact that it is being mentioned is very useful to me in directing our research program.

DR. ALBERTSSON: Yes. My name is Ann-Christine Albertsson, and my comment is the only reason that we, in the lab I came from in Stockholm, started in '60s in this area was the interest for nature -- for the interest between polymer and nature and that is the background for all this work with what should you do with polymers when you put them in nature.

So I think that the whole line with packaging material to discuss photodegradation, biodegradation and all this, that the reason is our concern for the nature.

And then here we also have our group of people that we are interested to use this material in the body and then again we are interested in that environment, what is happened, what is good. And when we started to come here we like to find out that these are very many similarities between these two approaches. But all the time we have in our mind we are only doing this because we take care of the nature. I mean, that is the background for everything, otherwise we haven't been here.

DR. WOOL: I agree with your comments. That is a very important, but it's also very general overriding consideration here. When we look at applications of degradable plastics and their analysis, we tend to tie our definitions into specific applications. When we look at degradation of plastics in solid waste fills by aerobic and particularly anaerobic degradation, that process needs to be characterized. The degradation products needs to be analyzed.

The environmental fate, it goes right along with that mode of analysis. When we look at development of agricultural mulch films for improvement of agricultural mulch films for improvement of agricultural yield, we also do the same analysis. We develop the materials, we test them in the field, we test them in the labs, we test the degradation products, we make sure that we don't have any toxic by-products. I can go on and on. With each method of analysis .. With each specific application, there is some address on the environmental fate issue. I don't think it's been ignored.

I would say in the interest of our one and half hour meeting or so last night in which most of it was spent in very interesting debate over the definition, that we perhaps didn't give equal weight to the environmental fate discussion, but that doesn't mean that it's not important.

DR. WILLIAMS: Two last points.

DR. ANDERSON: Jim Anderson, Case Western Reserve.

I don't want to be misunderstood here. Richard and Ramani did a very nice job in listing the research issues, and as a scientist I can appreciate the vast majority of those speak to ecological and environmental issues. But in looking at them they require scientific explanation. And our colleague from the government, the EPA, has told us how he could go back to his office and defend very will the actions of our group here in terms of addressing these issues.

That is not my point. My point is that if we look at this list and we need further explanation to explain to the general public that we really appreciate environmental, ecological issues, then we've missed the point.

All I want was a highlight where it would require no further explanation; that it would simply be listed there; that this group identifies ecological and environmental issues as being important and in need of much further study. Period.

And, Richard, everything that you've said I agree with. As a scientist we can appreciate that, but can we expect the fellow next door who is not a scientist to understand what we are trying to do? Let's tell him specifically.

DR. WOOL: This is part of the inter-relationship between science, research and policy.

DR. ANDERSON: State it.

DR. ENNIS: Bob Ennis, Plastigone Technologies.

I want to assure the gentleman from EPA Jim Anderson is not a voice crying in the wilderness, that there are many of us concerned with environmental fate and environmental issues. I think when you listen to what Tony Redpath has to say coming up next concerning the consensus opinions from the photodegradable session, you can see that this was certainly one of our primary concerns, and since there has been so much said here in the meeting concerning environmental fate which has not been, I agree, directly addressed by any of the speakers, I've asked the chairman for the opportunity to come up and spend a few minutes to speak to the group after the last presentation as co-chairman of the environmental fate section so ASTM where we have been working on developing some standardized test methods for toxicologic evaluation of both photo and biodegradable products which will be presented at the ASTM meeting next week in Orlando, Florida. But I'll give you a two-minute brief overview of what we've done after the last presentation.

DR. WILLIAMS: Okay. This is a very good time to move onto next presentation of the reports.

Tony and Sam will talk about the conclusion of the photochemical breakout session last night.

PRESENTATION BY TONY REDPATH and SAM MITRA:

DR. REDPATH: Well, I think in comparison to the other sessions, the photochemical session was sweetness and light, so to speak.

We covered several areas outlined here, definitions, test environments, physical property measurements, reasearch needs and issues we were instructed to do.

I think Sam and I are pleased to say we will be able to report to the group in the spirit of this conference some quantified data on the level of concern about environmental impact. We'll get to that at the end.

Definitions. What we did was to show the group members both of the definitions that were included in the handout here. We basically read them out, made certain everybody had absorbed them. Sam, if you could just flash those two, they are the ones you've seen in the handout. we then put it to a vote with three options.

Do you accept the first definition, and by that it's the coupling of the definition of degradable plastics with the subheading photodegradable plastics. Do you accept No. 1, do you accept No. 2 or do you accept neither?

And the results of the vote were six for the first definition, eleven for the second, nine for neither. I think it was clear then that there is no consensus on any one. You could view that as saying that 17 out 26 people voting -- which is about 75 percent -- could live with one or the other but that there is no one clear winner. So we then opened the floor for discussions.

We decided at the very outset -- or I guess the co-chairs decided at the very outset -- that we would not attempt to come out with a definition because we felt that we would spend the full time arguing about definition. Rather, we felt that we would take a quick vote on how people felt about these and then solicit opinions as to how these could be changed, what were the criticisms of these.

So we came up with a list -- and, again, these were suggestions from the floor. What we did is we simply had people state how they would change the definition in their view to improve it. We listed them all and then we took a vote as to how many people felt that particular change was an important one.

So we now have listed them in rank order. I guess the first one, to change -- to introduce the concept of induced degradation. For example, in definition one, where disintegration is induced under environmental conditions. Definition two, primary mechanism degradation is induced through the action of. Seventeen out of the 26 people felt that was an important concept to include.

Second point that was raised: Enhanced degradation. Again, that is under definition two, at a rate that is enhanced as compared to; definition one, conditions in a demonstrable period of time. Again, 17 out of the 26 people felt that was important to include that concept there.

The next one, Sam. To replace the word "reasonable." It was felt that the word reasonable was too soft and that should be replaced with "controlled." Twelve out of 26 in favour of that.

Replace "sunlight" by "light." Ten out of 26 voted to broaden the definition that way.

An alternative change, replace in favour of that.

If you combine Point C and E replacing "reasonable" with either "controlled" or "enhance," that tells me there is considerable objection simply to the word "reasonable" and that needs to be firmed up either by specifying that you mean it as a controlled degradation or it is in some fashion an enhanced degradation.

And then one last suggestion was to include the sense of initiation by sunlight. And, again, that ties in with the concept of induced degradation, initiation, induction. Six people in favour of that.

We did not go further and attempt to combine all of those and rewrite a definition. We felt it wiser to wait and see what the other groups were coming up with in terms of the general definition. But these are the suggestions then, and I guess the relative weighting of the importance of those suggestions as to how people feel the definitions should be modified.

We moved on from the into testing, and I'll turn the floor over to my co-chair, Sam, to talk about the decisions that we made on testing.

DR. MITRA: In the area of testing there are several procedures, some indoor accelerated tests and some outdoor tests. And there was general agreement that some of these tests, the four listed, some of them are pretty standard already with the ASTM standard procedures that they would be acceptable methods of exposure testing, followed by opening up the floor that we did last night to modifications or other suggestions as to how these test methods might be altered.

So the first one is the Xenon Arc indoor testing. It's got a G26 number in the ASTM. Then there's the accelerated natural sunlight, Fresnel mirror, ASTM 4364. These were explained, the ASTM methods, the numbers, were explained on the floor before they were discussed so that people -- everyone in the room, all 26, really had a little bit of a feel for what these tests were.

The fluorescent UVs at G53 on the ASTM method and outdoor is ASTM 1435. This was generally acceptable to most people.

If you move onto the next overhead you will see the modifications that have been suggested. And we have, again, ranked them by the number of votes we put each of these -- we first listed all this modifications that people suggested from the floor then we opened each one up for voting as to how many people thought that modification was an important thing to incorporate.

As you can see, we started off 25 votes out of 26 in the room, which was almost unanimous. And the use of a temperature report record along with the dose, because people felt if you just describe the light component to it and did not consider what the temperature of the specimen or the environment was under which it was done, that would cause some confusion as to how much degradation you got.

The next one is the angle option. Some people had some objections to a 45-degree angle specified, I think, in the ASTM method, but really wanted to have a full sky radiation on the sample so that the angle option, depending on where you are doing the test, especially the outdoor test, should be incorporated.

The third one is a choice of UV cycle. It depends on the material. The UV cycle that is used is also specified, I think, in some of the tests, and the UV cycle should really be opened up or about broadened so that it uses the most efficient one that you can try to do with the particular material that you are testing.

The fourth one was eliminate condensation cycles where appropriate because some of the testings are done in very dry areas. The use may be in very particularly dry areas. The use may be in very particularly dry areas where the condensation cycles really become irrelevant. They are some concerns also for the washing away of the degradation products from the samples by these condensation cycles and we wanted to -- there was really no test to collect them and look at those products to analyze them.

Five with 12 votes was calibrating the exposure set up for dose time because a lot of these exposures are really given in terms of the time that the exposure is done. But depending on the instrument or exposure set-up, how much energy that the sample actually saw during that period of time is sometimes not specified. So there was a strong suggestion from the floor that the dose be either calibrated with the time so that even when you are giving the time you are really mentioning how much energy you saw, or else give the dose instead.

The one with the 8, F, is a lower duty cycle, 4/4 or fluorescent UV test method. That is broadening out the UV cycle actually to suit the situation that is there, or else was really reluctance to speak with the 24, I think, cycle.

If you go to the next one. At this point we also want to make a point particularly about the characterization of the test samples. Yesterday and the day before there were a lot of concern that some people tend not to define the materials that are tested to the extent that everybody would like to see what the material is, exactly what it contains.

That is a very serious concern and, therefore, it is incumbent -- it was -- the language was that it should be incumbent upon the investigator to characterize their test materials to the fullest extent possible. That should be used -- most people, 21 out of the 26, felt that should be included and make it incumbent for them to describe that material in full detail, particularly the end-use formulation.

We move on to the next one. We move to discussing what properties you test for a polymer to see how it has degraded. There was several suggestions that Tony and I put on for discussion, put up for discussion. The general ones that had been discussed here and is commonly seen in the literature. We then opened the floor to discuss or propose other alternate methods of testing that we should consider.

We took this whole list that we had suggested and the ones that came from the floor, put every single item to a vote and now have prioritized or set up in a series by descending order of votes what those test methods ought to be.

Top on the list is mechanical tensile -- tensile test methods. And there are two ASTM standards on this, the 882 and the 638. And these two depend upon what the film thickness or the sample thickness is. Under a certain threshold, they switch over to a different method. So both have been included. And that was a completely unanimous, 26 out of 26 in the room. Everybody felt that should be definitely a test method to look at.

Molecular weight by GPC was something that most people felt was a very good method to follow these degradations.

IR spectroscopy inferred and included in there is the carbon index. That is a very common practice to follow the degradation process.

The tumbling friability is applicable only in certain samples but where it is, the tumbling friability was considered a very important test, but it is limited to certain kinds samples.

And the mechanical, the tensile had more support than the impact resistance in E with the 12 vote. But impact resistance was mentioned and it came from the floor and had a 12 vote on it.

Thermal analysis, both for induced phase separations used by the degradation or else the drop-off glass transitions or melting points. But generally glass transitions, thermal analysis could be used to look at how much it has degraded.

Molecular weight by solution viscometry. Molecular weight could be determined by GPC. That was a more absolute method people thought. But by solution viscometry, it's a simpler process and easier process to do. So that was also recorded.

You move on to the next one. Weight loss. The weight loss here is, of course, simple removal of volatiles by the degradation process. And so this is really -- we are talking about plain and pure metric measurements. You take the sample away before, and then after you exposure through the period of time, look at how much it weighs, how much it loss of weight you see. That was given an eight vote out of the 26.

The water absorption of the polymer. This, again, is only applicable in certain polymers, not necessarily will there be a significant increase in water absorption for all of them.

Extraction of the polymers to look at extractables practiced by several people as a method to look at the degradation of these materials.

The volatility, solubility and environmental stress cracking. These were methods that had been suggested from the floor and there was only, I think, one vote on them, that coming from person who had suggested it.

Compressive strength morphology by electromicroscopy, TGA, thermalography analysis. These were also methods suggested, but we listed them. We took them through the voting, and at that time people did not feel that it was really an important method that should be really included as a general method.

We then moved on to the research issues and research ideas.

DR. REDPATH: At this stage we got a little fancy. We opened the floor to discussion of scission and to what areas needed research. I think investigation needed further clarification. Seven main points came out. At the point we asked for any further concerns that should be tabled, and there were no further ones coming from the floor. So we put it to a vote.

We did things a little differently this time. People this time had two votes. They could vote for what they thought was the No. 1 issue that had to be addressed through R & D, then they also had a vote for the most second most important issue. We put five points to a No. 1 vote, three points to a No. 2 vote, to try and come up with a scale.

The No. 1 issue, as voted by this group, is to identify the by-products of degradation in different environmental areas. And when I said we would have some quantified information about the concern, that is an environmental question.

The numbers on the left-hand side, 56 is the ranking. In brackets, that had 10 votes as the No. 1 priority issue; two votes as the No. 2 priority issue.

Moving down then, the score of 42, defining photo-biomaterials. Really creating a category of materials that are both photo and bio recognizing that may be done by combining how we difine "photo" and "bio" but not knowing what was going in the bio session. That was left as something we wouldn't cover. It is viewed, though, as an important area. It's the second most important area that needs to be addressed. And that, again, six people thought that was the No. 1 issue for the No. 2.

Immediately after that specifying we considered was what meant by degraded. And I think we had that down as how degraded is degraded. That is really, I think, is the same comment as came up on the preceding session, what is meant by the end-point? We need to difine that. And, again, four people thought that was No.1; seven had that as their No. 2 priority.

A general catch description, environmental impact. I think we cought some of that concern in the No. 1 issue, but the general concept of environmental impact in its broadest sense; three No. 1; eight No. 2 votes, immediately behind then specifying of defining end-points.

Dropping then down to research in correlation between various exposure methods. There was three No. 1s, five No. 2s. By that we mean if you measure in a Q-panel, how do you correlate those results with results outdoors? How do you correlate results in one location outdoors with another location outdoors? There needs to be a lot of development work done to understanding that.

There was a significant drop down to the next issue which is studying the influence of spectral energy distribution. For example, location, geography, altitude, atmospheric conditions, what environmental factors influence light levels that will therefore influence performance of these materials. One No. 1; one No. 2.

And then the last, effect of atmospheric and/or maine conditions. Atmospheric or marine chemicals that can influence the photochemical processes that are taking place, that had two second place votes.

I think we have clearly identified the issues here. In summarizing the sense of consensus from the photochemical group, we may not know what we mean by a photodegradable material, but, we know how to measure it, we know how to expose it and we know how to test it.

We also have, I think, a fairly clear idea of where we should be going for the next session. If you take those top five points, they are all doable, they are all programs that can be initiated to elucidate information, and we have defined our homework for the next session.

At this point, I think I close our formal presentation and ask for questions.

DR FEIJEN: Thank you very much, Tony and Sam.

Who wants to ask the first question? Yes, please.

DR. WOOL: Richard Wool, University of Illinois.

There are several photodegradable technologies, some of which only lead to apparent fragmentation without any biodegradation thereafter, and then as pointed out several times during this meeting, there are those materials that begin to fragment via photo initiation and then proceed to either chemically react in such a way that the end products can also be biodegradable. Was this issue discussed?

DR. REDPATH: That was what was meant by the No. 2 issue in terms of research needs. We restricted our discussion to photodegradability and did not get into whether or not the products of photodegradability -- of photodegradation were biodegradable because I think we felt -- the chair certainly felt that we weren't sure what the definition of biodegradable was and that we would be coming up with a parallel definition that may have been different from the group focusing on that. We deliberately left that untouched which, I think, is why it appears as a high-level issue for further definition.

DR. ANDEWOLE: Akin Andewole, Himont.

In the area of characterization of materials, and especially that those end up in the environment, whether you sell it to a customer or you use it in research, you already recommended to the group to do the characterization to the fullest extent possible.

I think that is well taken, but I do have reservations as to how you can go ahead and make recommendations regarding particular technological methods while doing these characterizations.

For example, if you are going to link morphological information to, in fact, how we do arrive at mechanisms either for degradation or for making these materials, you cannot avoid using SEM -- it's actually SEM, not EM -- which is coupled directly with the whole concept of making these polymeric materials, especially in compounding activities.

So I do believe that the technology is always evolving for characterization, and we should probably leave it at that instead of making specific recommendations.

DR. REDPATH: I think we would agree with that.

DR. MITRA: In terms of the characterization, it was not intended to describe or make it incumbent upon the investigator to describe the entire characteristics of the material. I think what we intended to say in terms of characterization is what the components are in it. Because the degradation -- photodegradation of these materials, some of these polymers are very, very dependent upon what addivities it has and so on. So if you would have the photodegradation specified for a particular polymer under certain conditions, and then when you go out and really formulate it, you do add on extra things to it or take things out of it. Then it certainly affects the photodegradability of that material and does not serve the cause of defining and setting standards.

So what was really meant by characterization was mainly the composition of the materials as opposed to spectroscopic or other kinds of characterization.

DR. REDPATH: If I could just add one comment to that.

In terms of measurement techniques. We did not mean to exclude or rule out things such as SEM. I think what that list shows you is what people are most comfortable with in terms of interpreting results.

If you are going to present results on degradation, then I think it says that your peer group likes to see those particular measurements. It doesn't mean that other measurements are not perfectly valid. It's just that if you are going to use other techniques I think it indicates that there is going to have to be a little more justification as to how those techniques are valid. It certainly doesn't exclude them.

DR. FEIJEN: Any other questions?

I have one question myself.

You were mentioning the second important research item to make some kind of combination between the photodegradation process and, at a certain stage, the further biological degradation.

And my question is: How are you going to organize that? Because I understand that you are more or less working in a particular group focusing on the photodegradation within this workshop. Have you found any structure or any set-up to combine the efforts now on this important subject?

DR. REDPATH: Well, you see, what we were hoping is that we could define photodegradation and that the environmental biodegradation would come in with a clear consensus on what was meant by biodegradation there, and then we would simply be able to basically combine the two and say, fine, if the material has the following properties upon exposure to sunlight it's photodegradable. You then take the result of that, perform a specified test, and if it meets certain conditions then it fits biodegradable, ergo, it's a photo-biodegradable system.

I'm saying that a bit tongue-and-cheek because we didn't really expect we would be able to do that.

I would hope, though, it would be possible to do that, to define the two independently and then if a material happens to have both properties it's, therefore, a photo-biomaterial rather than create a separate classification adn then have a separate group trying to come up with a definition for that separate material.

DR. FEIJEN: You would like to make some comment?

DR. GRUBY: Walter Gruby, U.S. Environmental Protection Agency.

I wonder if your group considered that -- in one area of degradable plastics might be very simply a litter problem, a surface litter, and its decomposition which, in my mind, would be solved or impacted significantly by the photolytic degradation process.

Have you considered this aspect from the standpoint of research needed of variable exposure conditions, adequacy of currently available tests, formulation of materials perhaps not yet available that can be more accurately programmed for photolytic degradation after economic uses is ended? I wonder if you've addressed this, what I could call a surface litter area.

DR. REDPATH: I thought that was captured under the heading of definition of end-points when it comes to targeting removal of a litter item -- and I take that to be the gist your question. If you want a product to undergo photochemical degradation to cause litter products to disappear, then you need to have some data on how long it takes for that to happen, under what different conditions.

Is that the sense of your question or am I misreading it?

DR. GRUBY: I see where you have included that.

DR. REDPATH: So end-point was the way we thought of that. We simply listed it because that is almost a policy decision to decide how fast it should go. It's this group's job to determine how can you accurately measure whether or not it's doing that.

DR. GRUBY: Thank you.

DR. FEIJEN: Are there any other remarks or questions accoring to this?

You wanted to make a comment? Okay, yes. Comments by Robert Ennis:

DR. ENNIS: Good morning. I'm Bob Ennis for Plastigone Technologies.

I hadn't really expected to be up speaking to you this morning, so I'll try not to ramble.

You recall at the very beginning of the meeting Ramani was discussing ASTM, and he mentioned that one of the four subsections of the ASTM committee on environmentally degradable plastics was the environmental fate subsection.

This committee was given the mandate to develop and propose standard test procedures which might be used to evaluate as a unit the toxicologic impact of a given sample of photo or biodegradable plastic material.

I'm the co-chairman of that committee along with Steven Harper from Harper from Georgia Tech Research Institute. We initially divided our committee into two subgroups, one dealing with photodegradable toxicology or environmental fate; which I headed, the other groups was concerned with biodegradable environmental fate, which is headed by Dr. Robert Bailey from Dow Chemical who is here at the meeting.

We have developed some preliminary proposals which we hope to present for discussion and for a vote at the ASTM meeting which is coming up next week in Orlando, Florida, but what I'll try to do is give you a very rapid outline of what we've come up with. It is certainly beyond the scope of a full discussion at this point in our meeting.

We have taken certain assumptions. We first assumed that obviously there are as many environmental fates as there are environments. So that any discussion of the degradation products of a particular system must be related to the specific environment in which the product degrades.

We also made the assumption that the environmental fate or toxicology of individual breakdown products is really beyond the scope of the ASTM committee. This is more related to a research application, and certainly we realize and we recognize that a full evaluation of breakdown products in a different environmental condition will have to be done, but as we mentioned at the photodegradable section last nigh, this will keep a whole generation of Ph.D. chemists working for a long time to figure out all of those various things.

What we wanted to do, is to provide a standard test procedure which was both reliable, economical and simple.

Now, as far as the photodegradable section is concerned, we realize that the degradation products represent a continuum, so we took an end-point arbitrarily of degradation as being 95 percent loss of elasticity of the specimen under an accelerated weathering condition. We proposed an acute and chronic toxicity test measured at two days and seven days. We proposed a fresh water type of test, the way that one would do toxicology screening test for groundwater since groundwater contamination is one of the major concerns when you are talking about waste products from plastics as a liter phenomenon depend.

We proposed that we do an extraction, an aqueous extraction, and based upon annual rainfall amounts which would be an extractable material of approximately 500 ml per gram of degraded plastic material. This would be tested in a fresh water system such as the standard systems which are now used for water toxicology evaluation using something like syradaphnia and fathead minnow.

For the biological degradation systems, we recognize that there is an underlying baseline toxicity in any microbial system and therefore what we want to test is whether a system of microbial organisms in which plastic has been degraded adds by virtue of those degradation products a certain amount of toxicologic by-products or end-products to that system.

And for that situation, we'll have to do essentially a serial dilution test to measure the toxicology or toxic potential of the control solution over whatever added tocicity there may be in the test solution by virtue of the breakdown products of the plastic material.

That is a very rapid overview, and obviously a lot things have gone into these discussions. It also is not the final solution because this is a preliminary evaluation that we are going to be voting on next week. But I hope it will give you some idea of the fact that this is a concern; that we are trying to address it in a simple and straightforward manner and that there is still work to be done in this area.

Thank you very much.

DR. FEIJEN: Thank you very much for your contribution.

I will not re-open the discussion on this item. We have discussed this point very clearly, so we'll leave it at that, and I invite Ramani Narayan to give a summary statement on the basis of the discussion and the presentations we have had.

Ramani?

ISSUES AND SUMMARY BY RAMANI NARAYAN:

There were 165 attendees at this workshop from 14 different countries, representing 120 companies and/or research institutions, universities, governmental agencies and other institutions. It's truly a global representation and testifies to the importance of this subject area.

This is the first time to my knowledge that a biomaterials/biomedical group has interacted with the environmentlly degradable polymers group in a major fashion and I think that it is very good. If I may quote Michele Vert, a biomaterials researcher, --- he said that initially he wasn't too sure it was a great idea to bring the biomaterials and environmental group together and was very concerned that it would prove to be a terrible fiasco.

He, however, feels comfortable that this did happen and is pleased with the deliberations and interactions at this conference. That should be chalked up a a success for the workshop.

While the organizing committee never intended it, the distribution of attendees interested in these subject areas has been even. The workshop was not dominated by any one group. The number of participants in the breakout sessions, were almost evenly distributed among the biomaterials group, the environmental-bio group and the environmental-photo group.

So this was truly a global event -- I think every one of the people who participicted in this workshop did contribute in great measure to the success of it. The proceedings of this workshop will come out within 90 days of submission of manuscript, and therefore, all speakers who have not turned in their manuscript, must turn in their manuscripts by -- before Thanksgiving, the 25th of November, 1989. Please, do make an effort to do that because I think we have been very successful so far. We have done a great job. You have been very, very good in the getting these things done, and I think the impact of this workshop would be lost if we spent a year just trying to get this book out. The proceedings will be mailed to all the participants and registered attendees of this conference.

I am supposed to summarize the three consensus breakout session reports that were presented earlier and it's going to be a little hard. However, I will try to identify a few key points that figured prominently in all the breakout sessions.

The consensus breakout sessions in each of the three areas were supposed to address three key issues:

- Definitions/Terminology
- Testing of materials - specifically test methods, test environments, protocols and other related issues
- Research needs and issues and prioritize them.

It was indeed remarkable that the breakout sessions managed to address all of the issues and come to some sort of consensus in the short time period they met.

On the issue of definitions/terminology, the three breakout sessions managed to come to a consensus within their group to a major extent.

There were a lot of similarities across the board and it remains to be seen (probably at the next conference) whether we can tie them all together.

On the issues of testing and research needs, some common themes that figured across the board were:

- Fundamental Studies dealing with degradation mechanisms
- Fate of products whether in the body or in the environment/environmental fate.
- Detailed characterization of the starting materials used in the degradation studies.
- Molecular weight analysis is an important measure of degradability.
- General agreement amongst both environmental group breakout sessions on the use of mechanical property measurements to evaluate degradability.

In conclusion, I hope we can identify common discussion elements from the three breakout sessions as a starting point for the next consensus workshop.

So that summarizes it. I want to thank you for listening but please, before you go, Professor Bailey wants to present something to us, and it was missed out in the presentations yesterday because it was getting late and he was the chair person so he decided he wouldn't force it.

So we are going to let Professor Bailey make a few general comments and apresent some of his material.

---PRESENTATION BY WILLIAM BAILEY

DR. FEIJEN: I would like to invite Michele Vert to tell us something about the plans we have to continue this first workshop where we have learned a lot, and I think it is really necessary to proceed with what we are doing, the initial effort we have made here. Michele will tell us a little bit about the plans for the ongoing activities.

DR. VERT: Thank you.

Well, I can say it's decided that the next meeting will be organized in France. The date, it should be sometime in the fall of 1991. We have a small problem in France, is that it's a vacation country and July and August is a very crowded period of time, and so it shouldn't be good to organize a big meeting in this period.

There are two potential locations at the moment, Rouen, which is the town where I work, it's the capital of Normandy, it's close to the beaches and it's a nice area; but the meeting could be also organized in the Montpellier area, which is the capital of the southern region named Languedoc Roussillon, and it's in the South of France. It's one of the largest vineyard areas in France. I guess it could be also a good opportunity to check whether degradable material can affect the quality of wine or not.

An executive committee has already formed and it includes European scientists -- at least European scientists which have been contributing to the success of this first meeting, but also those who are participants of this first meeting. Indeed, we want to keep the spirit which is growing now.

We will focus our attention on the interdisciplinary aspect of our problems and we will, of course take advantage of all the comments, recommendations and conclusions of this meeting, and we will try to base something new on these conclusions. So we will focus on basic concepts trying to find out a way to better understand what are the mechanisms and so on, and will try to balance different disciplines.

The duration will be probably three days. It seems to be nice for all of us, not too long, not to short, and we will try to inform all the people soon enough to have them prepare carefully their presentation according -- as far as possible -- to the conclusions of this meeting. We would like to have people coming with data or information, scientific information, which can be usefull according to what we have talked about for these three days.

Well, I guess that's enough for the moment. Of course, we look forward to having you in Europe in 1991, and I wish you successful and fruitful work so that we can make the next meeting as good and as excellent as this one.

And, Jan, on your behalf I would like to thank very much the organizers of this meeting: John, Sam, Tony, Ramani and Sumner. I guess they did a very good job and it is appreciated and I hope we will do at least the same.

Thank you very much.

DR. FEIJEN: Thank you, Michele. You thanked the organizers already, that is what I wanted to do.

Also for my co-chairman, it has been an excellent meeting. We both learned a lot. It was a real experience for us to have the interaction with the other groups. We have a lot in common as has been outlined today, and I think it has been an excellent idea to put the groups together and to come to certain actions -- common actions together. We can learn a lot from each other. I did, and I really appreciate the time and effort the four of you have put in, Sumner

Barenberg and John Brash and Tony Redpath and Ramani Narayan. I know how difficult is is to put everything together, to organize everything. It's a very pleasant surrounding. I have slept very well in this excellent hotel.

So thanks again for the excellent organization. I hope we all can do the same. Michele will take care of it in Europe, but the flavour will be a little bit different. Thanks, again.

DR. REDPATH: I'm just going to interject one comment on behalf of organizing committee.

Every event, whether it's a degradation event or anything else, needs some form of catalyst or a initiator, and I think in the case of this conference that was Sumner Barenberg, and on behalf of, the entire group here, particularly the co-organizers, I would like to thank Sumner for taking the initiative to get this whole idea started, to get the thing going, to drive the rest of the co-organizers to make us to do our jobs and for really providing the spark to start this whole thing off. Thank you.

DR. BARENBERG: This was not meant to be a testimonial.

I'm here for closing remarks, and I do want to thank the local arrangements people, Tony, John and Ramani; the International Organizing committee, which was Ann-Christine, Jan, Michele, David, Steve Veng, who is in mainland China and who couldn't get out and Bob Lenz, for helping put this together. Additionally, I wish to thank corporate and federal sponsors, because their underwriting allowed for the low registration fee.

Lastly Melissa, who is here as a speaker, but who also did audio visual. We thank her very much for her assistance.

And lastly, thank the audience, because all we did as an organizing committee was to bring the participants together, have selected participants present, and the rest was up to you, to make this meeting a success.

So on behalf of the organizing committee, both local and international, I want to thank you, the participants.

Thank you very much for a very good meeting.

---Whereupon the Conference was concluded at 12:10 p.m.

## RESEARCH AND ISSUES

1. Reference Standards (+ & -)
2. Lab vs. Field Correlation
3. Standardization of Lab Procedures
4. Degradation Mechanism Elucidation
5. Research <-> Policy
6. Characerization of Working Materials
7. Characterization of Test Environments

## General Recommendations

Establish three committees to prepare consensus material for discussion in 1991 to address:

- Nomenclature
- "Reference" materials and "Standard" test methods
- Research need and long term planning

FIRST INTERNATIONAL SCIENTIFIC CONSENSUS WORKSHOP
ON DEGRADABABLE MATERIALS

2-4 NOVEMBER 1989
HILTON INTERNATIONAL HOTEL
TORONTO

REGISTRANTS

Akin Adewole
Himont USA
800 Greenbank Road
Wilmington, DE 19808

AnnChristine Albertsson
Polymer Technology
Royal Institute of Technology
10044 Stockholm
Sweden

Paul Allenza
Allied-Signal EMRC
50 E. Algonquin Road
Des Plaines, IL 60017

James Anderson
Case Western Reserve
Department of Pathology
University Circle
Cleveland, OH 44106

Tony Andrady
Research Triangle Institute
3040 Cornwallis Road
Raleigh, NC 27709

Tetsuzo Arai
Sumitomo Metal Industries Ltd
Tosho-Cho Shinjuku
Katori-Gunn
Chiba-Ken
Japan 289-06

Richard Austin
Exxon Chemical Company
5200 Bayway Drive
Baytown, TX 77922

Robert Bailey
Dow Chemical Company
1701 Building
Midland, MI 48674

William Bailey
Department of Chemistry
University of Maryland
Chemistry Building
College Park, MD 20742

Lucinda Ballinger
Rhone Poulenc
P.O. Box CN-5266
Monmouth Junction, NJ 0543

Alberto Ballisteri
University of Catania
Diprtimento di Chmica-Viale
A.Doria
6-95125
Catania, Italy 95125

Sumner Barenberg
Baxter Healthcare Corporation
Route 120 & Wilson Road
Round Lake, IL 60073

Janice Bennett
Polysar Limited
Plastics Group
475 Witmer Street
P.O. Box 3338
Cambridge, Ontario
Canada N3H 4T6

Steven Bloembergen
Polysar LTD
Vidal Street
Sarnia, Ontario
Canada N7T 7M2

A. Moushaka-Bodayav
University of Toronto
Center for Biomaterials
124 Edwards Street
Toronto, Ontario
M5G 1GL, Canada

Melissa Bouzianis
Belland Inc.
100 Burtt Road
Andover, MA 01810

Aldo Brancaccio
EniChem Anic SpA
Via Jannozzi, 1
20097 San Donato Milanese
Milan, Italy

John Brash
Mcmaster University
Chemical Engineering
Hamilton, Ontario
Canada L8S 4L8

Remi Briand
Union Carbide Canada LTD
10455 Metropolitan Blvd.
Montreal Quebec
Canada H1B 1A1

Stephen Bruck
Stephen Bruck Associates Inc.
P.O. Box 828
Rockville, MD 20851

Paul Buscemi
Scimed Life Sciences
6655 Wedgewood Road
Maple Grove, MN 55369

Peter Campbell
St. Lawrence Starch
P.O. Box 1050
Port Credit Station
Mississauga, Ontario
Canada M4B 1R3

David Carlsson
National Research Council
of Canada
Division of Chemistry
Montreal Road
Ottawa, Ontario
Canada, K1A 0R9

Andy Chang
Quantum Chemical
USI Division
1275 Section Road
Cincinnati, OH 45222

Peter Chang
Tredegar Industries
P.O. Box 1071
Terre Haute, IN 47805

Yu-Ling Cheng
University of Toronto
Department of Chemical
Engineering
200 College Street
Toronto, Ontario
Canada M5S 1A4

Emo Chiellini
Dept. of Chemistry
University of Pisa
Via Risorgimento 35
Pisa, Italy 56100

John Comerford
Chicopee Research Division
Dayton, NJ 08810-0940

Nick Curto
Owens Illinoise
13512 Venetian
Monroe Michigan 48161

Ervin Dan
Enviromer Enterprises
29 Fuller Street, Box 90
Leominster, MA 01453

Ervin Dan
Polysar Inc.
690 Mechanic Street
Leominster, MA

Daniel DiBasi
Mobil Chemical
Rt 31 Tech Center
Macedon, NY 14502

Pieter Dijkstra
Twente University
P.O. Box 217
Enschede 7500 AE
The Netherlands

Jim Dillon
FDA
12200 Wilkins Ave.
Rockville, MD 20852

Pete W. Dinger
Council-Solid Waste Solutions
1275 K Street NW, Suite 500
Washington, DC 20005

James Dixon
Canadian Standards Association
178 Rexdale Blvd
Rexdale, Ontario
Canada M9W 1R3

Yoshiharu Doi
Tokyo Institute of Technology
Imayado-Cho
Asahi-Ku
Yokoham City
Kanagawa-Ken
Japan 241

Timothy Draeger
National Corn Growers Association
1000 Executive Parkway
Suite 105
St. Louis, MO 63141

Daniel Drozdowski
U.S. Testing Company
1415 Park Avenue
Hoboken, NJ 07030

Phil Edwards
NOVA Corporation
1265 South Vidal Street
Sarnia, Ontario
Canada N7T 7M2

Terry Elliott
Polysar LTD
Suite 550 West Tower
Slipp Center
3330 Bleer Street West
Tornoto, Ontario
Canada M8X 2X2

Robert Ennis
Plastigone Technologies
P.O. box 165618
Miami, FL 33116

Ralph Ewall
Hercules Inc.
Research Center/8136 T8
Wilmington, DE 19894

Nancy Fair
University of Missouri
137 Stanley
Columbia, MO 65211

Daniel Fauchon
465 Lorne Street
Apt. 7
Burlington, Ontario
Canada L7R 2T2

Jan Feijen
Department of Chemcial Technology
University Twente
P.O. Box 217, 7500 AE
Enschede
Netherlands

Bruno Gaeckel
Dow Chemical
Box 1012
Sarnia, Ontario
Canada N7T 7K7

Tom Galvin
ICI Americas
Wilmington, De

Soraya Gavidia
Dept. Genie Chimique
Ecole Polytechnique de Montreal
Case Postale 6079-SUCC "A"
Montreal, Quebec
Canada H3C 3A7

Maurizio Germani
Ferruzzi Ricerca E
Technologia Fertec
Foro Buonaparte 31
Milano, Italy 20121

Don Gibbons
Biosciences Laboratory
3M Life Sciences Sector
3M Center
St. Paul, MI 55144-1000

Dan Gilead
Plastopil Corp.
Hazorea, Israel 30060

David Gilmore
Univ. of Massachusetts
Department of Biochemistry
Amherst, MA 01550

Sathya Gourisankar
Meadox Medical
103 Bauer Drive
Oakland, NJ 07436

Howard P. Greisler
Loyola Univ. Medical Center
j2160 South First Avenue
Maywood, IL 60153

J. Griffith
Colortech
8011 Dixie Road
Brampton, Ontario
Canada L6T 3V1

Walter Grude
U.S. EPA
26 W. M.L. King Drive
Cincinnati, OH 45268

Jim Guillett
Department of Chemistry
University of Toronto
Room 718, 80 St. George Street
Toronto, Ontario
Canada M5S 1A1

William Haffner
Kimberly Clark Corporation
1400 Holcomb Bridge Road
Roswell, GA 30076

Jun-ichi Hakozaki
Nippon Unicar Co.
2-6-1, Ohtemachi, Chiyoda-Ku
Tokyo, Japan 100

James Hammer
3M Company
3M Center 201 2W 17
St. Pual, MN 55144

Michael J. Hanner
Dow Chemical Co.
438 Building
Midland, MI 48667

George Harlan
Union Carbide Corporation
P.O. Box 450
Somerset, NJ 08875-0450

Jorge Heller
SRI
333 Ravenswood Ave.
Menlo Park, CA 94025

Michael Helmus
Harbor Medical
25 Drydock Avenue
Boston, MA 02210-3912

Alex Henderson
Ethicon
1421 Lansdowe West
Peterborough, Ontario
Canada, K9J 7B9

Marty Hitchcock
Dow Chemical
3825 Columbus Road, SW
Granville, OH 43023

Dennis Holtman
Personal Products Company
Johnson & Johnson
Van Liew Avenue
Milltown, NJ 08850

Norman holy
Rohm and Haas
Route 413 P.O. Box 219
Bristol, PA 19007

Ahmad Husseini
Canadian Standards Association
178 Rexdale Building
Rexdale, Ontario
Canada M9W 1R3

William Hutchinson
Esso Chemical
P.O. Box 69
480 S. Christina Street
Sarnia, Ontario
Canada

Gene Iannoti
University of Missouri
137 Stanley
Columbia, MO 65211

Michiomi Kabayama
Twinpak Inc.
910 Cnetral Parkway West
Mississauga, Ontario
Canada L5C 2V5

William Kissel
Amoco Chemical Co.
P.O. Box 400
Naperville, IL 60566

Kazuo Kitamura
Japan Steel Works America
200 Park Avenue
New York, NY

Peter Klemchuk
Ciba Geigy
Ardsley, NY 10502

Chester Kmiec
Union Carbide
1 Riverview Drive
Somerset, NJ 08875

Hiroaki Kobayashi
Toray Industries
Muromachi, Nihonbashi
Chuo-Ku
Tokyo, Japan

Naoki Kubo
Chuo Kagaku Co. Ltd
4021-6
Haraichi
Ageo-City
Saitama-ken
Japan 342

Denis Labombard
Millipore
80 Ashby Road
Bedford, MA 01730

Dean Laurin
Baxter Healthcare Corporation
Route 120 & Wilson Road
Round Lake, IL 60073

Jong Shin Lee
Sam Yang Co
339, 2GA Palbok-Dong
Chon-Ju, Cholla-Buk-Do
Korea

Ronald Liesemer
Council for Solid Waste Solutions
1275 K Street NW
Suite 500
Washington, DC 20005

Wayne Logan
DuPont
C&P Department
Chestnut Run 709
Wilmington, DE 19808

Gary Loomis
DuPont
E323/334
Wilmington, DE 19898

John MacLean
Union Carbide Canada
1210 Sheppard East
Suite 210
Box 38
Toronto, Ontario
Canada M2K 1E3

Wayne Maddever
St. Lawrence Starch
P.O. Box 1050
Port Credit Station
Mississauga, Ontario
Canada M4B IR3

Tetsuro Maeda
Research Center Deki Kagaku
Kogyo K.K.
6-1-1007
Atsugi-Cho
Atsugi-City
Kanagawa-Ken
Japan 243

Don Maschmeyer
Dow Chemical
Building B3826
Freeport, TX 77566

Mark Matlock
Archer Daniels Midland
1450 Westside Drive
Decatur, IL 62549

Mitsuo Matsumot
Biochemical Division
Basic Industries Bureau
MITI
2-40-RA31
Tagaara
Nerima-Ku
Tokyo Japan 179

Susan May
Polysar Ltd
1265 South Vidal Street
Sarnia, Ontario
Canada N7T 7M2

Toshio Mekaru
Borden Inc.
960 Kingsmill Parkway
Columbus, OH 43229

Robert Middlemiss
Mooney Chemical
2301 Scranton Road
Cleveland, OH 44113

Joree Miranda
Hart Chemical Ltd
256 Victoria Road South
Guelph, Ontario
Canada, N1H 6K8

Smarajit Mitra
3M Company
235-3E-06, 3M Center
St. Paul, MI  55144

Terry Mohoruk
B.F. Goodrich Canada Inc.
195 Columbia St. West
Waterloo, Ontario
Canada N2J 4N9

Frank Moos
Canada Cup
137 Bentworth Avenue
Toronto, Ontario
Canada M6A 1P6

Hiroshi Morimoto
Toray America
600 Third Avenue
New York, NY  10016

Edward Mueller
FDA
12200 Wilkins Avenue
Rockville, MD  20852

Charles Murray
AT Plastics
307 Lake Promenada
Long Branch, Ontario
Canada M8W 1A6

Kazuya Nakajo
General Affairs Department
3-27-11
Shimizu
Suginami-Ku
Tokyo Japan 167

Ramani Narayan
Michigan Biotech Institute
Michigan State University
East Lansing, MI  48909

Edwin Niemann
Arco Chemcial
3801 West Chester Pike
Newtown Square, PA  10703

Eiichiro Nishikaw
Tonen Sekyu Kagaku K.K.
862-37
Kitanagai
Sanho-Cho
Saitamo-Ken
Japan 354

John Nizio
Southeastern Reduction Company
P.O. Box 5366
Valdosta, GA  31603

Rogers Harry-O'Kuru
1501 E. Gardner Lane
No. 1510
Peoria Heights, IL  61614

Norihiko Oda
C&M Fine Pack Inc.
1200 North Holdale Drive
Fullerton, CA  92631

Masatoshi Oki
Polymer Research Laboratory
Nippon Unicar Company
8-1 Ukishima-Cho
Kawasaki-Ku
Kawasaki, Kanagawa 210
Japan

Dennis Olszannki
Mooney Chemical
2301 Scranton Road
Cleveland, OH  44113

Fidelis Onwumere
Becton dickinson
P.O. Box 1285
Dayton, OH  45401

George Ostapchenko
DuPont
E323 Box 80323
Wilmington, DE  19880

Reiko Otsuki
General Affairs Department
1-4-22
Minamiotsuka
Toshima-Ku
Tokyo Japan 170

Luba A. Pacala
Union Carbide Corporation
P.O. Box 670
Bound Brook, NJ  08805

Chung Park
Dow Chemical
P.O. Box 515
Granville, OH  43023

Satyen Patel
Pillsbury Company
Grand Metropolitan
311 SE 2nd Street
Minneapolis, MN 55414

Thomas Pfeiffer
Ampacet Corporation
250 South Terrace Avenue
Mount Vernon, NY 10550

Robert Pilliar
University of Toronto
Center for Biomaterials
124 Edwards Street
Toronto, Ontario
Canada M5G 1GL

Maurice Planchon
Solvay & CIE, S.A.
Rue de Ransbeek 310
B-1120 Brussels, Belgium

B.A. Ramsay
Dept. of Chemcial Engineering
Ecole polytechnique de Montreal
C.P. 6079 SUCCURSALE "A"
Montreal, Quebec
Canada H3C 3A7

Tony E. Redpath
EcoPlastics
518 Gordon Baker Road
Willowdale, Ontario
Canada M2H 3B4

Frank Ruiz
Heritage Bag Company
2051 Boston Drive
College Park, GA 30337

Dorin Ruse
Faculty of Dentistry
University of Toronto
124 Edward Street
Toronto, Ontario M5G 1G6

Simon Salame
International Paper
Long Meadow Road
Tuxedo, NY 19870

Hideo Sano
Kuraray Company
2-12-11-104
Namiki Kanazawa-Ku
Yokohama City
Kanagawa-Ken
Japan 221

Paul Santerre
Univ.of Ottawa Heart Institute
1053 Carling Avenue
Civic Hospital
Ottawa, Ontario
Canada K1Y 4E9

Yasoji Sasaki
Nippon Unicar Company
Asahi Tokai Building
2-6-1 Ohtemachi
Chiyoda-ku, Tokyo 100
Japan

Frederick Schwab
Mobil Chemical Company
P.O. Box 240
Edition, NJ 08818

Michael Sciaraffa
Kimberly Clark
2100 Winchester Road
Neenah, WI 54956

Gerald Scott
Department of Chemistry
University of Aston
Aston Triangle
Birmingham B4 7ET
England

Shalaby Shalaby
Ethicon Inc.
P.O. Box 151
Somerville, NJ 08876

Toshio Shimada
Daicel Chemical Industries
7-13-4
Koganehara
Matsudo-Shi
Chiba-Ken
270 Japan

Alan Sipinen
3M
3M Center Building 230-3F-06
St. Paul, MN 55144

Gary Smith
Abbott Laboratories
D-98W
Building AP-4A
Abbott Park, IL 60064

Duncan Smith
Ortech International
2395 Speakman Drive
Mississauga, Ontario
Canada L5K 1B3

Stephen Smith
Kimberly Clark
1400 Holcomb Bridge Road
Roswell, GA 33076

Ja-Lyang Song
Sun Hill Glucose Company
285 Ga Jwa-Dong Seo-Ku
Inchon, Kyong Ki
Korea

Mark Sova
Quantum Chemical
11500 Northlake Drive
Cincinnati, Ohio

Thomas Stevens
National Sanitation Foundation
3475 Plymouth Road
P.O. Box 1468
Ann ARbor, MI 48106

Thomas M. Sutliff
Boxter Healthcare Corporation
Route 120 & Wilson Road
Round Lake, IL 60073

Graham Swift
Rohm & Haas Company
Norristown Road
Spring House, PA 18477

Masami Tashiro
Sekisui America
780 Third Avenue
Suite 3102
New York, NY

Luanne Tilstra
N.I.S.T.
Polymer Division
Room B314
Gaithersberg, MD

Burnham Tinker
Mooney Chemical
2301 Scranton Road
Cleveland, OH 44113

Yutaka Tokiwa
Fermentation Research Institute
Agency of Industrial Science
and Technology
1-1-3 higashi
Tsukuba 305 Japan

George Upton
Ampacet Corporation
6500 Gaines Ferry Road
Unit C-4
Flowery Branch, GA 30542

J.M. Vallee
Baxter R&D
Rue du Progres,7
Nivelles, Brabant
1400 Belgium

Michele Vert
University de Rouen
LSM-ASNA
BP08 30 Mont Saint Aignan
Rouen France

John Vestek
Plastigone Technologies
10700 N. Kendall Drive
Miami, FL 33176

Richard Williams
Roy F. Weston
1 Weston Way
West Chester, PA 19380

David Williams
Institute of Medical and
Dental bioengineering
University of liverpool
P.O. Box 147
Liverpool
England L69 3BX

Richard Wool
Materials Science & Engineering
College of Engineering
University of Illinois
1304 West Green Street
Urbana, IL 61801

# INDEX

## A



## B

## C

**D**



**E**

## I



## K



## L



## M

## Q



## R



## S

**W**



**X**



**Z**